Metal Technology and Processes

John L. Feirer

John D. Lindbeck

Delmar Publishers

an International Thomson Publishing company I(T)P

Albany • Bonn • Boston • Cincinnati • Detroit • London • Madrid
Melbourne • Mexico City • New York • Pacific Grove • Paris • San Francisco
Singapore • Tokyo • Toronto • Washington

NOTICE TO THE READER

Publisher does not warrant or guarantee any of the products described herein or perform any independent analysis in connection with any of the product information contained herein. Publisher does not assume, and expressly disclaims, any obligation to obtain and include information other than that provided to it by the manufacturer.

The reader is expressly warned to consider and adopt all safety precautions that might be indicated by the activities herein and to avoid all potential hazards. By following the instructions contained herein, the reader willingly assumes all risks in connection with such instructions.

The publisher makes no representation or warranties of any kind, including but not limited to, the warranties of fitness for particular purpose or merchantability, nor are any such representations implied with respect to the material set forth herein, and the publisher takes no responsibility with respect to such material. The publisher shall not be liable for any special, consequential, or exemplary damages resulting, in whole or part, from the readers' use of, or reliance upon, this material.

Cover Design courtesy of Charles Cummings Art/Advertising, Inc.

Delmar Staff

Publisher: Alar Elken
Acquisitions Editor: Tom Schin
Project Editor: Cori Filson
Production Coordinator: Toni Bolognino
Art & Design Coordinator: Cheri Plasse
Editorial Assistant: Fionnuala McAvey

COPYRIGHT © 1999
By Delmar Publishers
an International Thomson Publishing company I(T)P®

The ITP logo is a trademark under license
Printed in the United States of America

For more information contact:

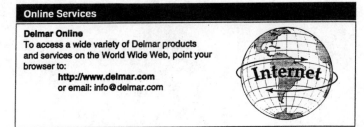

Online Services

Delmar Online
To access a wide variety of Delmar products and services on the World Wide Web, point your browser to:
   http://www.delmar.com
   or email: info@delmar.com

A service of I(T)P®

Delmar Publishers
3 Columbia Circle, Box 15015
Albany, New York 12212-5015

International Thomson Publishing Europe
Berkshire House
168-173 High Holborn
London, WC1V 7AA
United Kingdom

Nelson ITP, Australia
102 Dodds Street
South Melbourne,
Victoria, 3205 Australia

Nelson Canada
1120 Birchmont Road
Scarborough, Ontario
M1K 5G4, Canada

International Thomson Publishing France
Tour Maine-Montparnasse
33 Avenue du Maine
75755 Paris Cedex 15, France

International Thomson Editores
Seneca 53
Colonia Polanco
11560 Mexico D. F. Mexico

International Thomson Publishing GmbH
Königswinterer Strasse 418
53227 Bonn
Germany

International Thomson Publishing Asia
60 Albert Street
#15-10 Albert Complex
Singapore 189969

International Thomson Publishing Japan
Hirakawa-cho Kyowa Building, 3F
2-2-1 Hirakawa-cho, Chiyoda-ku,
Tokyo 102, Japan

ITE Spain Paraninfo
Calle Magallanes, 25
28015-Madrid, Espana

1 2 3 4 5 6 7 8 9 10 XXX 03 02 01 00 99 98

Library of Congress Cataloging-in-Publication Data

Feirer, John Louis.
   Metal technology and processes / John L. Feirer, John D. Lindbeck.
      p.   cm.
   ISBN 0-8273-7909-9 (alk paper)
   1. Metal-work.   2. Manufacturing processes.   3. Metals.
   I. Lindbeck, John D.   II. Title.
TS205.F38   1998
   671—dc21   97-30551
                                               CIP

Table of Contents

SECTION 4: MANUFACTURING METAL PRODUCTS **409**

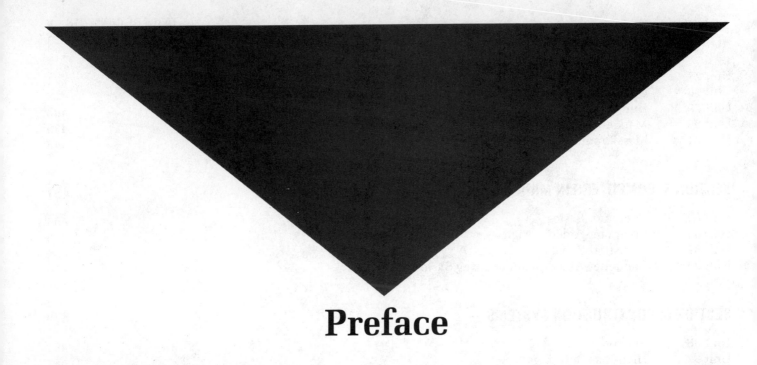

Preface

Our technological world requires that young people understand how products are designed and manufactured. Whether they make them, service them, or sell them, they will be better prepared to enter their careers if they have this knowledge.

The purpose of this book is to contribute to this goal by introducing the reader to the special field of metals and metal products. Because this material is so important in all product manufacture, it seems to be a logical place to begin.

Acknowledgments

The authors owe a debt to those many industrial and academic people who contributed their time, information, advice, resources, and fine illustrations in the preparation of this manuscript. They continue to be appreciated by everyone who has ever written a technical book. Special credit must be given to David Gregg for his excellent CAD drawings, and for his understanding in meeting impossible schedules; to Marilyn Lindbeck for her patience and encouragement, and for her word processing talents; to Lance Ferraro for his promptness in providing fine photographs; and to Pam Kusma for being a friendly and helpful editor. We are grateful for your kind assistance.

SECTION 1

Introduction to Metalworking

Are we living in the metal age, the electronic age, the atomic age, the jet age, or the aerospace age? Actually, we are living in all of these ages at once. However, the products of today's technology are dependent on parts made of metal.

Look up! The loud blast you hear could be a jet, a rocket, or a missile. These machines are made of metal. The attractive new buildings in your city also contain many metal parts. Did you ever stop to consider how these objects came to be made? The answer is that they are the products of the skill and cooperation of thousands of engineers, technicians, and skilled workers who make use of metals and other materials.

Unit 1

Metal—A Basic Material of Our Industrial Society

OBJECTIVES

After studying this unit, you should be able to:

- Know which areas of technology require metals.
- Understand the goals of metalwork.
- Understand the basic skills needed for working with metals.
- List the four major processes in manufacturing.
- Describe the different methods of cutting metals.
- List the metal-forming operations.
- Tell how products are assembled.
- Name the methods of finishing metals.

KEY TERMS

abrading	construction	milling	spinning
bending	cutting	pressing	texturing
brazing	drawing	punching	thermal cutting
buffing	drilling	riveting	thread fastening
casting	extruding	rolling	transportation
cementing	fastening	sawing	turning
chemical cutting	finishing	seaming	welding
coating	forging	shaping	
coloring	forming	shearing	
communication	manufacturing	soldering	

The United States is a huge workshop for metalworking. The metalworking industry employs a great number of workers, and the value of its finished products exceeds that of most other industry's. Metal is the raw material of this industry. Electricity is its source of power. Since so many products are made of metal, the field of metalworking is very large. In fact, the chances are more than one in ten that you will be working in the metalworking industry someday.

Metals in Industrial Technology

Metals are needed in four main areas of technology:

1. **Communications**—the devices and services that allow people to communicate (exchange information) with each other. Books, magazines, newspapers, telephones, radios, TVs, and computers are some of the means by which we communicate.

2. **Manufacturing**—the production of all goods and products people need to live; including everything from kitchen utensils to cars and airplanes.

3. **Construction**—the building of all kinds of structures; including: shelters, factories, high-rise buildings, and the nation's infrastructure (roads, bridges, tunnels, electrical and electronic systems, pipelines, etc.).

4. **Transportation**—the products and services needed to move people and goods from one place to another (Figure 1-1).

Goals of Metalwork

By studying the technology and processes involved in working with metal, you will begin to reach the goals that are important to success in the industry. In particular, you will:

- Gain an insight into American industry and its place in our industrial society.
- Acquire hand and machine metalworking skills (Figure 1-2).
- Appreciate the strength of the American free enterprise system.
- Discover and develop talents, aptitudes, interests, and potentials as related to industrial technology.
- Develop problem solving and creative abilities using tools, materials, processes, and products.
- Apply mathematics, science, and social science to metalwork.
- Explore and develop careers in metalwork.
- Build interesting products of metal.
- Prepare for advanced classes in metalwork, including avenues for skilled tradespeople, technicians, and engineers.

Skills Basic to Metalworking

Before you can begin to make a project in metalwork, you should learn how to work safely in the shop, how to select and design projects, how to read drawings,

Fig. 1-1 This modern helicopter is built of many lightweight materials, such as aluminum. (Courtesy of Sikorsky Aircraft Corporation)

Fig. 1-2 Workers in the metals industry, such as this welder, have a range of skills. (Courtesy of The Lincoln Electric Company)

how to plan a project, how to choose metals, and how to measure and make layouts. When you have learned these things, you are ready for the interesting experience of making something worthwhile out of metal. Though you may not have the time or skill to make projects in every area, you will have a chance to learn about them all.

Metal Processing

The tools, sporting goods, vehicles, and other articles we use every day are products of our metal manufacturing industries. These industries take raw materials and transform them into usable products. The techniques and methods used to bring about these transformations are called *manufacturing processes.* In this unit, we shall take a closer look at these processes to see how the skills and operations shown in this book relate to one another and the processes used in industry. You will note that many of these processes are used primarily in factory production systems and are not practical in the school shop. The reason for this is that they require massive, expensive, and complicated machinery and skills. However, the principles underlying these industrial processes can be learned and their relationships understood.

There are four main classifications of manufacturing processes—cutting, forming, fastening, and finishing. There are also other related techniques, such as tempering and heat treating, but the primary four processes are basic to the manufacture of most products (Figure 1-3).

Cutting

The process of removing or separating metal from a workpiece is called **cutting,** and there are a number of different types (Table 1-1).

Sawing

Sawing is a separation method whereby tools with teeth along their edges are moved against the workpiece. A good example is hacksawing. Metal pieces are cut to size by sawing.

Shearing

Shearing is another separation process. In shearing, a workpiece is forced between two cutting edges crossing one another. Using the tin snips to cut sheet metal is an example of shearing.

Punching

Punching is a removal process similar to shearing. Here, a shaped cutting edge is forced through a workpiece, such as when you use the hollow punch in sheet metal work. Circular shapes can be punched, and almost any shape can be removed from a piece of sheet metal. An example of this process is punching out an irregular shape for an ashtray or a seat for a metal chair.

Drilling

Drilling is a removal process wherein a cylindrical tool with angled cutting edges is forced through a workpiece to produce a hole. Drilling can be done on drill presses and on lathes. Reaming and tapping are related processes.

Abrading

Abrading is a removal process wherein cutting takes place by chipping away a softer material with a harder material such as mineral particles. Grinding and polishing, using abrasive paper, cloth, or wheels, are examples.

Shaping

Shaping is a removal cutting process wherein a single-edge tool is forced against a workpiece in a straight-line cutting path. This is a common machine shop operation. Planing, a type of shaping, is used primarily on very large workpieces. On a metal planer, the tool is fixed and the workpiece moves against it.

Milling

Milling is another example of removal cutting commonly used in the machine shop. Milling cutters are tools with sharpened teeth equally spaced around a cylinder and can have a variety of shapes. Metal filing is related to milling in that the filing tool is, in effect, a flat, stretched out milling cutter. Filing is commonly done by hand.

Turning

Turning is a removal process using an engine lathe. This type of removal cutting involves the revolving of a cylindrical workpiece against a single-edge tool. You can cut threads this way and also give workpieces various shapes.

Fig. 1-3 Military vehicles are made of cast, forged, formed, machined, and welded metal parts. They are tough and reliable pieces of equipment. (Courtesy of U.S. Army)

Table 1-1   Metal-Cutting Operations

Type	Equipment Used	Industrial Applications
Sawing 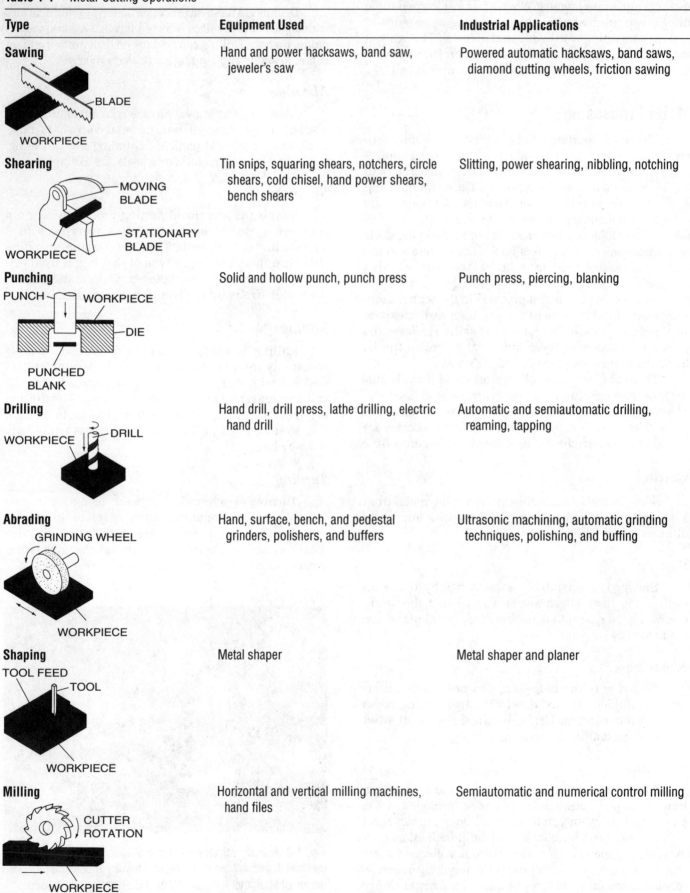	Hand and power hacksaws, band saw, jeweler's saw	Powered automatic hacksaws, band saws, diamond cutting wheels, friction sawing
Shearing	Tin snips, squaring shears, notchers, circle shears, cold chisel, hand power shears, bench shears	Slitting, power shearing, nibbling, notching
Punching	Solid and hollow punch, punch press	Punch press, piercing, blanking
Drilling	Hand drill, drill press, lathe drilling, electric hand drill	Automatic and semiautomatic drilling, reaming, tapping
Abrading	Hand, surface, bench, and pedestal grinders, polishers, and buffers	Ultrasonic machining, automatic grinding techniques, polishing, and buffing
Shaping	Metal shaper	Metal shaper and planer
Milling	Horizontal and vertical milling machines, hand files	Semiautomatic and numerical control milling

Table 1-1 Metal-Cutting Operations—*Continued*

Type	Equipment Used	Industrial Applications
Turning	Engine lathe	Automatic lathe turning
Thermal Cutting	Oxyacetylene and carbon-arc cutting torches	Automatic contour cutting, oxyacetylene, carbon arc, powder, air carbon arc, laser cutting, electrical discharge machining (EDM), electrical discharge grinding (EDG)
Chemical Cutting	Etching and pickling tanks	Chemical milling (CHM), electrochemical milling (ECM), chipless machining

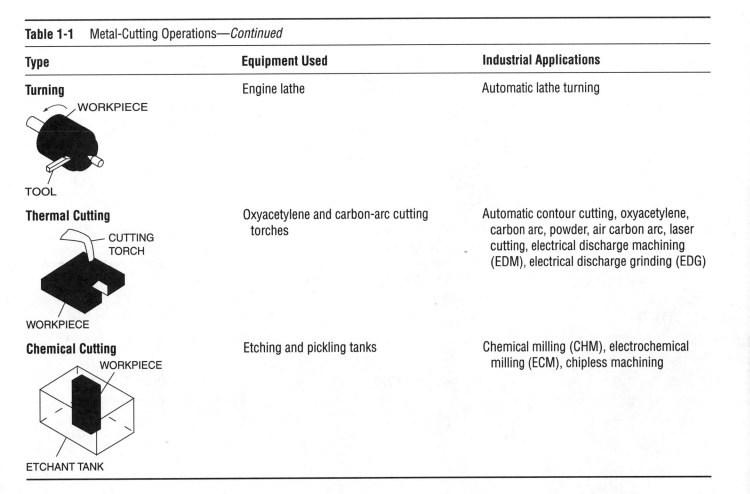

Thermal Cutting

Thermal cutting is a process wherein pieces are separated by the action of heat melting through the workpiece. Oxyacetylene and carbon-arc metal-cutting are examples of thermal cutting.

Chemical Cutting

Chemical cutting is a removal process using the action of chemicals which eat away the surface of metals. Etching is an example of this process.

You can see from the discussion above and by studying the table on metal-cutting operations that there are many different ways of cutting metals. The table also lists a number of related industrial applications. They are similar to the processes used in the school metal shop except that the equipment and skills are different.

Forming

The **forming** processes are techniques used to give objects shape or form without adding or removing any materials (Table 1-2). Some of these processes are referred to as *hot forming* or *cold forming* because the metal is worked at different temperatures. Forming provides an efficient and economical way to produce products or parts of products.

Casting

In the **casting** process, forming is achieved by pouring melted metal into a hollow cavity and allowing it to set, or harden. Sand casting is generally the method used in the school foundry, although investment casting is sometimes used. Casting is a hot-metal-working process.

Bending

Bending is a process by which metal is uniformly strained or stretched around a straight axis, which results in a product having a linear or straight shape. Typical operations include bending on either the bar folder, forming rolls, or a bending machine.

Forging

In **forging,** forming is achieved by hammering or applying steady pressure to a workpiece, forcing it to take the shape of a die. Forging may be done either hot or cold depending upon the size of the object and the end results desired. The methods are basically the same as those used by blacksmiths. The metal is heated to a high temperature and shaped by the pressure applied by the hammer as the metal rests on another surface (anvil). Huge forging hammers and forging presses are used in industrial operations.

Table 1-2 Metal-Forming Operations

Type	Equipment Used	Industrial Application
Casting 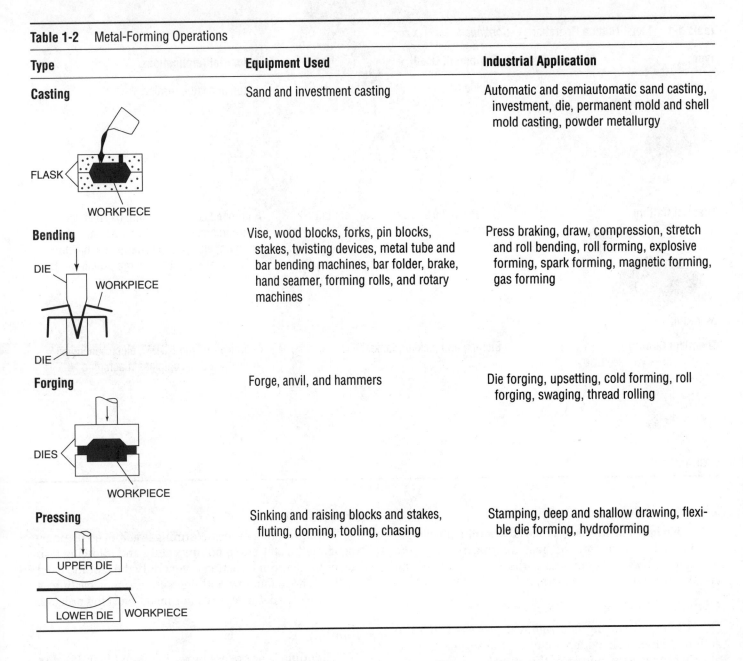	Sand and investment casting	Automatic and semiautomatic sand casting, investment, die, permanent mold and shell mold casting, powder metallurgy
Bending	Vise, wood blocks, forks, pin blocks, stakes, twisting devices, metal tube and bar bending machines, bar folder, brake, hand seamer, forming rolls, and rotary machines	Press braking, draw, compression, stretch and roll bending, roll forming, explosive forming, spark forming, magnetic forming, gas forming
Forging	Forge, anvil, and hammers	Die forging, upsetting, cold forming, roll forging, swaging, thread rolling
Pressing	Sinking and raising blocks and stakes, fluting, doming, tooling, chasing	Stamping, deep and shallow drawing, flexible die forming, hydroforming

Pressing

Pressing is the process by which sheet metal is shaped in several directions at one time by forcing it between two dies. Typical examples include the making of shallow dishes, trays, and covers of all kinds. Pressing differs from bending in that a single operation shapes the material in more than one direction. For example, compare the forming of a simple dish with the bending of a box.

Drawing

Drawing is giving shape by pulling a metal rod or ribbon through a die to reduce it to a wire or to form a tube. Drawing a wire generally involves several operations in which the wire is gradually reduced to the desired size.

Extruding

Extruding involves forcing metal through an opening or die which gives it a desired cross-sectional shape. When you squeeze the sides of a toothpaste tube, you are extruding a ribbon of toothpaste onto your brush. Typical shapes which are extruded include tubes and a variety of long pieces of metal. Extruding is primarily an industrial process.

Rolling

Rolling is one of the most commonly used industrial hot-forming methods. Hot rolling is mainly a roughing operation. The chief use is to shape large ingots of rough metal into a form that can be easily

Table 1-2 Metal-Forming Operations—*Continued*

Type	Equipment Used	Industrial Application
Drawing 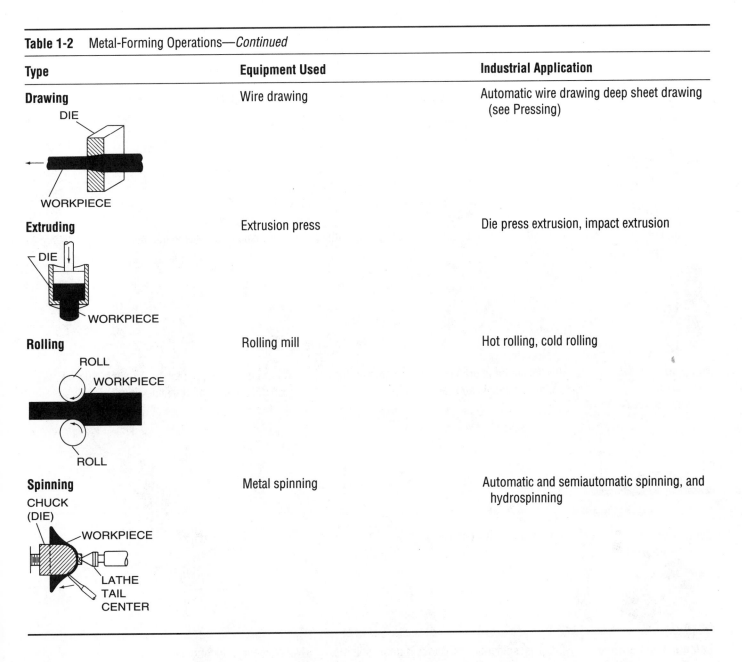	Wire drawing	Automatic wire drawing deep sheet drawing (see Pressing)
Extruding	Extrusion press	Die press extrusion, impact extrusion
Rolling	Rolling mill	Hot rolling, cold rolling
Spinning	Metal spinning	Automatic and semiautomatic spinning, and hydrospinning

worked. Rolling is performed on huge rolling mills wherein the metal is squeezed between two revolving rolls that are closer together than the starting thickness of the metal being rolled. Rolling mills can produce a variety of shapes, including: sheets, round rods, flat bars, and angles.

Spinning

Spinning is used to obtain shapes similar to those produced by deep drawing. Parts produced by spinning can be made with contoured sides which cannot be readily produced by other processes. This process involves pressing a rotating blank of metal around a specially contoured chuck. Very large shapes can be

produced this way, and a variety of thicknesses in material can be used.

Fastening

Fastening is the process of joining two or more pieces of metal by mechanical, fusion (heat), or adhesive means (Table 1-3). Many metal products are made up of a number of individual parts which must be joined in some way in order to make a usable product. These individual parts are made by cutting and forming and then assembled into a completed structure. Mechanical fastening systems include: rivets, lock seams, nuts and bolts, special clips, and retaining rings. Fusion, or heat, systems include welding, brazing, and soldering. Welding provides a much stronger joint than

Table 1-3 Metal-Fastening Operations

Type	Equipment Used	Industrial Application
Riveting 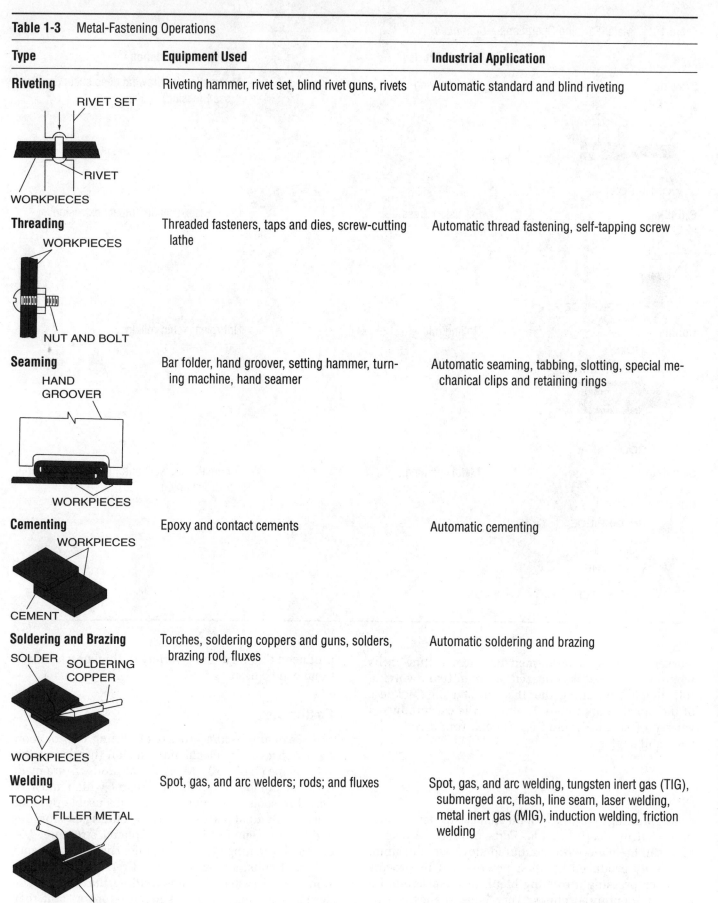	Riveting hammer, rivet set, blind rivet guns, rivets	Automatic standard and blind riveting
Threading	Threaded fasteners, taps and dies, screw-cutting lathe	Automatic thread fastening, self-tapping screw
Seaming	Bar folder, hand groover, setting hammer, turning machine, hand seamer	Automatic seaming, tabbing, slotting, special mechanical clips and retaining rings
Cementing	Epoxy and contact cements	Automatic cementing
Soldering and Brazing	Torches, soldering coppers and guns, solders, brazing rod, fluxes	Automatic soldering and brazing
Welding	Spot, gas, and arc welders; rods; and fluxes	Spot, gas, and arc welding, tungsten inert gas (TIG), submerged arc, flash, line seam, laser welding, metal inert gas (MIG), induction welding, friction welding

does brazing or soldering, although the processes all involve the use of heat. Adhesive cementing is a third important metal joining system.

Riveting

Riveting is a mechanical fastening process wherein metal pins with a formed head on one end are inserted through holes in the parts to be joined. A head is then formed by force at the other end in order to lock the parts firmly together.

Thread Fastening

Thread fastening is similar to riveting except that workpieces are locked together using nuts and bolts. Riveting is a permanent system, but thread fastening is considered semipermanent and permits disassembly of the product. Sheet metal screws, setscrews, and self-tapping screws are variations of threaded fasteners.

Seaming

Seaming is a fastening system using self-locking joints. Standing seams, grooved seams, and folding tab locks are good examples. Many of these seams are used in conjunction with soldering and adhesives.

Cementing

Cementing is permanently joining metal parts by using chemical adhesives. This process is the same as gluing two pieces of wood together except that special high-strength cements are used. Adhesives are generally used with sheet metals.

Soldering and Brazing

Soldering and **brazing** are closely allied to welding in that they are bonding systems. They provide permanent fastening and differ from welding in that the metals are not fused (melted) together but are joined by using filler metals.

Welding

Welding is the process of permanently joining metals by fusing them together with molten metal or pressure. Oxyacetylene, carbon arc, and spot welding are the most common systems used.

The fastening processes described above may be permanent or semipermanent. They may be used to join sheet metals or heavy structural members, or they may be used whenever a smooth attractive appearance is desired or great strength is required.

Finishing

Finishing is the process of treating or working the surface of the metal to protect it or to make it more at-tractive (Table 1-4). For example, metal can be painted to protect it or to make it more attractive.

Buffing

Buffing is achieved by working a surface with abrasive action using buffing (polishing) wheels and compounds. For example, a copper ashtray can be buffed to a bright or satin finish on a buffing machine. Industry uses many automatic machines to accomplish this purpose.

Texturing

Texturing is another important way of finishing metal. A simple technique is producing a hammered finish on a piece of metal by striking it with the ball peen of a hammer. Textured finishing is usually done to make a product more attractive, but it also has functional uses. For example, a knurled surface makes a handle easier to grasp and turn. Textured metal may also be further treated by coating.

Coloring

Coloring metal is achieved by using chemicals or heat to give a permanent color to a material. This involves placing a clean metal workpiece in a chemical, or subjecting it to an open flame. These colored workpieces may then be used as they are or covered with a coating of lacquer or wax.

Coating

Coating is the process wherein a metal surface is protected or made more attractive. You can spray or brush a coat of lacquer on a product and it will retain its luster longer because it is protected from fingerprints and corrosive fumes present in the air.

KNOWLEDGE REVIEW

1. Name the areas of technology in which metalworking plays a vital role.

2. List ten reasons why you should take a course in metalwork.

3. List the four main classifications of the manufacturing processes.

4. List three types of cutting operations used in school metal shops.

5. List three types of forming operations used in school metal shops.

6. List three types of assembly operations used in school metal shops.

7. List three types of finishing operations used in school metal labs.

Table 1-4 Metal-Finishing Operations

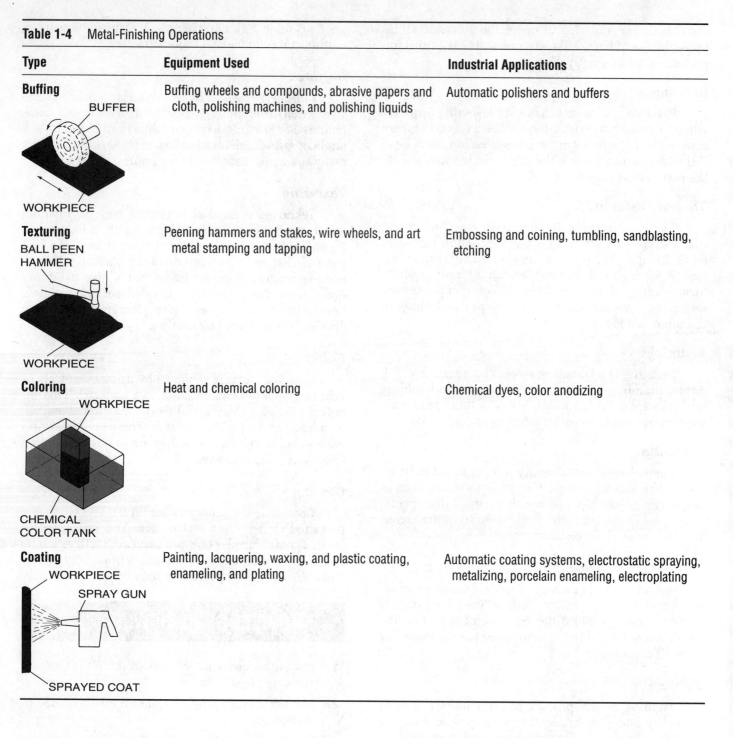

Type	Equipment Used	Industrial Applications
Buffing	Buffing wheels and compounds, abrasive papers and cloth, polishing machines, and polishing liquids	Automatic polishers and buffers
Texturing	Peening hammers and stakes, wire wheels, and art metal stamping and tapping	Embossing and coining, tumbling, sandblasting, etching
Coloring	Heat and chemical coloring	Chemical dyes, color anodizing
Coating	Painting, lacquering, waxing, and plastic coating, enameling, and plating	Automatic coating systems, electrostatic spraying, metalizing, porcelain enameling, electroplating

8. How do processes used in industry compare with those used in the school metal shop?

9. Give an example of a metal removal cutting operation and a metal separation cutting operation.

10. What is the difference between bending and pressing?

11. Find out all you can about one of the following people, and report on that person's activities in the metalworking field: Paul Revere, Eli Whitney, Thomas Edison, Henry Ford, Walter Chrysler, William Knudsen, Henry Bessemer, Jan Matzeliger, or Elijah McCoy.

12. Talk to a friend or neighbor who works in the metalworking industry. Ask the person about the job and about training for it. Report on this in class.

13. Select a product found in the home or the shop. Give an oral report on the processes used in making it.

14. Write a short research report on one of the industrial manufacturing processes.

Unit 2

Manufacturing Technology

OBJECTIVES

After studying this unit, you should be able to:

- Explain the words *science, technology,* and *industry.*
- Explain the relationships between science, technology, and industry.
- Describe the benefits which technology brings to society.
- Describe the problems of a technological society.
- Understand your responsibilities in a technological society.

KEY TERMS

industry   science   system   technology

The period in which we now live is often called the computer age, the age of automation, and the age of technology. These are accurate descriptions, for around us we see remarkable advances in communications, transportation, construction, materials science, manufacturing, and medicine. One of humankind's greatest technological achievements was carrying men to the moon, landing them on it, and then safely returning them to earth. The development of the Internet permits us to communicate with people all over the world from the comfort of our homes. Sleek, fast, and maneuverable jet fighters are an important part of America's air defense (Figure 2-1). New materials and power systems lead to new kinds of recreational products, such as the jet boat in Figure 2-2.

These kinds of achievements in technology have indeed changed the way we live and work, but along with the benefits have come many problems, some of which are described later in this unit.

It is helpful to know three important terms in order to understand how these advances have come about. **Science** is the broad field of human knowledge concerned with the study of the laws of the physical and material world. Scientists are people who discover and test these laws. For example, the scientists Charles Townes and Arthur Schawlow developed the working laser. Technologists took this working theory and applied it to the development of laser machines used, for example, to cut and weld metal. **Technology,** therefore, can be defined as the application of science for some practical purpose, such as creating useful laser welders. Technology is the organized study of the practical or industrial arts. A manufacturing **industry,** in turn, uses technology to make the many products we use every day. Similarly, technology is used by the communications, transportation, and construction industries to improve and expand their functions.

The relationships between science and technology are interesting and important. In the above example, laser theory had to be perfected before technologists could develop laser machines to process materials. However, early humans learned to extract crude copper from ores and to use it to make simple tools long before there was a science of metallurgy. In other words, these early people made metal tools without knowing anything about the properties of copper, such as its melting temperature. They discovered that by beating the soft copper into a knife blade or an axe head it became hard and brittle. Through trial and error, they learned that the axe head could be softened (or annealed) by heating it, so that it could be further shaped and improved. Science and technology continue to work together to develop new processes and products. Technology provides the link between scientific discovery and industrial production.

Fig. 2-1 This jet fighter is one example of technical advances in the aerospace industry. (Courtesy of Aluminum Company of America [ALCOA])

Technology Systems

Technology can be further described as a series of systems. A **system** is any combination of basic parts which form a usable plan or program. For example, an automobile has an ignition system and a fuel system, among others. Fuel is added to the tank, is drawn from there to the carburetor, and is fired in the engine cylinders to create the energy which causes the vehicle to move. The automobile is also part of a greater transportation system which is made up of vehicles, highways, bridges, tunnels, and service stations. (Railways, aircraft, and ships are also part of this transportation system.) A common method of describing these technological arrangements is through a universal systems model, which includes input, process, output, and feedback (Figure 2-3).

In your home can be found plumbing systems, electrical systems, and heating systems. Each is needed in order for the home to function properly as a place to live. In early America, the usual way to heat a living room was to load a potbelly stove with wood, set fire to the wood, and soon the room was warm and pleasant. When the people in the room began to feel cold, more fuel was added. The *inputs* in this heating system were the stove, the wood, and the people to tend the fire. The *process* was the burning of the fuel, and the *output* was the heat. The *feedback* was provided by the occupants of the room who were either warm enough, or felt cold and required more heat.

Fig. 2-2 This jet boat is a recreational product resulting from modern technology. (Courtesy of Arctco, Inc.)

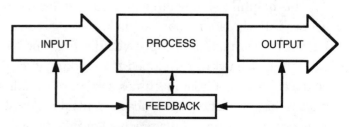

Fig. 2-3 The universal systems chart.

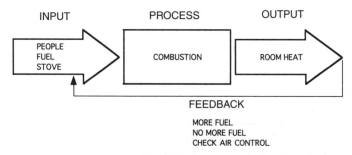

Fig. 2-4 The systems chart applied to a home heating system.

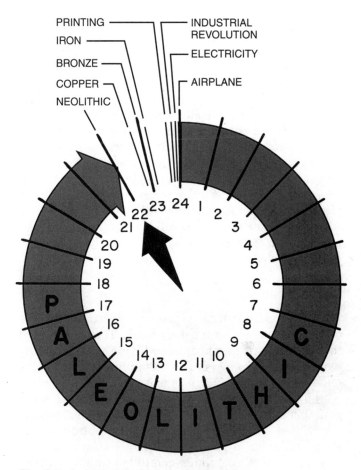

Fig. 2-5 A chart of the history of technology.

In a modern home heating system, a fuel oil tank automatically feeds to the furnace where the oil is burned, thereby heating the house. A room thermostat set at a certain temperature will automatically signal the furnace to turn on when the temperature falls below the thermostat setting, and to turn off when the setting has been reached. In this case, a series of electronic and mechanical devices have taken over the duties of people as furnace tenders (Figure 2-4).

Because this book deals with the area of metals technology, much of the discussion will center around metal product manufacturing systems. While people still continue to weld and drill and otherwise work metal, many computer-controlled machines are now in use. Examples of these machines and their special manufacturing systems will be described and illustrated throughout this book.

Humans and Technology

It would seem that there is very little that humans cannot do if they set their minds and hands to the task. In a sense, this has always been so; humans have long been inventing and finding better ways of doing things. Today, however, changes take place much faster than in past years. This is so because humans are better trained and educated, and have the use of better tools and materials. In other words, their technology is better.

It is estimated that human toolmakers have been on earth for roughly 240,000 years. If we take this number and set it to the scale of a twenty-four-hour clock, we get the illustration that appears in Figure 2-5. Note that humans spent twenty-two of the twenty-four "clock-hours" in the Paleolithic, or Old Stone Age. During this time they lived in caves and used the first crude stone tools. Just a short two hours ago, they entered the Neolithic, or New Stone Age. At this time, humans made polished stone tools of a more advanced design (they added tool handles, for example), began to farm and raise animals for food, and invented the all-important wheel.

One hour and twenty-four minutes ago, they discovered copper. One hour ago, they learned how to make and use bronze. This allowed these early people to make metal tools which were much better than the cruder stone implements of their ancestors. They were stronger and sharper, lighter and easier to use, and simpler to make. Forty-eight minutes ago, they discovered iron, and learned to cast and shape it into even stronger tools. With the discovery of metals, humans could make more efficient and dependable machines, and a variety of useful and aesthetic products. They invented printing just six minutes ago.

Two minutes ago on our twenty-four-hour clock, the Industrial Revolution began. During this period appeared some of the most significant inventions: Watt's steam engine, Arkwright's spinning machine, Whitney's cotton gin, and Maudslay's screw-cutting lathe. The factory emerged at this time, and machines became more efficient because of steam power. People learned to produce and use electricity only one minute and eighteen seconds ago. About twenty-five seconds ago, a man flew the first airplane. Atomic power, computers, and space flight occurred just a few seconds or fractions of seconds ago.

As you can see from studying this chart, the pace of technological progress is great, and you must learn to move with it. Your experiences in the classroom and the shop will help you to do this.

Technology Problems

Humans no sooner learn to use a machine or a material than a new one comes along for them to master. Along with new materials, skills, and ideas, a vast array of new tools has also been perfected to do work more quickly, accurately, and efficiently. From simple clay pots made by hand, people have moved to dishes mass produced by machines. And, whereas workers once had to hew logs by hand in order to make boards, they now do so more easily with the aid of woodworking machinery. The steady improvements in technology have made our work easier, given us better food and clothing, and made us more comfortable and healthier. We live longer and better due to our technologies. However, there are some serious problems which come to us with technology.

Detachment

Technology can cause people to become more separated or detached from their work. This means that they have less physical contact with the materials being worked into products. For example, cabinetmakers once hand-carved chair legs and arms, studying their workpieces and taking advantage of interesting wood grain patterns and knots. It was hard work, but satisfying in that a human with a tool was controlling the shape of the wood product. In a modern furniture factory, a person at a computer-controlled spindle-carving machine operates buttons and levers, and the machine does the work. Sometimes metal-casting machine operators do not even see the material, which may be in another part of the building. In these situations, while the operators are working in a safer, more comfortable environment, it can be difficult for them to become personally involved and interested in their work.

Reduction of Jobs

Technology demands a very high degree of accuracy in the mass production of goods. Very often a machine can do things that humans cannot possibly do, because machines are in some ways more accurate and reliable. Using a laser welder to join very thin metal parts is a good example. This causes changes in the types of human jobs that are available, as the machines take over the tasks that humans once did. The result is that many people must learn to live with the fact that they will often have to change jobs, or at least be retrained, several times during their working years.

Division of Labor

The efficiency of technology is directly related to repetitive actions, or operations which are done over and over. In other words, technology is most effective in mass production situations, where automatic devices are readily used in the manufacture of goods. Not only does this result in great quantities of identical products, but the production line jobs become very specialized. This is called *division of labor,* and while it makes for production line efficiency, it can become very boring.

Pollution

Technology-based societies consume great quantities of raw materials and cause serious pollution problems. Wherever people live together, such problems occur. However, in technology-based societies they are much more common and serious because too many products are designed to be discarded after use, such as plastic shavers and ball-point pens.

Living with Technology

In this unit you have learned of some of the achievements and problems of technology. No one can predict the future, but it seems very likely that our society will continue to become more automated and specialized. This means that the work force must have more education and training in order to work in modern factories (Figure 2-6).

Some people worry that machines may change our environment and economy so much that we will be ruined. However, humans are capable of solving these problems because they are greater than any machine. Your task in building a better, more humane society is to understand and appreciate technology for what it is and can do. To achieve this, you must be informed. Machines cannot think; humans can. The more you under-

Fig. 2-6 Living in a technical world requires new skills. This technician is using a coordinated measuring machine to check a part. (Courtesy of The Conard Corporation)

stand about technology, the less reason you will have to worry about it. This metals technology course you are taking will help prepare you to live, work, and play in a technical society.

KNOWLEDGE REVIEW

1. Ask your grandparents about their early years and how different their lives are today. What kinds of technological advances can you and they identify?

2. Write your own descriptions of the words *science, technology,* and *industry.*

3. Go to the library and prepare an historical research report on a product such as the bicycle, airplane, or typewriter.

4. Visit a local industry and interview some of the workers to learn about the modern machines and equipment they use in their work. Also, ask them about the education and training they needed to prepare them to do this work. Prepare an oral report of your findings. Your teacher can help you to arrange this industrial visit.

Unit 3

Safety

OBJECTIVES

After studying this unit, you should be able to:

- Identify the meaning and purpose of OSHA.
- Demonstrate a positive attitude toward safety.
- Know why never to violate safety rules.
- List important safeguards when using machines.
- Describe the general safety policies to follow in a lab.

KEY TERMS

OSHA   regulations   safety

You will get more out of shop activities and your future career if you avoid accidents. It is smart to be careful, for an accident can change your life.

Safety is important in all industrial education activities. When you get a job in industry or business, you will be required to observe the safety standards that were established by the U.S. Congress in 1970 with the passage of the **Occupational Safety and Health Act (OSHA).** These strict regulations provide protection for both the employee and the employer in all aspects of work. Many states have passed laws similar to the federal law. More and more, these regulations apply to both industries and schools.

Safety standards established by law are very comprehensive. These standards include not only the obvious, such as safety glasses or goggles, but also standards for guards, noise and air pollution, electrical hazards, and every other aspect of working conditions (Figure 3-1). Each employer must establish and maintain conditions of work that are safe and healthy. All employees must follow all safety **regulations** (Figure 3-2). No employee may willfully remove, displace, damage, destroy or carry off any safety device or safety item.

It is important in your school activities to develop proper attitudes toward health and safety. By the time you take a job in industry, you will have developed safe working procedures. The basic purpose of a comprehensive safety program is to provide you and your fellow students with a safe and healthy working environment. It is mandatory that you follow the safety regulations outlined in this unit and in each of the sections where safety regulations apply.

It is important for you to learn how to be both a good and a safe worker. The best way is to do each part of a job carefully and correctly as described in this book or as demonstrated by your instructor. Accidents usually happen when you do the wrong thing "just this once" or when you fail to follow proper methods (Figure 3-3). The *right* way is the *safe* way.

Fig. 3-1 In industry, dangerous work areas are surrounded by safety fencing. (Courtesy of Cincinnati Milacron, Inc.)

The metal shop is planned to be a safe place in which to work. All machines are guarded and in good working order. You must help maintain the tools and machines so that the shop will remain a safe place in which to work. Let your teacher know if you find any unsafe tools or machines in the shop.

To have an accident-free shop, remember the ABC of safety: *Always Be Careful.* Here are some important safeguards:

1. *Remember that the shop is a place for work, not horseplay.*

   a. Any tricks or pranks are dangerous to you and your friends.

   b. Do not be responsible for sending a fellow student to the hospital by playing a practical joke. For one moment of laughter, you will pay with a lifetime of regret.

2. *Whenever possible, use the buddy system when working at a difficult task.*

3. *Dress safely.*

   a. Roll up your sleeves, tuck in or remove your tie, and wear a shop coat or an apron (Figure 3-4).

   b. Remove all jewelry, including rings and watches. A watchband, for example, can catch in moving machinery (Figure 3-5).

OSHA Safety Checklist

Hand and Portable Power Tools

1. All hand and portable power tools are in good operating condition: no defects in wiring; equipped with ground wires.   Yes   No
2. All protable equipment is equipped with necessary guarding devices.   Yes   No
3. All compressed air equipment used for cleaning operations is regulated at 30 psi or less; chip guarding and personal protective equipment are provided.   Yes   No

Machine Guarding and Mechanical Safety

1. Every production machine has been inspected as to the following items and found to be in satisfactory operating condition:
   a) Cleanliness of machine and area   Yes   No
   b) Securely attached to floor   Yes   No
   c) Operations guarded   Yes   No
   d) Illumination   Yes   No
   e) Effective cutoff devices   Yes   No
   f) Noise level   Yes   No
   g) Adjustment   Yes   No
   h) Material flow   Yes   No

Material Hazards

1. All hazardous gases, liquids, and other materials are properly labeled and stored.   Yes   No
2. Areas where hazardous materials are in use are fire-safe and restricted to authorized employees.   Yes   No
3. Where X-ray is used, the area is properly shielded and dosimeters are used and processed for all authorized employees.   Yes   No
4. Protective clothing is worn by employees when oxidizing agents are being used.   Yes   No
5. All hazard areas are posted with NO SMOKING signs.   Yes   No
6. All areas where caustics or corrosives are used have been provided adequately with eye fountains and deluge showers.   Yes   No

Fig. 3-2 This OSHA checklist provides a good guide to school shop safety.

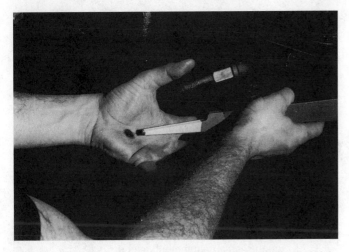

Fig. 3-3 Failure to use a handle on a file can result in a painful wound.

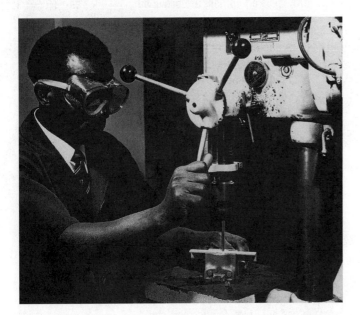

Fig. 3-4 This is the proper way to operate a drill press. How many safe practices can you identify?

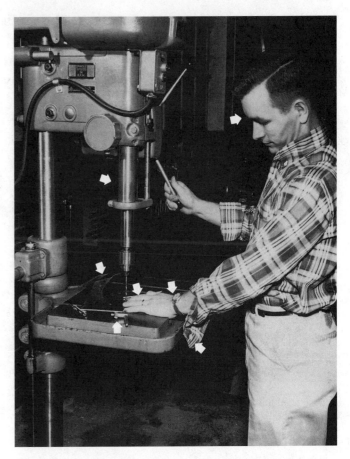

Fig. 3-5 The drill press can be dangerous. Find the following unsafe practices: quill extended too far; sharp edges on metal; workpiece being held by hand; chuck key, drill, and other tools lying on table; loose sleeves on operator; operator not wearing apron or goggles; and operator wearing watch and ring.

   c. Keep your hair cut short or out of the way. Long hair around moving parts is dangerous and must be covered with a net or cap.

   d. Always wear special protective clothing when working in the welding, forging, and foundry areas.

4. *Protect your eyes and ears.*

   a. Wear goggles or a shield whenever there are sparks. You only have two eyes. Protect them.

   b. Wear safety glasses when grinding or buffing or whenever there is danger of flying chips (Figure 3-6).

   c. Wear special goggles or a shield for gas and arc welding.

   d. Wear goggles and protective clothing when pouring hot metal in the foundry or when working with acids.

   e. Wear ear protection whenever shop noise is loud enough to damage ears (Figure 3-7).

5. *Take proper care of hand tools.*

   a. Most accidents are caused by incorrect use of hand tools or poor tool maintenance. Leaving scraps of metal lying around is always a hazard.

   b. Dull tools are dangerous. Always keep tools sharp.

   c. Make sure that hammer heads and screwdriver blades are fastened tightly to their handles.

   d. Always put a handle on a file before using it.

   e. Grind "mushroom" heads and all burrs off cold chisels, center punches, and other small hand tools (Figure 3-8).

   f. Always keep pliers, screwdrivers, and metal shears in good working condition.

Fig. 3-6 Machine tools can be dangerous if you are not careful in your work. Practice safety as this operator is. (Courtesy of Norton Company)

Fig. 3-7 This headset electronically couples low frequency noise waves with shop noises to cancel sounds which may damage the ear. (Courtesy of illbruck, inc.)

6. *Use tools correctly.*

   a. Choose the right tool or equipment for the job. The work can then be done faster and more safely.

   b. There is always a right and a wrong way to use a tool. Learn to use a tool the right way (Figure 3-9).

   c. Never carry sharp tools in your pockets.

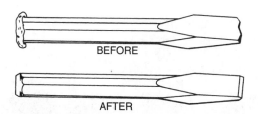

Fig. 3-8 The cold chisel at the bottom is sharp and in good condition. The one at the top is dull and has a "mushroom" head. Take care of your tools.

Fig. 3-9 This student is holding the workpiece with a file while soldering. This is a good safety practice.

   d. You can pinch your fingers with pliers or snips. You can get burned with hot metals or hot tools. Always be careful.

7. *Use portable electric hand tools correctly.* These tools operate on 110 volts. This voltage can kill or cause a serious shock or burn under certain conditions.

   a. Always check the electric hand tool before using it. Make sure the cord is in good condition and that it does not have a broken plug or switch.

   b. Always keep the cord away from oil or hot surfaces.

   c. Never use electric tools around flammable vapors and gases. This could cause an explosion.

   d. Always be sure that your hands are dry when using an electric hand tool.

8. *Observe safety rules when running machines.*

   a. Always follow the safety rules given in each unit on each machine.

   b. Stop the machine before oiling, lubricating, or adjusting.

Fig. 3-10 A neat, well-organized tool cabinet contributes to shop safety.

Fig. 3-11 Which kind of housekeeper are you?

   c. Never feel the surface of metal while it is being machined.

   d. Clean chips off with a brush—never with a rag or your hand.

   e. Never allow anyone to stand near the machine you are using.

   f. Never use measuring tools on metal while it is being machined.

   g. Keep the guards in place. They were put there for your protection.

9. *Be a good housekeeper.*

   a. Do your part in keeping the shop clean and in order.

   b. Clean the tool or machine after using it.

   c. Put away all tools and accessories (Figure 3-10).

   d. Wipe up any oil or grease on the floor.

   e. Get rid of waste materials.

   f. Put your work away at the end of the period.

   g. A clean shop is likely to be a safe shop. Do not wait for someone else to clean up (Figure 3-11).

10. *Ask for first aid every time you need it.*

   a. Do not laugh off a small injury or burn. Get first aid no matter how slight the injury. Infection may start many days after you scratch your hand.

   b. Report every accident to your instructor. If necessary, he or she will send you to the school nurse or to a doctor.

   c. A small burn, a metal sliver in your finger, or a cut can easily cause blood poisoning. A piece of metal in the eye can cause blindness. Don't believe it never happens; each year there are over 153,000 eye injuries to students. Do not add yourself to this number.

   Whenever you are involved in any shop activity, remind yourself of these basic rules of safety, and practice them: (1) dress properly, (2) know your job, (3) and do your job correctly.

KNOWLEDGE REVIEW

1. What is the ABC of safety?

2. What is the proper way to dress in the shop?

3. In what ways can the eyes be protected?

4. What is the right way to use a file?

5. Why should the right tool be used for a job?

6. What should you do in case of an accident in the shop?

7. What are the three basic rules of safety?

8. Write to an insurance company or to the National Safety Council for information about school shop and industrial safety practices. Give a report.

Unit 4

Careers

OBJECTIVES

After studying this unit, you should be able to:

- Name the four basic technologies in metalwork.
- Find where detailed occupational information can be obtained.
- Name some of the skills needed in metalworking.
- Identify what a diemaker does.
- Explain how to become a skilled metalworker.
- Describe the work of a technician.
- Name several kinds of technicians.
- Identify what an engineer does.
- Describe the work of some of the engineers involved in metals.
- Tell what kind of education is needed to become an engineer.
- Describe what a scientist does.
- Name three areas in which scientists work.
- Describe some of the qualifications needed to become a metalworker.

KEY TERMS

designer	engineer	patternmaker	technician
diemaker	machinist	research	toolmaker
drafter	millwright	scientist	welder

The metalworking industry contributes to the production of goods in many ways. It produces metal raw materials, tools, machines, and structures for other industries. It also delivers a wide variety of goods and services directly to consumers. As described in unit 1, the metalworking industry can be divided into four basic areas: communication, manufacturing, construction, and transportation. Each group includes many different occupations in which metalworking skills are essential (Figure 4-1). The purpose of this unit is to help you learn more about these careers and occupations.

The field of metalworking offers many exciting careers and occupations. Engineers and technicians are needed to plan and supervise the work. Skilled workers are needed to perform the difficult operations. Semiskilled workers are required to operate production machines. Routine jobs are done by unskilled workers. Detailed information on occupations in the metalworking field is available in *Occupational Outlook Handbook,* published by the U.S. Government Printing Office, Washington, D.C.

General job categories and a number of specific jobs are discussed below. The skilled workers, technicians, engineers, and scientists serve on technical teams in the metalworking industry. Each job has certain educational and training requirements.

Skilled Metalworkers

The metalworkers of modern industry are trained to do skilled work and to use precision tools and machines to build, operate, and maintain many different products. These workers include the toolmakers, diemakers, wood and metal patternmakers, millwrights, electricians, welders, plumbers, sheet metal workers, and many others (Figure 4-2).

A brief description of some of the skilled occupations and a list of typical jobs done in each follow. These are not all the skilled trades. Several of the skills of one trade may be common to a number of other trades. Nonetheless, the occupations described below represent the types of skill generally found in the metalworking industry.

All-Around Machinist

The all-around machinist is a skilled worker who uses machine tools to make metal parts. Machinists can set up and operate most machine tools. A wide knowledge of shop practice and the working properties of metals, plus an understanding of what the various machine tools can do, enable the machinist to turn a block of metal into an intricate part meeting precise specifications (Figure 4-3).

Variety is the main characteristic of the job of the all-around machinist. This worker plans and carries through all operations needed in turning out machined products, sometimes switching from one kind of product to another. A machinist selects the tools and material needed for each job and plans the cutting and finishing operations in order to make the product according to blueprint or written specifications. A machinist makes standard shop computations relating to dimensions of work, tooling, feeds, and speeds of machining. A machinist often uses precision measuring instruments, such as micrometers and gages. After completing machining operations, a machinist may finish the work by hand, using files and scrapers, and then as-

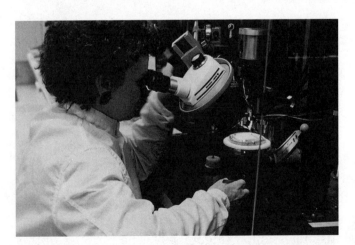

Fig. 4-1 This technician is using a microscope in the manufacture of electronic parts. (Courtesy of Rockwell's Collins Businesses)

Fig. 4-2 These engineering technicians are preparing metal sheet samples for testing. (Courtesy of Aluminum Company of America [ALCOA])

Fig. 4-3 The skilled machinist must be able to use many different kinds of machines. Here Machinist Mate Third Class Jeffrey Frederick is at work in the machine shop aboard the aircraft carrier USS John F. Kennedy. (Courtesy of U.S. Navy; photograph by Airman Jones)

semble the finished parts with wrenches and screwdrivers. The all-around machinist may also heat-treat cutting tools and parts to improve their machinability.

Diemaker

The diemaker is a skilled machinist who specializes in making the molds and dies used for die casting, stamping, and pressing out metal parts. This worker also repairs damaged or worn dies and sometimes makes tools and gages. This trade requires a very high degree of manual skill as well as a good knowledge of shop mathematics, blueprint reading, and the properties of many kinds of metal. Imagination and ingenuity are also important, for the diemaker must be able to visualize the completed die or part before starting to work.

Toolmaker

The work of the toolmaker is closely related to that of the diemaker, since both must be skilled machinists. The toolmaker also constructs precision gages and measuring instruments for checking the accuracy of parts.

Patternmaker

The patternmaker constructs the wood or metal patterns and core boxes that are used to make molds in casting parts from metal or plastic. This worker must have a high degree of manual skill because of the many hand operations involved in pattern building. Accuracy is crucial in patternmaking because any imperfection in the pattern will show up in the casting. The patternmaker must also have the artistic ability to visualize the

parts the patterns will be used to make. A knowledge of metal shrinkage and foundry practices is essential in both wood and metal patternmaking.

The wood patternmaker, working from blueprints, first lays out each section of the pattern to be made. Next, this worker uses woodworking machines and hand tools to shape the pieces. The patternmaker then fits the pieces together to form the final pattern or core box.

Metal patternmakers use metalworking machines and equipment to make patterns. This skill is similar to that of the toolmaker and diemaker. Unlike the wood patternmaker, the metal patternmaker makes patterns that are normally used many hundreds of times.

Machine Maintenance Worker

The machine maintenance worker makes certain that the machinery of the plant is kept in good condition. Such workers must be able to repair breakdowns as quickly as possible. A complete knowledge of the structure and operation of every machine used in the shop is needed. It is necessary to know all types of bearings and the lubricating requirements of the machines serviced. Doing routine maintenance work, making replacement parts, and doing repairs for presses, lathes, mills, and other machines are also parts of the job.

Sheet Metal Worker

The sheet metal worker must be able to read blueprints to lay out complicated metal shapes and surfaces. The job requires the skillful use of hand and power tools for cutting, forming, and fabricating sheet metal. The job may involve constructing aircraft parts, hoppers, air-conditioning ducts, paint-spraying booths, furnace piping, or safety shields for the shop machines. In the automobile industry, the sheet metal worker makes and assembles the parts for the handmade metal bodies of experimental and preproduction automobiles.

Millwright

Millwrights install, dismantle, move, and set up the machinery and industrial equipment used in factories. They often prepare the platforms and concrete foundations on which the machines are mounted. They also frequently help plan the location of new equipment.

The millwright must have a thorough knowledge of the structure of the buildings and the operation of the machines. A good knowledge of blueprint reading is also needed. The millwright must be skillful in welding and many other trades in order to install and service overhead tracks and cranes, conveyors, bins, chutes, and structural members.

Welder and Oxygen Cutter

Welders join metal parts together by first applying gas flames or electricity to cause melting, then either

Fig. 4-4 Setting up and controlling a robotic welding operation is an important industrial metalworking skill. (Courtesy of Arctco, Inc.)

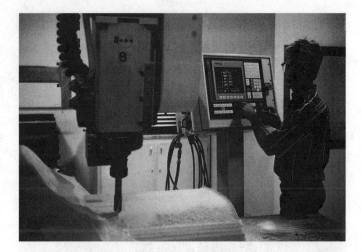

Fig. 4-5 This technician is using a rapid prototyping machine to prepare a material forming mold. (Courtesy of Arctco, Inc.)

letting the parts fuse by themselves or pressing them together. Welders must know how to weld different metals. They must also know how to select and operate welding equipment. The ability to read blueprints is important. Welders are employed in making and repairing thousands of metal products. Some welders specialize in setting up welding work for others. Other welders work with engineers on designing new welding equipment (Figure 4-4).

An oxygen cutter's job is like gas-flame welding, except that instead of joining the metal parts together, the cutter cuts them apart.

Learning to Become a Skilled Metalworker

There are several ways you can learn the skills of a metalworker. A craft or trade can be learned in a vocational or trade school or in an industrial apprenticeship program. Many vocational courses, such as machine shop, welding, foundry, and drafting, can be taken in high schools. These are useful if you enter a trade apprenticeship program. The length of the program varies with the trade. Tool-and-die work, machine operation, and patternmaking frequently take four years. Some maintenance trades can be completed in three years. In every case, you must satisfactorily complete a certain number of work hours set down in the training schedule for each skill and operation in your chosen trade.

While on the job, you will be under the guidance of the supervisor of the department in which you work. Complete records will be kept on all phases of your training. Your shop supervisors and instructors will review your progress from time to time. They will comment on your learning ability, skill, initiative, and attitude toward the job.

In addition to your work in the shop as an apprentice, you may also be required to attend classes in shop mathematics, blueprint reading, mechanics, and other subjects related to your trade. Here your practical courses and shop training in school will be a great help. The more you have learned in school, the less extra study you will have to do to complete your apprenticeship.

Vocational programs in high schools, trade schools, and community or junior colleges often can prepare you for jobs in small local industries. You can then receive extra training on the job in the special skills needed in your particular craft.

Technicians

There are many different scientific and technical careers available to technicians. In industry, technicians apply the science and mathematics learned in a technical institute or a community or junior college. They use specialized and highly accurate measuring devices. They build special machinery to test new devices. They collect data and compile reports (Figure 4-5). In short, they work with engineers and scientists, helping them change ideas into reality.

Many technicians work in the metalworking industry. Some of these are drafters and designers, others are mechanical, metallurgical, building-construction, and instrumentation technicians.

Drafters and Designers

Drafting and design technicians are the graphic communicators on the technical team (Figure 4-6). They translate scientific and engineering sketches and notes into accurate drawings. Prototypes or other kinds of experimental equipment are then made from these drawings. These technicians have a background in technical

Fig. 4-6 The design technician must learn the art of graphic reproduction. Both computer and manual drafting skills are needed. (Courtesy of Grob Werke, GmbH)

Fig. 4-7 Technicians use electron microscopes to inspect and analyze everything from metal grain structures to air samples. (Courtesy of NASA)

drawing, mathematics, and processes. This knowledge enables them to produce graphical and tabular data that the skilled worker uses. Technical sketching, rendering, and writing are parts of this job.

Mechanical Technician

Mechanical technology involves machines, mechanisms, and industrial processes, and is an important part of industrial activities. The work done by the mechanical technician is a part of, or is related to, research and engineering activities. The job requires a broad understanding of engineering design and production.

The work of the mechanical technician involves two general areas: design and production. The design area includes such jobs as designer, drafter, tool designer, research assistant, or engineering assistant. The production area relates to mechanical manufacturing functions. It includes activities such as quality control, production planning, methods analyzing, and job estimating.

Metallurgical Technician

The metallurgical technician is an important member of a technical team engaged in product or process research and development or in production. The work requires special knowledge of the principles of chemistry, physics, and mathematics as applied in the metallurgical industry. The work also requires a broad knowledge of laboratory procedures, metallurgi-

cal processes, and laboratory analyses. This background enables the technician to work closely with scientists and engineers. As a member of the metallurgical research team, the technician obtains test data on processes for developing new products or improving production methods (Figure 4-7).

Building-Construction Technician

The construction technician works closely with civil engineers. The job involves detailed design work, materials testing, surveying, estimating, construction, supervision, report writing, and other engineering work. Many tasks previously done by a civil engineer are now performed by the building-construction technician. The technician is not chiefly responsible for the design or construction of engineering projects and remains accountable to the professional engineer. Nevertheless, the building-construction technician may head a surveying crew or supervise the materials-testing laboratory as well as hold other responsible positions.

Instrumentation Technician

Scientific instruments extend human senses and control in space exploration, weather prediction, communications, environmental control, industrial research, automated processing and production, and many other areas of applied science.

Instrumentation technicians must be able to work closely with instrumentation engineers and scientists. They must also be able to supervise and coordinate the efforts of skilled workers and instrument maintenance workers. Technicians are members of the scientific team that plans, assembles, installs, calibrates, evaluates, and operates instruments in metalworking processes or systems. Because instrumentation technicians are employed in so many different and specialized situations, they must have a broad range of scientific knowledge and technical skills.

Engineers

Engineers who work in the metalworking industry are trained in mathematics and the physical sciences. They apply their knowledge, experience, and judgment to develop economical ways of controlling and using the materials and forces of nature. They transform raw materials into useful devices and harness energy to help people do work. They are builders or inventors who apply scientific knowledge to solve the practical problems of life, to build better systems, and to create new products. Engineering could be called *the art of making science useful.*

Although there are many types of engineers, only those closely related to the metalworking industry will be described here. These are civil, mechanical, metallurgical, mining, aerospace, and industrial engineers.

Civil Engineer

Civil engineers are chiefly interested in the relationships between people and the physical environment. They design and construct structures and systems that add to comfort, convenience, and safety. They are trained in soil mechanics, building materials and structural analysis, hydrodynamics, fluid mechanics, and environmental health. Civil engineers help develop commercial buildings, expressways, pipelines, waterways, sanitation systems, urban renewal projects, and bridges of all types.

Mechanical Engineer

Mechanical engineers develop the hardware for the generation, transmission, and application of heat and mechanical energy. Mechanical engineers design power sources, such as reactors, gas turbines, internal-combustion engines, and heat exchangers and transmitters. They also design the machines and tools used in manufacturing operations. Their training in applied mechanics, strength of materials, thermodynamics, and fluid mechanics gives them broad knowledge that can be applied to many industries.

Metallurgical and Mining Engineers

These engineers employ new techniques to design systems for locating and extracting increasingly scarce minerals from the ground and for converting them into useful raw materials or consumer products. Mining engineers use their knowledge of geology, chemistry, surveying, materials handling, and mining operations in working with solid minerals such as iron ore, coal, and limestone. Metallurgical engineers help extract impurities to refine the ores and help produce useful metals and alloys.

Aerospace Engineer

Aerospace engineering deals with the application of many areas of engineering to all types of flight vehicles, from aircraft to missiles. Aerospace engineers study aerodynamics, hydraulics, electronics, structural design, materials strength, and flight mechanics. They develop and test guidance and propulsion systems for airplanes and spacecraft. This field demands a broad base of scientific and engineering knowledge and rapid adjustment to changes and discoveries.

Industrial Engineer

Industrial engineers help make manufacturing and assembly processes more efficient by combining people, materials, machines, and methods in new and better ways. Industrial engineers need a broad knowledge of engineering and manufacturing and an understanding of plant operations, cost analysis, personnel relations, and applied statistics. They must be familiar with human as well as technical concerns when designing and managing systems for the production of quality products at a reasonable cost.

Becoming an Engineer

If you are interested in engineering, your objective in high school should be to lay a firm foundation for further training in college. To qualify for most engineering schools, you must take several basic subjects, including mathematics, science, and communication. These subjects provide the groundwork for any field of engineering and are the basic tools of the engineer. If you do not enter an engineering school directly after high school, you may attend a two-year technical institute or a community or junior college and then transfer to a university to complete your engineering degree. All engineering programs are at least four years long. After graduating from such a program and, perhaps, working for a while, you may wish to study for an advanced degree.

Scientists

The scientist on the technical team is the discoverer, the person who seeks new knowledge. The scientist is an inquirer—someone trained in the basic laws of nature who investigates the physical world and its secrets to learn more about why and how things behave as they do. Although the various branches of science demand different skills and training, scientists in every field possess several similar characteristics. Perhaps the most important trait is curiosity—a driving need to know. It is a desire for knowledge that compels the scientist to search until a reasonable explanation has been found for something not understood before. This curiosity forces the scientist to question even accepted facts and established theories. Though there are many specialized fields in science, almost all scientific careers are based in some part on studies in the basic sciences—chemistry, physics, biology, and mathematics. After gaining a fundamental knowledge of these subjects, most prospective scientists advance to higher levels of study and select a field in which to concentrate.

Careers in science cover a wide range of activities that are difficult to define in specific terms. In various industries, the same type of scientific work may be known by several different names. There are, however, a few basic descriptions that cover the work of most scientists.

Basic Research

Scientists engaged in basic research are trying to gain a better understanding of the natural and physical world and are seldom concerned with the immediate use of their discoveries.

Applied Research

Scientists in applied research usually seek new knowledge with a well-defined goal and a specific application in mind. Applied research scientists often work closely with engineers in developing new products or techniques.

Teaching

Scientists who are teachers help fill the need for future qualified scientists. In addition, college and university instructors often do individual research on projects supported by their school or by business or government. Like most scientists, science educators usually hold at least a master's degree.

Learning to Become a Scientist

If you are interested in science, high school courses such as English, mathematics, science, and social studies will help you fulfill the basic requirements for most college programs in science.

Getting Started in a Metalworking Career

There is no certain way of telling whether you will make a good metalworker or engineer until you actually try. However, aptitude tests may help you find out what skills you already have. As in any occupation, success depends a good deal on your attitude, temperament, and willingness to work and learn.

There are, however, a few qualifications you must have in order to be on the technical team. You must

Fig. 4-8 Every industrial occupational level, from skilled crafter to scientist, requires computer literacy. (Courtesy of U.S. Department of Labor)

have acquired mechanical ability. You must like working with materials, machines, and ideas. If you have ever built a birdhouse, put an alarm clock back together, or fixed a bicycle, you probably have some idea of your aptitudes. Your school metals class will give you an excellent opportunity to find out whether you will enjoy this type of work as a career.

A knowledge of blueprint reading and drafting is essential in nearly all technical jobs. Your school courses in these subjects will be a big help to you. Take as many science courses as you can, particularly physics. Your shop classes will give you a good foundation in shop practice and theory. They will also give you a chance to use many of the hand and machine tools skilled workers use. English, speech, history, and computer training courses will round out your education and aid you in expressing your opinions (Figure 4-8). The information and skills gained from them may be helpful in future promotions to supervisory positions.

But most of all, you will have to answer this question: Do I think I have the ability and the desire for a career in metalworking?

KNOWLEDGE REVIEW

1. What are the four main classes of metalworking industries?

2. Which four types of workers serve on technical teams in the metalworking industry?

3. List the eight types of skilled metalworkers.

4. What six kinds of technicians are employed in the metalworking industry?

5. What six kinds of engineers work in the metals industry?

6. What are three kinds of careers open to scientists?

7. Interview a friend or relative who works in the metalworking industry. Write a short description of the job.

8. Discuss a career in metals with your teacher and counselor.

Unit 5

Metal Properties and Applications

OBJECTIVES

After studying this unit, you should be able to:

- Describe the meaning of materials as universal system inputs.
- List and describe the various material properties.
- Sketch and explain the different mechanical property forces.
- Describe the basic differences among plastic, wood, ceramic, and composite materials.
- List and define the special mechanical properties of metal.

KEY TERMS

acoustical	fusibility	plasticity	thermoset
brittleness	hardness	property	torsion
chemical	machinability	shear	toughness
compression	magnetic	strength	
ductility	malleability	tension	
elasticity	melting point	thermal	
electrical	optical	thermoplast	

Materials are substances from which products are made; they are the *inputs* of the universal systems model described in unit 2. Designers must select from the many kinds of metals, plastics, woods, and ceramics when planning things to be manufactured. These input materials are *processed* or changed by various machines into a product, or *output.* The relationships between materials and product design are described in unit 11. The information presented here will help you to learn more about materials in general, and metals in particular. This will help you to select the best material for a project you may wish to make. Modern industrial designers also make use of this information to make their material selections. This was also true in the early history of metalcrafting. For example, the small sword in Figure 5-1 was made of steel for strength and a keen edge. The other materials add to its function and appearance.

Fig. 5-1 Aubry Le Jeune, French, Small-sword, steel, gold, silver, wood, leather, c.1775–85, 1.: 105.7 cm, George F. Harding Collection, 1982.2156 view 1. (Photograph © 1996, The Art Institute of Chicago, all rights reserved)

Material Properties

Materials have several features or qualities, called **properties,** which product designers study to aid in their material choices. *Physical* properties describe material melting points, density, and moisture content. For example, steel has a higher melting point than aluminum. *Chemical* properties relate to the material's resistance to corrosion and deterioration, such as the rusting of steel. *Thermal* properties are measures of how temperature affects materials. For example, ceramics can generally withstand higher temperatures than can pieces of wood. *Electrical* properties deter-

mine material conductivity and resistance to electrical charges. Copper is a good conductor of electricity, and plastics are better insulators. *Acoustical* properties indicate reactions to sound, and *optical* properties reactions to light.

Mechanical properties are especially important because they are indicators of material strength, workability, and durability. For example, *tension* is a mechanical force that stretches a material; *compression* is a force that applies squeeze pressure; *torsion* is a twisting or torquing force; and *shear* involves two opposing forces tending to fracture a material, as in the shearing of a piece of paper or metal. The drawings in Figure 5-2 illustrate the actions of these mechanical forces. A knowledge of how these forces act on materials is valuable in determining which material to use. These forces provide measures of material hardness, toughness, abrasion resistance, brittleness, and ductility, among others. Special testing machines, described in unit 63, are used to measure these qualities.

Finally, the designer must consider characteristics such as appearance, odor, feel, and general impression. These can help the designer to discover the special uses of materials for both functional and aesthetic purposes. The striking Aeron chair in Figure 5-3 has a base of recycled aluminum and a frame of recycled plastic polymer. These materials were selected to ensure that the chair structure is strong, safe, and durable. The seat and back material is a breathable membrane known as *pellicle.* This is a woven, specially-engineered plastic material. The chair also is designed to adjust mechanically to the human frame for comfort. It is a truly unique design created from several materials.

The tools, machines, furniture, vehicles, and sports equipment people use every day are often made from several materials. Plastic, wood, ceramic, and composite materials are briefly described below. This is followed by a broader discussion of metal materials.

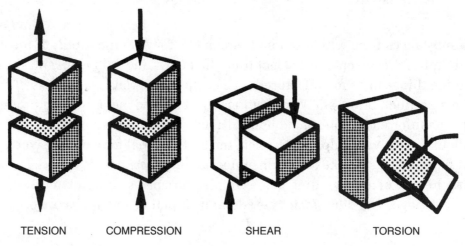

TENSION   COMPRESSION   SHEAR   TORSION

Fig. 5-2 Four common kinds of forces.

Fig. 5-3 This attractive Aeron chair is engineered for comfort and appearance. (Courtesy of Herman Miller, Inc.; photographs by Nick Merrick and Hedrich Blessing, Chicago, IL)

Plastic Materials

Plastics are synthetic (made by humans) materials made from substances such as coal, petroleum, and wood byproducts. They are available as rods, bars, sheets, and films. Plastics can be machined, cast, molded, bent, blow-formed, and worked in many other ways. One of the best reasons to use this material is because of its light weight. For example, a typical cast aluminum air intake manifold for a small gasoline engine can require costly casting operations, and weighs 1.75 pounds. The replacement plastic manifold is easier to manufacture, and weighs only a half pound (Figure 5-4). A special process called *two-shell overmold technology* was used to produce this part. In this process, two half-shells of the hollow part are injection-molded in separate cavities. The shells are then brought together and placed in a second mold for overmolding, which permanently joins the two parts. Plastic injection molding is similar to metal die casting described in unit 51. It is interesting to note that the molds and the molding machines, or *tooling,* are made of metal. This is true of most products. Regardless of the material used, the tooling is usually made of metal.

What makes these materials so desirable is that no matter what the need is, one is sure to find a plastic to meet that design need. Plastics are either heat-resistant or easily melted. They are available as structural shapes approaching the strength of steels, or as tough, pliable films. They are also excellent adhesives, lubricants, and

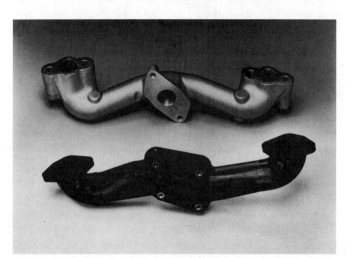

Fig. 5-4 The aluminum engine manifold, above, was replaced by a lighter weight plastic model, below. (Courtesy of Bayer Corporation, Polymers Division)

finishes, and are waterproof, corrosion-resistant, odorless, colorful, and easy to use. However, one weakness is that plastic products generally cannot be used in high-temperature conditions.

There are two basic groups of plastics. Thermoplastic materials, or **thermoplasts,** become soft when heated and harden when cooled, no matter how often the process is repeated. (This feature makes them easy to recycle.) They are made up of long, chain-like molecules, and undergo a physical, not a chemical, change as they are being formed. Thermosetting materials, or

Fig. 5-5 These Sled Dogs snow skates feature a lightweight polyurethane boot with a ski-like base made of a tough elastomeric plastic. The replaceable boot sole, or base, has four steel rails molded in place. (Courtesy of Bayer Corporation, Polymers Division. Sled Dogs is a trademark of The Sled Dogs Company, St. Paul, MN)

thermosets, harden into permanent shapes when heat and pressure are applied to them. Reheating will not soften them. They are of a cross-link molecule structure, where the individual chain segments are chemically attached to one another. Regardless of forms or properties, all plastics fall into one of these two groups. Plastics are important materials in a broad range of products (Figure 5-5).

Wood Materials

Historically, perhaps no other substance has enjoyed the popularity of wood as a design material. From the earliest of times, it has been used in craft, construction, and furniture applications because it was readily available and easy to work. Many technical improvements have contributed to its continued use as an important product material. These improvements include plywoods, laminated beams, and the many modified woods where plastic is forced into the wood cells under pressure.

There are two classes of woods. *Softwoods* have needle-like leaves (pine needles, for example), and almost all are evergreen. Softwoods are used for cabinetwork, home building, furniture, and windows and doors. *Hardwoods* (oak, for example) have broad leaves which they shed in autumn. Hardwoods are generally harder and tougher than softwoods, and are used for products such as baseball bats, tool handles, fine furniture, and turned-wood bowls. As shown in Figure 5-6, wood can be combined with metal structures to create interesting tables.

Fig. 5-6 An aluminum angle structure supports a top and shelf of hard maple. Note the interesting use of Allen-head capscrews to hold the pieces together.

Ceramic Materials

Ceramic products are made from nonmetallic, inorganic materials such as sand and clay. These raw materials are worked at high temperatures to create a variety of important industrial and consumer products. These include *brickwares,* such as building bricks and sewer pipes; *concretes,* such as road-paving and building materials; *glass,* for windows and camera and telescope lenses; and *whitewares,* such as sinks

Fig. 5-7 This lovely covered pot is an example of porcelain ceramicware. (In honor of Marc Hansen)

Fig. 5-8 These tough, flexible ski poles are made of a plastic-ceramic fibrous composite. (Courtesy of SAC/GOODE)

and toilets, fine dining room china, and earthenware pottery (Figure 5-7).

Refractory ceramics are important in the steelmaking industries for blast furnace lining, casting crucibles, and firebrick. Other industrial ceramics include many kinds of *abrasives,* such as grinding wheels and sandpapers.

Composite and Fiber Materials

When two or more materials are combined, but do not mesh completely into each other to become a single material, the result is a **composite.** Such materials have different structures (or compositions, hence the name), and have properties different from the individual materials of which they are made. Fiberglass is a good example. It is made from woven glass or carbon fibers, bedded in liquid plastic to form a strong, durable sheet with many uses. Many automobile bodies are made from fiberglass. The glass and plastic used to make the fiberglass composite are called *constituents.* The properties of the composite are influenced by the properties of the constituents. Thus, a composite can have the strength of one constituent and the temperature properties of another. Any number of property combinations are therefore possible. Other composite products include laminated kitchen countertops, automotive brake shoes, tennis rackets, and ski poles (Figure 5-8).

Metal Materials

There are two basic classes of metals: *ferrous,* which are primarily made of iron, such as steel and cast iron; and *nonferrous,* which contain little or no iron, such as copper and aluminum. These will be described in units 6 and 7. Metals are seldom used in their pure states. In-

stead, they are combined with other metals or chemicals to form *alloys.* This is done to secure a metal material with the properties needed for a specific product.

Mechanical Properties of Metal

As stated earlier, mechanical properties relate to how a material will behave when forces or loads are applied to it. Various combinations of tension, compression, shear, and torsion tests will determine, for example, how easily a metal can be forged, rolled into thin sheets, or drawn to form a metal beverage can. Some of these special properties are described below.

1. **Strength** is the ability of a metal to resist applied forces. Bridge girders, elevator cables, and building beams all must have this property.

2. **Hardness** is the ability of a metal to resist penetration, or piercing. The harder the metal, the less likely it is to change in shape. Hardness can be increased by work-hardening or heat-treating (explained in later units).

3. **Brittleness** is the tendency of a metal to break easily. Certain kinds of cast iron are brittle and break if dropped. Hardness and brittleness are closely related, since hard metals are more brittle than soft metals.

4. **Malleability** is the ability of metal to be hammered or rolled out without breaking or cracking.

5. **Ductility** is the ability of a metal to be drawn out thin without breaking. Copper is very ductile and can be easily made into wire. Deep-formed automobile bodies and fenders, washing machines, and other stamped and formed products are made possible by this property.

6. **Elasticity** is the ability of a material to return to its original shape after bending. The steel used to make springs is very elastic.

7. **Plasticity** is the ability of a material to keep its new shape after being formed.

8. **Fusibility** is the ability of a material to become liquid easily and join with other metals. Metals that readily weld usually have this property.

9. **Machinability** involves several properties. Some of these are: the rate at which the material can be removed in machining, the kind of chip produced, the amount of tool wear, and the kind of surface finish that can be obtained.

10. **Melting point** is the temperature at which the material becomes liquid (Figure 5-9). Few pure metals are used in product manufacturing. Instead, metal *alloys* are used; an alloy being any material containing two or more metallic elements. Alloys are more usable because they have better properties. Alloys do not have a melting point, but instead a melting range, meaning that they lose their solidity between upper and lower temperature limits. This melting range is important for several reasons:

    a. It helps to determine the useful maximum working temperature of an alloy. When a metal approaches the temperature at which it will melt, its strength rapidly decreases.

    b. Metal technicians must choose a temperature below the melting range in order to perform such operations as softening a metal by annealing it, or heating an alloy for further hot working such as forging.

    c. Metallurgists must know the melting range in order to learn to use the alloy more effectively, or to improve it and produce better alloys.

11. **Toughness** is the ability to absorb mechanically applied energy. Strength and ductility determine a material's toughness. Tough metals are needed for such things as railroad cars, automobile axles, hammers, rails, and similar products.

Strength Properties of Metal

Strength may refer to any one or all of the properties defined below that enable metal to resist different loads. When just one property of strength is meant, only that property should be named.

1. *Bending strength* is a measure of the load-carrying capacity of horizontal structural parts between two supports. Examples of parts subjected to bending are joints, girders, and stringers.

2. *Compressive strength* (endwise) is a measure of the ability to carry loads along the length of a member without deformation. An example of an endwise compression member is a column.

Metal	°F	°C
Pewter	420	216
Tin	449	232
Lead	621	327
Zinc	787	419
Aluminum	1218	659
Bronze	1675	913
Brass	1700	927
Silver	1760	960
Gold	1945	1063
Copper	1981	1083
Cast Iron	2200	1204
Steel	2500	1371
Nickel	2646	1452
Wrought Iron	2700	1482

Fig. 5-9 Chart of common metals' melting points.

Columns are calculated not to exceed a certain length in comparison to cross section (usually eleven times the thickness or width, whichever is less). Both stiffness and compressive strength affect the load capacity.

3. *Stiffness* is a measure of resistance to deflection. The stiffness of a beam is proportional to its width and also to the cube of its depth.

4. *Shear strength* means resistance to forces tending to cause one part of a piece to slide over the other.

5. *Torsion strength* refers to the ability to resist twisting.

Strength usually refers to a force acting on a piece of metal and tending to change the shape of the material. Stress may be tensile, compressive, shear, or torsion. Tensile stress, or **tension,** occurs when forces tend to elongate the piece. Compressive stress, or **compression,** is the result of squeezing or crushing. **Shear** stress is caused by opposing forces that meet almost head-on (like the blades of a pair of shears). **Torsion** stress is the result of twisting.

Stress is also related to the internal strength by which a material resists external force. For example, tensile strength refers to the ability of a material to withstand tensile stress—in other words, its resistance to being pulled or stretched apart.

Other Properties of Metal

In addition to the mechanical and strength properties of metal, there are several other properties to be aware of for material selection.

1. **Magnetic** properties of metals are those which have to do with reactions to magnetic or electrical forces. Some metals are attracted by magnets, some are not. Certain high carbon and alloy steels

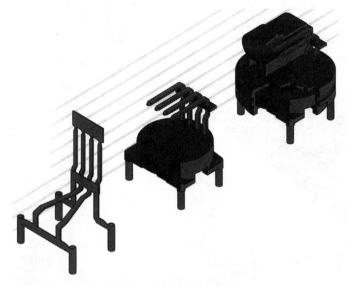

Fig. 5-10 Copper is a good material to use in this plastic insert molded electronic connector. It is a good electrical conductor and is ductile. (Courtesy of Tricon Industries, Inc.)

retain their magnetism and are used to make permanent magnets.

2. **Chemical** properties of metal pertain to metal's resistance to corrosion and oxidation. Rust is one form of corrosion. Iron and steel rust when exposed to weather. Many metals are coated to increase corrosion resistance.

3. **Thermal** properties pertain to the effect of temperature upon metals. Most metals are good conductors of heat. Heat also causes metal to expand.

4. **Electrical** properties determine conductivity and resistance to electrical charges. Copper, for example, is a very good conductor of electricity (Figure 5-10). A semiconductor is a metal that has properties of both a conductor and an insulator.

5. **Acoustical** properties are related to sound transmission and sound reflection. Acoustics is the science of sounds; sounds wanted or unwanted. Metals are often covered with insulating materials to reduce the reflection of sounds.

6. **Optical** properties relate to the way metal reacts to light. Materials such as polished aluminum reflect light. Materials that you can see through, such as glass, are transparent. Materials that you cannot see through, such as metal, are opaque.

The information you have learned in this unit should help you to select materials for your metalworking activities.

KNOWLEDGE REVIEW

1. In your own words, define the meaning of the term *material property.*

2. Write definitions of these properties: physical, chemical, thermal, electrical, acoustical, optical, and mechanical.

3. Sketch a diagram of the four mechanical forces.

4. Explain the meanings of the terms *ferrous, nonferrous, thermoplast, thermoset, hardwood, softwood, whiteware, glass,* and *composite.*

5. List and define five of the mechanical properties of metal.

6. Select a simple product, such as a tool or household utensil. Study it and write a report on the materials used to make it, and why you think they were used.

7. Prepare a test piece of 22-gauge sheet steel, one inch wide and eight inches long. Clamp the piece at its midpoint in a vise. Bend the piece 180 degrees from side to side. How many bends does it take to break the piece? Prepare a copper test piece of the same size and repeat the experiment. What conclusions can you draw from these experiments?

Unit 6

Ferrous Metals

OBJECTIVES

After studying this unit, you should be able to:

- Tell the difference between ferrous and nonferrous metals.
- Describe a base metal.
- Explain how to produce an alloy.
- Identify where most iron ore comes from in the United States.
- Describe a blast furnace.
- Name three methods of making steel.
- Describe an ingot.
- Name the kinds of cast iron.
- Describe wrought iron.
- Tell how carbon steels are classified.
- Name five metals that are used in alloy steels.
- Describe the use of tool-and-die steel.
- List three kinds of rolled steel.
- Name three methods of identifying steels.
- List the standard shapes of metals.

KEY TERMS

alloy	flats	nickel	squares
base	hexagons	nonferrous	tungsten
blast furnace	iron	octagons	vanadium
carbon	manganese	rounds	
chromium	metallurgy	silicon	
ferrous	molybdenum	steel	

Metals are desirable materials for product manufacturing because of their many outstanding properties. For example, they are excellent conductors of heat and electricity; they are opaque and are easily shined; they are strong, tough, and ductile; and they are easily machined.

Metallurgy is the science and technology of metals and their behavior. A scientist who specializes in metals and their use is called a *metallurgist.* In this unit, you will become acquainted with some basic information on metals and metallurgy.

Metals and Alloys

Metals are among nature's most common elements. Iron, copper, and aluminum are examples. An **alloy** is a mixture of two or more metals. Usually, it consists of a **base** metal and a smaller amount of other metals. Brass, for instance, is an alloy of copper, the base metal, and zinc. In the metal shop, metals and alloys are both called metals. Metals are divided into two groups: the **ferrous** metals, which have a large percentage of iron, and the **nonferrous** metals, which have little or no iron.

The Ferrous Metals—Iron and Steel

Iron as a pure metallic element is not suitable for use in industry. However, a small amount of the non-metallic element **carbon** can be added to iron to produce cast iron, wrought iron, and steel. **Steel,** which is very tough and useful, is an alloy composed mainly of iron and carbon. Other alloying elements, such as manganese and chromium, give steel other properties, such as corrosion resistance and strength. Steel is the most important metal known to people. Most of the metal used in the world today is steel (Figure 6-1).

Fig. 6-1 The space shuttle Discovery is rolled to the launch pad on a mobile launch platform and crawler transporter. These are huge and complex steel structures. (Courtesy of NASA)

Production of Iron and Steel

The making of iron and steel begins with the mining of high-grade iron ore, either in underground or in open-pit mines. A lower-grade ore, called *taconite,* also is used. Taconite is crushed and screened, and then processed into small, round pellets of a higher iron content. There are large iron ore deposits in the states of Michigan, Minnesota, and Wisconsin. The iron in these ores is removed by a *smelting* process which takes place in a blast furnace. The material taken from these furnaces is an impure metal called *pig iron.* Figure 6-2 illustrates the entire steelmaking process, from ore to finished product.

Blast Furnace

A blast furnace is basically a huge steel cylinder lined with firebrick and charged with iron ore, coke, limestone, and hot air. The ore provides the iron; coke supplies the heat needed to melt the ore, and provides carbon monoxide to remove the oxygen from the iron oxide; limestone mixes with such impurities as sulphur and silicon dioxide to purify the iron; and hot air provides the oxygen needed to burn the coke.

In blast furnace operations, measured amounts of iron ore, coke, and limestone, called the *charge,* are loaded into skip cars, carried to the top of the furnace, and dumped. Hot air from the stoves is then blown into the bottom of the furnace, causing the coke to burn at about 3,000°F and melt the ore. As the ore melts, it settles to the bottom of the furnace, and more charge is added at the top in a continuous process. The hot air moves upward, forming the carbon monoxide which helps to remove the oxygen from the ore by turning into carbon dioxide. The limestone (which acts as a flux) melts and joins with impurities to become the substance *slag.* The slag, being lighter than the molten iron, floats on top of it. Slag is drawn off the melt regularly and can be used to make cement blocks.

Removing the molten iron from the furnace (called *tapping*) is done every four or five hours. From 150 to 300 tons of iron can be drawn off with each tap. The iron is cast into 100-pound forms called *pigs,* or it can be transported as molten pig iron to the steel mill.

The raw materials of steelmaking must be brought together, often from hundreds of miles away, and smelted in a blast furnace to produce most of the iron that goes into steelmaking furnaces. Air and oxygen are among the most important raw materials in iron and steelmaking.

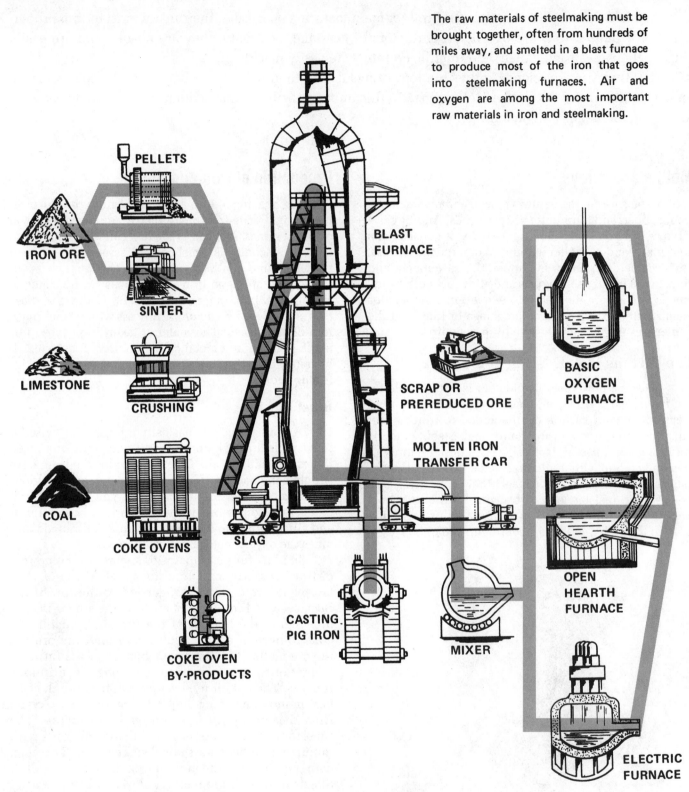

PELLETS

IRON ORE

SINTER

LIMESTONE

CRUSHING

COAL

COKE OVENS

COKE OVEN BY-PRODUCTS

SLAG

CASTING PIG IRON

BLAST FURNACE

SCRAP OR PREREDUCED ORE

MOLTEN IRON TRANSFER CAR

MIXER

BASIC OXYGEN FURNACE

OPEN HEARTH FURNACE

ELECTRIC FURNACE

Fig. 6-2 The flowchart for steelmaking. (Courtesy of American Iron & Steel Institute)

Molten steel must solidify before it can be made into finished products by the industry's rolling mills and forging presses. The metal is usually formed first at high temperature, after which it may be cold-formed into additional products.

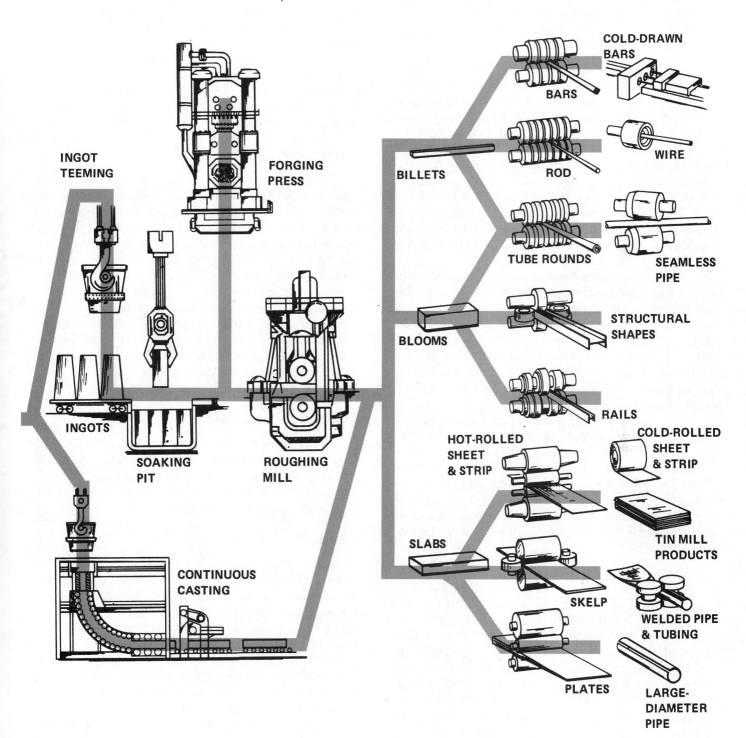

Fig. 6-2 Continued

It is interesting to note that it takes about two tons of ore, one ton of coke, nearly half a ton of limestone, and four tons of air to make one ton of pig iron. Pig iron contains about 3 to 4 percent carbon, 0.06 to 0.10 percent sulphur, 0.10 to 0.50 percent phosphorous, 1 to 3 percent silicon, and traces of other impurities.

Steelmaking Methods

To change pig iron into steel, the impurities must be removed in one of three kinds of steelmaking furnaces: open hearth, basic oxygen, or electric. The *open hearth furnace* has a shallow basin (hearth) open to the sweep of flames across it to convert molten pig iron, iron ore, scrap iron, and limestone into steel. A single furnace load ranges in size from one hundred to three hundred tons, and is called a *heat*. Limestone acts as a purifying flux, and the iron ore oxidizes and refines the melt. Between eight and twelve hours of processing is required before purified molten steel can be run from the furnace into a ladle. This furnace is used primarily to make carbon steel.

In the *basic oxygen furnace*, eighty tons of scrap and molten iron are changed into steel in about fifty-five minutes. In operation, the furnace is tipped on its side and charged with iron scrap, and then moved to an upright position where the molten iron is poured in. Oxygen is blown into the furnace at high speed to burn out the impurities. Limestone acts as the flux to collect impurities and form slag. About fifty percent of American steel is produced by this method.

The *electric furnace* is used to produce high grade carbon and alloy steels. Powerful electric currents are sent through three large rods (electrodes) that pass through the top of the furnace. When sparks from these electrodes strike the iron charge, they generate an intense heat. This heat is sufficient to melt the charge and burn out any impurities. Since no oxygen is present to support combustion, the purity of the metal can be better controlled.

Ingot Processing

The steel that emerges from the furnaces described above is formed into *ingots* for further processing. An ingot is a large steel casting that has solidified into a workable size and shape, and is produced by three operations. *Teeming* is pouring the molten steel into ingot molds. A ladle of freshly made steel is held by a crane above a row of ingot molds. A valve in the bottom of the ladle is opened and the steel flows into each mold, filling it to the top. *Stripping* is removing the ingots from the ingot molds by means of a stripper crane with large tongs or grippers. The ingot is the first solid form of steel. *Soaking* is placing the ingots in a furnace called a soaking pit where the ingots are slowly brought to a uniform temperature. If the ingots were not soaked, their outer surfaces would solidify before their insides, resulting in a faulty workpiece. After the soaking is completed, the ingots are moved from the soaking pit to the rolling mill.

The white-hot steel ingots are fairly soft and can be squeezed into various shapes by passing them between the powerful steel rolls of a rolling mill. Ingots first go to the primary mill where they are rolled into heavy shapes called *blooms, slabs,* or *billets.* These three semifinished shapes then go to finishing mills where they are rolled, either hot or cold, into thick plates, thin sheets, strips, rods, tubes, and bars. A number of different steels are used in product manufacture (Figure 6-3).

Continuous Casting

It was pointed out earlier that molten steel can be used directly to cast blooms, slabs, or billets without going through the expensive intermediate step of ingot casting. This process is called *continuous casting,* where molten steel is carried from the furnace and poured into a reservoir (tundish) above a special water-cooled containment mold (Figure 6-4). The molten steel entering this mold is chilled and develops an outer shell, or crust, having the same shape as the mold. The shell is gradually pulled down by the rolls, is cooled and straightened, and then cut to the desired length. The resulting blooms, slabs, and billets are then further reduced to finished products in the rolling mills.

Kinds of Iron and Steel

Most of the projects you will make will be of carbon, alloy, or tool-and-die steel. In addition, you will probably use some of these other common ferrous metals: wrought iron, galvanized steel, tin plate, and perforated and expanded metal.

Cast Iron

Cast iron is used for the heavy parts of many machines. It is the most common material for making castings. It contains two to four percent carbon. The basic kinds of cast iron are white, gray, and malleable iron. *White iron* and *gray iron* are low in cost and wear well. They are very brittle, however, and cannot be hammered or formed. *Malleable iron* is cast iron that has been made more malleable by annealing (this process will be discussed in later units). Malleable-iron castings are not so brittle or hard. They can take a great deal of hammering. Many plumbing fixtures are made of malleable iron. Cast irons are difficult to weld.

Wrought Iron

Wrought iron is almost pure iron. It contains only minor amounts of carbon. It is often used for orna-

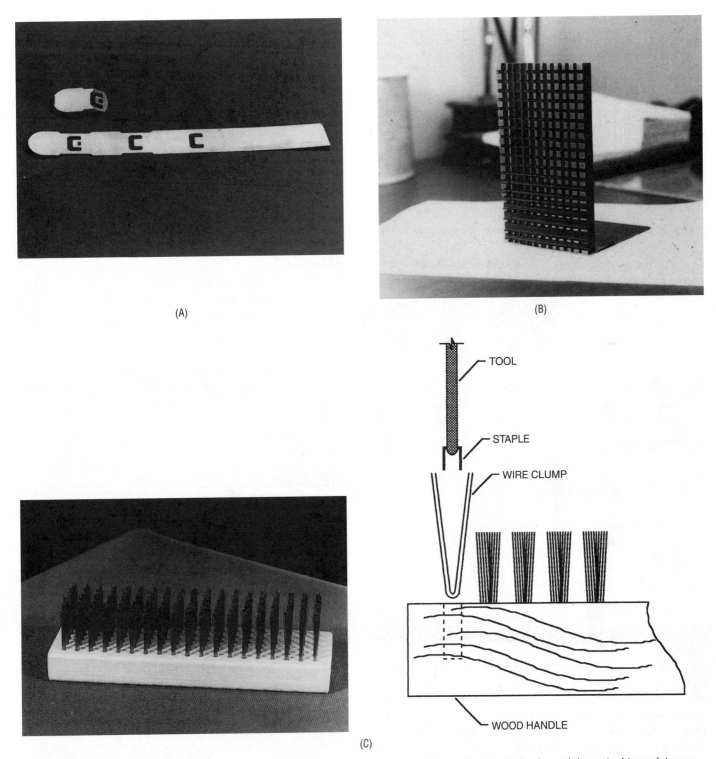

Fig. 6-3 Different steels for different products: (A) the filler flap on an automobile gas tank is made of a stainless steel to resist corrosion; (B) this bookend is made of 3/8-inch mild steel—its face pattern was cut on a metal shaper; and (C) the wire clumps on this brush are made of tough, sharp spring steel. Note the installation diagram.

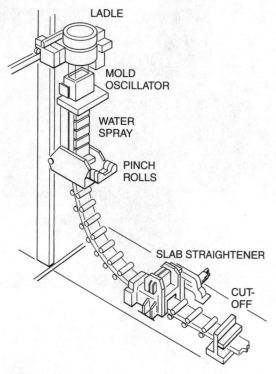

Fig. 6-4 The continuous-casting process. (Courtesy of American Iron & Steel Institute)

mental ironwork. However, wrought iron is seldom used in the school shop because of its high cost. Wrought iron forges well, can easily be bent hot or cold, and can be welded.

Carbon Steels

Carbon steels are classified by the amount of carbon they contain. This amount is given in *points* (100 points equals 1 percent) or by *percentage.*

Low-Carbon Steel

Low-carbon steel, often called *mild* or *soft steel,* contains 0.1 to 0.3 percent carbon (10 to 30 points). It does not contain enough carbon to be hardened. This type of steel is available as black-iron sheet, band iron, bars, and rods. Because it is easily welded, machined, and formed, low-carbon steel is suitable for products and projects in which an easily worked metal is needed. It is used for most bench-metal and ornamental ironwork.

Medium-Carbon Steel

Medium-carbon steel has 0.3 to 0.6 percent carbon (30 to 60 points). It is used for many standard machine parts. In the school shop, it is used for projects like hammer heads and clamp parts.

High-Carbon Steel

High-carbon steel contains 0.6 to 1.7 percent carbon (60 to 170 points). The best kind for school shops

contains 75 to 95 points. It is used for making small tools or for any item that must be hardened and tempered.

Alloy Steels

Alloy steels have special properties determined by the mixture and the amount of other elements, particularly metals, added. Each alloy steel has a "personality" of its own. Some of the common alloying elements are described below.

Nickel

Nickel is added to increase strength and toughness. It also helps steel resist corrosion.

Chromium

Chromium adds hardness, toughness, and resistance to wear. Gears and axles, for example, are often made of chromium-nickel steel because of its strength.

Manganese

Manganese is used in steel to produce a clean metal. It also adds strength to the steel and helps in heat treating.

Silicon

Silicon is often used to increase the resiliency of steel for making springs.

Tungsten

Tungsten is used with chromium, vanadium, molybdenum, or manganese to produce high-speed steel, used in cutting tools. Tungsten is said to be *red-hard,* or hard enough to cut even after it becomes red-hot.

Molybdenum

Molybdenum adds toughness and strength to steel. It is used in making high-speed steels.

Vanadium

Vanadium improves the grain of steel. It is used with chromium to make chrome-vanadium steel, from which transmission parts and gears are manufactured. This type of steel is very strong and has excellent shock resistance.

Tool-and-Die Steels

Tool-and-die steels are a large group of steels used when careful heat-treating must be done. You will use these steels to make tools that must have a cutting edge.

Rolled Steel

Rolled steel, which includes bar, rod, and structural steels, is produced by rolling the steel into

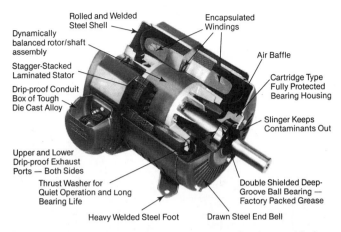

Fig. 6-5 This industrial electric motor is made of many kinds of steel parts. Notice the perforations on the drawn steel end bell. (Courtesy of The Lincoln Electric Company)

shape. *Hot-rolled steels* are formed into shape while the metal is red-hot. The metal passes through a series of rollers, each one a little closer to the next. As the steel passes through the last rollers, hot water is sprayed over it, forming a bluish *scale*. The steel produced by this method is fairly uniform in quality and is used for many kinds of parts. Hot-rolled bars of the best quality are used to produce cold-finished steels. *Cold-finished steels* are used when great accuracy, better surface finish, and certain mechanical properties are needed. There are several ways of producing cold-finished bars. The most common results in what is called *cold-worked steel.* First, the scale is removed from the hot-rolled bars. Then, the bars are either *cold-drawn,* drawn through dies a little smaller than the original bar, or *cold-rolled,* rolled cold to the exact size.

Coated Steel

Galvanized, or *galvannealed,* steel is mild sheet steel coated with zinc to keep it from rusting. *Tin plate* is mild steel coated with tin.

Patterned Steel

Metals are available in many different patterns. *Perforated metal* has a design stamped through it (Figure 6-5). *Expanded metal* is made by cutting slits in the metal and pulling it open to expand it. *Embossed metal* has the design pressed into its surface.

Identifying Steels

All steels look very much alike. Thus, it is difficult to identify the type of steel merely by looking at it. There are three primary methods of identification: the number system, color coding, and the spark test.

Number System

A number system, or series, to identify carbon and alloy steel has been developed by the American Iron and Steel Institute (AISI) and the Society of Automotive Engineers (SAE) (Table 6-1). You will find complete information about these numerical systems in *The New American Machinist's Handbook* (New York: McGraw-Hill) or *Machinery's Handbook* (New York: The Industrial Press). The systems are based on the use of numbers composed of four digits:

- The first number tells the kind of steel: 1 indicates carbon steel, 2 is nickel steel, 3 is nickel-chromium steel, 4 is molybdenum steel, and so forth.

- The second number in alloy steel indicates the approximate percentage of alloy elements. For example, 2320 indicates a nickel steel with about three percent nickel.

- The last two (and sometimes three) numbers indicate the carbon content in points (100 points equals 1 percent). For example, SAE 1095 is a carbon steel with 95 points of carbon.

Another letter-and-number, or alphanumeric, system is used to identify tool-and-die steels.

The American Iron and Steel Institute has also developed a system for identifying the kind of furnace in which the steel was made:

- B = acid Bessemer carbon steel
- C = basic open-hearth or basic electric-furnace carbon steel

The letter is placed before the number of the steel. For example, AISI C1018 is a low-carbon, general-purpose steel suited to project work in wrought metal.

Color Code

Most manufacturers paint each different kind of high-carbon alloy and tool-and-die steel a different color. Some paint only the ends. Others paint all along the bar. This is done to avoid confusing the steel bars when they are stored on racks. If a certain steel is painted red, it may mean that it is a high-carbon steel. Each company has a different color code.

Spark Test

The spark test method of identification is rather inaccurate. The test is performed by watching the sparks given off when the metal is ground. The kind, frequency, position, and color of the sparks are all considered in making the identification. Figure 6-6 shows the sparks given off by various metals. Spark tests are often used in choosing a particular type of steel from among a group of unidentified steels (Figure 6-7). Always wear eye protection when doing this test.

Table 6-1   AISI Chart for Identifying Steels

Steel Type	Series	% Nominal Alloy Content						
		Ni	Cr	Mo	Mn	V	Si	W
Carbon-Plain	10xx				1.00			
—Free Cutting	11xx							
—Free Cutting	12xx							
—Plain	15xx				1.00–1.65			
Manganese	13xx				1.75			
Nickel	23xx	3.50						
	25xx	5.00						
Nickel-Chronmium	31xx	1.25	0.65; 0.80					
	32xx	1.75	1.07					
	33xx	3.50	1.50; 1.57					
	34xx	3.00	0.77					
Molybdenum	40xx			0.20; 0.25				
	44xx			0.40; 0.52				
Chromium-Molybdenum	41xx		0.50; 0.80; 0.95	0.12; 0.20; 0.25; 0.30				
Nickel-Chromium-Molybdenum	43xx	1.82	0.50; 0.80	0.25				
	47xx	1.05	0.45	0.20; 0.35				
	81xx	0.30	0.40	0.12				
	86xx	0.55	0.50	0.20				
	87xx	0.55	0.50	0.25				
	88xx	0.55	0.50	0.35				
	93xx	3.25	1.20	0.12				
	94xx	0.45	0.40	0.12				
	97xx	0.55	0.20	0.20				
	98xx	1.00	0.80	0.25				
Nickel-Molybdenum	46xx	0.85; 1.82		0.20; 0.25				
	48xx	3.50		0.25				
Chromium	50xx		0.27–0.66					
	51xx		0.80–1.05					
Chromium (Bearing) C 1.00 Min	50xxx		0.50					
	51xxx		1.02					
	52xxx		1.45					
Chromium-Vanadium	61xx		0.60; 0.80			0.10; 0.15 Min		
Tungsten-Chromium	72xx		0.75					1.75
Silicon-Manganese	92xx		0; 0.65		0.65; 0.82; 0.85		1.40; 2.00	

Atom Symbols:
C   Carbon   Mo   Molybdenum   V   Vanadium
Ni Nickel   Cr   Chromium   W   Tungsten
Mn Manganese   Si   Silicon

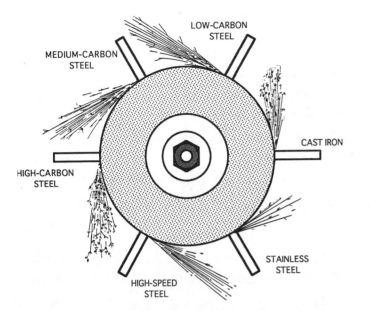

MEDIUM-CARBON STEEL

LOW-CARBON STEEL

HIGH-CARBON STEEL

HIGH-SPEED STEEL

STAINLESS STEEL

CAST IRON

Fig. 6-6 Ferrous metal spark patterns. Notice how different clusters form for different workpieces as the pieces touch the grinding wheel.

Fig. 6-7 The spark test. What kind of metal is being ground?

Standard Shapes and Sizes

Metals can be purchased in many shapes and sizes. Standard lengths range from 10 to 24 feet (3 to 8 m). Not all shapes, sizes, and kinds of metal are available in every length. The best way to purchase metals is to use a stock list or reference book available from the supplier. You will find that most suppliers have the following common shapes and sizes or their metric equivalents in stock (Figure 6-8).

1. **Rounds,** or shafting, in diameters from 3/16 to 9 inches (Table 6-2).

Fig. 6-8 Common shapes of bar stock.

Table 6-2   Round, Square, and Hexagon Bar Sizes

inch	mm*
1/8	3
1/4	6
3/8	10
1/2	12
5/8	16
3/4	20
1	25
1-1/4	30
1-1/2	40
1-3/4	45
2	50

*Metric replacement sizes are based on ANSI B32.4-1974.

2. **Squares** from 1/4 by 1/4 inch to 4 1/2 by 4 1/2 inches (Table 6-2).

3. **Hexagons,** or six-sided shapes, from 1/4 to 4 inches measured across opposite flat sides (Table 6-2).

4. **Octagons,** or eight-sided shapes, from 1/2 to 1 3/4 inches measured across opposite flat sides.

5. **Flats,** or rectangular shapes, from 1/8 by 5/8 inch to 3 by 4 inches (Table 6-3).

Table 6-3 Flat, or Rectangular, Bar Shapes

Thickness		Width	
inch	mm*	inch	mm*
1/16	1.6	3/8	10
1/8	3	1/2	12
3/15	5	5/8	16
1/4	6	3/4	20
5/16	8	1	25
3/8	10	1-1/4	30
7/16	11	1-1/2	40
1/2	12	1-3/4	45
9/16	14	2	50
2/8	16		
11/16	18		
3/4	20		
7/8	22		
1	25		
1-1/8	28		
1-1/4	30		
1-5/16	32		
1-3/8	35		
1-1/2	40		
1-3/4	45		
2	50		

*Metric replacement sizes are based on ANSI B32.3-1974.

KNOWLEDGE REVIEW

1. What is the difference between a metal and an alloy?
2. What is the difference between ferrous and non-ferrous metals?
3. Name the mechanical properties of metals.
4. What are the four raw materials for making iron?
5. What are three ways of making steel?
6. How do cast iron and wrought iron differ in composition?
7. What is mild steel?
8. Name seven elements used in making alloy steels.
9. What are the three methods of identifying steel?

Unit 7

Nonferrous Metals

OBJECTIVES

After studying this unit, you should be able to:

- List three groups of nonferrous metals.
- Describe the characteristics of copper.
- Name several kinds of brass.
- Explain what metals are used to make bronze.
- Give another name for nickel silver.
- Identify the uses of zinc.
- Describe lead.
- Explain how tin is used as an alloy.
- Name the metals used in aluminum alloys.
- Describe some of the common aluminum alloys.
- Describe the contents of sterling silver.
- Give another name for pewter.
- Name three space-age metals.

KEY TERMS

aluminum	copper	nickel silver	titanium
beryllium	gar-alloy	pewter	zinc
brass	lead	sterling silver	
bronze	molybdenum	tin	

There are many kinds of nonferrous metals (metals containing little or no iron). They are grouped as follows:

1. *Base metals:* copper, lead, tin, nickel, zinc, and aluminum
2. *Alloys:* brass, bronze, nickel silver or German silver, pewter or Britannia metal, and gar-alloy
3. *Precious metals:* sterling silver, gold, and platinum

Many of the nonferrous metals listed above are not used very often in the school shop. Some are too expensive. Others do not have good working qualities.

Common Metals and Alloys

Some common nonferrous metals and their characteristics and applications are described below.

Copper

Copper is a warm reddish brown metal used for much art metalwork. It is easy to work and is an excellent conductor of electricity (Figure 7-1). Copper becomes hard when it is worked. However, it can easily be softened by heating it to a cherry-red color and then cooling it. Copper can be joined with rivets or by silver soldering, bronze brazing, or soft soldering. It will take an excellent polish and finish and is used for many decorative pieces, such as bowls, trays, and containers. It is easy to form by stretching, but it does not machine easily. For example, it is more difficult to cut threads on copper rod than on mild steel. Copper becomes covered with a green tarnish, which is easily removed by brushing and polishing or by chemical means.

Brass

Brass is one of the more important of the copper alloys. Brass is very ductile and can be made into wire. It is also a good base for plating. There are many kinds of brass, each with different working qualities. The principal alloying element in most brass is zinc. *Gilding brass,* a rich bronze in color, is five percent zinc. It is used for jewelry, coins, and plaques. *Commercial bronze,* actually a brass, contains about ten percent zinc. The *red,* or *colonial,* brass that contains fifteen percent zinc is somewhat softer and is used for jewelry. It has a gold color that is very similar to real gold. *Cartridge brass,* which is thirty percent zinc, is used to make artillery and ammunition. *Yellow brass,* thirty-five percent zinc, is characterized by its bright yellow color. *Leaded brass* and *forging brass* each have a little lead added to the copper and zinc.

Bronze

Bronze, another copper alloy, has a reddish gold color. Bronze may or may not contain some zinc. Common alloying elements used in bronze include:

1. *tin,* to increase strength and resistance to corrosion;
2. *aluminum,* to increase resistance to corrosion; and
3. *nickel,* to make the bronze whiter. Nickel bronze is a beautiful metal for castings, plaques, and similar articles.

Nickel Silver

Nickel silver, sometimes called *German silver,* is usually made of about sixty-four percent copper, eighteen percent nickel, and eighteen percent zinc. The more nickel it contains, the whiter it is. Nickel silver can be soldered, formed, and annealed, much like brass. Its color varies from a light pink to a silver white. Nickel silver is a good imitation of sterling silver. However, it is quite brittle, cannot be hammered into shape, and discolors easily. Nevertheless, it is easy to solder, bends quite well, and is very good for etching. Nickel silver is used for bracelets, rings, and other jewelry. It is used as a base for all good silver-plated tableware.

Zinc

Zinc is most often seen in the form of a protective coating on sheet metal. Zinc is also used as an alloying element in making brass and some bronze. Sheets of pure zinc can be used for protecting other materials.

Lead

Lead is one of the heaviest metals. The surface of lead is quite gray, but a scratch in the surface looks white. Lead is very soft; thus, a lead block is a good thing to use as backing when punching holes with a hollow punch or when bumping or hammering sheet metal. Lead is most often used as an alloy with some other metal. When lead is alloyed with tin, it becomes soft solder.

Tin

Tin is seldom used except as an alloying metal. It is used with copper to produce bronze and with lead to make soft solder. Tin is nonpoisonous. A coating of tin is put on mild-steel sheets used for making food containers.

Aluminum

Aluminum has become one of the leading metals of industry and everyday life (Figure 7-2). Embossed

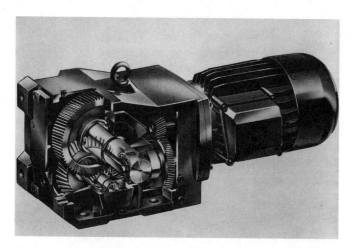

Fig. 7-1 This industrial bevel gear motor contains copper alloys for the motor windings and bronze bushings for the gears. (Courtesy of Nord Gear Corporation, Waunakee, WI)

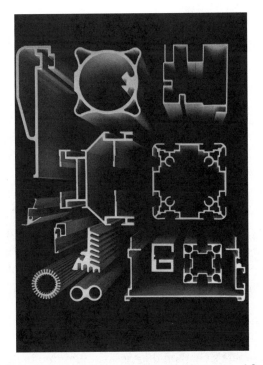

Fig. 7-2 Extruded aluminum bar shapes. (Courtesy of Cardinal Aluminum Company, Louisville, KY, USA)

Fig. 7-3 These high-strength aluminum alloy drumstick shafts are press-fitted with flexible polyurethane covers. Hard nylon tips screw over the covers to hold the units together. (Courtesy of Bayer Corporation, Polymers Division)

sheet aluminum is used in building construction, furniture, and appliances. It is being used more and more because it is lightweight, is easy to work, and has a pleasing appearance. Aluminum resists corrosion and is a good conductor of electricity. Pure aluminum is soft. Making it into alloys adds strength and provides other desirable characteristics (Figure 7-3). Elements added to aluminum include:

1. *copper,* to add strength and hardness and to make it easier to machine;
2. *magnesium,* to improve ductility and resistance to impact;
3. *manganese,* to increase strength and hardness;
4. *silicon,* to lower the melting point and improve castability; and
5. *zinc,* to improve strength and hardness.

There are well over one hundred different alloys of aluminum. Some of the more common ones you might use include:

1. *wrought alloys,* which are used for rolling, pressing, and hammering. These are indicated by a capital *S.* Wrought alloy 1100 is pure aluminum;
2. *alloy 3003,* which has some manganese in it. This can be purchased as 3S-O, which is dead-soft or annealed, or in the forms 3S-1/2H, 3S-3/4H, or 3S-H, which vary in degree of hardness;
3. *alloy 5052,* which has a little magnesium in it; and

4. *casting aluminums* used in sand molds, including 43, which contains a little silicon, and 113, which contains some copper and silicon.

Sterling Silver

Sterling silver is used in jewelry and tableware. It has a warm silver luster. Pure silver is soft. Sterling silver, although not so ductile as copper, can be formed and shaped well. It is suited to chasing, etching, spinning, and all other surface decorations. It also hard-solders well. To be marked *sterling,* an article must contain at least 0.925 parts of silver, with 0.075 parts or less of copper and other alloying elements to give it hardness.

Pewter

Pewter, or *Britannia metal,* has a pleasing gray color. It is ninety-two percent tin, six percent antimony, and two percent copper. Because of its high tin content, it is quite expensive and often hard to obtain. It was used a great deal during colonial days. Some of the early pewter contained lead, which made it poisonous in combination with acid foods. Pewter is very soft and easy to work cold. It is difficult to solder, however, because of its low melting point.

Gar-Alloy

Gar-alloy is bluish gray. It is a zinc-based alloy, with copper and silver added to give it strength. It has good cold-working qualities for drawing, spinning, and hammering. It resists corrosion well and buffs to a high polish.

Space-Age Metals

When people began to approach the threshold of space, they needed new metals that could withstand great heat and cold, were light and tough, and were easy to fabricate. Three new metals are particularly important in the aerospace industry and in other areas of new technology.

Titanium

Titanium is a very desirable metal because it is light and strong and resists corrosion. Though titanium is heavier than aluminum, titanium alloys are much stronger than aluminum alloys.

Titanium alloys are much lighter than steel alloys but are comparable to them in strength. Titanium is very expensive because it is difficult to produce and fabricate. It melts at a very high temperature and must be worked in an atmosphere of inert gases.

Beryllium

Beryllium, which is lighter than aluminum, is as strong in proportion to its weight as high-strength steel. Beryllium conducts heat well and is slow to heat up. Thus, it can be used where other metals would melt. Beryllium is frequently used as an alloy with copper, nickel, or aluminum.

Molybdenum

Molybdenum has a high melting point and a high strength-to-weight ratio. It is a good conductor of heat and electricity.

KNOWLEDGE REVIEW

1. What is the base metal in brass and bronze? What are the chief alloying elements?

2. What elements are alloyed with aluminum?

3. In what products is aluminum commonly found?

4. What properties does titanium have which make it such a desirable metal in aerospace work?

5. What materials are used in beryllium alloys?

6. Describe some typical uses for molybdenum.

7. Find out all you can about one of the following metals: tungsten, molybdenum, nickel, chromium, cadmium, titanium, magnesium, or beryllium.

8. Write a report on Charles Martin Hall and his work in the aluminum industry. Any of the aluminum companies will supply you with information about this great inventor.

9. Prepare a chart on the uses of aluminum in industry.

10. Select a simple product, such as a hand tool or household utensil, and analyze it as to the appropriateness of the materials used in its construction. List any faults and materials specification changes you would suggest.

11. Write a research report on new uses for space-age metals in industry.

SECTION

2

INDUSTRY FUNDAMENTALS

The exchange of information is a critical element in any manufacturing operation. Computer technicians create engineering designs and data bases to guide the production control specialists in factories (Figure S2-1). Managers speak to plant supervisors either in person, by telephone, or on the Internet. Beyond the factory, airplanes are guided on their flight paths by communications control centers, as are trains on their track routes from city to city (Figure S2-2). The purpose of this book section is to introduce both the theory of communications and its practical methods, and to describe their many applications in industry, transportation, business, and society.

Fig. S2-2 Modern onboard communications equipment is a vital part of every railroad system.
(Courtesy of Rockwell's Collins Businesses)

Fig. S2-1 Computers are a vital link in industrial communications. (Courtesy of Western Michigan University)

53

Unit 8

Communication Systems

OBJECTIVES

After studying this unit, you should be able to:

- Explain the meaning of *communication.*
- Identify the various forms of communication.
- Describe the uses of communication in the manufacturing industries.
- Describe the meaning of graphic reproduction technology.
- Describe the major processes of offset lithography, gravure, flexography, and screen.
- Give examples of the major printing processes as they relate to metal product technology.

KEY TERMS

flexography	gravure	planographic	screen printing
graphic reproduction technology	intaglio	plate	substrate
	offset lithography	relief	telecommunication

Very simply, communicating is the act of sending and receiving information by speech or sign. A spoken (oral) language is one kind of communication; through it we can tell others what we feel, think, or want. Any time two people speak to one another, they are sending and receiving information and are therefore communicating (Figure 8-1). We can also communicate by sign (graphics) through drawings, photographs, or printed words.

Oral and graphic communication can be improved by electronic systems to make it simpler and more efficient, and to permit it to take place over greater distances. Examples here include radios, television sets, telephones, satellites, and computers.

Before these technical advances, the written message was the main way to communicate over long distances. A person wrote a letter by hand, addressed, stamped, and mailed it, and a mail carrier delivered it to the receiver. With the invention of the telegraph, the message could be sent more quickly by wire signals; and with the development of the telephone, two people could have an immediate conversation. The computer and modem now make it possible for people to converse over a worldwide network via electronic mail, or *e-mail.* The act of writing a book (such as this textbook you are reading) is now much simpler than typewriting it as was done in the past (Figure 8-2). By using the word processor, an author keyboards in

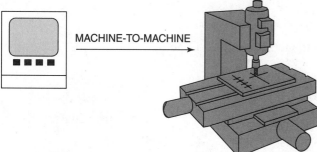

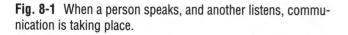

Fig. 8-1 When a person speaks, and another listens, communication is taking place.

Fig. 8-2 The computer word processor is a modern and efficient means of communication.

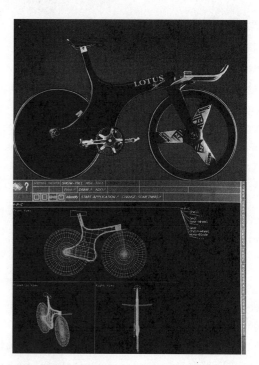

Fig. 8-3 This striking carbon-fiber structure bicycle is a product of a computer-aided design system.
(Courtesy of Concentra Corporation, and Lotus Engineering)

the words of a chapter, rearranges and corrects them, and sends a computer disc containing the completed chapter to the publisher.

The same capacity holds true with the drawings for a book. These are prepared with a computer-aided drafting (CAD) program, any necessary changes or corrections are made by the drafter, and these graphics discs are then sent to the publisher. Modern publishing houses complete the necessary editing and proofing of this book material, and the printing is then done from these computer data. Book writing and publishing have now become speedier, more accurate, and sometimes paperless operations due to technology. In other industries, design engineers use this CAD system to create totally revolutionary new products, such as the carbon-fiber bicycle in Figure 8-3.

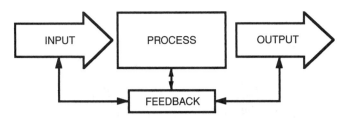

Fig. 8-6 The universal systems chart is useful in describing communications systems. Can you apply this to computer word processing?

Fig. 8-4 Avionics (aviation electronics) is vital to the guidance and control of aircraft.
(Courtesy of U.S. Navy; photograph by Photographer's Mate Third Class John S. Lerblance)

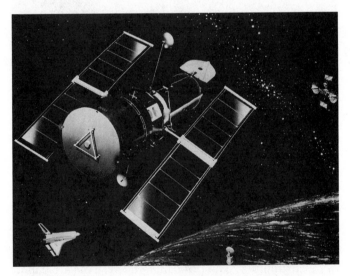

Fig. 8-5 This concept drawing of the space telescope is an illustration of the importance of telecommunications in the aerospace program.
(Courtesy of NASA)

Telecommunications

Communication affects our lives in many other ways. **Telecommunication** involves the sending of information over long distances. Examples include: the landing instructions sent by a control tower operator to the pilot of an airplane (Figure 8-4); the telephone calls people make everyday; the audio and visual programs we listen to and watch on radio and television; the weather reports sent to us by earth-orbiting satellites; and the research information retrieved from libraries via the computer ter-

minal. These systems employ technical devices to facilitate communication. For example, fiber-optic cables are used to send telephone messages by pulses of light generated by a laser. A single glass fiber is smaller than the diameter of a human hair, yet it can carry over five thousand voice messages. They have several advantages over copper wires and cables: they are less expensive, lighter in weight, more durable, and more reliable.

Efficient and reliable worldwide telecommunication is made possible by sophisticated earth satellites (Figure 8-5). These communications devices are placed into earth's orbit by the now-familiar NASA space shuttle. Complete newspapers are written and edited in one city, and are transmitted and automatically printed in another hundreds of miles away. People use computers to exchange information as a part of a national and worldwide communication network. The future of communication technology is impossible to predict, but it will have a tremendous impact on all our lives.

Industrial Communications

You were introduced to the universal systems model in unit 2 of this book and learned that it included input, process, output, and feedback. This same theory can be applied to a *communication system* because it has these same elements, as shown in Figure 8-6. Here the *inputs* are the people that create messages and the resources needed to record these messages. The act of recording the message is the *process,* such as using the computer word processor to type a letter. The finished letter prepared by the computer printer is the *output.* Errors, such as misspelled words or incorrect spacing, may be found before or after printing and can be corrected. This is an example of *feedback.*

Industrial communication deals with the transmission of the data necessary to make industry work. It takes many forms. Shop supervisors have to give machinists oral instructions regarding the jobs they are working on. Machinists must study an operations manual in order to check the proper settings for a computer lathe control. Human resources people post work and vacation schedules on bulletin boards for employees to read. Product designers must communicate their plans

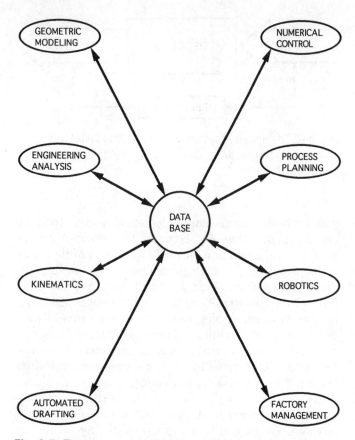

Fig. 8-7 The computer data base provides an important computer link among all functions of a modern manufacturing operation.

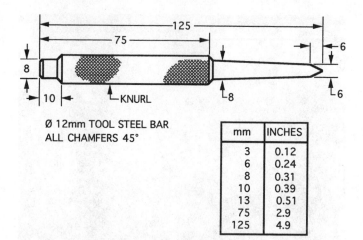

Fig. 8-8 CAD drawing of a center punch.

Fig. 8-9 The decals on aluminum baseball bats are a product of the printing industry.
(Courtesy of Worth, Inc.)

to the people who will make the products. This can be done manually by preparing engineering drawings on paper or electronically by creating computer (CAD) images. Paper drawings can be hand-delivered to the machinists and technicians who will build the product. Electronic images and instructions can be transmitted directly to a computer-controlled machining center, without any human intervention. In any of these cases, needed information moves from the sender to the receiver so that some action can take place (Figure 8-7).

An example of the systems model is the product plan for a new center punch, which includes the product design (Figure 8-8), the necessary materials, and the tools required to make it. These are all examples of *input.* The *process* is cutting the steel workpiece on a lathe to shape it; and the *output* is the completed punch. The act of creating the tool could not have occurred unless the product information was communicated to a person or a machine. This systems model operates in all communication situations, whether it involves oral or written instructions, or product design information.

Graphic Reproduction Technology

Graphic communication includes the many visual messaging methods, such as printing, drafting, and photography. Newspaper photographs, technical drawings,

product labels, and pictures used to illustrate magazine articles and decorate product cartons are all examples of these graphics. The print imaging systems are called **graphic reproduction technology,** a major industry. Printing is very important in metal product manufacture. The warning labels on chainsaws and the user manuals for lathes are printed by the lithography process. The decalcomania (or "decal") found on an aluminum baseball bat is produced by that same process (Figure 8-9). Wooden bats have the name burned in with hot letter dies. The company name on many wrenches is added as part of the forging operation. Study the tools and machines you use in your shopwork and you will find many other examples.

Printing theory is rather simple: (1) an image is prepared on a printing surface called a **plate;** (2) ink is applied to the plate; and (3) paper, plastic, or some other material (called a **substrate**) is pressed against the plate to receive the inked image. The primary printing methods are, in order of importance: offset lithography, gravure,

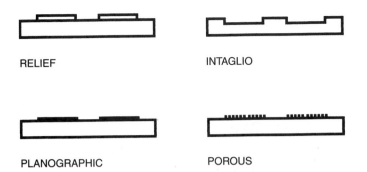

RELIEF

INTAGLIO

PLANOGRAPHIC

POROUS

Fig. 8-10 Image surface characteristics of the four major printing processes.

flexography, and screen. These methods differ in the way the image is carried to the plate, as shown in Figure 8-10.

Offset Lithography

Offset lithography is a **planographic** method wherein the image is transferred from a plane surface. It is based upon the principle that oil and water do not mix, which fairly defines the term *lithography*. The word *offset* comes from the fact that the plate cylinder never touches the paper (Figure 8-11A). In this process, a photographic image is acid-etched on a plate cylinder. The ink rolls carry the ink to the plate cylinder at the same time water rolls carry moisture to the plate. The plate accepts ink and repels moisture in the etched image areas, and attracts moisture and repels ink in the non-image areas. Next, the plate image offsets (transfers) to the blanket cylinder, which transfers it to the paper under pressure from the impression cylinder. One reason for the wide use of lithography is that the blanket cylinder allows very clear images to be printed onto smooth or textured substrates. It is generally used for printing newspapers, magazines, advertising inserts, and books.

Gravure

The **gravure** method involves direct printing from an engraved copper-plated cylinder to a substrate. Note that while the lithography plate image is etched or burned in with acid, the gravure plate image is engraved or cut in with a diamond stylus tool. The gravure cylinder turns in an ink pan to fill the engraved image areas. A doctor blade wipes away all the excess ink, leaving ink only in the engraved image areas. The image is then transferred to the substrate by contact with the impression cylinder (Figure 8-11B). The gravure process creates exceptionally high quality color images. However, the etched cylinder costs are high, and so the process is restricted to large volume runs such as the Sunday newspaper's glossy color inserts. Other items include wallpaper, product labels, postage stamps, foil packages, and magazines. Gravure is an **intaglio** process, which means that the print image is incised or cut into the plate, a precision and expensive process.

Flexography

Flexography is a **relief** printing method wherein the raised image is in the form of a rubber mat wrapped around a plate cylinder. In this process, a rubber fountain roll picks up ink from a pan and passes it onto a steel anilox roll (Figure 8-11C). A doctor blade causes a thin ink film to adhere to the anilox. This film is passed to the plate cylinder, and the image is transferred to the substrate. The notable feature of flexography is that it transfers images to long, continuous substrate webs, making it a very efficient and cost-effective method. The individual pieces are later cut from this web. It is primarily used for packaging materials, such as plastic and foil bags, milk cartons, boxes, and labels.

Screen

Screen printing is a process wherein a porous mesh material covered with a stencil (partially cut) design is stretched over the substrate (Figure 8-11D). It is often called the *silk-screen* process, although metal screens are commonly used in production work. A moving rubber squeegee forces ink through the fine screen openings, transferring the stencilled image to the printing surface. This method is used to print images on bottles, cartons, banners, fabrics, and the graphics on the sides of huge semitrailers. Look at the speed and feed charts on lathes and other machines; many of these are screen printed with an enamel, which is then oven-baked for durability.

There are many other printing processes important to the manufacturing industries. For example, the graduations on measuring tools such as steel rules and micrometers are photoengraved (Figure 8-12). Also, hard copy CAD drawings are produced by dot matrix, inkjet, laser, or plotter printers (Figure 8-13).

KNOWLEDGE REVIEW

1. Write a short essay on the important uses of printing in the metal product manufacturing industries.

2. Explain what communication is.

3. Define the term *telecommunication* and give some examples of it.

4. Explain what graphic reproduction technology is.

5. Explain what printing is.

6. Describe the four major printing processes and give some product examples.

7. Select some hand tools and prepare an oral class report on the method used to produce the printing found on them.

8. Visit a local printing shop and report on the printing methods used there.

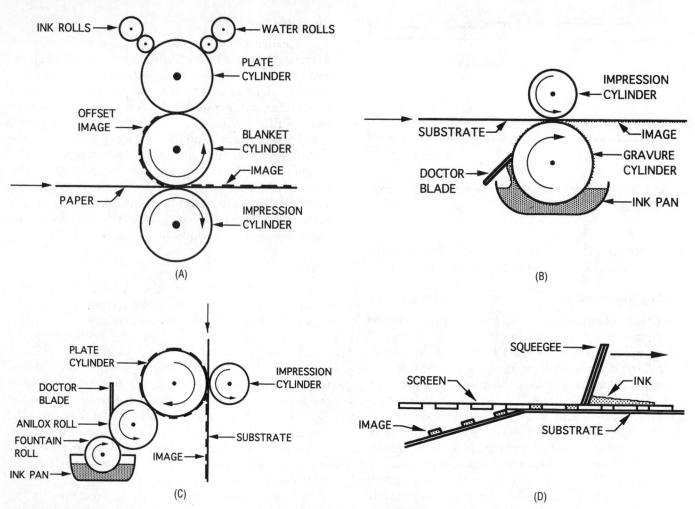

Fig. 8-11 Diagrams of the four major printing processes: (A) offset lithography; (B) gravure; (C) flexography; and (D) screen.

Fig. 8-12 The corporate name and button control labels on this digital electronic micrometer were printed on adhesive metal tabs. The barrel and thimble graduations were photoengraved. (Courtesy of The L. S. Starrett Company)

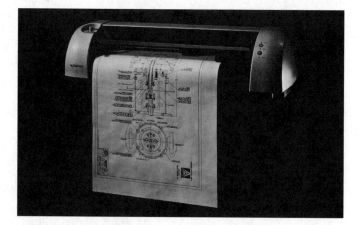

Fig. 8-13 An example of a computer plotter drawing. (Courtesy of CalComp)

Unit 9

Measurement

OBJECTIVES

After studying this unit, you should be able to:

- Give some reasons why the United States should become a metric nation.
- Define the terms *measurement* and *standard units.*
- State some customary units and give their metric replacements.
- List the seven base metric units.
- Explain what derived units are.
- Explain the difference between hard and soft conversion.
- Know how to measure with a metric rule.

KEY TERMS

base unit	kilogram	prefix	speed
Celsius	liter	pressure	standard unit
derived unit	meter	second	symbol
energy	power	SI metric system	

Measurement is one of the most common yet most important activities in our lives. We thrill at a football player running ninety-eight yards for a touchdown. A baseball hit four hundred feet is a home run in any ball park. And every household has a ruler and a tape measure to determine a person's height or the size of a rug.

Most everyday measurements are satisfactory if they are off by a small fraction of an inch. However, in the mass production industries where things are made by the millions, measurements have to be more precise. Here, many dimensions have to be accurate within thousandths of an inch. Why? Because mass production is based on interchangeability, which demands that all product parts fit together perfectly. The only bridge between the engineers who design things and the skilled workers who make them is the blueprint that contains the product dimensions, or measurements.

Fig. 9-1 This steel tape is graduated in both inches and millimeters. Some tapes show millimeters only, and others inches only.

Unit	Symbol
meter (length)	m
kilogram (mass)	kg
second (time)	s
kelvin (temperature)	K
ampere (electric current)	A
candela (luminous intensity)	cd
mole (amount of substance)	mol

Fig. 9-2 The seven base units of the SI metric system. All other units are derived from these.

What Is Measurement?

When we measure something, we compare it to a **standard unit.** The foot is a standard unit in the United States. If you wish to measure the length of a piece of copper tubing, you must lay a steel tape alongside it and count the number of foot-units. Each of the measures mentioned above—a person's height, the size of a rug, the length of a touchdown run, the distance the baseball was hit—is an example of this comparison. We use such standard units as inches, pounds, and gallons to measure things. To measure, we must find the number of units there are in the object we are measuring.

Two kinds of measurement are used in the United States: the customary (inch-pound) system and the metric system. You have grown up using feet, inches, miles, pounds, quarts, and gallons, and are familiar with them. Most of the world uses the metric system only. All of the automobiles, and much of the scientific equipment and machine tools, made in America for export are produced in metric units. In the years ahead, everyone will need to know how to use this system and its special measuring tools, such as steel tapes (Figure 9-1). Therefore, it is important to learn about metrics and to build a project to metric measures.

The SI Metric System

The metric system began in France in 1790, as a replacement for the variety of confusing measures which were then being used in Europe. The basis for this new system was the distance from the North Pole to the equator through Paris, France. They selected this distance because it was fixed and measurable, and all scientists could agree upon it. This distance is so great that one ten-millionth of it was chosen as the new base unit for length, and it was called the *meter*. Other measures for mass, volume, and temperature also were developed. This system was used and improved, and became the law in France in 1840. After that, the metric system spread throughout the world.

As you can see, the metric system has been in use in many countries for a long time. In recent years, it has been replaced by a newer, more accurate, and more usable method to meet the needs of modern technology. Named the International System of Units, it carries the symbol "SI" for convenience. This **SI metric system** consists of the seven **base units,** shown in Figure 9-2, plus two supplementary units and many derived units. The supplementary units are the *radian* and the *steradian* to measure plane and solid angles. **Derived units** are the combinations of base, supplementary, or other derived units obtained by multiplication and division. For example, the derived unit for area is the square meter (m^2). It is calculated by multiplying length by width in meters.

Some derived units have been given special names and symbols. They are used to express the derived unit in a simpler way than in terms of the base units themselves. For example, the derived unit for force is called the *newton*. Expressed in base units it is $kg \times m/s^2$. Obviously, it is much easier to say "newton" than to state the more complex formula. These derived units will be discussed later.

The metric system, like the monetary system of the United States, is based on multiples of ten, simply called a *base-ten system*. For example, a meter is divided into ten decimeters, one hundred centimeters, and one thousand millimeters by using the **prefixes** *deci-, centi-,* and *milli-*. A meter multiplied by one thousand equals a kilometer. Therefore, a kilometer is one thousand meters, a kilogram is one thousand grams, and a kilowatt is one thousand watts. As you can see in Figure 9-3, the same prefixes are used for all units.

Multiple or submultiple	Prefix	Symbol	Pronunciation*	Means
$1\,000\,000\,000 = 10^9$	giga	G	*jig'a* (*a* as in *about*)	one billion times
$1\,000\,000 = 10^6$	mega	M	as in *megaphone*	one million times
$1\,000 = 10^3$	kilo	k	as in *kilowatt*	one thousand times
$100 = 10^2$	hecto	h	*heck'toe*	one hundred times
$10 = 10^1$	deka	da	*deck'a* (*a* as in *about*)	ten times
base unit $1 = 10^0$				
$0.1 = 10^{-1}$	deci	d	as in *decimal*	one tenth of
$0.01 = 10^{-2}$	centi	c	as in *centipede*	one hundredth of
$0.001 = 10^{-3}$	milli	m	as in *military*	one thousandth of
$0.000\,001 = 10^{-6}$	micro	μ	as in *microphone*	one millionth of
$0.000\,000\,001 = 10^{-9}$	nano	n	*nan'oh* (*an* as in *ant*)	one billionth of

*The first syllable of every prefix is accented.

Fig. 9-3 The SI metric unit prefixes.

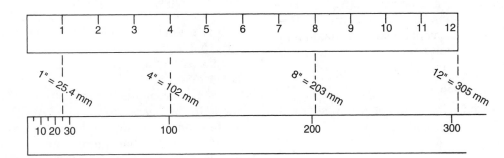

Fig. 9-4 Note that the common 300-mm rule is a little shorter than the standard 12-inch rule.

Common Metric Measures

Most people will use only three of the seven base units in everyday life: the **meter** for length, the **kilogram** for mass or weight, and the **second** for time. While the minute, hour, day, and year will continue to be used, they are not really SI units because they are not base-ten measures. Another common measure for which there is a base unit is temperature. However, the degree **Celsius** is the temperature measure rather than the Kelvin, which is the base unit. Let's take a closer look at the primary units of measurement.

Length

The metric unit of length is the meter, and its symbol is *m*. Note that this is a **symbol,** and not an abbreviation. All metric units and prefixes have special symbols. Use them correctly. The meter is equal to about thirty-nine inches, and is therefore longer than a yard, which is equal to thirty-six inches, or three feet. The meter can be divided into one hundred parts called *centimeters* (cm), and one thousand parts called *millimeters* (mm).

The meterstick is a little longer than the yardstick. A 300-millimeter rule is therefore slightly shorter than

a 12-inch rule (Figure 9-4). The terms *rule* or *scale* are used in place of the word *ruler* in technical work. The millimeter (about the thickness of a dime) is the smallest division on most metric rules. It and the meter are the preferred units. Centimeters are never used in technical measuring. They are restricted to clothing and sporting goods sizes.

Volume

The fundamental volume measure is based upon the meter. It is the cubic meter (m^3), which is about thirty percent more than a cubic yard. It is generally used for large quantities of air and water, and for solid substances such as earth and concrete.

The cubic decimeter (dm^3), typically used to measure liquids, is called the **liter** (L). One liter of water has a mass of one kilogram. The liter is about three tablespoons larger than the quart, and is the metric unit for liquids such as paint, oil, soft drinks, and milk. Since these are normally packaged in quart and pint (half-quart) sizes, the metric equivalents are the liter and half-liter (five hundred milliliters). A four-liter can of paint replaces the present gallon size. (Gasoline is sold by the liter.) Since the liter and half-liter are a little larger than the quart and pint, the contents of the metric containers will be more.

Mass

The metric unit of mass is the kilogram (kg), equal to one thousand grams. For everyday purposes, mass is the same as weight. The kilogram equals 2.2 pounds, or slightly more than twice the pound in weight. There are one thousand kg in a metric ton, which equals about 1.1 U.S. tons. If you weigh one hundred pounds, your metric weight will be about forty-five kilograms. There are about twenty-eight grams in one ounce. Butter and coffee will be sold in five-hundred gram packages, slightly more than a pound.

Temperature

Metric temperature is measured in degrees Celsius (°C). On this scale, water freezes at 0°C and boils at 100°C, and normal body temperature is 37°C. The easiest way to convert Fahrenheit (°F) temperatures to Celsius (or vice versa) is to use your pocket calculator. Subtract 32 from the Fahrenheit temperature and multiply the result by 0.56. The formula is

$$°C = (°F - 32) \times 0.56$$

For example, the conversion of a normal room temperature of 68°F to °C would be:

$$°C = (68 - 32) \times 0.56$$
$$°C = (36) \times 0.56$$
$$°C = 20.16$$

This means that a temperature of 68 degrees Fahrenheit is equal to about 20 degrees on the Celsius scale. Many shop and classroom bulletin boards have temperature conversion charts to make these calculations even simpler.

This brief overview will help you to better understand and use metrics in your daily living and purchasing. You will find a drawing of the metric sizes of some common objects in Figure 9-5 to familiarize you with these measures. Now we shall learn how metric measures and tools are used in the metalworking shop.

Technical Metric Measures

The most common metric length units for shop use are the meter (m) and the millimeter (mm). Metal stock lengths are measured in meters, and project drawings in millimeters. The area of a piece of sheet metal is given in square meters (m²).

Metric rules are usually available in 150 mm and 300 mm lengths in addition to the meterstick. Use a rule that is graduated in millimeters (not centimeters), with every tenth line marked 10, 20, 30, and so on (Figure 9-6).

Measurements in the customary system are often made to the nearest 1/16 inch. Working in the metric system and measuring to the nearest millimeter is therefore somewhat more accurate. For even greater accuracy, values may be rounded off to a half millimeter (0.5 mm). A half millimeter is about 1/50 inch in size, or about halfway between 1/32 inch and 1/64

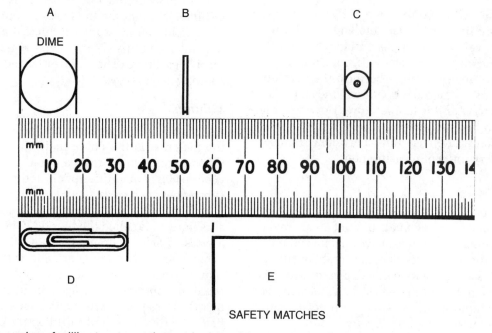
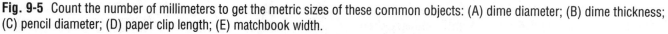

Fig. 9-5 Count the number of millimeters to get the metric sizes of these common objects: (A) dime diameter; (B) dime thickness; (C) pencil diameter; (D) paper clip length; (E) matchbook width.

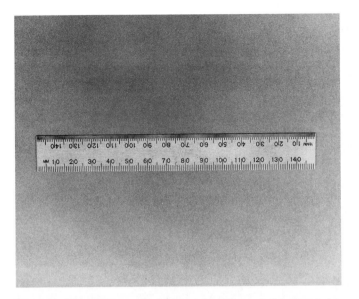

Fig. 9-6 The 150-mm rule replaces the common 6-inch rule in metric measuring.

Measurement	Name	Symbol	Formula
frequency	hertz	Hz	l/s
force or weight	newton	N	$kg \cdot m/s^2$
pressure or stress	pascal	Pa	N/m^2
energy	joule	J	$N \cdot m$
power	watt	W	J/s
electric charge	coulomb	C	$A \cdot s$
electric potential	volt	V	W/A
capacitance	farad	F	C/V
electric resistance	ohm	Ω	V/A
conductance	siemens	S	A/V
magnetic flux	weber	Wb	$V \cdot s$
magnetic flux density	tesla	T	Wb/m^2
inductance	henry	H	Wb/A
luminous flux	lumen	lm	$cd \cdot sr$
illuminance	lux	lx	lm/m^2

Fig. 9-7 SI metric derived units with special names.

inch, the smallest division on a precision machinist's rule or scale.

The derived units with special names are shown in Figure 9-7. Note that these refer to technical and scientific measures. In the metric system, there is only one unit for any physical quantity. In our customary system there are many, each dealing with a different area. For example, the *joule* is the metric unit for energy. It replaces the British thermal unit (Btu) and the therm, which are heat measures; the erg, which is a unit of work; the calorie, which is a measure of the energy content of food; and four others. One of the reasons to adopt the metric system is that it will make measuring easier by doing away with this confusion. All technical people will speak the same measurement language. Several common technical units you may use in metalwork are described below, associated with the physical quantity they measure.

Speed

Speed is given in *meters per second* (m/s), for use in determining machine cutting speeds. Meters per minute (m/min) also may be used. Highway speed limits are shown in kilometers per hour (km/h). Even though the minute and hour are not SI units, the above measures will be used.

Pressure

The *pascal* (Pa) is unit for **pressure**. Because this is such a tiny unit, pressures are usually given in kilopascals (kPa), or one thousand pascals. For example, a dollar bill lying flat on a table exerts a pressure of about one pascal on the table. One pound per square inch (psi) is equal to about seven kilopascals. A tire pressure of 25 psi is therefore about 172 kPa.

Energy

The unit of **energy** or work is the *joule* (J). This also is a very small unit, so the kilojoule (kJ) is most often used. The joule replaces a number of older units, as described earlier.

Power

The unit for all **power**—not just electrical—is the *watt* (W) or kilowatt (kW). A popular size of electric lamp consumes one hundred watts. One horsepower equals about 0.75 kW. Therefore, a 100 horsepower engine would also be a 75 kilowatt engine.

Tools, Machines, and Materials

The drill press has a convenient inch rule attached to the frame to indicate the depth of the hole being drilled. Other machines have similar scales. These machines can be converted to metric measurement by placing an adhesive metric tape alongside the customary gage or over it (Figure 9-8). On machine tools such as lathes, shapers, and grinders, dual reading dials must be added that show both thousandths of an inch and hundredths of a millimeter. On the lathe, a separate set of gears is needed to cut metric threads. All measuring tools, such as rules and micrometers, must be available in metric units. Metric drills, taps and dies, and wrenches are also needed.

Metal materials in metric sizes are now available. A sample chart of metal bars is shown in Figure 9-9. Note that these numbers indicate a change in physical size (hard conversion). A bar 1" wide now becomes a bar 25 mm wide. A "soft conversion" would be merely showing

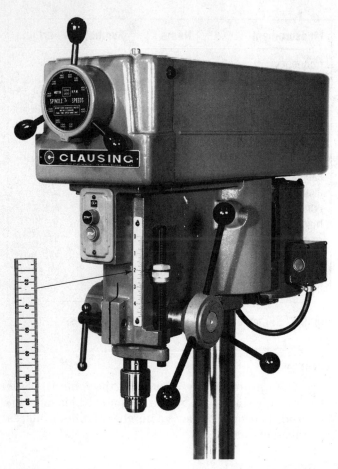

Fig. 9-8 A metric tape can be fixed to a drill press as a depth gage for hole drilling.

the 1″ bar as a converted 25.4 mm size. Study this chart to become familiar with these size changes. More information on metric materials can be found in unit 6.

KNOWLEDGE REVIEW

1. Why must measurements in industry be very accurate?

2. List the seven base metric units and indicate those which you will use in daily life.

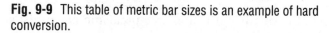

Thickness		Width	
inch	mm	inch	mm
1/16	1.6	3/8	10
1/8	3	1/2	12
3/16	5	5/8	16
1/4	6	3/4	20
5/16	8	1	25
3/8	10	1 1/4	30
7/16	11	1 1/2	40
1/2	12	1 3/4	45
9/16	14	2	50
5/8	16		
11/16	18		
3/4	20		
7/8	22		
1	25		
1 1/8	28		
1 1/4	30		
1 5/16	32		
1 3/8	35		
1 1/2	40		
1 3/4	45		
2	50		

Fig. 9-9 This table of metric bar sizes is an example of hard conversion.

3. Describe the metric technical measuring units important in metalwork and give examples of their use.

4. What is the difference between "soft" conversion and "hard" conversion?

5. Visit a hardware store and report to the class the metric tools you found there.

6. Make a poster-size conversion chart for a selection of customary and metric units.

7. Write a research report on the history of the metric system.

Unit 10

Technical Drawing

OBJECTIVES

After studying this unit, you should be able to:

- Understand the purpose of drawings in the metal shop and in industry.
- Explain the differences among drafting, drawing, and sketching.
- Prepare and read pictorial and mulitview sketches.
- Recognize and know how to use the different types of lines.
- Dimension a drawing or sketch.
- Prepare a conversion chart of a metric sketch.
- Know how to read and use drafting scales.
- Recognize drafting instruments.
- Explain the purpose of detail and assembly drawings.

KEY TERMS

cabinet	drafter scale	metric scale	protractor
cavalier	drafting	oblique	sketching
compass	drawing	orthographic	triangle
divider	engineer scale	perspective	T square
drafter	isometric	pictorial	

In metalwork, you must prepare a plan or drawing for any project you wish to make. You also must be able to read and understand a drawing made by someone else. Accurate plans are important because you will transfer measurements from the drawing to the material when you are making the project. If you have had a drawing course, this unit of work should be easy for you. If you have not, read this unit very carefully to learn what good drawings are and how they are made. Drawing is a technical language skilled workers must understand if they are to be successful in their work.

The word **drawing** was used several times in the above paragraph. A drawing is a project or product plan made with instruments such as T squares and triangles for making straight lines, and compasses for making circles. Most of the line illustrations in this book are therefore drawings. They are neat and accurate, but they take time to make. **Sketches** are freehand plans made without the use of instruments. They can be made quickly and accurately by using grid papers, and are quite suitable for shop use. Much if not all of your project planning will be done with sketches. These will be described later in the unit. The process of making drawings and sketches is called **drafting,** and the technical people who do this work are called **drafters.**

As was mentioned in unit 8, drafting is a method of graphic communication. It is a visual language which uses lines, numbers, words, and symbols to create plans for products. In industry, a designer plans a product, a drafter makes a drawing of it by hand or computer, and skilled crafters make the product from the drawing.

Types of Lines and Their Meanings

In the language of drafting, different lines have different meanings, and these lines are used to make a drawing. In a sense, this is an alphabet of lines, much as the alphabet of letters is used to form words. As you make and read drawings, it is important that you understand the meaning of each. The illustration in Figure 10-1 will help you to do this. Note how the lines are used, and that they differ in both weight, or heaviness, and in structure.

1. *Object* lines are heavy and solid, and show all the visible edges and outlines of a thing. These lines describe the shape of an object.

2. *Hidden* lines are thin and dashed, and show edges and details that cannot be seen.

3. *Center* lines are thin and made of short and long dashes. They locate the centers of arcs and circles, and divide an object into symmetrical, or equal, parts.

4. *Extension* lines are thin and solid, and are continuations of the object lines of the drawing. They are used to locate the ends of dimension lines.

5. *Dimension* lines are thin and solid, and show the measures or distances between extension lines. Note the use of neat arrowheads.

6. *Leaders* are thin and solid, and serve as a special dimension line to direct a measure or note to some detail of a drawing.

7. *Break* lines show parts of objects which have been removed (broken away) to reveal some hidden details. Thick, freehand, irregular lines are used for short breaks, and thin lines with V notches for long breaks.

8. *Cutting plane* lines are heavy and solid, generally made of long and short dashes, and indicate an imaginary cut (section) through an object.

9. *Section* lines are thin, solid, and parallel, drawn at an angle to show the cut surfaces in a section view.

Study the many drawings in this book to learn more about these lines and how they are used.

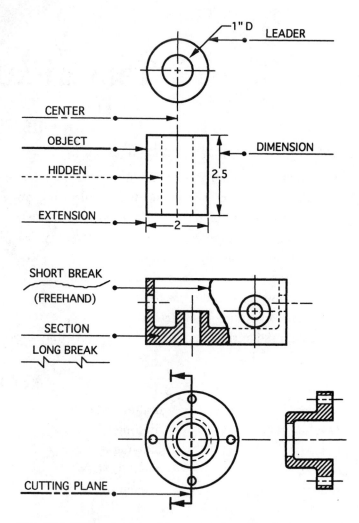

Fig. 10-1 Different types of lines, showing what they mean and how they are used.

Kinds of Drawings

Many books and magazines contain realistic drawings that look much like photographs (Figure 10-2). These are called **pictorial** drawings, and there are three kinds of them.

1. A **perspective** drawing accurately shows the object as it would appear to the eye. They are drawn

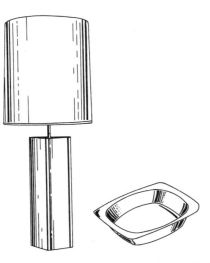

Fig. 10-2 Pictorial sketches such as these are realistic illustrations of products. Simple shading improves their appearances.

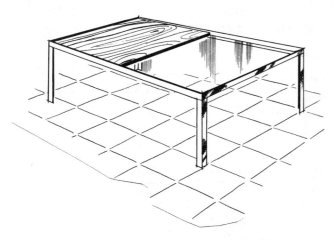

Fig. 10-3 A perspective sketch of a coffee table with an angle iron frame.

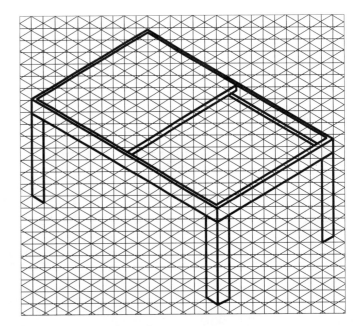

Fig. 10-4 An isometric sketch of the coffee table.

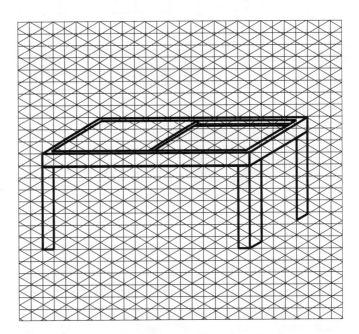

Fig. 10-5 A cabinet sketch of the coffee table. The foreshortened side lines make the sketch appear less awkward.

by using vanishing points. If you were to stand between two railroad tracks and follow them with your eyes, they would appear to meet on the horizon. In like manner, if you extended the lines of the side rails of the table top in Figure 10-3, they too would meet. While this is a good method of preparing project drawings, it is not very often used in metal shopwork unless you wish to make an illustration for a report.

2. An **isometric** drawing looks something like a perspective drawing except that vanishing points are not used. Instead, the sides of the object are shown 120 degrees apart (Figure 10-4). Note the use of grid paper for this sketch. This method is used mostly for drawing rectangular objects, because isometric circles and curves are difficult to prepare.

3. A **cabinet** drawing is the easiest kind of pictorial to make. A front view of the object is drawn true to size, and the sides are inclined back from the front view at 30-degree angles. These angular lines are not true length. They are foreshortened by one half so the view will not look distorted (Figure 10-5). If you made these angular lines true length, you would have a **cavalier** drawing. The cabinet and cavalier are types of **oblique** drawings.

The multiview, or **orthographic,** is the second major kind of drawing used for making project plans. It is the working drawing which is followed as a project is

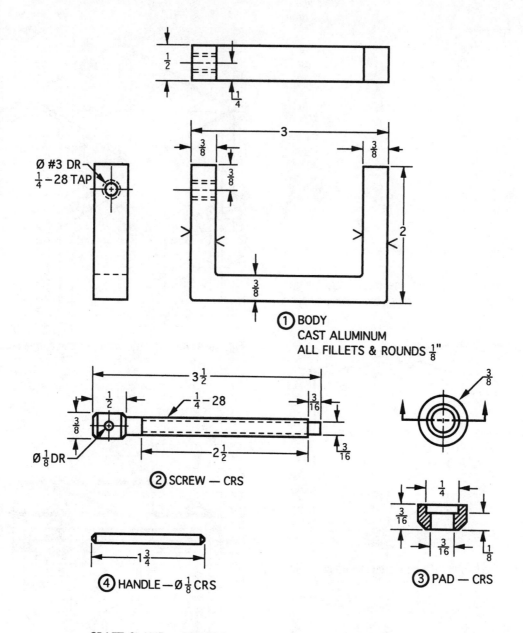

① BODY
CAST ALUMINUM
ALL FILLETS & ROUNDS $\frac{1}{8}$"

② SCREW — CRS

④ HANDLE — Ø$\frac{1}{8}$ CRS

③ PAD — CRS

CRAFT CLAMP — DETAILS
ALL CHAMFERS 45°

Fig. 10-6 A three-view drawing of the body is shown on this detail drawing of C clamp parts. One- and two-view detail drawings also are shown.

being built. Multiviews usually show two or three views of an object. A standard three-view drawing is made with the front view in the lower left-hand corner of the page. The top view is in line and directly above, and the end view is in line and usually to the right.

Study the illustration of the C clamp body in Figure 10-6 for an example of a three-view drawing. Note here that the end view is to the left because it better shows the details of the screw hole. The views are lined up with one another and you get a good idea of what the piece looks like. Note also that the pad requires only two views, and the screw and handle only one view. Can you explain why?

The drawings in Figure 10-6 are *detail* drawings, which show the individual project parts and all the information needed to make them. An *assembly* drawing shows how the parts fit together to make the project (Figure 10-7). For some simple projects, both the detail and assembly drawings are shown on the same sheet. Larger projects require separate detail and assembly drawings.

Dimensions

Whereas lines are used to describe the *shape* of an object, dimensions describe the *size* of that object. A drawing that is correctly made contains all the dimensions and

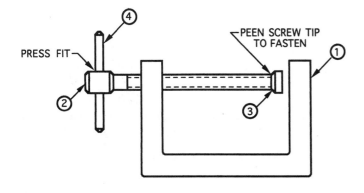

CRAFT CLAMP — ASSEMBLY

Fig. 10-7 An assembly drawing of the C clamp clearly shows how the parts fit together. The numbers refer to the detail drawing in Figure 10-6.

information (or notes) necessary to make the part. There are four common ways to dimension a drawing so that the product can be built to customary and/or metric measures.

1. *Customary.*
   A customary-inch drawing can be dimensioned in inches and fractions of inches when the part to be made does not require a fine degree of accuracy. These dimensions are used for many of the project drawings in this book, such as the part details shown in Figure 10-6.

2. *Decimal.*
   For more precise parts, dimensions are always given in inches and decimal parts of an inch, such as 1.25 inches or 0.25 inches. Most machine drawings use

decimals. This method also alerts a machinist to use precision measuring tools, such as micrometers or verniers. These tools are graduated in decimal units. Dimensions that require the use of precision measuring tools are given in thousandths (0.001) or tenthousandths (0.0001) of an inch. Also, many machine dials and gages are calibrated in decimal inches. An example of decimal dimensioning is shown in Figure 10-8. This is an industrial computer drawing of a part of the exercise machine described in unit 12. Look at this drawing and you will see that a .225 inch dimension is shown. It is better to place a zero (0) in front of the decimal point (.), so that the dimension will read 0.225, in case the decimal point is lost when making prints or copies of a drawing.

3. *An all-metric drawing.*
   Such a product plan shows metric dimensions only.

4. *A metric drawing with readout chart.*
   Many manufacturers are designing new products in metric, and use only those dimensions on the drawing. A small conversion chart is then added to the drawing, often as a paste-on. This chart indicates the drawing dimensions in millimeters, and their equivalents in inches. Therefore, a skilled crafter can make the product to either metric or customary measures (Figure 10-9).

Some drafters also use dual dimensions on drawings, as illustrated in Figure 10-10. This method is suited to very simple plans only. On more complicated drawings, these dimensions clutter the sheet and can become confusing. It is better not to use these; use the conversion chart system instead.

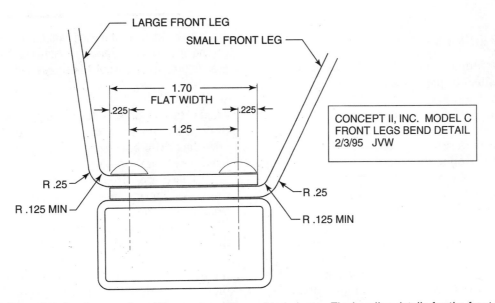

Fig. 10-8 Decimal dimension drawings such as this are generally used in industry. The bending details for the front legs of an exercise machine are shown.
(Courtesy of Concept II, Inc.)

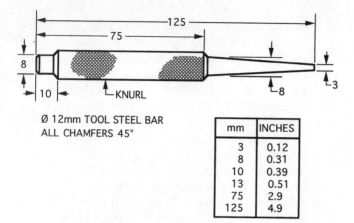

Ø 12mm TOOL STEEL BAR
ALL CHAMFERS 45°

mm	INCHES
3	0.12
8	0.31
10	0.39
13	0.51
75	2.9
125	4.9

Fig. 10-9 Metric drawing of a metal punch, with the conversion chart shown.

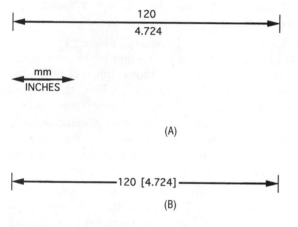

DIM. IN [   ] ARE INCHES

Fig. 10-10 Two systems for dual dimensioning: (A) line method and (B) bracket method. The conversion chart method shown in Figure 10-9 is preferred.

Drawing Instruments

Special instruments make project drawing easier. For example, a sheet of paper may be taped to a drawing board and a **T square** used to draw straight lines (Figure 10-11). **Triangles** are used to draw both vertical and angular lines (Figure 10-12). The two common types of triangle are the 30°–60°, and the 45°, which can be used in combination to draw angles as shown. Other angles can be laid out with a **protractor** such as you used in your mathematics classes. A **compass** (Figure 10-13) is used to scribe circles and arcs, and **dividers** are used to transfer measurements from the scale to the paper, or from a drawing to a metal workpiece.

Very often the project you are drawing is so large that it will not fit on the sheet of drawing paper. In this case you will have to draw it half size or quarter size by using a *drafting scale* (Figure 10-14). These are special measuring devices, about twelve inches long, used to

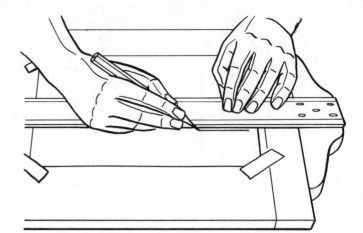

Fig. 10-11 Using the T square to rule straight lines.

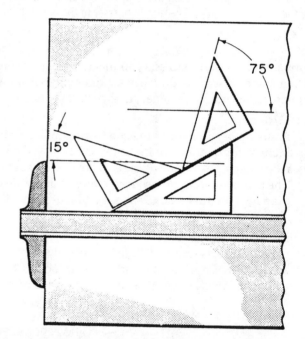

Fig. 10-12 Triangles are used with the T square to draw vertical and angular lines. Shown here are combinations of triangles used to produce angles of 15 degrees and 75 degrees.

"scale down" objects to a desired size. You may also draw a small object double size, or "scale up." Each edge has different graduations so that one tool can be used for several operations.

There are three scales commonly used for shop drawings. The **engineer scale,** used for accurate decimal drawings, is graduated in inches and tenths of an inch. Civil engineers also use this for making maps. The **drafter scale** is graduated in inches and sixteenths of an inch, much like an ordinary shop rule. This makes it the best tool for ordinary shop drawings based upon the inch and its divisions, such as 1/16, 1/8, 1/4, etc. The **metric scale** is graduated in millimeters. The drafter and the engineer scales are used for making customary-inch

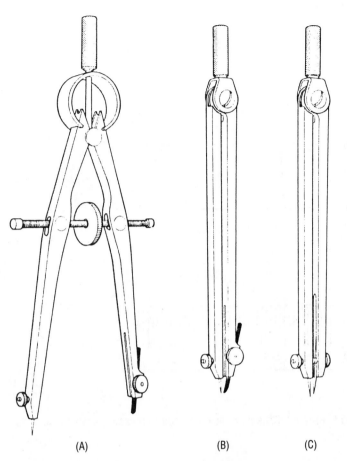

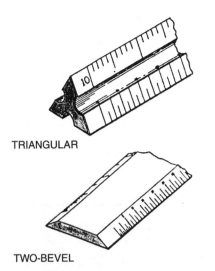

TRIANGULAR

TWO-BEVEL

Fig. 10-14 Two styles of drafting scale.

(A)   **(B)**   **(C)**

Fig. 10-13 Circles and arcs are drawn with a compass. Two types (A and B) as well as a pair of dividers (C) are shown here.

drawings. The metric scale is used for making metric drawings.

To use the drafter scale, study the illustration in Figure 10-15. The "1/4" marking indicates that this is a 1/4 size scale, meaning 1/4" = 1". The distance to the left of the zero mark is divided into sixteen equal parts, each part representing 1/16". The numbered divisions to the right of the zero represent one inch each. To measure 4 5/16" on this scale, you must first count four numbered divisions to the right of the zero. Then count five divisions to the left of the zero. Add the two distances to get the desired 4 5/16" measurement.

Sketching

As stated earlier, shop sketches can be most easily made by using grid paper. This allows you to conve-

niently organize views, sketch straight lines, and determine sizes by counting grid squares. If grid paper is not available, you can prepare satisfactory freehand sketches with practice. The most common grids are the 1/8-inch or 1/4-inch square, the combination square and isometric, and the metric sheets.

Begin your work by sketching short, "wiggly" horizontal, vertical, and diagonal practice lines (Figure 10-16). Circles and arcs can be made with layout guide markers. When you feel comfortable with your sketching skills, proceed to use this new skill in preparing a project plan according to the following outline.

1. *Decide on the views you will need for the plan.*
   A simple project may need only one or two views.

2. *Select the scale.*
   If the paper has eight squares to the inch, each square can represent 1/8 inch, 1/4 inch, 1/2 inch, or any other suitable fraction to fit the sketch to the page.

3. *Sketch the project.*
   Figure 10-17 shows a one-view plan for a drill gage. This simple sheet metal tool acts both as an angle guide when sharpening a drill bit, and as a measuring device to check common drill diameters. The properly placed dimensions complete the plan. As you can see, the sketch is a quick and easy way to prepare a project work plan.

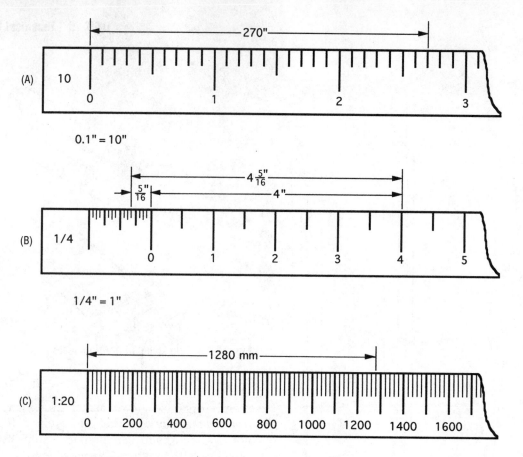

(A) 10

270"

0.1" = 10"

(B) 1/4

$4\frac{5}{16}$"

$\frac{5}{16}$"

4"

1/4" = 1"

(C) 1:20

1280 mm

1mm = 20 mm

Fig. 10-15 Three types of drafting scale: (A) engineer, (B) drafter, and (C) metric. Note how they are used to measure scaled distances.

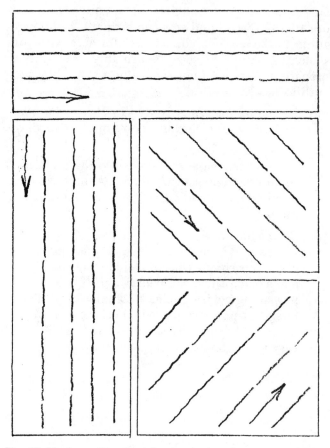

Fig. 10-16 A practice sheet for developing sketching skills.

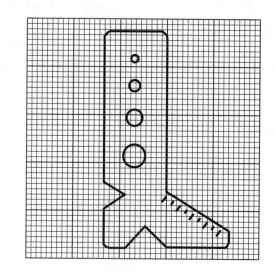

Fig. 10-17 A working sketch for a drill gage.

KNOWLEDGE REVIEW

1. What is a pictorial drawing?

2. Explain the differences among the several kinds of pictorial drawings.

3. Prepare isometric, cabinet, and cavalier sketches of a simple bookcase. Include dimensions.

4. Describe the purposes of the six types of lines used to convey information in a drawing.

5. What are multiview drawings?

6. Explain the reasons for one-, two-, and three-view drawings.

7. Why is grid paper so helpful in sketching?

8. Get a print of a simple metal part from a manufacturer of metal products or from a magazine. Show the print to the class and explain how to read it.

9. Select a simple object and use a scale to increase or decrease its size. Prepare a sketch of that object.

Unit 11

Project Design and Planning

OBJECTIVES

After studying this unit, you should be able to:

- Understand the purpose of product design in industry.
- Explain the meaning of the word *design*.
- Apply the three design requirements to your project planning.
- Understand the design elements and principles.
- Apply the Design Analysis Method to your personal design projects.

KEY TERMS

appearance	function	proportion	variety
balance	lines	shapes	
elements	materials	surface qualities	
forms	principles	unity	

Products are the things made by manufacturing processes or operations. They are the tools and tennis rackets and motorcycles people use each day. Every product starts from a plan or design. In the beginning, the design is only an idea in a person's mind. Then it becomes a sketch on a piece of paper, and next an experimental model. Finally, it is made into a finished product.

The next time you use a tool in school or at home, take the time to think about it. Is the tool usable? Does the wrench hold the nut safely and securely, or the hacksaw cut as it should? Have the proper materials been used to make the tool strong enough? Is it comfortable to hold and well-balanced? If the answers are yes, the object is probably well-designed. Tools and other products are planned and built to be useful, usable, and safe. Study the mountain bike in Figure 11-1. It is a good example of a sport vehicle built on a strong, lightweight, welded aluminum frame with a high-performance rear shock absorber. The bike's features include: a comfortable saddle, a handlebar with conveniently placed control levers, a durable and reliable shift mechanism, safe and sure pedals, and tough wheels and tires. It looks like a rugged, usable machine—and it is.

Fig. 11-1 This mountain bike is a well-designed piece of sport equipment.
(Courtesy of GT Bicycles, Inc.)

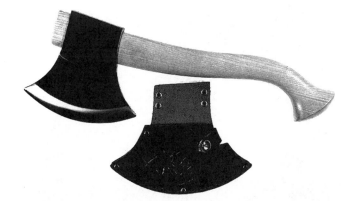

Fig. 11-2 Proper materials and a unity of line and form make this hatchet a functional and attractive cutting tool.

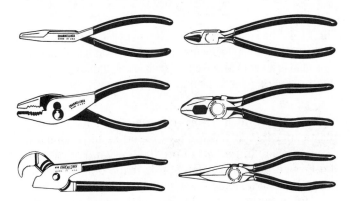

Fig. 11-3 The shapes and structures of these pliers' jaws are different because they are used for different kinds of holding, turning, and cutting operations.
(Courtesy of CHANNELLOCK, Inc.)

The people who plan these products are the industrial and the engineering designers who work in research and development departments of manufacturing industries. Their jobs are to design new products and to improve old ones. If the products work as they should, are made of the right materials for strength and durability, and have a pleasing appearance, they are well-designed. If they do not meet these conditions, they are poorly designed. It is important that you learn something about design so that you can plan and make your own personal projects. Knowing something about design can also help you to make wise decisions when purchasing things.

Design Requirements

Very simply, designing is creative planning. In order to be of high quality, the thing being designed should meet three requirements: it should work properly (functional requirement), it must be made of the correct materials (material requirement), and it should be pleasing to look at (visual requirement).

Look at the hatchet in Figure 11-2, which was designed for light chopping and chipping of wood. To function correctly, there must be a correct balance between the weight of the head and the length of the handle. The wooden handle absorbs the shock of cutting. Note the curve at the end of the handle for a sure grip and good blade control. The blade is made of tough steel to withstand stress and to hold a sharp edge. The gentle, graceful curves are pleasing, as is the contrast of color and texture. The sturdy leather sheath protects both the blade and the fingers. This hatchet is functionally, materially, and visually correct. Let's take a closer look at these three requirements.

Function

A product is **functional** if it works as it is supposed to work. This is the most important feature of product design. Regardless of how carefully the materials have been chosen, or how attractive the piece is, if the thing does not work it is a failure. Chairs must be of the proper height for easy working and lounging. A stool used at a drafting table must be higher than the chair you sit on while using a computer. The controls for a machine tool, such as a metal lathe, must be located for safe and convenient operation. Plier jaws must be designed variously for holding delicate parts, for cutting wire, or for heavy gripping and turning (Figure 11-3). The jaws may be curved or straight, long or short, depending on the use. The examples described here all relate to function. Once a problem has been identified, the designer must plan the product so that it works.

Material

A product must be made from the proper kind and amount of **material.** If you design something to be used outdoors, it must withstand water, wind, and sun. The mountain bike discussed earlier must be strong and durable for rugged outdoor use and abuse. Materials obviously must be chosen carefully according to product use. These are the inputs of product manufacture as

Fig. 11-4 The French horn is an elegant coiled structure of brass tubing leading to a flared bell. It has a warm and mellow tone, and is an attractive piece of metal sculpture. (Courtesy of Dr. Aline C. Lindbeck)

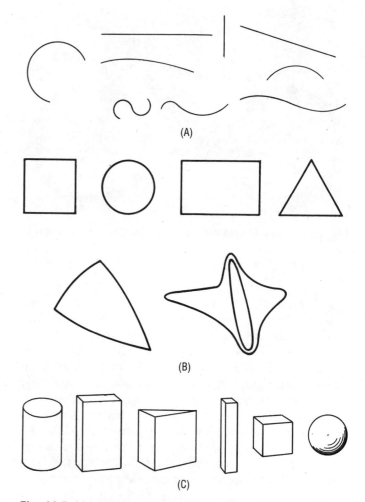

Fig. 11-5 Lines (A) are used to create shapes (B) and solid forms (C).

described in unit 5. Every material, be it metal, wood, plastic, or ceramic, has unique properties and characteristics which are important in product design. For example, the French horn in Figure 11-4 is made of a thin coil of brass about sixteen feet long, which ends in a wide, flared bell. Brass is the material of choice because it is easy to form and is durable, easy to maintain, and attractive.

Appearance

A good product must have a pleasing **appearance.** Everyone prefers things which are beautiful to those which are ugly. Designers must keep this in mind as they design products. This is the most troublesome of the design requirements because taste and beauty vary from individual to individual. For guidance in designing attractive things, we can look to fundamental design elements and principles.

Design Elements

A designer must understand and use certain visual features of products. These "building blocks" of design are called **elements.** They are the lines, the shapes, the forms, and the surface qualities (color and texture) that are present in every object.

Lines can be light graceful curves, rigid arcs, or straight and strong. **Shapes** are created when lines are combined into two-dimensional planes. These familiar circles, rectangles, triangles, and free forms are the beginnings of recognizable objects. Plane shapes are, in turn, joined and changed to make **forms** which define the appearance of solid objects (Figure 11-5). All of the products you use daily are designed by using these ba-

sic elements. Finally, all the products which emerge from these shapes and forms have natural **surface qualities** of color and texture. Steel products are gray in color and are smooth. Bricks are red and rough. Products also can be painted or chromeplated. Others can be sandblasted to give a rough surface texture. Sometimes texture is added for functional reasons, such as knurling or roughing a metal tool handle for a better grip in an oily fist. Designers must determine how a product is to be finished for purposes of appearance and protection. The forged aluminum automobile wheels in Figure 11-6 are a good example of the elements applied to product design. Lines and planes are used to create a visually pleasing wheel form. The bright chromeplating is attractive and also protects the wheel from rain and dirt.

Design Principles

There are four basic **principles** which are the guides to good visual organization (or appearance) of the elements. The different parts of a product should look as though they belong together. This is called

Fig. 11-6 Note the good use of line and shape, and decorative plastic color inserts, in these attractive automobile wheels. (Courtesy of Aluminum Company of America [ALCOA])

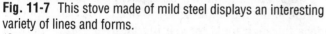

Fig. 11-7 This stove made of mild steel displays an interesting variety of lines and forms.
(Courtesy of Peter Haythornthwaite Design Limited, New Zealand)

unity. The hatchet shown earlier in Figure 11-2 has unity because the gently shaped handle and curved cutting edge of the blade flow together cleanly and pleasantly. The hatchet also has **variety** since the blade is made of dark tool steel, the handle of a lighter ash wood, and the protective sheath of warm brown leather. The different colors and textures give this product an interesting appearance. A variety of form is also present in the clever stove in Figure 11-7. Here the designer played with common shapes to create a most interesting yet functional heating device.

Product parts should be in proper **proportion** to one another. The handle and the blade of the hatchet de-

Fig. 11-8 This target pistol is well-balanced for accurate shooting. (Courtesy of Daisy Manufacturing Company, Inc.)

scribed above have a good size relationship. The handle is just long enough to control the chipping action of the blade. Now study the air pistol in Figure 11-8. You can see that the handle grip is proportional in size to the barrel. The structure does not look awkward, as it would if the barrel were very heavy and supported by a thin, spindly grip. Because the parts of the pistol are well-proportioned, the product has the necessary **balance** for accurate target shooting. Similarly, tables and lamps also must have proportional parts and balance so as not to look as though they will topple over.

In summary, unity, variety, proportion, and balance are called *principles* of design because they deal with basic ways of organizing the lines and shapes of products. These control the form and appearance of every product. Look for the applications of these principles in the many product examples which appear in this book.

School Shop Project Design

Many different kinds of projects can be designed and made in the school shop. *Craft* objects, for example, are very popular because they present an opportunity to combine art and skill. Their purpose is to make us feel comfortable by serving us politely and helpfully, as well as to thrill us. The pieces result from experiments with materials and forms and processes. They generally are a one-of-a-kind project, not designed for mass production. For example, attractive metal sculptures can be created by bending and winding wire (Figure 11-9). Scrap lengths of copper electric wire may be used. A basic structure (armature) can be shaped with 12-gage wire, and lighter wire of 14 or 16 gage then bent and twisted to shape the body. Epoxy glue or solder drops at the joints will give strength to the craft piece. Short strips of lightweight sheet metal can be bent into a variety of forms. Metal sculpture is an inexpensive, fast, and rewarding craft activity.

Fig. 11-9 Three clever penguin wire sculptures. (Courtesy of Aluminum Company of America [ALCOA])

Fig. 11-10 A vase made of scraps of automotive muffler tailpipes. Gun bluing and grinding were used to finish the piece.

Flower vases, similar to the one shown in Figure 11-10, can be made from scraps of steel tailpipe tubing you can get from your neighborhood muffler shop. Fit a piece of mild steel in the bottom of the tube and braze it in place. Design a shorter vase to hold pencils and pens. A taller one of a larger diameter can hold kitchen utensils such as ladles, spatulas, and spaghetti tongs. They can be given a gun-blue, sandblasted, high-heat, or a color spray finish. These finishes can be "distressed" by etching or light grinding. The factor that allows these holders and vases to be called craft pieces is that they are creative, clever, and unique. No two are alike.

Fig. 11-11 A lightweight camp saw made of bent aluminum tubing.

Fig. 11-12 An attractive and useful turned-metal hammer.

Utility projects are the tools and machine parts that are so useful in the home workshop. Hammers, punches, vises, and saws are typical of these. The camping saw illustrated in Figure 11-11 is simply and cleverly made from a length of aluminum tubing. It is of the proper size and is lightweight for convenience in cutting wood for the campfire. The saw can be easily flexed when inserting the blade, yet it will spring back to hold the blade tightly and safely.

Many different kinds of hammers can be designed and made as utility projects. Ball-peen hammers can be functional and interesting, and made in many different styles. Special lightweight models can be used for cracking nuts. The one shown in Figure 11-12 has a knurled handle and a hardened grooved head. You can get other ideas for such tools by visiting hardware and sporting goods stores.

Home furnishing projects are the stools, tables, lamps, and bookcases which can be used in the home. The example shown in Figure 11-13 is a side table made of 1/2-inch angle iron parts brazed together. Note the two different methods of joining the legs to the top. The tabletop can be made of perforated sheet metal, wood, or a plywood sheet covered with ceramic tiles. You can get ideas for many other tables by visiting furniture stores or studying catalogs. Use your imagination to create a special table for your bedroom.

Whatever kind of project you set out to design, remember that it should be functional and attractive. Search for ideas, and then create your own personal plans.

As stated earlier, designing is a kind of planning. You should approach project planning as you would

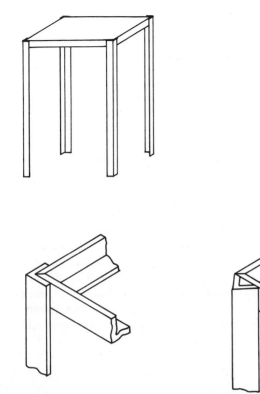

Fig. 11-13 A simple angle iron table structure. Note the different techniques for attaching the legs to the top.

any other kind of detailed work. The project that you wish to design should be thought of as a problem to be solved, much as you would solve a problem in mathematics or science. It is helpful to separate the problem into a series of steps to make it easier to solve.

Design Analysis Method

A good way to begin your personal designing is to attack the problem as professional designers would. They use many different approaches, but a typical one is called the *Design Analysis Method*. It is similar to the scientific method used in your science classes. Look at the chart in Figure 11-14. You will see there are five steps to this process.

1. *State the problem.*
    What do you wish to design, what is it for, and where and how is it to be used? Assume that you wish to design a hanging candleholder, or sconce, to give to your parents or a friend as a gift. You now know exactly what problem you are trying to solve.

2. *Analysis and research.*
    You must consider the questions of function and materials in this step. What are the candle sizes, how strong must the holder be, and how must it be made to contain dripping wax? What is wrong with some candleholders you have seen or used?

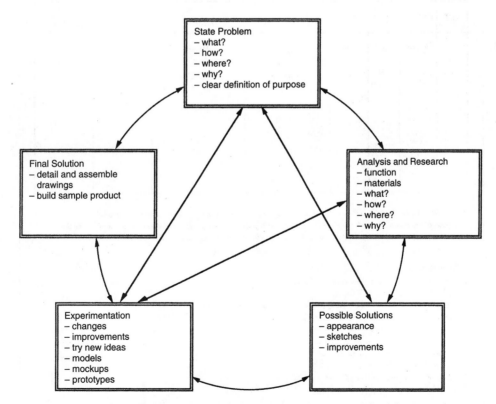

Fig. 11-14 A diagram of the Design Analysis Method. Note that you must move back and forth between the steps to arrive at good design solutions.

3. *Possible solutions.*
   Sketch several ideas for this project, and consider its visual requirements (appearance). Make certain that you consider available candle sizes. Rework and improve the sketches. Share your ideas with classmates and get some feedback.

4. *Experimentation.*
   Make a cardboard mockup (or prototype) of a selected design. Try it out, see if it works, and make any necessary improvements.

5. *Final solution.*
   Prepare detail and assembly drawings or sketches of your best idea. Make a sample product, and change it to make it perfect.

This example of how the Design Analysis Method is used can be a helpful guide to planning your own metalworking projects. It will also help you to design projects for use in unit 81, "Mass Production in the School Shop."

Project Planning

After you have designed a project and made the drawings, you must decide how you are to make the project. This is necessary in order to avoid making mistakes and wasting materials. Industry does this through the careful tooling and control procedures described in section IV of this book. You must do this in the shop by preparing a planning sheet for your work. A sample sheet for the candleholder designed above is shown in Figure 11-15. There are four main parts to the planning sheet.

1. *Bill of materials.*
   List all of the items that are needed to build the project. This should include all fasteners such as nuts and bolts. Study the units of this book dealing with metals, fasteners, and finishes in order to use the correct names for each item. This materials bill includes the numbers of parts and their sizes, the names of each part, the kinds of materials, and the costs. Materials price sheets are usu-

---

Name *Martinez, Maria*   Grade *9th*

Candle Sconce	*October 5*	*October 19*
(Name of project)	(Date started)	(Date completed)

Bill of Materials

No.	Size T	Size W	Size L	Name of part	Material	Unit cost	Total cost
1	*18 Gage*	*2"*	*14"*	*hanger*	*Black iron sheet*		
1		*3/4"*	*1"*	*candle cup*	*Pipe, mild steel*		

Tools, Machines, and Supplies: Combination square, layout fluid, scriber, dividers, round and flat files, grinder, disk sander, drill and drill press, center punch, hammer, sheet metal brake, hacksaw, brazing equipment, steel wool, C clamp, flat black paint, paint thinner, brush.

Outline of steps

1. Cut hanger to size shown in bill of materials.
2. Apply layout fluid to hanger.
3. Lay out patterns for each end; lay out bend lines and center punch holes.
4. Cut hanger to shape with hacksaw, disc sander, and files. Remove all burrs.
5. Bend hanger to shape.
6. Cut pipe to length; remove all burrs.
7. Clamp pipe in position on hanger.
8. Braze pipe to hanger.
9. Clean sconce with paint thinner and steel wool; wipe dry.
10. Apply two coats of flat black paint.

Fig. 11-15 An example of a planning sheet for the candleholder project.

ally posted on the shop bulletin board. Study these in order to compute the cost of your project.

2. *Tools, machines, and supplies.*

All needed hand tools, power and manually operated machines, special equipment, and supplies should be included in this list. You must study the project plans and survey the available shop equipment in order to determine this. For example, how are you to cut the sheet metal hanger? You have specified a 19-gage (thickness) metal sheet. If your shop has a squaring shear heavy enough to cut 18-gage material, this would be the best machine to use. If not, you must use a hacksaw or a steel-cutting bandsaw. Don't forget to include the supplies needed to apply a finish to your project.

3. *Outline of steps.*

This is simply a list of the operations or steps you would follow in order to build the project. Shearing, drilling, grinding, and filing are typical metal shop operations. A good way to identify these is to once again study the project drawing or the mockup. Then make an outline or list of all the things you would have to do in order to make it. Once you have done this, you must rearrange the outline to the proper order or sequence, as shown in the sample planning sheet. You begin with layout operations, and end with finishing operations.

4. *Working drawing.*

You must refer often to an accurate drawing or sketch when making the project. This plan must include both detail and assembly drawings, and must contain any corrections or changes made in the process of preparing the planning sheet. The working drawing of the candleholder appears in Figure 11-16.

When the planning sheet has been prepared, discuss it with your instructor for his or her approval. You are now ready to build the project that started with your creative design ideas.

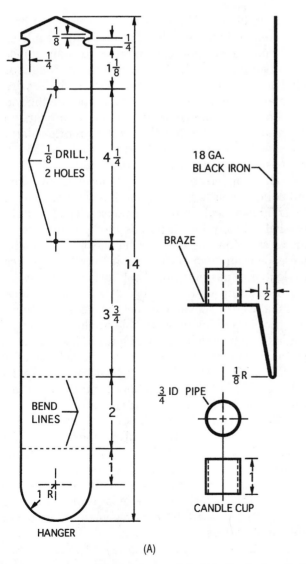

(A)

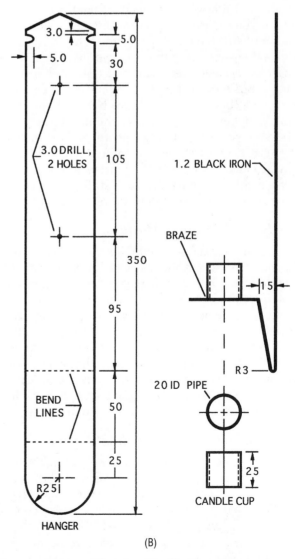

(B)

Fig. 11-16 Two drawings of the candleholder. Study the relationships between customary (A) and metric (B) dimensions.

IN FOCUS: *Young Inventors*

Jamie Lynn Villella, fourteen, is a ninth grader at Centennial Elementary School in Fargo, North Dakota. She was selected to be a member of the "Super Mario All-Stars Team," which developed the Nintendo Kids' Platform. The organization addresses issues of concern to America's youth and presents those concerns to elected officials in Washington, D.C.

She has been an inventor since she was in kindergarten, and usually got her ideas by listening to what other people complained about, or from something that bothered her. When she heard about or experienced a problem, she thought of ways to fix it. Jamie also kept a list of all of her new ideas for her school's yearly Invention Convention. Here are descriptions of some of her inventions and how they came about.

Grow-A-Size Shoe

Invented at age five. This is a shoe that has an insert in it that is a size smaller than the outer shoe. When you outgrow the size of the insert, you pull it out, and still have the same shoe. This way you get twice the wear out of the same pair of shoes. She got the idea because her sister had outgrown a pair of shoes, but they still looked new. They were too big for Jamie, but she slipped into her own shoes, and then put her sister's on over them. Although they were not very comfortable, they fit, and the idea for a two-size shoe was born. She entered it in her school's invention contest, won first prize there and then placed first at the tri-city contest. She received two blue ribbons and ten dollars for her first invention.

Child Safe Cabinet Alarm

Invented at age seven. This device has a switch that attaches to a cabinet, and an alarm that is hidden inside a picture frame in another room. When someone opens the cabinet, the alarm goes off. This idea came about when her little sister broke the "childproof" latch on a cupboard, opened the door, and began playing with things in the cupboard. Jamie thought there should be an alarm that would sound if someone too young opened a cupboard. This invention won first prize and grand prize at her school, as well as first prize at the tri-city contest.

E.Z. Tools (E.Z. Vac and E.Z. Shovel)

Invented at age eight. An invention that won the state contest, the E.Z. Vac and E.Z. Shovel have handles like those on a lawn mower. This makes it possible to use both hands, instead of one, when vacuuming or shoveling (Figure 11-17).

Jamie's mother and aunt both have arthritis, and have complained about how much their hands hurt when they vacuum. Jamie noticed that a lawn mower handle lets you use both hands, and thought it might be easier to vacuum if the machine had lawn mower-type handles. She designed such a handle, and her mom said it really helped. She wrote to some doctors and to the Arthritis Foundation in Fargo about her invention, and they wrote back to tell her they thought it was a great idea. She later decided that a snow shovel could also use the E.Z. Handle. Jamie is trying to get a manufacturer interested in the E.Z. Vac idea. She was the third grade Invention Convention winner for the state of North Dakota, and won a two hundred dollar U.S. savings bond as a prize.

Electronic Neighborhood Watch

In her own words, this is what Jamie has to say about her latest idea:

I am busy working on my newest invention. I have come up with an idea that will help people escape from fires, floods, or burglars in their homes. The invention is called "The Electronic Neighborhood Watch." It is a weathervane attached to a wireless alarm system. The weathervane goes on the roof of a house. If there is a fire, flood or burglar, the alarms inside the house will set off the rooftop alarm. A strobe light and siren will turn on, so the neighbors or a passing motorist can call for help.

I have already written to eight local fire marshals and police chiefs to get their opinion of my invention. So far, I have received three responses. The Fargo fire marshal said my invention has the "potential to reduce loss of life," because "if the person were unable to escape from the home or unable to call authorities, a neighbor or passerby could call for help quickly."

I came up with this idea after hearing of local fires that were discovered by people passing by them. The people passing by were able to alert the people inside the house that they had a fire. This invention will also work great if you're not at home and a fire starts. You wouldn't be home to hear your smoke detector go off, but your neighbors would know there was a fire and could call the fire department.

The Electronic Neighborhood Watch was awarded several trophies and ribbons. Jamie has these final words about her successful inventing: My parents have helped me a lot with my inventions because they always help me figure out how to make my ideas work. But the most important thing my parents help with is encouragement. They never laugh at my ideas.

IN FOCUS: *Young Inventors* (continued)

Sometimes kids have a way of looking at things in a different way, and we come up with ideas that others may not have thought of yet. I think we're willing to do a lot of things, because we do them for fun, and we're not afraid to fail. I was told that everyone who invents something is a winner because they tried. I think that's what kids should remember. If you don't try, you won't know if you could have been the inventor of the next hula hoop. No idea is a bad idea, because it is an idea. Every idea leads to the next idea.

You learn a lot of things when you invent something. Art, science, music, marketing, economics, and a lot of other things are part of inventing. But the best part of it is, you learn these things without knowing you're learning them—because it's fun!

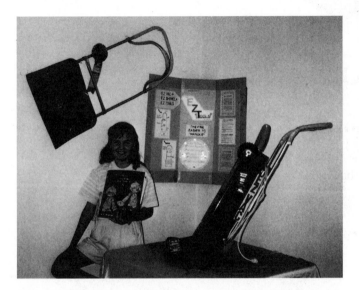

Fig. 11-17 Jamie Lynn Villela, inventor of the E.Z. Tools. (Excerpted from *Girls and Young Women Inventing* by Frances A. Karnes, Ph.D., and Suzanne M. Bean, Ph.D., © 1995. Used with permission from Free Spirit Publishing Inc., Minneapolis, Minn.) All rights reserved.

KNOWLEDGE REVIEW

1. Explain the meanings of the three design requirements.
2. Select a product which you have found to be especially functional and useful. Prepare an oral report on your selection to present to the class.
3. Explain the relationships among the four design elements.
4. Select some examples of products that show unity and variety, and some that do not.
5. Prepare a bulletin board display of pictures of some well-designed products.
6. Your class instructor will arrange the class members into design teams of four people. Each team will use the Design Analysis Method to plan a metal rack to hold pliers, screwdrivers, and wrenches. Select the best solution to this problem for use as a possible mass production project.
7. Describe the four major parts of a planning sheet.
8. Ask some of your friends what their favorite products are and why they like them. Evaluate this information in terms of the design requirements. Prepare an oral report of this market research to present to the class.

Unit 12

Industrial Product Design

OBJECTIVES

After studying this unit, you should be able to:

- Understand the purpose and methods of design in industry.
- Know what industrial and engineering designers do.
- Describe the meaning of *collaborative design.*
- Describe how the terms *serviceability, safety, human factors, producibility, handicapped,* and *environment* relate to product design.
- Understand how and why industry redesigns products.

KEY TERMS

anatomy	human factors	producibility	reuse
anthropometry	industrial design	rebuilding	safety
availability	maintainability	reconditioning	serviceability
engineering design	maintenance	recycling	source reduction
ergonomics	physiology	reliability	

The people who design things to be mass manufactured use practically the same methods as you did in your project design activities. They all use some design analysis system to guide and direct their work.

Manufactured products are planned by industrial designers and engineering designers. **Industrial designers** usually are concerned with planning consumer goods such as furniture, sporting goods, recreational vehicles, toys, and household appliances. The familiar snowmobile shown in Figure 12-1 is typical of such products. This rugged, reliable vehicle is designed for travel under cold and snowy conditions, and is safe and user-friendly. Its designers had to study both material and human factor needs in order to make it an excellent product.

Fig. 12-1 Snowmobiles are designed for safe, dependable travel over rough winter terrain.
(Courtesy of Arctco, Inc.)

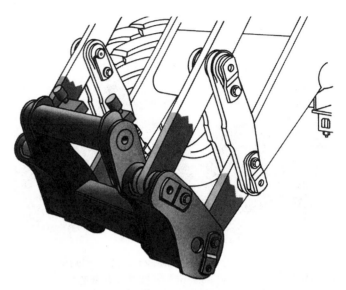

Fig. 12-3 Details of the Quick Coupler mechanism for the tractor toolcarrier.
(Courtesy of Caterpillar Inc.)

Fig. 12-2 The pallet fork on this tractor is conveniently attached to its toolcarrier with the Quick Coupler. The fork is lifting a wooden pallet frame loaded with rolls of grass sod.
(Courtesy of Caterpillar Inc.)

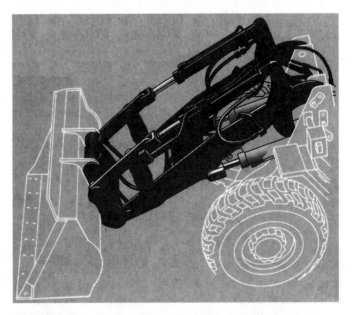

Fig. 12-4 The tractor toolcarrier linkage, showing the Quick Coupler in place.
(Courtesy of Caterpillar Inc.)

Engineering designers plan machines and equipment for the manufacturing, transportation, communications, and construction industries. They are specialists in materials science and mechanisms. For example, the wheeled tractor in Figure 12-2 is a common engineering design product. These vehicles can be found on the farm, in the forest, and on construction sites. They can be equipped with tools such as buckets, fork-lifts, and clam jaws. The designers at one equipment company created a special Quick Coupler (Figure 12-3) for changing these tools safely and conveniently. When this Coupler is attached to the tractor's toolcarrier (Figure 12-4) the operator can exchange a bucket for a fork

in less than thirty seconds without leaving the tractor cab. A lever in the operator's compartment is used to control a hydraulic device for positive tool attachment and detachment. This integrated toolcarrier is a safe, rugged, and reliable piece of farming equipment. The machines and equipment found in the metalworking laboratory are other engineering design examples.

One of the best examples of a product resulting from the joint efforts of industrial and engineering designers is the famous Harley-Davidson motorcycle

Fig. 12-5 The famous Harley-Davidson motorcycle results from the efforts of a capable, dedicated design team. (Courtesy of Harley-Davidson Motor Company)

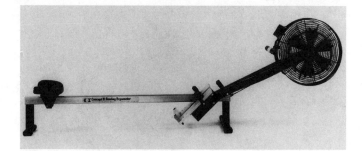

Fig. 12-6 The Model B Rowing Ergometer, before redesign. (Courtesy of Concept II, Inc.)

Fig. 12-7 The redesigned Model C Rowing Ergometer with its many improvements. The numbers on the photograph refer to the product changes listed in the text. (Courtesy of Concept II, Inc.)

shown in Figure 12-5. A combination of custom-engineered metal parts and tough plastic components makes this a safe, rugged, powerful, and attractive driving machine. Industrial designers were largely responsible for the motorcycle's human factors, such as comfort, safety, conveniently placed controls, and ease of operation. The engineers designed the engine, drive train, and the control systems. Such product excellence is achieved through the efforts of well-educated, dedicated, and creative design teams.

Product Redesign

In addition to creating something new, designers also must improve existing products. All manufacturers must regularly evaluate the things they make in order to improve them and stay competitive. Opportunities for product redesign come from new materials and production methods, quality control tests, new safety regulations, and from user feedback. With this new information, designers can create products safer, more usable, easier to service, and less expensive to make.

The rowing machine shown in Figure 12-6 was one such candidate for product redesign. This first model was well-accepted by physical fitness enthusiasts and performed well. However, after seven years, the manufacturer decided to do a complete product analysis leading to a new design. Sales records, maintenance and production reports, and consumer evaluations were all studied. This information resulted in a product redesign. The new, improved model features a number of product changes which are listed below (Refer to Figure 12-7).

1. Increased user load, or exercise resistance, range, with load change lever conveniently located on wind damper. (This refers to the amount of force the user is working against while exercising.)

2. Safer flywheel fan enclosure.

3. Larger, higher, more comfortable seat.

4. More stable and durable seat carriage and track system.

5. Easy disassembly into two pieces for convenient storage. This feature also makes it possible to package the machine in one box to reduce shipping costs.

6. Machine monitor panel moved closer to user for easy reading and ability to change settings from seat. The new panel can also be tilted to reduce glare.

7. More stable frame design reduces flywheel imbalance. The legs are more sturdy both front-to-back and side-to-side, and all four feet sit firmly on the floor with no forward tipping if either fan enclosure or arm is leaned upon.

8. Rollers attached to front feet make it easier and safer to move the machine to a new location.

9. New pull-cord handle holder for accessibility.

10. Single sprocket design permits the use of an anti-slip device for safety; also eliminates the need to handle the greasy chain when changing between large and small sprockets.

11. Improved return mechanism, featuring larger diameter pulleys, longer shock cord, and better spacing to reduce possibility of interference between moving parts.

12. Easier to clean and maintain: fan cover, fan, and axle are removable without special tools; return mechanism visible and accessible from under front arm.

13. Improved quality control to eliminate assembly errors.

14. Quieter and smoother operation.

15. Cleaner, more attractive appearance.

The technical, visual, and human factor improvements can be easily seen by comparing the two rowing machines. All manufacturers must practice such redesign activities to maintain product quality and sales.

Collaborative Product Design

All too often, a designer plans a new product and then gives the plan to a manufacturing engineer who has to figure out a practical way to make the thing. Product features such as safety, ease of maintenance, and usability also have to be checked to assure product quality. Because these inspections are often made after the product has been designed, much time is lost and errors made. Unfortunately, many industries still work this way. A more effective and newer approach to product design is a technique variously called *design for manufacture, design/build groups, simultaneous engineering, team engineering, concurrent product development,* or *collaborative design.* The term *collaborative design* will be used in this book.

The collaborative method involves assembling a team of experts to work together (collaborate) on a new product. A simple screwdriver generally does not require a team design effort. One person can do it. More complex products such as machines and vehicles do require collaboration. Team members include industrial and engineering designers; maintenance, safety, and human factors specialists; and materials and production technicians. The aim of this group is to create a product that is as near to perfect as it can be. They do this by studying the important product issues at the very beginning of the design process, and solving at that time any problems which may arise. These issues typically include: serviceability, safety, human factors, producibility, design for the handicapped, and environmental concerns.

Design for Serviceability

The word **serviceability** refers to the performance and availability of products in use. If a machine breaks down, it does not perform and is therefore not available to a user. At best, the design goal is to develop a product which needs little or no service. Barring this ideal situation, any needed service should be able to be done easily and at low cost. Several important terms relate to this serviceability concept.

Maintenance

Maintenance is the act of keeping products in, or restoring them to, an operating condition. It involves preventive actions such as regularly lubricating a machine to keep it running (or prevent it from breaking down). Maintenance is also corrective, as in repairing a broken machine (or correcting a breakdown).

Maintainability

Maintainability is a design feature which assures that any necessary maintenance can be performed with a minimum of cost, inconvenience, and effort. Designing a milling machine so that all test, adjustment, and service points are easy to reach is an example of maintainability.

Reliability

Another related issue is **reliability**, which simply means that a product will continue to perform its intended function over a prescribed useful lifetime. A ball-point pen is a reliable product. It usually works until it runs out of ink. There is an important relationship between reliability and serviceability. A product cannot continue to perform (be reliable) if it has not been adequately maintained, or if it is abused by the user.

Availability

Availability is operation upon demand, and is related to both reliability and maintainability. It implies a readiness for use. People assume that a product such as a copy machine will be usable (available) when it is needed. If it is not, they usually blame the manufacturer even if the failure was due to faulty maintenance or to operator abuse. Regardless of the reason, any time the copy machine breaks down, the user suffers a loss because he or she must find another machine to use, or call a repair technician to fix it. In any event, the user was unable to make the copy when it was needed because the machine was not available.

A good example of a maintenance-free product is the derailleur cable guide for bicycle shifting mechanisms (Figure 12-8). This plastic piece is placed under the bottom bracket of a bicycle frame. The shifting cable rides in it. The guide prevents cable wear, is self-lubricating,

Fig. 12-8 This protective plastic derailleur cable guide prevents shifter cable maintenance problems.
(Courtesy of AMOCO Polymers, Inc.)

and provides a smooth shifting action. Because of it, cable maintenance problems are reduced or eliminated.

Design for Safety

There is perhaps no single issue of greater concern to manufacturers than that their products will not endanger the people who use them. This can be better realized if safety is built into a product at the design stage, and not tacked on later. **Safety** can be defined as the freedom from those conditions that can cause injury to people, or damage to equipment. As such, safety relates both to work environments and to products. A person can be injured either by having to work in a messy, unsafe shop or factory, or by using an unsafe machine or tool. Both are problems in the home, in the school shop, and in industry. The information found in unit 3 of this book can help you to learn more about safety and to practice it in the metalworking shop.

The key used to lock a drill bit in a drill press can be a hazard. If you forget to remove it before you turn on the machine, the key can be thrown into your face and cause an accident. An improved chuck key designed for safety is shown in Figure 12-9. It has a built-in spring that permits chuck-tightening with normal finger pressure. However, it will slowly eject when the pressure is released. This simple improvement has eliminated a potential hazard.

Aside from the important issue of protecting users from product hazards, manufacturers also are concerned with the matter of product liability. Strict federal product safety laws have set legal obligations for the makers of defective or faulty products. Such laws further encourage industry to make things which are safe. A manufacturer who makes an unsafe product can

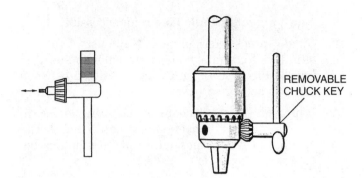

REMOVABLE CHUCK KEY

Fig. 12-9 The standard drill press chuck key, shown on the right, may cause an accident if not removed before starting the machine. The key on the left will safely self-eject.

be held liable and sued by the person injured while using that product.

Design for Human Factors

Products must be created for human use. This can only be done by recognizing the abilities, needs, and limitations of people as products are being planned. **Human factors** is the study of the interactions between people and the products they use, and is important in industrial design. Another word for human factors is **ergonomics,** which is a combination of two Greek words: *ergon* (work) and *nomos* (study of). Either term implies convenience, ease of use, and satisfaction, as contrasted to confusion, difficulty, and aggravation. Usable (user-friendly) things give people great pleasure; unusable products are annoying. Several fields of science are important to the issue of human factors.

Anthropometry

Anthropometry is the science of human measurement, and it deals with body sizes and proportions. Products must be designed to "fit" human beings. Examples include the proper work heights for computer chairs and keyboards, or arranging the controls of a lathe, washing machine, or motorcycle for convenient and safe operation. Designers make use of special anthropometric data charts to determine these human measures.

Physiology

Physiology is the science dealing with the human senses of sight, sound, and touch. Sight, or visibility, factors involve the clear and obvious placement of machine controls. Users should not have to guess or engage in trial and error in order to determine which button to press next. This is both dangerous and inefficient.

Sound, or auditory, factors take advantage of the fact that humans can hear different sound signals. Sound can provide information which is difficult to convey by any other means, or which an operator may

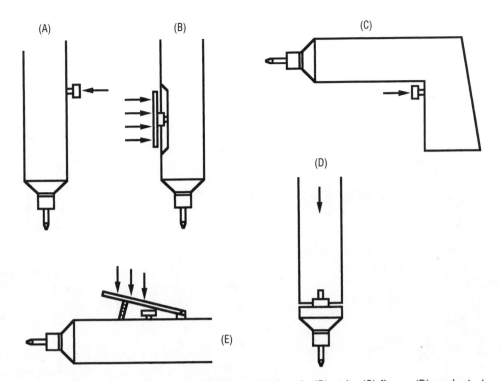

Fig. 12-10 Trigger designs for automatic assembly tools include: (A) thumb, (B) strip, (C) finger, (D) push start, and (E) lever. Which of these would be the most tiring to use for long periods of time?

not be aware of otherwise. Some examples include the rattle or the rushing air noise which indicates that a car door is not closed; the change in pitch caused by a dull band saw blade; or the satisfying purr of a properly tuned motorcycle engine.

Touch, or tactile, factors relate to feel. A skilled woodworker can "feel" the sharpness of a chisel shearing through a piece of walnut, or of the blade of a tablesaw while ripping a piece of oak. In both cases there may be a feeling of unusual resistance, and for the tablesaw a change in sound, which are clear indications of tools which need sharpening.

Anatomy

Anatomy is the science of the structure of the human body. It deals with the bones and muscles and organs which make up our bodies. Designers use this information in many ways. For example, they know that people on assembly lines can tire easily and can suffer muscle damage if they are using poorly designed tools. Several kinds of triggers are used on assembly tools such as automatic screwdrivers. The type used depends upon the force needed, how long the hand or finger must press the trigger, and the kind of work being done. Study the trigger designs in Figure 12-10. See if you can figure out which would be the most comfortable to use for long periods of time.

The term *user-friendly* was mentioned earlier. A product is said to be user-friendly when its design is based upon people and how they use and respond to things. For example, ball-point pens are reliable and easy to use. In contrast, public address systems are often unreliable and hard to use. A computer can be said to be user-friendly, or user-unfriendly, if it is easy or confusing to operate.

A very usable socket tool is shown in Figure 12-11. The typical socket wrench is a reliable tool. The only problem is that one needs a separate socket for each nut size, and this increases the cost and inconvenience of the set. The tool shown is convenient, reliable, and less costly than a conventional socket set. To use it, the adjustable socket is placed over the nut or bolt head, and the knurled collar is turned to tighten. Special triple-action jaws close and lock on the nut to provide an anti-slip grip. This is a clever tool; well-designed and usable.

Design for Producibility

The word **producibility** means the ease with which materials can be processed in order to manufacture something. It is the art of making parts easily, and then assembling them with equal ease into a product. For example, study the mounting brackets in Figure 12-12. The top bracket (A) is made of nine separate metal pieces, each of which must be made and then spot-welded together. It is expensive and time-consuming to make. Special fixtures must be designed to hold the parts during welding. The other model (B) is made of one piece of sheet metal, which is cut, punched, and bent in a single

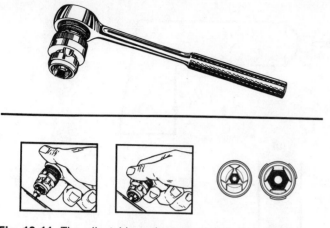

Fig. 12-11 The adjustable socket wrench is both convenient and less expensive than ordinary socket wrench sets. (Courtesy of CHANNELLOCK, Inc.)

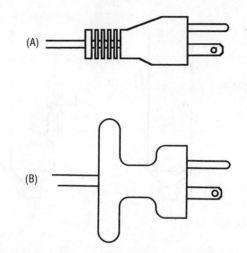

Fig. 12-13 The small electric plug (A) is difficult for people with finger strength problems to hold. The special plug (B) has a large fin at its end for easier removal.

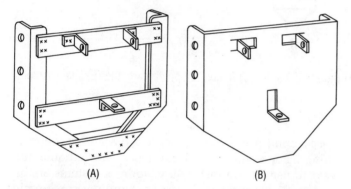

(A)   (B)

Fig. 12-12 The original bracket (A) is made of many parts and is more costly and difficult to make. The redesigned model (B) is more producible.

series of operations performed on one machine. No assembly is required. It is far more producible than the first example because it is simpler and less expensive to make. Industry designs for producibility by studying the product plan to simplify the manufacturing process and to reduce the numbers of parts in a product.

Design for the Handicapped

Industrial designers must consider the product needs of handicapped and elderly people. This is one of the fastest growing population groups. It is estimated that by the year 2000, a third of the United States population will be over sixty-five years of age, or be disabled in some way. These persons may have special emotional and mental problems. They may also have physical disabilities, including: (1) physical or motor handicaps, such as the loss of use of one or more limbs, which may require the use of wheelchairs, crutches, or other devices, or (2) sensory handicaps, such as sight, hearing, or feeling impairments, which generally require special aids and devices.

Such handicaps can lead to special product needs. For example, electric cord plugs are a problem for people with fingers weakened by arthritis or some other muscle problem. Many ordinary plugs are too small to permit a good finger grip. The bottom plug (B) in Figure 12-13 features a large rear projection which can be easily and effectively grasped and pulled from an electric outlet. Contrast this with the other plug (A). The finned projections are also a gripping aid, but not nearly as effective as the other. This is one example of the unique products needed for this population group. Others include: kitchen cooking tools with fatter, softer handles for better gripping; easy-to-open medicine bottle caps; and barrier-free bathroom fixtures for wheelchair users. The hook and loop self-gripping fasteners described at the end of this unit are very user-friendly to all users, but especially those who have physical problems lacing shoes.

The attractive and functional bank checkstand in Figure 12-14 was designed to provide access for people in wheelchairs. This furniture piece meets the requirements of the American Disabilities Act, and will accommodate both disabled and able-bodied bank customers. Especially notable is the fact that the disabled are not segregated by having to have a separate banking area. They remain part of the mainstream of bank customers.

Design and the Environment

The environment also is an important product design matter. Whereas ergonomics is the study of the relationships between humans and their products, environmental issues are concerned with the interactions of humans with the world in which they live. Very simply, the problem is that Americans produce over 160 million tons of solid waste each year, or about three pounds per person per day. This is the highest per capita trash

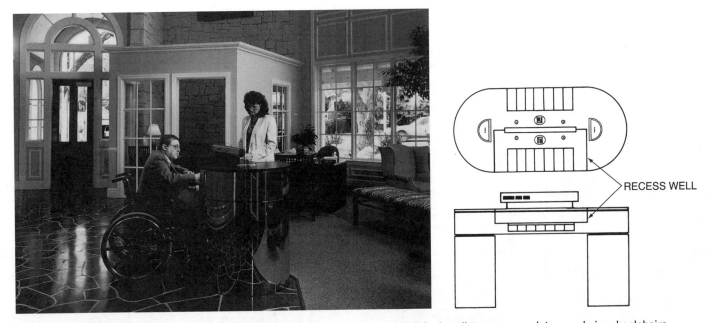

RECESS WELL

Fig. 12-14 This attractive and functional checkstand is designed with a special desk well to accommodate people in wheelchairs. (Courtesy of Matel Manufacturing, Inc.)

rate among the world's industrial nations. There are a number of ways to resolve this problem.

Source reduction means decreasing the amount of waste at the point of origin. Manufacturers can use fewer materials, or use them more wisely, in their products. **Reuse** is returning products to the manufacturing stream, such as returning glass beverage bottles for refilling. **Rebuilding** and **reconditioning** are approaches whereby a basic structure, such as a toaster, is kept, and replacement parts are used to restore it to a usable condition. Excellent examples here are the many older buildings which have been restored to active use rather than being razed. (It costs about one kilowatt of energy to tear down each kilogram of building material.) Collecting discarded or outdated equipment and reconditioning it for reuse is another example, such as rebuilding industrial machinery.

Recycling is especially important because it reduces the amount of material entering the waste stream. It also gives people a feeling of accomplishment because they are actively engaging in a solution to the problem of trash. Aluminum cans are lightweight and valuable, are easy to collect, crush, transport, and reprocess, and are therefore the most popular recycling material. Recycling is encouraged by soft drink container deposit laws, reverse vending machines, and buy-back centers, and has resulted in a scrap aluminum recovery rate of about 65 percent in nondeposit states and over 90 percent in deposit states. The chart in Figure 12-15 illustrates the efficiency of the recycling process.

Designers can contribute to environmentally correct product manufacture by specifying the use of recycled materials. Wildlife signs, benches, tables, docks,

lighting posts, planters, bird feeders, and barricades can be made from a plastic lumber. This material is made from recycled polystyrene foam coffee cups, food containers, and milk jugs. It can be worked like wood but requires no finishing and withstands weather and decay. The design challenge here is to create useful and attractive products from waste materials which would normally end up in a land fill. Recycling waste materials can also be profitable. The Chicago Board of Trade now includes recyclables on its exchange.

Each of these approaches must be met cooperatively by designers, makers, and users in order for environmental problems to be resolved.

Collaborative Design in Action

Complicated products such as air racing planes result from the work of design teams. The exciting award-winning racer in Figure 12-16 is made of space age materials, has a twenty-two foot wingspan, and is designed to sustain about twelve G's. Called the *Nemesis,* the plane with engine and fuel weighs about five hundred pounds. Many people with many design specialties were needed to create this aircraft, including: industrial designers to plan the human factors requirements of the pilot compartment; aero-engineers to plan the engine and structure; materials technicians to develop the carbon-reinforced plastic components and metal structure; computer technicians for CAD work; production engineers to plan manufacturing methods; and safety specialists to plan a hazard-free product. It was a truly collaborative team effort with dramatic results.

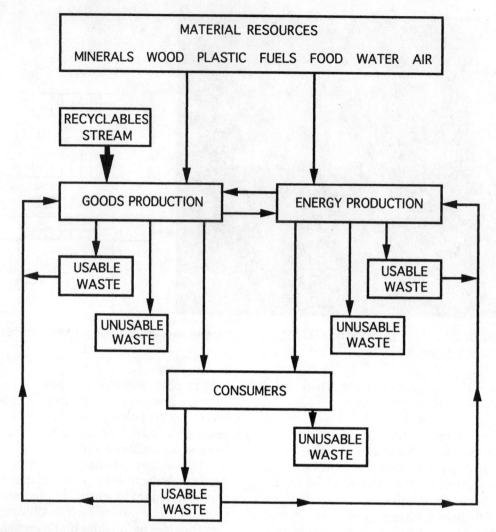

Fig. 12-15 Chart of the recycling system.

Fig. 12-16 Exciting racing planes such as this require the efforts of a collaborative design team.
(Courtesy of National Instruments Corporation, and Nemesis Air Racing)

IN FOCUS: *Hook and Loop Fastening System*

A good example of using plastic fibers in a creative way is the familiar self-gripping plastic hook and loop fastener, commonly called "Velcro." The term comes from two French words, *velour* (loop) and *crochet* (to hook). This product was invented in 1958 by a Swiss hiker who, after a walk, had to pick off the cockleburrs which were clinging to his trousers. This experience led him to the idea of developing a fastening system which could hold two pieces of material securely, but which could be easily separated by gently tearing the pieces apart. The microphotograph of this plastic fiber structure (Figure 12-17) shows how the tiny hooks grip the loops as they are pressed together.

The hooks can be molded into an endless variety of shapes and sizes to meet the specific grip-strength and cycle-life needs of products. For example, a loop tape with a soft backing for a surgical mask has both low strength and short cycle-life requirements. It must only hold a

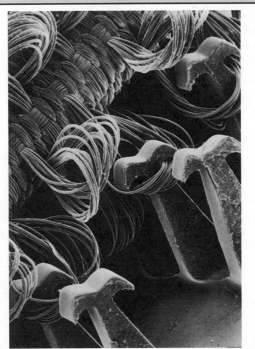

Fig. 12-17 A microphotograph of the hook and loop fastening system. (Courtesy of Aplix Inc.)

light weight mask to the face, and is used only once and then discarded. On the other hand, the fastening strip for a tennis shoe must be stronger and more durable. It must provide a tougher grip, and must be used many times. Each of these examples have a different hook and loop structure design, because the product needs differ. Other systems are made to meet the high strength, durability, and temperature demands required, for example, by an aircraft's removable fire-retardant wall panels.

The most outstanding feature of these clever fasteners is that they are easy, quick, and convenient to use. Handicapped and elderly people with finger dexterity problems often have difficulty tying the laces of their tennis shoes. With the hook and loop fastener, they only need to press the tabs together to "tie" the shoes. The fasteners also are used on helmet, pocket, and jacket straps; on the backing panels of television sets and appliances; and for automotive upholstery. The designers of these systems can create hook and loop structures to meet the self-grip fastener demands of many products.

KNOWLEDGE REVIEW

1. Describe the difference between the work of industrial and engineering designers.

2. Describe how the terms *serviceability, safety, human factors, producibility, handicapped,* and *environment* relate to product design.

3. What is collaborative design?

4. What three sciences are important in human factors design?

5. The ball-point pen was identified as an example of a reliable product. Select another such product and tell why it is reliable.

6. Visit a cycle shop in your community and prepare a report on the human factors features you find on some of the bicycles or motorcycles.

7. Visit your local pharmacy and find some examples of products which are designed for handicapped people. Prepare an oral report of your findings to your class.

Unit 13

Measuring and Marking Out

OBJECTIVES

After studying this unit, you should be able to:

- Define the terms *measuring* and *marking out.*
- Demonstrate the use of the common layout tools.
- Understand basic project layout procedures.

KEY TERMS

center head	hook rule	square head	steel tape
circumference rule	protractor head	steel rule	square
combination set	scale	steel square	

After you have designed a project and are ready to build it, you must transfer the dimensions to the metal workpiece or stock. This operation is called *measuring and marking out,* and is an important step in project work. In this unit you will be introduced to the common layout tools and their correct usage. These instruments must be handled carefully and used properly. Avoid bending or damaging them.

Fig. 13-1 The common six-inch steel rule, or scale.

Customary Rules

Customary rules are available in different lengths and configurations to accommodate the specific size and shape of the object being measured or marked out. The six- or twelve-inch **steel rule** (or **scale**) (Figure 13-1) is generally used for taking and laying out measurements in metalwork. Most of these rules are marked on all four edges, front and back. The first edge is divided into eight parts to the inch, with each small division representing one-eighth of an inch. Every second division represents one-quarter of an inch, and every fourth division represents one-half of an inch. The second edge is divided into sixteenths of an inch, the third edge into thirty-seconds of an inch, and the fourth edge into sixty-fourths of an inch. Each sixty-fourth of an inch represents a little less than sixteen-thousandths (actually 0.0156) of an inch. In measuring, the experienced machinist can judge half or one-fourth of this amount. He or she can therefore read the rule as close to the exact size as 0.003–0.005 (three- to five-thousandths) of an inch. The six-inch rule with decimal graduations also is available. Fractions of an inch and their decimal and millimeter equivalents are shown in Figure 13-2.

In measuring or marking out distances, the rule must be held on edge to be accurate. If the end is worn, start measuring from the one-inch mark. Be sure to practice reading a rule. It is surprising how many people have difficulty finding the 7/16 or 13/32 division.

Inches		in.	mm	Inches		in.	mm
	1/64	0.01563	0.397		33/64	0.51563	13.097
1/32		0.03125	0.794	17/32		0.53125	13.494
	3/64	0.04688	1.191		35/64	0.54688	13.890
1/16		0.06250	1.587	9/16		0.56250	14.287
	5/64	0.07813	1.984		37/64	0.57813	14.684
3/32		0.09375	2.381	19/32		0.59375	15.081
	7/64	0.10938	2.778		39/64	0.60938	15.478
1/8		0.12500	3.175	5/8		0.62500	15.875
	9/64	0.14063	3.572		41/64	0.64063	16.272
5/32		0.15625	3.969	21/32		0.65625	16.669
	11/64	0.17188	4.366		43/64	0.67188	17.065
3/16		0.18750	4.762	11/16		0.68750	17.462
	13/64	0.20313	5.159		45/64	0.70313	17.859
7/32		0.21875	5.556	23/32		0.71875	18.256
	15/64	0.23438	5.953		47/64	0.73438	18.653
1/4		0.25000	6.350	3/4		0.75000	19.050
	17/64	0.26563	6.747		49/64	0.76563	19.447
9/32		0.28125	7.144	25/32		0.78125	19.844
	19/64	0.29688	7.541		51/64	0.79688	20.240
5/16		0.31250	7.937	13/16		0.81250	20.637
	21/64	0.32813	8.334		53/64	0.82813	21.034
11/32		0.34375	8.731	27/32		0.84375	21.431
	23/64	0.35938	9.128		55/64	0.85938	21.828
3/8		0.37500	9.525	7/8		0.87500	22.225
	25/64	0.39063	9.922		57/64	0.89063	22.622
13/32		0.40625	10.319	29/32		0.90625	23.019
	27/64	0.42188	10.716		59/64	0.92188	23.415
7/16		0.43750	11.113	15/16		0.93750	23.812
	29/64	0.45313	11.509		61/64	0.95313	24.209
15/32		0.46875	11.906	31/32		0.96875	24.606
	31/64	0.48438	12.303		63/64	0.98438	25.003
1/2		0.50000	12.700	1		1.00000	25.400

Fig. 13-2 This table shows the fractions of an inch and their decimal and millimeter equivalents.

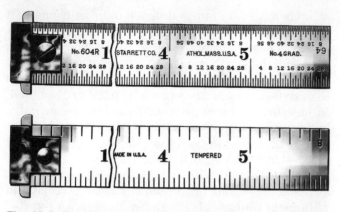

Fig. 13-3 The hook rule, with front and back faces shown. (Courtesy of The L. S. Starrett Company)

Many errors are made through incorrect measuring and layout.

Another customary rule, the **hook rule,** has a hook on the end for holding the rule to the workpiece (Figure 13-3). It measures inside and outside hole diameters, and holes which go all the way through a workpiece. The hook serves as the starting point of the measurement.

A thirty-six-inch steel **circumference rule** is a sheet metal layout tool (Figure 13-4). As its name implies, it is used to measure the circumference, or rim, of a circle. Along one edge of this tool there is a regular rule divided into sixteenths for measuring diameters. Along the other edge there is a scale that shows the circumferences for those diameters. For example, if you want to lay out a piece of sheet metal that is to be rolled into a two-inch-diameter cylinder, first locate the 2 inch mark on the rule. Then look across to the mark on the scale, which indicates about 6 1/4 inches.

A fourth customary layout tool, the **steel tape,** can be bent around curved or circular objects. You can use it to measure on the inside or outside of openings. The tape has a little hook on the end that holds it in place for making long measurements (Figure 13-5).

Metric Rules

Metric rules are available in 150 mm, 300 mm, and 1 m lengths. The rules are divided into full millimeters (1 mm) for most work and into half millimeters (0.5 mm) for more precise measures. Steel tapes have dual (inch-mm) or metric-only graduations. The relationship among customary-inch, decimal-inch, and millimeter rules is shown in Figure 13-6.

Squares and Protractors

In addition to customary and metric rules, squares and protractors are also important layout tools with specific purposes.

The toolmaker hardened steel **square** has a sturdy steel beam and blade to help hold it to the work. It is used to check the squareness of stock, to make a line square to an edge, and to square-off a line when measuring stock to length (Figure 13-7). A large **steel square** is very useful for large layouts, such as in sheet metal work (Figure 13-8).

The **combination set** (Figure 13-9) is one of the most versatile and useful measuring tools. It has four parts: (1) a a twelve-inch steel rule, with a groove running lengthwise along one side. Any one of the three heads (described below) can slide along the groove on the rule and be locked in place with a knurled nut; (2) a **square head,** which forms a 90° angle on one side and 45° angle on the other. There also is a spirit level in the head, for leveling up a machine or workpiece, and a scribing tool. This head can be used for laying out lines at 45° or 90° angles, for checking the squareness of stock, for measuring depths, and for other operations; (3) a **center head,** used to locate the center of the end of a cylinder or rod; and (4), a **protractor head,** used to lay out and measure any angle up to 180°. The head is rotated to the desired angle and held in place by a lock screw.

Examples of the many uses of the combination set are shown in Figure 13-10.

Sheet Metal Gages

Sheet metal gages are round disks of hardened steel with slots and holes cut around the outer edge. They provide a convenient and accurate way to measure sheet metal thicknesses and wire diameters in inches. Each slot is numbered, and the number represents a certain thickness or diameter in decimals of an inch. The gage of the sheet metal or wire is the number

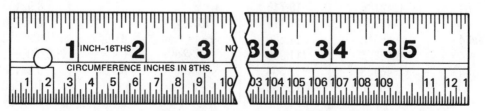

Fig. 13-4 The circumference rule is used in sheet metal layout. It indicates the circumference for a given diameter.

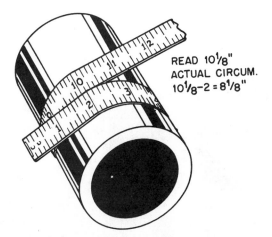

Fig. 13-5 A steel tape can be used to measure the diameter of a pipe or round bar.

READ 10⅛"
ACTUAL CIRCUM.
10⅛−2 = 8⅛"

of the slot into which it fits. Gages are not common in the metric system. Instead, a metric micrometer is used to measure thickness or diameter in millimeters.

There are two common customary gage systems used for metal. The United States Standard (USS) gage is used for black and galvanized mild-steel sheets, steel plates, and steel wire. The Brown & Sharpe, or American, gage is used for measuring nonferrous metals such as copper, brass, and aluminum. To measure a piece of galvanized sheet, select a gage that is stamped USS. Then try the metal in the various slots until you find the one in which it just slips (Figure 13-11). If the metal is 16 gage, it will be 0.0598 inch thick. To measure brass, again use the Brown & Sharpe gage. If it is 16 gage, it will be 0.050 inch thick. A table of sample gage sizes appears in Figure 13-12.

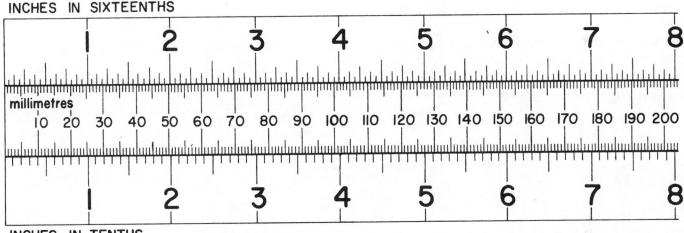

INCHES IN SIXTEENTHS

millimetres

INCHES IN TENTHS

Fig. 13-6 The relationships between customary and metric measures are shown in this diagram. Note the alternate spelling of the word *millimetre*. Some American industries use this spelling.

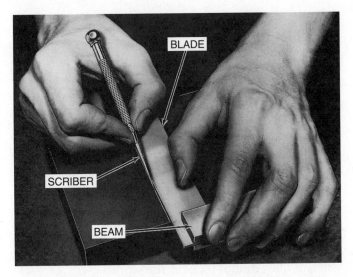

Fig. 13-7 The use and parts of a hardened steel square. The scriber will scratch a visible line onto a metal workpiece. (Courtesy of The L. S. Starrett Company)

BLADE
SCRIBER
BEAM

Fig. 13-8 The twenty-four-inch steel square is used for layout work on larger workpieces.
(Courtesy of Stanley Tools, Division of the Stanley Works, New Britain, CT)

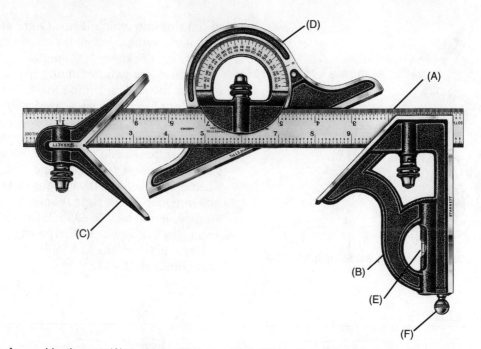

Fig. 13-9 The parts of a combination set: (A) steel rule; (B) square head; (C) center head; (D) protractor head; (E) spirit level; (F) scriber. (Courtesy of The L. S. Starrett Company)

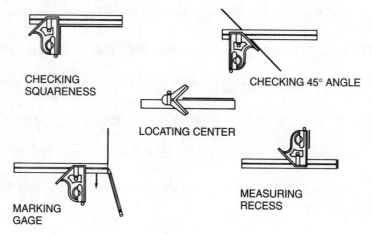

CHECKING SQUARENESS

CHECKING 45° ANGLE

LOCATING CENTER

MEASURING RECESS

MARKING GAGE

Fig. 13-10 Some uses of the square and center heads.

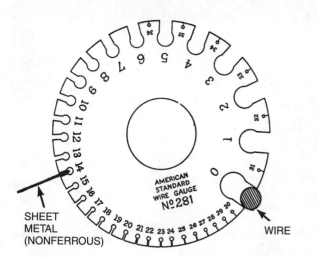

SHEET METAL (NONFERROUS)

AMERICAN STANDARD WIRE GAUGE N⁰ 281

WIRE

Fig. 13-11 The sheet metal gage can measure both thickness and diameter.

Calipers

Simple, adjustable, spring-joint calipers are used with scales to measure diameters and distances. Outside calipers for outside diameters have bowed legs, and inside calipers for measuring inside diameters have straight legs (Figure 13-13).

Caliper legs are moved together or apart by turning a finger nut. To adjust an outside caliper, hold one leg over the end of the scale. Move the second leg until it exactly fits the correct measurement line (Figure 13-14). To use an outside caliper to determine the size of an object, hold it squarely to the work. When you check the measurement, the leg should just drag lightly over the work (Figure 13-15). Hold the caliper lightly so that you can feel when it touches the work.

To adjust an inside caliper, place one leg of the caliper and the end of a rule against a machined surface. Adjust the caliper until the other leg exactly fits the measurement line. To use the inside caliper, make sure that the points of the legs touch the work so that the largest diameter is measured (Figure 13-16). Hold the ends at right angles to the work, and drag lightly against it for proper fit.

When transferring a measurement from one caliper to the other, place one leg of the inside caliper against the leg of the outside caliper. Use this as the pivot point. Now adjust the outside caliper until the other leg just drags across the second leg of the inside caliper (Figure 13-17). With practice, you will be able to use these calipers to make accurate measurements.

Measuring and Marking Out Heavy Stock

There are no strict rules for measuring and marking heavy stock. However, here are a few suggestions.

Gage	Ferrous		Nonferrous	
	inch	mm	inch	mm
16	0.0598	1.60	0.050	1.20
18	0.0478	1.20	0.040ᵃ	1.00
20	0.0359	0.90	0.032ᵇ	0.80
22	0.0299	0.80	0.025ᶜ	0.65
24	0.0239	0.65	0.020ᵈ	0.50
26	0.0179	0.45	0.015	0.40
28	0.0149	0.40	0.012	0.30
30	0.0120	0.35	0.010	0.25
32	0.0097	0.22	0.007	0.18
ᵃ32-ounce ᵇ24-ounce ᶜ20-ounce ᵈ16-ounce				

Fig. 13-12 Table of common sheet metal gage sizes and their metric equivalents.

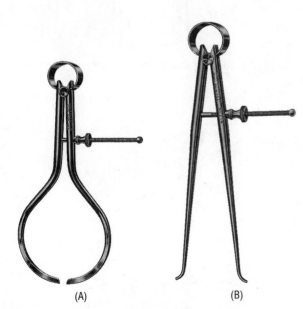

Fig. 13-13 Outside (A) and inside (B) spring calipers. (Courtesy of The L. S. Starrett Company)

1. Make sure the end of the workpiece from which the marking is done is square.

2. Hold the rule on edge and in contact with the surface of the workpiece (Figure 13-18). If the end of the rule is worn, hold the 1-inch or 10-mm mark over the edge of the workpiece.

3. Make sure the rule is kept parallel along the workpiece so that the exact length can be obtained.

4. Mark the length shown on the rule with a **scriber.** The scriber is a marking tool with a sharp point used to scratch a line or mark on metal. Use it carefully.

5. Hold a square firmly against the side of the workpiece. Mark a line across the workpiece (Figure 13-19). Turn the scriber at a slight angle so that

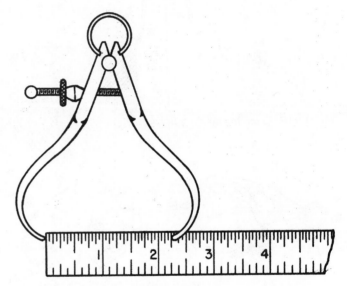

Fig. 13-14 Setting the outside calipers for a measurement.

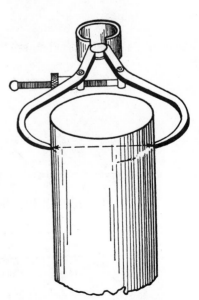

Fig. 13-15 Measuring the diameter of a bar.

the point will draw smoothly along the lower edge of the rule.

6. To measure the thickness of a heavy metal bar or the diameter of a pipe, hold the rule as shown in Figure 13-20.

Marking Out Sheet Stock

For marking out sheet stock, follow the procedure below.

1. Make sure that an edge and an end of the sheet are at right angles to each other.

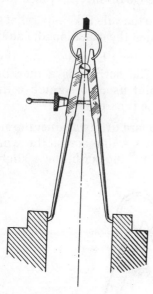

Fig. 13-16 Using the inside calipers.

Fig. 13-19 Using a scriber and the square head of a combination set to mark a line on a workpiece.

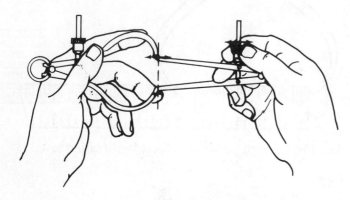

Fig. 13-17 Transferring a measurement between the two calipers.

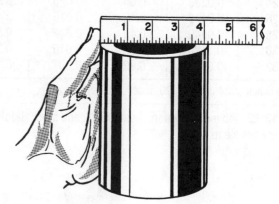

Fig. 13-20 A method of measuring the diameter of a heavy pipe.

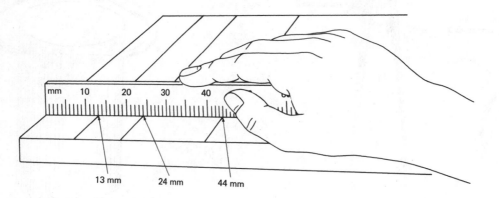

Fig. 13-18 Hold the rule on edge for more accurate measuring.

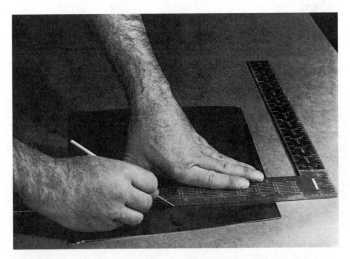

Fig. 13-21 When laying out on a flat sheet, mark several points to indicate the desired size. Then hold the square over these points and scribe a line.

2. Hold a rule over the end of the workpiece, and mark out a point that shows the length of stock to be cut.

3. Move the rule over, and mark out another point for the length.

4. Use the large steel square or a straightedge to join these two points with a scribed line.

5. Mark several points to indicate the width.

6. Lay a square along these points, and scribe a line (Figure 13-21).

KNOWLEDGE REVIEW

1. What are the common inch divisions on a steel rule?

2. List some uses for the hook rule.

3. What is the circumference of a thirty-five-inch circle? Find the answer by reading the scale in Figure 13-4.

4. What are the common divisions on a metric rule?

5. What are the four parts of a combination set?

6. List two uses for a sheet metal gage.

7. Study the chart in Figure 13-12 and list the inch and millimeter sizes for 22-gage ferrous and non-ferrous metal sheets.

8. Obtain a six-inch rule, try square, protractor head, and center head, and explain to the class how they are used.

9. List some ways of measuring the thicknesses of stock.

Unit 14

Making a Simple Layout

OBJECTIVES

After studying this unit, you should be able to:

- Recognize and be able to use common layout tools.
- Know how to prepare a metal surface for layout.
- Know how to enlarge a pattern.
- Know how to transfer a pattern onto a metal workpiece.
- Know how to make a layout directly onto a metal workpiece.

KEY TERMS

angle plate	cross peen	layout fluid	template
ball peen	dividers	mallet	toolmaker clamp
bench vise	hermaphrodite caliper	prick punch	V blocks
center punch	layout	straight peen	

As described in unit 13, marking out is the process of transferring measurements from a project drawing to the material from which the project is to be made. The resulting flat pattern made directly on the metal is called the **layout.** It shows the shape and size of the object, the location of all holes or openings, and the areas to be machined or otherwise removed. A layout is similar to a working drawing laid out on a metal workpiece. Accuracy is very important. If you make an error, your job can be ruined before you even start it. To make a good layout, you must be able to (1) read and understand drawings and prints, (2) use layout tools correctly, and (3) transfer measurements accurately from a drawing to the material itself.

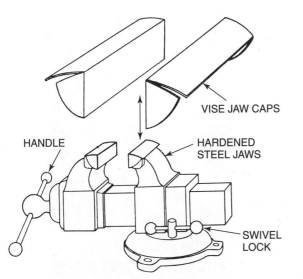

Fig. 14-1 A common bench vise used by metalworkers. The vise jaw caps protect the workpiece from scratches and dents.

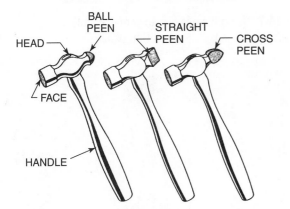

Fig. 14-2 Three types of metalworking hammers. The cross and straight peens are designed to reach into tight corners.

Layout Tools

Some of the tools you will use for making layouts include: the bench vise, special types of hammers and punches, dividers, hermaphrodite calipers, an angle plate, toolmaker clamps, and a V block.

The **bench** or **machinist, vise** is the type of work holder most often used in the shop. It has two hardened jaws with faces that are lightly grooved to hold workpieces securely. The solid jaw is fixed, and the other moves in and out when you turn the handle. Vise jaw caps of copper or aluminum can be slipped over the hardened jaws to protect a metal workpiece (Figure 14-1).

Hammers are used for striking, driving, and pounding. The most common are the **ball peen, straight peen,** and **cross peen** (Figure 14-2). The ball-peen hammer is used most often. The flat face of the hammer is used for general work, and the rounded end for riveting, or for hammertracking (decorating) art metal projects. The straight-peen hammer and cross-peen hammer are

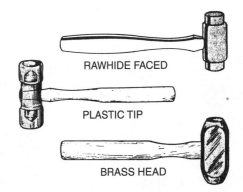

Fig. 14-3 These mallet heads are softer than steel and will not mar a workpiece.

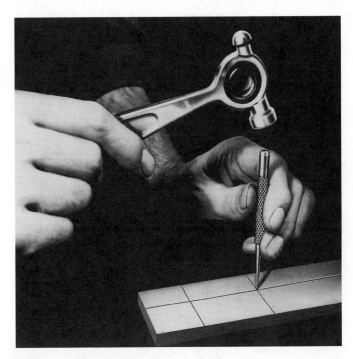

Fig. 14-4 Using the prick punch.
(Courtesy of The L. S. Starrett Company)

used for hammering in tight corners and making sharp bends in metal.

The size of a hammer is determined by the weight of the head, which varies from four to forty-eight ounces. A five-ounce hammer is common for shop use. Special hammers, often called **mallets,** have heads of lead, brass, rawhide, plastic, rubber, or wood. These are used when a steel hammer might mar the metal surface (Figure 14-3).

Two kinds of punches are used in layout work. The **prick punch** is a short piece of hardened steel rod with a knurled handle and a point ground to a sharp angle of about 30 degrees (Figure 14-4). It is used to make the first marks locating holes for drilling. In transferring patterns from paper layouts to sheet metal, the prick

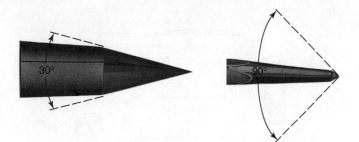

Fig. 14-5 Note the correct point angles of the prick punch, above, and the center punch.

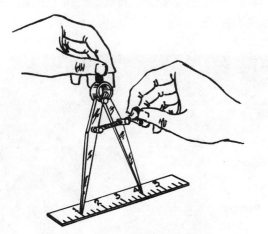

Fig. 14-6 Setting the dividers to the proper dimension.

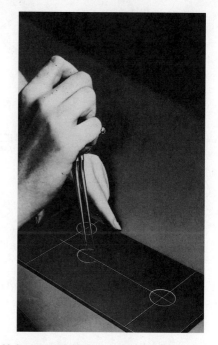

Fig. 14-7 Using the dividers to scribe a circle. (Courtesy of The L. S. Starrett Company)

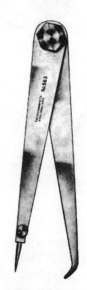

Fig. 14-8 An hermaphrodite caliper. (Courtesy of The L. S. Starrett Company)

punch is used to make small dents to locate lines and corners. It is important to keep the point sharp. The **center punch** is similar to a prick punch except that the point is usually angled at about 90 degrees. It is used to enlarge prick-punch marks so that the drill will start easily and correctly. The properly ground punch tip angles are shown in Figure 14-5.

Dividers are used to lay out arcs and mark the size and location of holes to be drilled or cut. Divider legs must be sharp and of equal length. As shown in Figure 14-6, the tool is set by placing one point on a rule line and adjusting the other to the desired radius. To use the tool, place the point of one leg into the prick-punch mark, tip the dividers, and turn it clockwise (Figure 14-7).

An **hermaphrodite caliper** has one outside-caliper leg and one divider leg (Figure 14-8). It is used to locate the center of an irregularly shaped workpiece, or to lay out a line parallel to an edge.

In addition to the bench vise, there are various other tools used to hold workpieces. An **angle plate** (Figure 14-9) has two finished surfaces at right angles to each other. It is a bracket used to hold workpieces in making a layout, as shown in Figure 14-10. A **toolmaker** or **parallel, clamp** (Figure 14-11) holds parts together when making a layout. The **V block** is a rectangular steel block in which deep V-shaped grooves have been

cut to hold round workpieces when making a layout. They also can be clamped in a vise to hold round workpieces when drilling (Figure 14-12).

Layout Surface Preparation

Special fluids and other materials can be used to coat the surface of metal so that scribed layout lines will show more clearly. Ordinary white chalkboard chalk or poster paint can be used for simple layouts, such as locating drilling hole centers. There are also special **layout**

Fig. 14-9 A typical angle plate.

Fig. 14-10 Using the angle plate in layout work.
(Courtesy of The L. S. Starrett Company)

Fig. 14-11 A pair of toolmaker clamps.
(Courtesy of The L. S. Starrett Company)

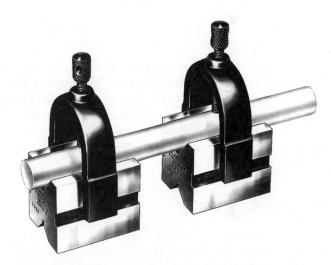

Fig. 14-12 These V blocks are holding a round workpiece during layout work.
(Courtesy of The L. S. Starrett Company)

fluids available, and they are the best and most convenient to use. They are spread on the metal with a brush or spray can, and leave a dark blue coating on which scribed lines show up sharp and clear (Figure 14-13).

Enlarging a Pattern

Occasionally, you may find an illustration in a magazine or catalog which you wish to use for a project. It generally must be enlarged to full size before you can make the layout. The easiest way to do this is to enlarge the illustration by using a copy machine. Experiment with increasing the size by 100 percent, and then repeating the process until you have the desired size.

Another way is to use the squared-paper method described below.

1. Cover the original drawing with ruled squares. The size of the squares will depend on the scale of the drawing. For example, if the drawing is one-fourth full size, lay out quarter-inch squares on the original.

2. Next, lay out a pattern sheet with one-inch squares. Letter across the bottom and number up the left side on both the original drawing and the pattern.

3. Locate a number-letter point on the original drawing and transfer it to the same point on the enlarged

Fig. 14-13 Spreading layout fluid on a metal sheet. Can you identify the layout tools in the foreground? (Courtesy of The L. S. Starrett Company)

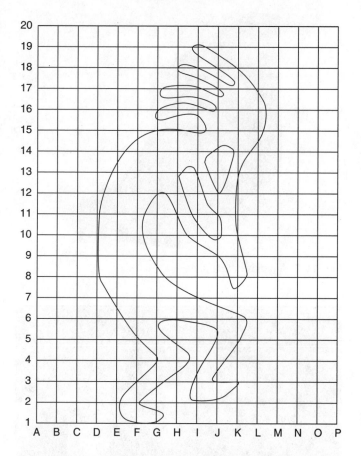

Fig. 14-14 An illustration of pattern enlargement. Note the square numbering and lettering system. This is a sketch of Kokapelli, the mythical Navaho Indian hunchback flute player.

grid. Continue this until you have transferred enough points to draw the full-size pattern layout.

4. Connect these points with a pencil, using a straightedge for straight lines and a plastic drafting curve or bent wire for the curved lines. Instead of working with the curves and straightedges, you may wish to sketch the design using the points as guides. Touch up and improve the pattern as necessary (Figure 14-14).

Transferring a Pattern

Enlarged patterns or other sketches can be transferred to a sheet of 32-gage copper to create an interesting tooled art metal project. These designs also can be copied onto a heavier piece of sheet steel to make a cutout for a bookend, or for a wall plaque or candleholder. There are several ways to do this.

1. Draw the design on tracing paper and glue it to the thin metal sheet with rubber cement, or hold it in place with masking tape (Figure 14-15). Metal tooling can be done through the paper by using a tracing tool.

2. Apply a coat of white poster paint to the metal. Place a piece of carbon paper (face down) and then the paper design on the metal and clip or tape them together. Next, outline the design with a pencil or tracing tool (Figure 14-16).

3. If you are making parts that have the same shape, you can cut a **template,** or pattern, out of thin hardboard or sheet metal (Figure 14-17). Hold the template on the metal and trace the pattern with a scriber, as shown in Figure 14-18. This method is

Fig. 14-15 Design transfer by taping the paper to the metal sheet.

especially useful when a class is mass producing a project, such as the candle sconce in Figure 14-19.

Making a Layout Directly on Metal

For some projects, such as a sheet metal box, it is easier and more efficient to draw (lay out) the object di-

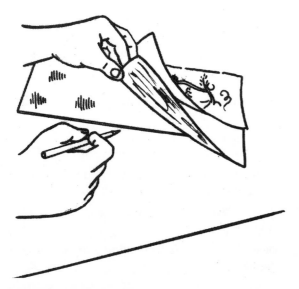

Fig. 14-16 The carbon paper design transfer method.

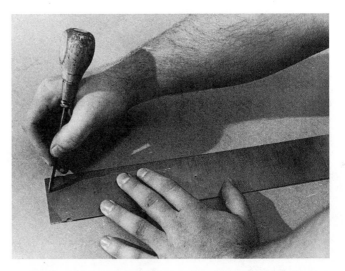

Fig. 14-18 Tracing around the template with a metal scriber to produce the workpiece for a candle sconce.

Fig. 14-17 This sheet metal template is a tracing guide.

Fig. 14-19 The completed metal candle sconce is meant to be hung on a wall.

rectly onto the metal workpiece. This saves the trouble of first drawing the project on paper, and then transferring the drawing to a workpiece. Since the layout will vary for each project, only general suggestions can be made for this procedure (Figure 14-20). Before starting, be sure to have the correct size and kind of metal.

1. Square the workpiece so that the left and the bottom edges are at right angles to one another. These two edges will be the reference lines from which all measurements will be made.

2. Apply layout fluid to the workpiece.

3. Use a square and scriber to lay out all straight horizontal and vertical lines.

4. Lay out all angular and irregular lines.

5. Locate and mark all arcs, circles, and holes to be drilled.

6. Lay out all internal lines.

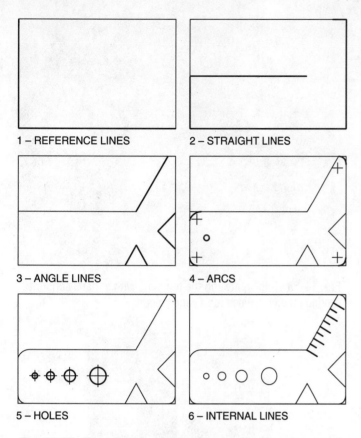

1 – REFERENCE LINES

2 – STRAIGHT LINES

3 – ANGLE LINES

4 – ARCS

5 – HOLES

6 – INTERNAL LINES

Fig. 14-20 Steps in laying out a drill gage directly onto a metal sheet.

KNOWLEDGE REVIEW

1. What is a layout?
2. Name four kinds of hammers and describe their uses.
3. Name some materials that can be applied to metal to make layout markings show up better.
4. Write a brief description of the process of enlarging a design.
5. Write a brief description of the process of transferring a design.
6. Describe the steps in making a layout directly on metal.

SECTION 3

MANUFACTURING PROCESSES

Manufacturing consists of enterprises that create economic goods for the basic needs of people. This technology is divided into two major groups—*durable goods,* which are long-lasting products such as automobiles, airplanes, and trains; and *nondurable* goods such as clothing, food, and other products that are consumed in a relatively short time.

Metal machines are needed to produce both types of products. Metals are also the primary materials used to produce most durable goods. Each industry is based on the need for qualified workers who use special machines. For example, in manufacturing cars, a company needs designers, engineers, technicians, skilled workers, and assembly workers. Manufacturing is an essential conversion process in which raw materials are changed into finished products. In the United States, about seventeen percent of the labor force is employed in manufacturing industries.

In this section of the book, you will learn about the manufacturing processes needed to produce durable goods. You will use more hand tools than machines; however, in industry, more machines than hand tools are used. Skilled metalworkers must know how to use both hand tools and machines.

Unit 15

Introduction to Bench Metal

OBJECTIVES

After studying this unit, you should be able to:

- Describe other terms used for bench metal.
- Name five tools used in bench metal.
- Name two professions that require a knowledge of hand tools.
- Name four shapes of soft metal used in bench work.
- Describe the four basic processes done in bench metal.
- Name another term for ornamental metalwork.
- Name four products that can be made of wrought metal.
- Define the term *machine* and know its basic task.
- Name the six types of simple machines.

KEY TERMS

bench metal	lever	resistance	wrought-metal work
effort	machine	screw	
fulcrum	ornamental metalwork	wedge	
inclined plane	pulley	wheel and axle	

The technical progress of prehistory was directly related to the ability of humans to make and use hand tools. Even today, the skilled machinist must rely upon such tools to aid in the most complex machining operations. Stock is often cut to length with a hacksaw, and deburred with a file or grinder prior to installing the workpiece on a product. Similarly, many workpieces are finished with hand polishing operations. Completed parts are frequently assembled with metal fasteners which require hand tapping and threading operations. Such work is done with hand tools, and the operations are commonly called *benchwork*. The range of hand tool skills is important, and must be mastered as a part of industrial machining training and practice.

One of your first experiences in the metal shop will probably be in the area of **bench metal.** This is because bench-metal work is basic to all other metalwork. Bench-metal work is sometimes called *elementary machine shop, benchwork, cold-metal work, metal fitting,* or *hand tooling.* Whatever the name, the purpose is to introduce you to common metal tools. You will soon learn how to use and to care for such tools (Figure 15-1).

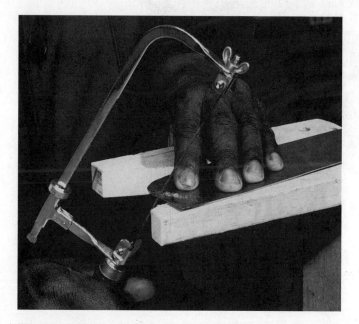

Fig. 15-1 Metal sawing is one of the bench-metal processes you will learn in this course.

Tools

You will learn to use hand tools for cutting, forming, fastening, and finishing. These tools include hammers, drills, wrenches, hacksaws, vises, and benders. Using them can be more difficult than operating some machines. All metalworkers and mechanics must know how to handle bench-metal tools. Even people in the professions of medicine and dentistry must have a knowledge of precision, or very accurate, tools.

No matter how you plan to earn a living, learning to use these hand tools and machines is a rewarding accomplishment. You can make many interesting projects. You can also do many small maintenance jobs in the shop and home. These might include repairing a bicycle or automobile, or putting a new lock in a door.

Materials

In bench-metal operations, the metal is usually worked cold. It is therefore important to use soft metals. One common material is mild steel in the form of flats, rounds, angles, and heavy sheets. Aluminum and copper are also used, especially for ornamental metalwork.

Processes

The basic bench-metal processes are shown in Table 15-1. You will need to master them in order to make your projects. Both hand and machine processes are used in bench metal.

Table 15-1 Bench-metal processes

Type	Equipment used
Cutting	
sawing	hand and power hacksaws, bandsaw
shearing	cold chisel, bench shears, bolt cutter
drilling	hand and electric drills, drill press, reamers, taps (related dies)
abrading	grinders and polishers
filing	files
Forming	
bending	vises, forks, bending machines
Fastening	
riveting	rivets and sets
threading	threaded fasteners, wrenches, and screwdrivers
Finishing	
polishing and buffing	polishing and buffing wheels and compounds
texturing	hammers

An interesting part of bench metal is **ornamental metalwork,** sometimes called **wrought-metal work.** It involves some cutting operations plus a few others such as bending curves and scrolls and twisting metal. As you master your bench-metal skills, keep in mind the things you have learned about safety. Know how you are to do something before you do it. Keep your tools in good condition. Avoid unsafe working habits.

Your bench-metal experience may help you to a career in metalworking. Bench-metal work may also become a satisfying hobby. There are many things that can be made of wrought metal—floor lamps (Figure 15-2), boot scrapers, house number plaques, porch furniture, and sports equipment, to name just a few.

What is a Machine?

You probably know that complicated devices, such as the drill press and lathe, are machines. However, you might be surprised to learn that the hand tools you will use in the shop are also machines. Chisels, hammers—even an object like a screw—are machines, operated with hand power. Each of these

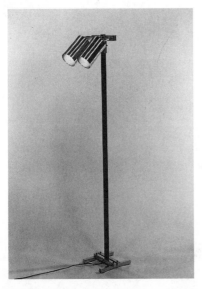

Fig. 15-2 This interesting floor lamp is an excellent example of a metal project.

simple items is just as truly a machine as the more complicated ones. As a matter of fact, machines such as the lathe are made up of several very simple machines fitted together in different ways.

A **machine** is a device used to make work easier. As you use hand tools, materials, and machines you will be doing work. What do we mean by *work?* Work is done when a force moves through a distance to make something move or stop moving. You are "working" when you strike a ball with a bat or when you ride a bicycle.

There are six simple machines: the lever, the wheel and axle, the inclined plane, the wedge, the screw, and the pulley (Figure 15-3). A **lever** is a long bar with something to place it on for support. The place where the lever is supported is called a **fulcrum.** This is where the lever turns. The fulcrum may be at either end of a lever or anywhere between the ends. The **effort** is the amount of push or pull you have to use to move the lever. The **resistance** is what you are moving with the effort. Levers are called *first class, second class,* or *third class* depending on where the fulcrum is.

A wheel on an axle is a kind of lever. A doorknob, the steering wheel of a car, and an egg beater all use the **wheel and axle.** When you turn the crank on a pencil sharpener, you are using a wheel and axle.

Inclined means sloping or slanted. A plane is a flat surface, such as a floor or the top of a desk. So an **inclined plane** is a sloping or slanted flat surface. A board used to slide a box or barrel from the ground onto a truck is an inclined plane. Two inclined planes placed back to back form a **wedge.** A chisel is a wedge; so is a knife. A **screw** may be defined as a wedge that winds around a cylinder. You are familiar with the screws used as fastening devices.

A **pulley** is a simple wheel that is free to turn on an axle. Usually it is a grooved wheel over which a rope is run. Several pulleys may be combined to increase the effort, such as with a block and tackle. When you raise a flag by pulling down on rope, you are using a pulley at the top of the flagpole.

KNOWLEDGE REVIEW

1. List some bench-metal operations.
2. What kinds of workers must know how to handle bench-metal tools?
3. Why are soft metals used in bench-metal work?
4. What is wrought-metal work?
5. What are the six simple machines used to make work easier?
6. Trace the history of one of the common hand tools from its beginning. Prepare a written or oral report for presentation to your class.

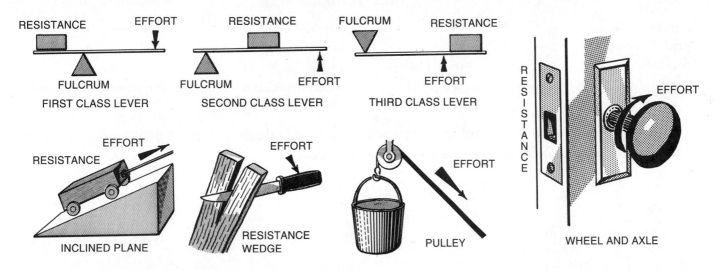

Fig. 15-3 The effects of effort and resistance on simple machines.

Unit 16

Cutting Heavy Metal

OBJECTIVES

After studying this unit, you should be able to:

- Know the safety rules when using the hacksaw.
- Describe four factors used to identify hacksaw blades.
- Describe the word *set*.
- Tell why the jaws of a vise are covered with caps.
- Describe the four kinds of chisels.
- Explain how to cut thin metal with a hacksaw.
- Name the device that should never be used as a back plate when cutting with a chisel.
- Explain how to shear metal with a flat cold chisel.
- Identify the two kinds of bench shears.
- Explain the purpose of a bolt cutter.
- Name two kinds of power metal saws.
- Name two kinds of portable power saws.

KEY TERMS

band saw	cape chisel	flat cold chisel	saber saw
bench lever shears	cutoff saw	hacksaw	set
bolt cutter	diamond-point chisel	kerf	slitting shears
burr	end-cutting nippers	roundnose chisel	throatless bench shears

There are several ways of cutting heavy metals. Sawing is done with hand saws and power saws. Shearing is done with cold chisels, bench shears, and bolt cutters.

Safety

Always observe the following safety guidelines when cutting heavy metal.

- When using a hacksaw, make sure the blade is properly tightened.

- Always wear safety goggles when striking with tools such as hammers and chisels.

- Remember that the **burr,** or rough area along a cut edge, is sharp and can cause a serious injury.

- Never use a dull chisel or one with a "mushroom," or flattened, head. Make sure to sharpen chisels and to remove mushroom heads using the grinder.

Hacksaws

A **hacksaw** has a fine-toothed blade held under tension in a frame. Hacksawing is probably the most common method of metal cutting (Figure 16-1). The hand hacksaw has a U-shaped frame with a handle and replaceable blades. The frame itself may be either solid, to take one length of blade, or adjustable for different lengths (Figure 16-2). The posts that hold the blade can be adjusted to four different positions. For ordinary cutting, the blade is placed in line with the frame.

Blades

Here are some facts you should learn about hand hacksaw blades:

- *Size.*
  Blades are available in lengths of eight, ten, and twelve inches, a width of 1/2 inch, and a thickness of 0.025 inch.

- *Kind of material.*
  Blades may be made of carbon steel, tungsten alloy steel, molybdenum steel, molybdenum high-speed steel, or tungsten high-speed steel.

- *Types of blades.*
  Three variations of blade type are commonly used: all-hard, semiflex, and flexible back. The flexible back has a hard edge that makes it good for general sawing.

- *Number of teeth per inch* (Figure 16-3):
  1. 14 teeth—for cutting soft steel, brass, and cast iron.
  2. 18 teeth—for cutting drill rod, mild steel, light angle iron, and tool steel; for general work.
  3. 24 teeth—for cutting brass tubing, iron pipe, and metal conduits.
  4. 32 teeth—for cutting thin tubing, thin sheet metal, and channels (Figure 16-4).

- *Tooth set.*
  Set refers to the way the teeth are bent to one side or the other and the amount of bend. The set makes the **kerf,** or width of the cut made by the saw, wider than the blade itself. Thus, the blade will not bind or stick. In an *alternate,* or *raker set,* the teeth are bent alternately to right and left, with a straight tooth between. In a *wave set,* several teeth are bent in one direction, and then several are bent the other way. Blades with fourteen and eighteen teeth are made in the alternate set. Those with twenty-four and thirty-two teeth are made in the wave set.

- *Special blades.*
  A tough steel blade with carbide grains bonded to its cutting edge is called a *carbide grit blade.* It can

Fig. 16-1 Hacksawing is a simple and common method of cutting heavy metal.

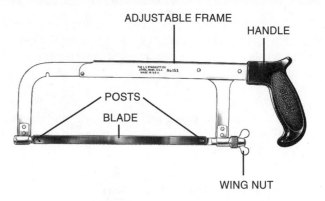

Fig. 16-2 Parts of an adjustable hacksaw frame that can be used with blades of different lengths. (Courtesy of The L. S. Starrett Company)

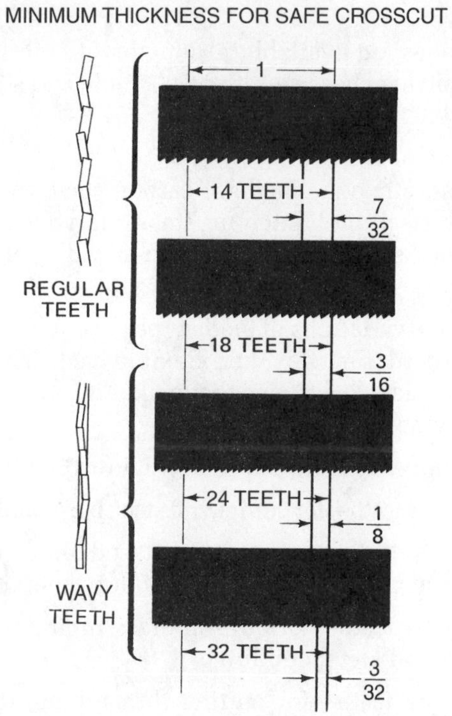

MINIMUM THICKNESS FOR SAFE CROSSCUT

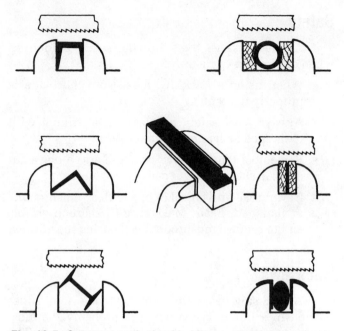

REGULAR TEETH

14 TEETH — $\frac{7}{32}$

18 TEETH — $\frac{3}{16}$

24 TEETH — $\frac{1}{8}$

WAVY TEETH

32 TEETH — $\frac{3}{92}$

Fig. 16-3 This guide will aid in hacksaw blade selection.

Fig. 16-4 To cut thin tubing, use a wave-set blade with thirty-two teeth per inch. Notice the simple method of holding the tubing. A hole is drilled in a piece of wood about the size of the tubing, and then the piece is cut in two.

cut ceramics, glass resin plastics, metals, and other very hard materials which would destroy an ordinary blade.

Cutting with a Hand Hacksaw

Follow the procedure below for cutting with a hacksaw.

1. Select the correct blade, making sure that the blade will have at least *three teeth* in contact with

Fig. 16-5 Common methods of holding workpieces.

the metal at all times. Place it in the frame with the teeth pointing *away* from the handle.

2. Tighten the wing nut on the handle until the blade is tight. After a few cuts, retighten it with a turn or two.

3. Cover the jaw of the vise with jaw caps. Fasten the workpiece with the layout line as close to the end of the jaws as possible. Figure 16-5 shows the right ways to fasten different metal shapes.

4. If thin sheet stock is to be cut, sandwich it between two pieces of scrap wood.

5. Start the cut by guiding the blade with the thumb of the left hand and taking one or two light strokes with the hacksaw. Then, grasping the end of the frame firmly with the left hand, take full-length strokes. Move both hands in a straight line. If you do not, the blade will twist, making the cut uneven.

6. Apply pressure on the forward stroke. Release the pressure on the return stroke. Do not allow the teeth to drag over the metal. Use a uniform motion, with about forty to sixty strokes per minute.

7. When the cut is about completed, hold the end to be removed with your left hand (Figure 16-6). Make the last few cuts by holding the saw with the right hand only.

8. For making deep cuts, turn the blade at right angles to the frame after you have cut as deep as you can with the blade in the regular position (Figure 16-7).

9. Check your work. Did you twist the blade? Is the cut crooked? Does it follow the layout line?

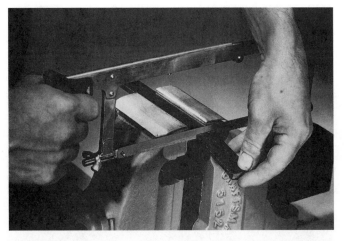

Fig. 16-6 Hold the workpiece with your left hand as you make the last few cuts.

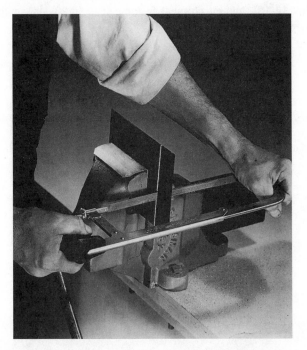

Fig. 16-7 Turn the blade at right angles to the frame for deep cuts.

Chisels

A chisel is a wedge-shaped cutting tool used to cut, shear, and chip metal. One end is hardened and sharpened to make a good cutting edge. There are four kinds of chisels used in the machine shop (Figure 16-8). The **flat cold chisel** is a plain flat chisel used for cutting or chipping metal and for splitting nuts or rivets. The common widths are 1/8 to 3/4 inch. The **diamond-point chisel** has a square end and a cutting edge at one corner. It is used for cleaning out sharp corners or for cutting a sharp bottom groove. The **cape chisel** has a narrow blade for cutting keyways and grooves. The **roundnose chisel** is used for cutting grooves or for moving a drilled hole that was started wrong.

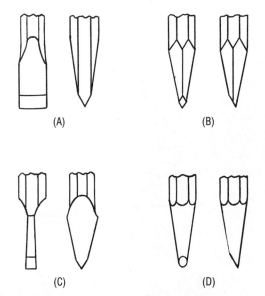

Fig. 16-8 Four common kinds of chisels: (A) flat cold chisel, (B) diamond-point chisel, (C) cape chisel, and (D) roundnose chisel.

Cutting with a Flat Chisel on a Plate

Follow the procedure below for using a flat chisel to cut on a soft metal plate.

1. Lay out the area to be cut with a chisel. Put on safety glasses.

2. Place the workpiece over a soft metal plate. *Never use a surface plate, or an anvil, or the finished surface of a machinist's vise.*

3. Hold the chisel firmly in your left hand with the thumb and fingers around the body.

4. Place the cutting edge on the layout line and strike the tool with a hammer. Keep your eye on the cutting edge, not on the tool.

5. Go over the layout line lightly (Figure 16-9), then start back and strike the tool with a firmer blow to cut through the metal. If necessary, turn the stock over and cut from the opposite side.

Shearing in a Vise

For using a chisel to shear in a vise, follow the procedure below.

1. Clamp the work in the vise with the layout line just above the edge of the jaw.

2. Hold the chisel at an angle of about 30 degrees with the side of one cutting edge parallel to the layout line. If the chisel is held too high, it will dig into the vise jaw. If it is held too low, it will tear the metal.

3. Strike the chisel firmly to shear off the metal (Figure 16-10).

4. The chisel is also used to shear off a rivet or bolt.

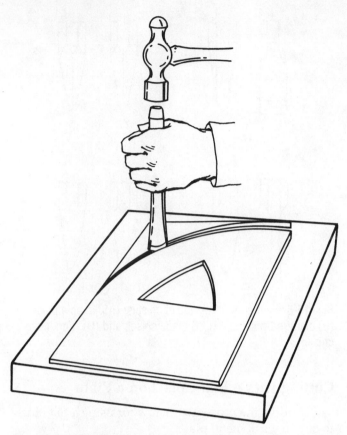

Fig. 16-9 Hold the chisel firmly. Make light indentations along the layout line first, then strike with heavier blows to cut through the metal.

Cutting an Internal Opening

To cut an internal opening with a chisel, follow the steps below.

1. Drill a series of small holes close together just inside the waste material.

2. Cut out the opening with a chisel (Figure 16-11).

3. File to the layout line.

Bench Shears

There are two kinds of bench shears. The **bench lever,** or **slitting, shears** have straight blades. They are used for cutting band or strap iron and black-iron sheet. *Never cut nails, rivets, or bolts with these shears.* To operate, open the blades and insert the metal with the layout line directly over the shearing edge. Then carefully lower the handle to cut the metal (Figure 16-12).

The **throatless bench shears** have curved blades. They are used to make curved or irregular cuts as well as straight cuts on sheet stock. When cutting an irregular shape, constantly move the stock as you cut, following the layout line (Figure 16-13).

Fig. 16-10 Hold the chisel at an angle of about 30 degrees to the vise when shearing metal in a vise.

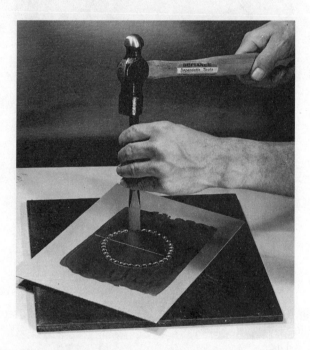

Fig. 16-11 Cutting an internal opening. A series of holes is drilled just inside the waste stock. This makes it easy to cut out the opening with a chisel. Finish the inside edge by filing.

Bolt Cutters and Nippers

These special tools come in many sizes and styles, and are used to cut mild steel or nonferrous rivets, rods, wires, and bolts. The **bolt cutter** is used for heavy-duty work (Figure 16-14). **End-cutting nippers,** or end cutters, are designed for lighter work (Figure 16-15). Be careful not to pinch your fingers when using cutters and nippers.

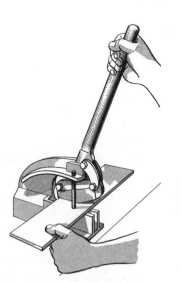

Fig. 16-12 Using the bench lever shears.

Fig. 16-14 Bolt cutters are used to cut heavy workpieces.

Fig. 16-15 This nipper is one of several types of tools used to cut metal rods, wires, and bolts. (Courtesy of CHANNELLOCK, Inc.)

Fig. 16-13 Using throatless bench shears. Make sure that the metal is inserted as far into the shears as possible. Keep your fingers away from the sharp cutting blade.

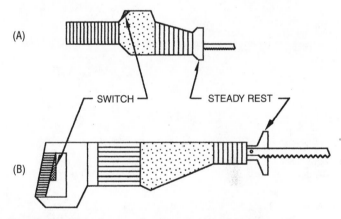

Fig. 16-16 Typical power handsaws: (A) saber saw and (B) cutoff saw.

Bandsawing

Metal also can be cut to size and shape quickly and accurately with horizontal and vertical **band saws.** Information on this equipment can be found in units 52 and 54.

Power Handsaws

Portable electric power sawing tools are convenient because they can be carried to the metal workpiece. Such tools are commonly found in the home. They also are used in industry and on metal building construction job sites. Some require electric cords, while others are powered by rechargeable batteries. They generally have short, tough, replaceable blades and operate with a reciprocating (back-and-forth) cutting action. Two common styles are the **saber saw** and the heavy-duty **cutoff saw** (Figure 16-16).

For efficient and safe cutting, follow these directions:

1. Be sure to use a safe, grounded electric cord, or check the battery pack to be sure that it is fully charged.

2. Keep the blade off the workpiece until the motor has been started.

3. Start the cut on the surface where the most teeth will be in contact with the workpiece.

4. Place the steady-rest against the workpiece and lower the blade into the cut.

5. Do not bear down while cutting. The weight of the tool will generally supply enough pressure.

6. When completing the cut, hold the tool firmly so that it will not fall against the workpiece or into your body. For additional safety, keep both hands on the tool at all times.

KNOWLEDGE REVIEW

1. List some important safety rules to be followed when cutting heavy metals.

2. When a blade is installed in a hacksaw frame, which way should the teeth point?

3. What should you do when a "mushroom" forms on the head of a chisel?

4. Explain the different uses for the two types of bench shears.

5. List two common types of power handsaws. Should you bear down on the blades when using them?

6. Prepare a research report on cutting heavy metals in industry.

Unit 17

Drilling

OBJECTIVES

After studying this unit, you should be able to:

- Describe drilling.
- Know the safety rules when using a drill press.
- Name the major parts of a drill press.
- Name two other tools that can be used to drill holes.
- Name the parts of a twist drill.
- Describe the three sets of drills.
- Name three ways of determining the size of a drill.
- Describe three holding devices used when drilling.
- Define drill *speed* and *feed.*
- Explain why cutting fluid is used when drilling.
- Describe how to locate the correct position for drilling.
- Name two types of punches used to locate the center for drilling.
- Tell how to do countersinking.
- Explain the purpose of reaming.

KEY TERMS

chuck	drill gage	fly cutter	portable electric drill
countersink	drill press	hand drill	reaming
cutting fluid	drill speed	hole saw	twist drill
drill drift	feed	micrometer	

Drilling is cutting round holes in a material. Drilling machines, including the drill press, are used for this purpose. They are used also for countersinking, reaming, boring, and tapping (Figure 17-1).

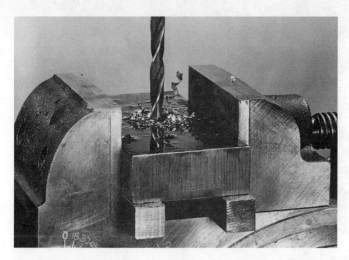

Fig. 17-1 Drilling is the process of generating round holes in workpieces. Most workpieces can be held safely in a drill press vise for drilling.

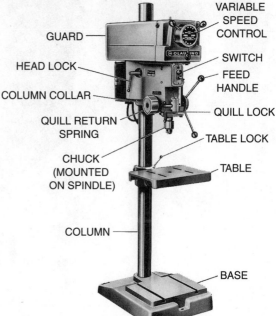

Fig. 17-2 The parts of a floor-model drill press. (Courtesy of Clausing Industrial, Inc.)

Drill Press Safety

Always observe the following safety guidelines when using a drill press.

- Wear safety glasses.
- Make sure the drill is tight and straight in the **chuck,** or special clamp.
- Remove the chuck key before starting the drill.
- Never force the drill.
- Do not try to hold workpieces by hand when drilling. Make sure your work is firmly clamped to the table or is held tightly in a drill press vise.
- Never leave the drill press when it is moving.
- Never try to stop the drill with your hand.

Drilling Machines

Drill presses include bench models and floor models (Figure 17-2). The major parts are the *base, column, table,* and *head.* The head contains all the operating parts. There is usually a four-step pulley on the spindle drive and the motor for setting different speeds. Some machines have a convenient variable speed control instead of the step pulley. The *spindle* moves the cutting tool up and down. The size of a bench or floor drill press is equal to twice the distance between the drill and the column. Thus, if the distance between drill and column is seven inches, the machine is called a fourteen-inch drill press.

Another drilling machine, the **portable electric drill,** can be used for many jobs. The size is indicated by the largest drill it will handle. Common sizes are 1/4, 3/8, 1/2, 3/4, and 1 inch. The **hand drill** is used for drilling smaller holes.

Twist Drills

Twist drills with two flutes, or grooves, running around the body are the type most often used for drilling. They are made with either *straight* or *taper shanks.* The straight shank is the most common type, especially on drills 1/2 inch or smaller.

Drills are made of either high-speed steel or carbon steel. High-speed-steel drills are stamped HS or HSS on or near the shank. Though a little more expensive than carbon-steel drills, high-speed-steel drills are much better. They wear much longer and do not break so easily. The main parts of a twist drill are shown in Figure 17-3.

Drills larger than 1/2 inch are available in fractional or metric sizes only. Three sets of customary drills 1/2 inch or smaller are in common use:

1. Wire-gage sizes from 80, the smallest, to 1, the largest.

2. Letter sizes from A, the smallest, to Z, the largest. Lettered drill sizes begin where numbered drill sizes end.

3. Fractional-size drills from 1/64 to 1/2 inch, increasing by 1/64 inch.

Metric drills are available in all equivalent sizes. A conversion chart for drills can be found in the appendix to this book. Note that all three sets are made so that each drill differs by only a few thousandths of an inch from the next size. No two drills are exactly the same size, except the 1/4-inch and the E drill, which are both 0.2500 inch. The size of the drill is stamped on the shank. If the mark is worn off, you can check the size

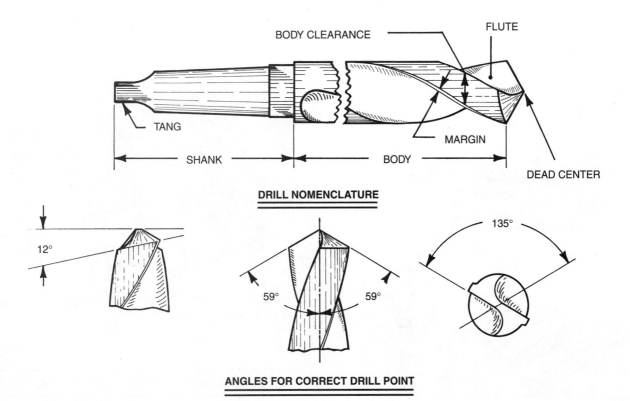

DRILL NOMENCLATURE

ANGLES FOR CORRECT DRILL POINT

Fig. 17-3 Parts of a taper-shank drill bit. Drill point angles also are shown.

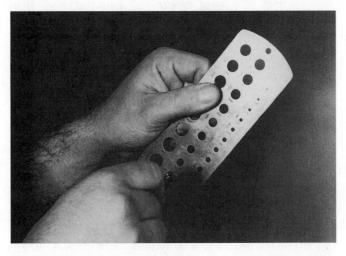

Fig. 17-4 Use a drill gage to find the size of a customary drill. Insert the drill into the holes until you find one it just fits. There is a separate gage for each set of drills—number, letter, fractional, and metric.

with a **drill gage.** There is a drill gage made for each different set of drills (Figure 17-4). The drill size can also be checked with a **micrometer** (see unit 53).

To cut properly, the drill point must meet these requirements:

1. The included angle must be correct. For most work, this angle is 118 degrees, or 59 degrees on either side of the axis.

2. The lips of the drill must be the same length.

3. There must always be proper relief, or clearance, behind the cutting edge. This lip relief must be 8 to 12 degrees for ordinary drilling.

4. The chisel edge must be ground at the proper angle. See unit 22 for information on sharpening twist drills.

Holding Devices

There are several ways to hold workpieces safely for drilling (Figure 17-5). A *drill press vise* is best for holding rectangular or heavy flat pieces for drilling. A *V block* holds round pieces. On some V blocks, a U-shaped clamp holds the workpiece in place. *C clamps* can be used to fasten the work to the table. Put a piece of scrap wood underneath the work, or center it directly over the table hole. Special *hold-down clamps* can be used to hold some irregularly shaped workpieces. For other irregular, small, or special-case workpieces, a *monkey wrench* or a pair of *pliers* can sometimes be used as a holding device. However, this must be done with extreme care.

Speeds, Feeds, and Lubricants

Drill speed is the distance the drill would travel in one minute if it were rolled on its side. Generally, the

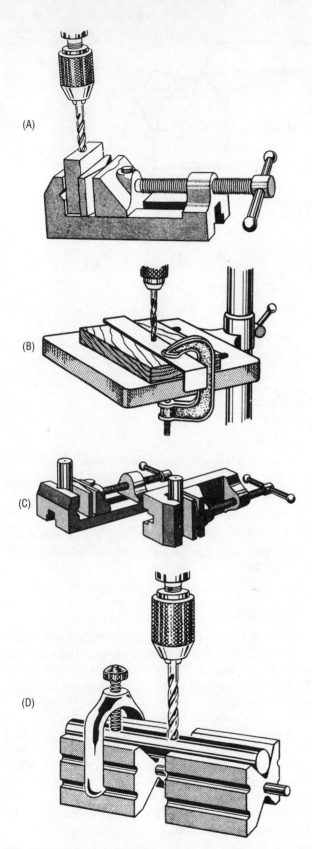

Fig. 17-5 Workpiece holding devices: (A) drill press vise, (B) C clamp, (C) drill press vise for round stock and end-drilling, and (D) V block with clamp.

BELT POSITION FOR:
FASTEST SPEED

SLOWEST SPEED

Fig. 17-6 Various drilling speeds can be obtained by changing the position of the belt on the pulley.

Table 17-1	Drill speeds in revolutions per minute*		
Diameter		**Aluminum,**	**Cast iron,**	**Mild**
in.	**mm**	**brass, bronze**	**hard steel**	**steel**
1/8	3.18	9170	2139	3057
1/4	6.35	4585	1070	1528
3/8	9.53	3056	713	1019
1/2	12.7	2287	535	764
1	25.4	1143	267	282

*Rpm should be reduced one-half for carbon-steel drills.

larger the drill, the slower the speed; the softer the material, the higher the speed. Using a coolant or cutting fluid allows higher drill speeds.

The drill press with belt and pulley has only four to eight speeds. Set the belt on the pulley below the highest speed for a 1/4-inch drill (Figure 17-6). Increase the speed for smaller sizes, and decrease it for larger ones (Table 17-1). To adjust the speed on a variable-control machine, dial the desired speed while the machine is running. The dial mechanism generally has a range of 450 to 5,000 revolutions per minute (rpm) (Figure 17-7).

Feed is the distance the drill moves into the stock with each complete turn. Apply just enough pressure to make the drill cut the metal. Too much pressure will cause the drill to burn or break. Too little will produce a scraping or dulling action. You can easily learn how much pressure to use by looking at the chips.

Cutting fluid is used in drilling as a lubricant to reduce friction and to keep the drill cool (Table 17-2).

Fig. 17-7 A drill press with variable speed pulleys. (Courtesy of Clausing Industrial, Inc.)

Table 17-2   Cutting fluids

Material	Drilling	Reaming
aluminum	kerosene kerosene and lard oil soluble oil	kerosene soluble oil mineral oil
brass*	soluble oil kerosene and lard oil	soluble oil
cast iron*	air jet soluble oil	soluble oil mineral lard oil
copper*	soluble oil mineral lard oil kerosene	soluble oil lard oil
malleable iron*	soda water	soda water
mild steel	soluble oil mineral lard oil sulfurized oil lard oil	soluble oil mineral lard oil
tool steel	soluble oil mineral lard oil sulfurized oil	soluble oil lard oil sulfurized oil

*Also drilled and reamed dry, without the use of cutting fluids.

Drilling on the Drill Press

Follow these basic steps when performing drilling operations on a drill press.

1. To locate a hole for drilling, scribe two lines at right angles to show the center of the hole. Mark this center with a prick punch.

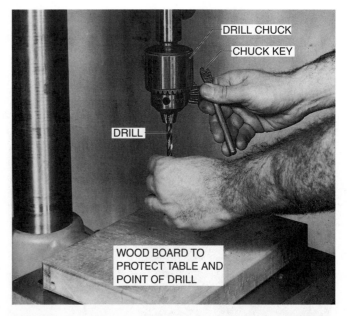

Fig. 17-8 Inserting a drill into a drill chuck. This chuck takes a drill to 1/2 inch (12.7 mm) in diameter. Be sure to remove the chuck key when the drill is tight.

2. Enlarge the mark with a center punch, so the drill will start easily.

3. Select the correct size of drill. Your drawing or sketch shows this. Insert the drill in the chuck and tighten it with a key (Figure 17-8). Always remove the key immediately.

4. If a taper-shank drill is used, insert it directly in the spindle or first into a drill sleeve and then into the spindle. The drill is removed with a **drill drift** (Figure 17-9).

5. Turn on the power and check to see that the drill is running straight. If it wobbles, it may be bent, or it may be in the chuck the wrong way. The shank may also be worn. Turn off the power.

6. Adjust for the correct speed.

7. Fasten the workpiece securely in a vise or to the table.

8. Move the drill down with the hand feed to see if the point is exactly over the punch mark. If necessary, move the workpiece or the table slightly. Also check to see that the drill can go all the way through the work. Raise the table if necessary. Make sure that there is a piece of scrap wood or a clearance space between the workpiece and the table.

9. Turn on the power and bring the point of the drill down to the workpiece. Feed the drill in to enlarge the punch mark. Now release the drill slightly and apply a little cutting fluid, if needed.

10. Begin the drilling. Apply even pressure and let the drill do the cutting. When it is cutting correctly on

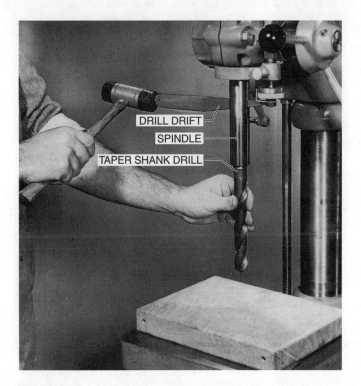

Fig. 17-9 Removing a taper-shank drill with a drill drift. A board is placed on the table to keep the point of the drill from striking the metal of the table if the drill drops.

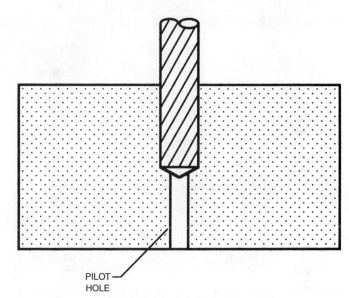

Fig. 17-10 Drill a pilot hole before drilling a large hole in metal.

Fig. 17-11 A hole saw is used for cutting holes in metal sheet stock. (Courtesy of Stanley Tools, Division of the Stanley Works, New Britain, CT)

steel, thin ribbons of metal rise from the flutes in the hole. Brush away the shavings as they pile up. Use a stick or brush for this. Watch carefully as the drill begins to go through the other side of the workpiece. Release the pressure slightly. Do not bear down on the feed handle as this may cause the drill to catch, breaking the workpiece or twisting it.

11. After the drilling is finished, release the pressure on the feed handle and turn off the power.

Drilling Larger Holes

To drill a hole larger than 3/8 inch, it is a good idea to drill a *pilot hole* first. If this is done, the larger drill will cut with less friction. The pilot hole should be equal to, or slightly larger than, the web of the larger drill (Figure 17-10).

Hole saws can be used to cut large holes in cast iron, mild steel, and nonferrous metals (Figure 17-11). They range in size from 5/8 to 6 inches, in increments (or steps) of 1/8 inch. The **fly cutter** is used to cut large holes in heavy sheet and plate (Figure 17-12). Note that the tool can be set to cut various diameters. A starter hole is first drilled to mark the hole center. The rod is guided into this hole to control the cutting. To use the hole saw and fly cutter, fasten the workpiece firmly and locate the hole to be drilled. Use a slow speed, as indicated on the use charts. Feed slowly until the saw teeth

engage the workpiece. Increase the feed and proceed with the cutting. The workpiece may chatter if you drill too fast or if it is loose. Add cutting oil as needed, and lessen the feed as the cut is finished.

Drilling Blind Holes

A *blind hole* is one that is cut only part way through the workpiece. Move the workpiece aside, then lower the drill until the right depth is reached. Then set the depth gage to stop the drill. Place the workpiece under the drill, and drill as described in the basic procedure above.

If the hole is to be deep, release the drill several times and apply a cutting fluid. This helps clean out the hole and keeps the drill cool.

Countersinking

Countersinking is machining a cone-shaped recess at the outer end of a hole. This recess is needed when using countersunk rivets and flat-head bolts or screws. An 82-degree countersink is used (Figure 17-13).

1. Insert the countersink in the drill chuck.

2. Adjust for a slower speed—about half that of drilling.

3. Feed the countersink slowly until about the right amount of material is removed. Check the size of

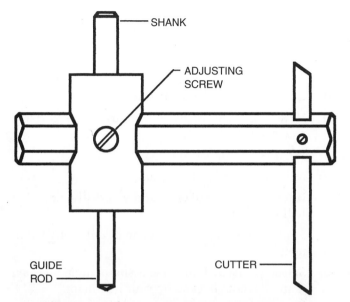

Fig. 17-12   The shank of the fly cutter fits into the drill press chuck. The guide rod fits into the starter hole drilled into the workpiece. The adjusting screw allows for the setting of the hole sizes.

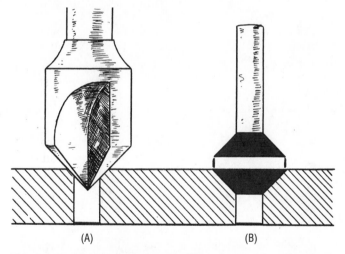

Fig. 17-13   (A) A countersink. (B) The countersink hole should match the size of the flat-head fastener.

the hole by turning the rivet, screw, or bolt upside down over the hole.

4. When countersinking several holes, set the stop on the depth gage.

Reaming

Reaming is smoothing the surface of a hole and finishing it to a standard size. This process is also called *sizing the hole.* When a very accurate, smooth hole is needed, the hole is first drilled a little undersize and then reamed to exact size. Hand reamers are identified by the square on the end. Machine reamers have either a straight or a taper shank.

There are two common kinds of reamers: hand and adjustable hand. *Hand reamers* are solid, straight reamers with either straight or spiral flutes (Figure 17-14).

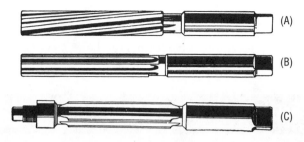

Fig. 17-14   Hand reamers with (A) spiral flutes and (B) straight flutes. The adjustable reamer (C) also is shown.

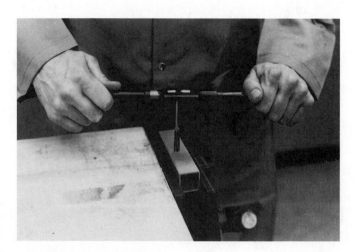

Fig. 17-15   Hand reaming.

They are made of carbon or high-speed steel. A tap wrench is used on the square of the shank to rotate the tool. Hand reamers are available in sizes from 1/8 to 1 1/2 inches. The cutting end is ground with a starting taper to make sure the reamer starts easily in the hole.

Adjustable hand reamers have blades inserted in tapered slots along their threaded bodies. They are used to ream odd-sized holes, as in repair work. The size of these reamers can be adjusted from 1/32 to 5/16 inch. About twenty reamers are needed for all hole sizes from 1/4 to 3 inches.

Reaming is done as follows:

1. Drill the hole 1/64 inch (0.5 mm) undersize.

2. Clamp the workpiece in a vise, with the hole vertical. (If the workpiece is small, the reamer is sometimes clamped in the vise and the workpiece rotated.)

3. Fasten a double-end tap wrench to the square end of the reamer.

4. Hold the reamer at right angles to the surface of the workpiece and apply slight pressure to it. Turn the reamer slowly and evenly, making sure it aligns itself with the hole.

5. Feed the reamer into the hole steadily. In one turn, it may go into the hole as much as one-fourth the length of its diameter (Figure 17-15).

6. Use cutting fluid if needed.

7. Continue turning the reamer clockwise until the hole is reamed. Continue rotating the reamer forward as you remove it from the hole. *Never turn a reamer backward.*

KNOWLEDGE REVIEW

1. Name the major parts of a drilling machine.

2. What kind of shank is most common on drills 1/2 inch (12.7 mm) or smaller?

3. Why is a high-speed-steel drill better than a carbon-steel drill?

4. Where on a drill is the size marked?

5. What is used to hold rectangular or flat pieces for drilling?

6. What is drill *speed?*

7. Define *feed.*

8. How do you know when a drill press drill is cutting correctly on steel?

9. How are round pieces held for drilling?

10. How might you break off the drill on a portable electric drill?

11. What is a countersink?

12. Name the two kinds of reamers.

13. How much undersize should you drill a hole that will be reamed?

14. Find out how square holes can be cut or drilled in metal.

15. Select a twist drill that needs sharpening. Sharpen it and have it checked by your instructor.

Unit 18

Filing

OBJECTIVES

After studying this unit, you should be able to:

- Name the parts of a file.
- Describe how files differ.
- Name three classifications of files.
- Describe file care and safety.
- Explain how to do cross-filing.
- Describe draw-filing.

KEY TERMS

cross-filing

draw-filing

rasp-cut file

single-cut file

double-cut file

Filing is one of the most useful of bench-metal operations. It is used to take the burrs off a piece of hacksawed metal, for smoothing a ground workpiece, or for cutting intricate shapes. Filing is a form of metal milling. If you were to take a milling cutter and stretch it out, you would have a file (Figure 18-1).

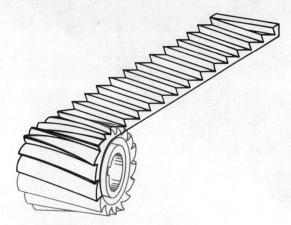

Fig. 18-1 Filing is a form of metal milling. This illustration shows how the file and the milling cutter are similar.

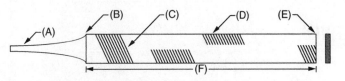

Fig. 18-2 Parts of a file: (A) tang, (B) heel, (C) face, (D) edge, (E) point, and (F) length.

Files

A *file* is a hardened steel tool that forms, shapes, and finishes metal by removing small chips. A good metalworker has a variety of files and uses each for its own particular purpose. The parts of a file are shown in Figure 18-2. Note that the length is measured from the *point,* or *tip,* to the *heel,* or *shoulder.* The length does not include the *tang.* Files differ in four primary ways: length, shape, cut, and courseness.

Length

File lengths vary from 4 to 18 inches. In general, files increase by 1 inch from 4 to 8 inches and by 2 inches from 8 to 18 inches. The most common lengths are 6, 8, 10, and 12 inches.

Shape

Files are made in four common geometric shapes: square, triangular, round, and rectangular (Figure 18-3).

Cut

The various cuts of files are shown in Figure 18-4. A **single-cut file** has one row of teeth cut at an angle (from 65 to 85 degrees) across the face. A **double-cut file** has two rows of teeth cut to form individual diamond-shaped cutting points. The first row is called the *overcut.* The second row is finer and deeper and is called the *upcut.* The angle of the two cuts is not the same. The double-cut removes metal

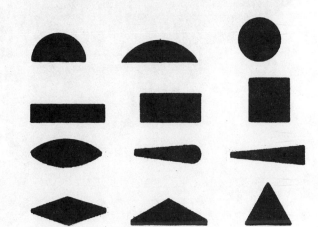

Fig. 18-3 Common shapes of files.

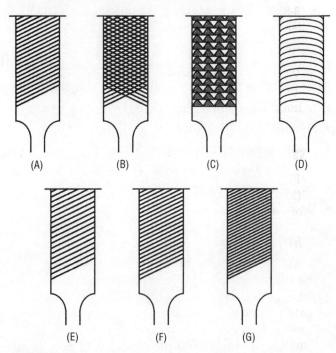

Fig. 18-4 Cuts of files: (A) single cut, (B) double cut, (C) rasp cut, and (D) curved-tooth cut. File coarseness also is shown: (E) heavy, (F) second-cut, and (G) smooth.

faster than the single-cut, but leaves a rougher surface.

The **rasp-cut file** has individually shaped teeth and is used only on softer metals. Some special files have *curved* teeth for fast, smooth cutting.

Coarseness

The common grades of coarseness are as follows:

- *Heavy*—quite widely spaced teeth for fast removal of stock.

- *Second-cut*—medium spacing of teeth for average cutting.

- *Smooth*—closely spaced teeth for fine finishing.

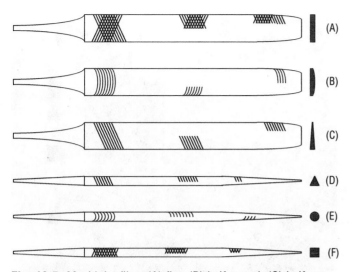

Fig. 18-5 Machinist files: (A) flat, (B) half-round, (C) knife, (D) three-square, (E) round, and (F) square.

With files of the same grade of coarseness, the teeth on a shorter file are closer together than the teeth on a longer file.

Classification of Files

Files are divided into three groups: (1) machinist files, (2) Swiss-pattern and jeweler's files, and (3) special-purpose files.

Machinist Files

Machinist files are the files most commonly used in the metal shop. They remove metal fast and are used when a very smooth finish is not needed. These files come in the shapes shown in (Figure 18-5). All are double-cut, except the very small round files and the back of the smooth-cut half-round files.

The *flat file* is rectangular and slightly tapered toward the point in both width and thickness. It is cut on both edges and sides. It is a good general-purpose file for fast removal of metal. The *hand file* is similar to the flat file, but it does not decrease in width. It tapers in thickness only. One edge is "safe" (uncut). It is used by machinists to finish flat surfaces.

The *half-round file* has one rounded side (back) and one flat side. The flat side is always double-cut. The rounded side is also double-cut, except in all-smooth and four- and six-inch second-cut files, which are single-cut. The half-round file is very useful for filing either flat or concave surfaces.

The *knife file* has a knife-blade shape and is used for filing sharp angles. The *three-square file* is double-cut with edges left sharp and cut. It is used for internal filing and for cleaning up square corners. The *round file* is made in a tapered shape and is generally used to file

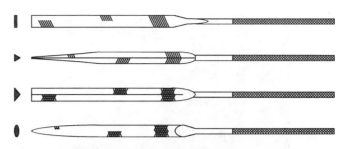

Fig. 18-6 Jeweler's files come in sets of different shapes for doing precise filing.

round and curved surfaces. The *square file* is used mostly for filing slots and for general surface filing.

In addition to these, the machinist often uses a *mill file*. This is a single-cut file used mostly in the smooth or second-cut grades. It is used for draw-filing and lathe filing, and produces a very fine finish.

Jeweler's Files

Swiss-pattern and jeweler's files are used for fine precision filing. They are sold in sets (Figure 18-6).

Special-Purpose Files

Special-purpose files include the many files designed for specific purposes. One commonly found in the metal shops is the *long-angle lathe file*. It has teeth cut at a much longer or shallower angle than those of a mill file. This file gives a good shearing action, cuts faster, and produces a very fine finish. It is used for both draw-filing and lathe filing. The *curved-tooth file* is also common, designed for filing aluminum and other soft metals.

File Care and Safety

Keep in mind the following safety and proper care guidelines for the use of all files.

- *Attach a tight-fitting handle.*
  It is very dangerous to use a file without a handle. The tang can pierce your hand. Choose a handle of the correct size. Slip the tang into the hole and tap the handle on the bench until the file and handle are tight. A screw-on handle can also be used. *Never strike a file with a hammer to push it into the handle.*

- *Keep the file clean.*
  Tap the handle on the bench after every few strokes to remove the loose chips. Brush the file in the direction of the teeth with a *file card* (Figure 18-7). If a small chip remains in the teeth, pick it out with a metal *scorer*. This is a piece of wire or soft metal sharpened to a point. *Never clean the teeth by striking the file against a bench. The teeth are brittle and break easily.*

Fig. 18-7 Keep the file clean by brushing it at regular intervals with a file card. Always follow the angle at which the teeth are cut (arrow).

- *Keep files separated.*
  Do not put files together in a drawer or on a bench. Bumping or rubbing files together can damage their teeth.
- *Chalk the file teeth.*
  Rub ordinary chalk across the face of the file to help keep chips from wedging between the teeth. These chips scratch the workpiece. Chalking is especially important when filing nonferrous metals, such as copper, brass, and aluminum.
- *Keep files dry so they will not rust.*

Cross-Filing

Cross-filing, the usual method of filing, is done in the following way:

1. Cover the jaws of the vise with jaw caps, then lock the workpiece in place. The top of the workpiece should be about as high as your elbow when your arm is bent.

2. Stand with your feet about 2 feet (0.6 m) apart, your left foot ahead of your right foot. You should be able to swing your arms and shoulders freely.

3. Grasp the handle of the file in your right hand. Your thumb should be along the top and your fingers curled around the handle.

   a. For heavy filing, place the palm of your left hand over the tip of the file and your fingers underneath. (Figure 18-8).

Fig. 18-8 The correct way to hold a file for heavy-pressure filing. Place the palm of the left hand on the file point with the fingers curved and pressed against the underside. Remember that the amount of pressure must vary with the hardness of the metal. Press just hard enough to make the file cut throughout the forward stroke.

   b. For light filing, place your thumb over the file and your other fingers underneath.

4. Files cut only on the forward stroke. Apply pressure to the point at the start of the stroke with your left hand, then with both hands, and finally with only the right hand. Move the file forward in a straight line. Do not rock it. Press only hard enough to make the file do the cutting.

5. Lift the file slightly on the return stroke. When filing soft metal, you might allow the file to drag slightly to help clean and position it for the next stroke. Beginners often make the mistake of moving the file back and forth in short, jerky strokes.

6. Work from one side of the workpiece to the other, filing evenly.

7. Every few strokes, tap the handle lightly on the bench to loosen the chips, then clean the file with a file card. It is good practice to chalk the teeth before doing the final filing.

8. When grinding, drilling, or cutting, a small burr usually develops. Remove this by running the file across the corner.

Draw-Filing

Draw-filing is done to get a very smooth, level surface. A mill file or a long-angle lathe file is generally used. If more metal must be removed, a second-cut, *bas-*

Fig. 18-9 Draw-filing. You should push the file sideways across the workpiece. Always use a new portion of the file after each stroke.

tard file (between second-cut and heavy), or a flat file may be used.

1. Grasp the handle in your right hand, the file in your left hand with only enough space between for the workpiece. In this position, push the file (Figure 18-9). Turning the file around (with the handle in your left hand), file on the draw stroke. Curl your fingers around the far edge of the file.

2. Hold the file steady. Use the part of the face near the tang. Start the stroke at the end of the workpiece.

3. Apply moderate pressure and push the file forward. Release the pressure on the return stroke.

4. After each stroke, move your hands to expose a new part of the file. This is important because any pin caught between the teeth will make a heavy scratch on the filed surface.

5. After the entire face has been used, clean the file.

6. Always remove the sharp wire edge by holding the file at an angle and making a light stroke across the edges of the workpiece.

Fine Filing

Be especially careful when working with small files, for they break very easily.

Fig. 18-10 Use jeweler's files for precision filing. If the metal is thin, sandwich it between two pieces of wood.

1. Fasten the workpiece in a vise. If the metal is thin sheet, place it between two pieces of plywood.

2. Choose a file of the correct shape. No handle is needed because the tang is large and round.

3. To file a concave or round opening, twist the file slightly during the forward stroke (Figure 18-10).

KNOWLEDGE REVIEW

1. Name the parts of a file.

2. By what four features can files be identified?

3. Why must the file be equipped with a handle?

4. What do you use to remove metal chips from file teeth?

5. In cross-filing, do you apply pressure on the forward or backward stroke?

6. In draw-filing, what must you do after each stroke?

7. Make a study of how files are made. Several manufacturers have booklets they will send you that describe the methods in detail.

Unit
19

Bending and Twisting Metal

OBJECTIVES

After studying this unit, you should be able to:

- Define *bending*.
- Describe how to make angular bends.
- Explain how to twist metal.
- Describe how to bend a scroll.

KEY TERMS

bending   scroll   twisting

Bending is a way of forming metal in one direction only. When you take a piece of wire in your hands and form a U shape with it, you are bending it. Metal can be bent to form round scrolls or angular V shapes (Figure 19-1). Sheet metals, rods, bars, and wires are easily bent. In this unit, you will learn how to bend and twist heavy metals.

Fig. 19-1 This mild steel trivet includes both angular bends and scrolls.

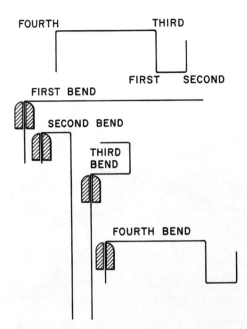

Fig. 19-2 Notice the order in which the bends are made.

Types of Angular Bends

Most metals 1/4 inch thick or less can be bent cold. The most common angular bends are made at *right angles,* or 90 degrees. However, it is sometimes necessary to bend other angles. An *obtuse angle* is greater than 90 degrees. An *acute angle* is less than 90 degrees.

Making Angular Bends

To make an angular bend in a workpiece, follow the general procedure below.

1. Make a full-size drawing of the part to be bent. Bend a piece of soft wire to the shape of the bend to find the length of material needed. In bending right angles, add an amount equal to one-half the thickness of the metal for each bend. For example, if you are using metal 1/4 inch thick and must make two right-angle bends, add 1/4 inch to the length of material needed.

2. If the piece has more than one bend, decide which bend is to be made first (Figure 19-2).

3. Fasten the metal vertically in the vise, with the bend line at the top of the jaws. The extra material allowed for the bend must be above the vise jaws. Use a square to check the workpiece.

4. Bend the metal by striking it with the flat of a hammer near the bend line. If the piece is long, apply pressure with one hand and strike the metal at the same time. Do not strike it so hard that you thin out the metal near the bend.

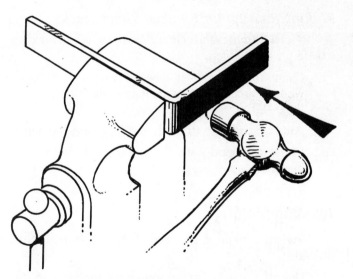

Fig. 19-3 Squaring off a bend. Notice that the workpiece is held in a vise with the edge parallel to the top of the vise jaw. This is the way to get an accurate right-angle bend.

5. To square off the bend, place it in the vise as shown in Figure 19-3. Strike directly over the bend. Make the other bends as needed.

6. To make an obtuse bend, use a monkey wrench, as shown in Figure 19-4.

7. To make an acute bend, first make a right-angle bend. Then place the bend between the vise jaws and squeeze the two sides together. Another method is to hold one side of the right-angle bend over the anvil and strike the other side with a hammer.

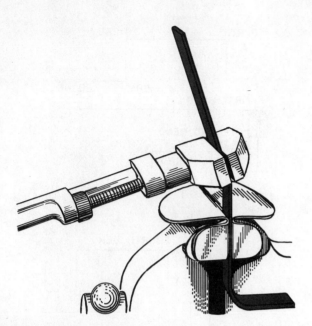

Fig. 19-4 Making a bend that is greater than 90 degrees, using a monkey wrench as a bending tool.

Making an Angular Bend on Sheet Stock

The following steps should be taken when making a bend on sheet stock.

1. Clamp the metal sheet between two pieces of hardwood or angle iron that are longer than the width of the metal.

2. Apply pressure with both hands to start the bend.

3. Finish the bend with a wooden, rawhide, or rubber mallet.

Twisting Metal

Twisting metal is accomplished by following this procedure:

1. Cut off a piece of metal somewhat longer than the finished piece will be. Metal decreases in length when it is twisted. You can check the amount of "shrinkage" on a particular length of metal by making a single twist on a piece of scrap.

2. Mark a line at the beginning and end of the twisted section.

3. If the piece is short, place it vertically in the vise with the top of the jaws at one end of the twist. Place a monkey wrench at the other end.

4. Clamp long pieces horizontally in the vise. If you wish, you can slip a piece of pipe over the section to be twisted to keep it from bending out of line.

5. Hold your left hand over the jaws of the wrench to steady the metal. Then apply pressure to the handle, making a definite number of twists (Figure 19-5).

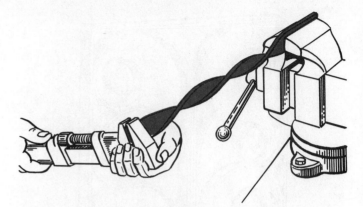

Fig. 19-5 Twisting metal. The twist may be either right-hand or left-hand, depending on the way you turn the wrench.

Bending a Scroll

A **scroll** is a strip of metal that forms a constantly expanding circle. It is similar in shape to an open clock spring. You bend a scroll in the following way:

1. Enlarge the drawing of the scroll to full size for a pattern (see unit 14).

2. Measure the length of stock needed. This can be done by forming the scroll with a piece of soft wire and then straightening it out.

3. Cut the stock to the needed length.

4. Flare the end of the stock, if desired, by holding the end flat on a metal surface and striking it with glancing blows.

5. Start the scroll by placing the metal flat on an anvil or bench block, with one end extending slightly beyond the edge. Use the flat of a ball-peen hammer to strike the metal with glancing blows to start the curve (Figure 19-6). Continue extending the metal beyond the edge, a little at a time, as the beginning of the scroll is formed. Check often by holding the metal over the pattern. If the curve is too tight, open it slightly by holding the curved section over the anvil horn and striking the edge.

6. When the first part of the scroll is complete, the remainder can be formed better on a *bending jig,* or *fork,* as shown in Figure 19-7. Most of these are adjustable to take different thicknesses of metal. If these are not available, a bending fork can be made by bending a rod into a U shape. The opening should be equal to the thickness of the metal being bent.

7. Lock the bending fork in a vise and adjust it to the proper opening. Slip the partly bent scroll into the fork to the point at which the scroll is already bent. Now, grasp the straight end of the metal in your left hand. Apply pressure with the thumb and fingers of your right hand to continue forming the curve.

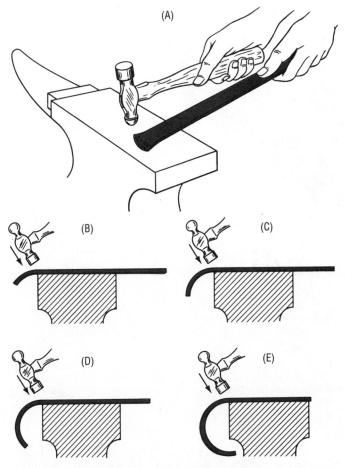

Fig. 19-6 Starting a scroll. Place the metal flat on an anvil and flare the end (A). Extend the end over the edge a little and strike with a glancing blow (B). Strike and move it alternately, a little at a time, until the curve begins to form (C, D, E).

8. Bend the scroll a little at a time as you feed the stock into the jig.

9. Constantly check the scroll as it is formed by holding it over the pattern (Figure 19-8). You will have to open the scroll a little if it is too tightly bent.

10. Two scrolls are often formed on the same piece, usually bent in opposite directions. There should be a continuous curve from one scroll to the other for the most pleasing appearance.

KNOWLEDGE REVIEW

1. What is bending?

2. What kinds and thicknesses of metal can be bent cold?

3. What allowance in length must be made for a right-angle bend?

4. With what tool do you strike metal to bend it?

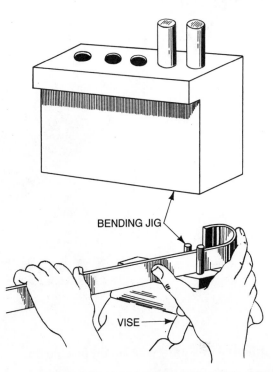

Fig. 19-7 A metal block with pins that can be set for different thicknesses of metal.

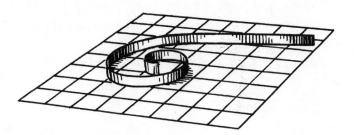

Fig. 19-8 Hold the partly bent scroll over the pattern to see if it is correct.

5. What has been done incorrectly if the metal is thinned out at the bend line?

6. What tool can you use to complete a bend in sheet stock?

7. When metal is twisted, what happens to its length?

8. What is a scroll?

9. After you have hammered part of a scroll, what tool can you use on the remainder?

10. Find out the exact amount by which a particular length of metal shortens when it is twisted once, twice, and three times. Does the metal shorten by the same amount if it is heated before twisting?

Unit 20

Machine Bending

OBJECTIVES

After studying this unit, you should be able to:

- Tell what a plate and bar bender is used for.
- Identify what kind of bender can be used to bend wire, flat metal, and tubing.
- Describe the Di-Acro bender.
- Explain how to do draw, or rotary, bending.
- Indicate another name for stationary-die bending.
- Describe roll bending.
- Tell why sharp bends in tubing are difficult.
- Calculate the length of material needed for a bend.

KEY TERMS

bar bender	Di-Acro bender	roll bending	stretch forming
compression bending	draw bending		

In unit 19, you learned a number of different ways to bend metal by hand. This unit explains the methods used in machine bending. Machines are used to make bending easier and more accurate and to make interesting projects. Several different kinds of bending machines are found in the school shop. Some of these are discussed in this unit.

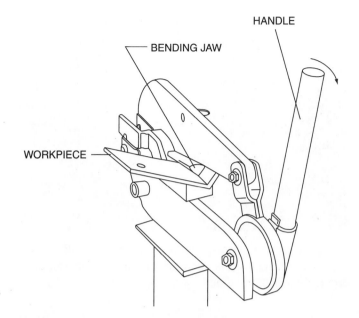

HANDLE

BENDING JAW

WORKPIECE

Fig. 20-1 A typical plate and bar bender.

Bar Bender

Bar benders are mounted on the floor or on a bench. They can be used to bend flat bar stock, usually up to 1/4 inch thick. Some nonferrous soft metals of a slightly larger gage can also be bent on them. The capacity of the machine is usually stamped near the jaws. Do not exceed this capacity or you may damage the machine. To use the bender, carefully mark the bend line with chalk so that it can easily be seen. Place the workpiece between the jaws, and close them just enough to hold the workpiece. Then, grasp the bending lever with both hands and pull hard to make the bend (Figure 20-1). Next, carefully open the jaws and remove the workpiece. Some of these bending machines are hydraulically operated, especially for industrial operations.

Di-Acro Bender

The **Di-Acro bender** makes it possible to bend a variety of metals and shapes (Figure 20-2). It consists of a form that has the same shape as the bend to be made, and a nose that moves around the form to shape the metal. Since all metals are somewhat elastic, they spring back a little after they have been formed. Because of this, the bending form is usually made with a smaller radius than that needed for the bend. Here is the correct way to use this bending machine:

1. Make a full-size drawing of the part to be bent on a piece of wrapping paper.

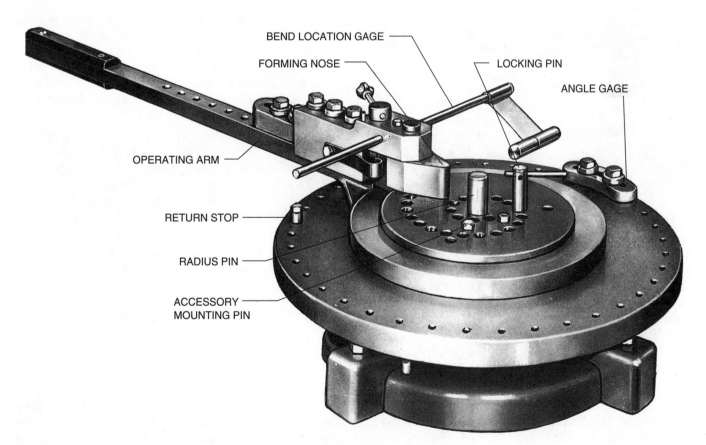

BEND LOCATION GAGE

FORMING NOSE

LOCKING PIN

ANGLE GAGE

OPERATING ARM

RETURN STOP

RADIUS PIN

ACCESSORY MOUNTING PIN

Fig. 20-2 Parts of a Di-Acro bender. (Courtesy of Strippit, Inc.)

NUMBER OF DEGREES IN BEND

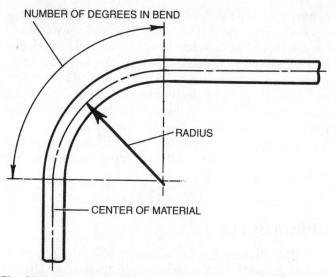

RADIUS

CENTER OF MATERIAL

Fig. 20-3 Measure both the radius to the center of the material and the angle of the arc before figuring the amount of material needed.

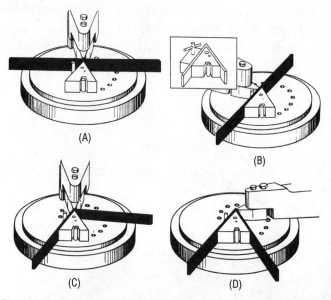

(A)

(B)

(C)

(D)

Fig. 20-4 (A) Adjust the forming nose so that the material will fit snugly between the nose and the point of the radius block. (B) Clamp the material close to the bending edge, using a locking pin or holding block as shown. (C) Move the operating arm until it strikes the gage. You will thus obtain the exact degree of bend you want. (D) The completed bend. (Courtesy of Strippit, Inc.)

2. Use a piece of soft wire to find the length needed, then cut a piece of metal to this length.

3. Another method of determining the workpiece length is shown in Figure 20-3. First find the radius of the arc from its center to the center line of the material. Then use this simple formula: *The length of the circular part equals the radius times the num-*

Fig. 20-5 Making a right-angle bend in a piece of angle iron. This type of bend is made when constructing the frame for a wrought-metal table. (Courtesy of Strippit, Inc.)

ber of degrees in the arc times 0.0175. For example, if the radius is 2 inches and you wish to bend a 90-degree angle with a rounded corner, the length of material needed will be, in customary units: $2 \times 90 \times 0.0175 = 3.150$, or about 3 5/32 inches.

4. Mark the start of the bend on the metal with a piece of chalk.

5. Place the metal in the bender with the mark against the form or against the radius collar. For example, in bending a sharp right angle with a 0-radius form (a form designed to bend sharp corners), the mark should be to one side, not at the center of the forming nose. Clamp the material close to the bending edge, as shown in Figure 20-4.

6. Move the operating arm until it starts to bend the metal. After it has bent a short distance, take the metal out and check it over the drawing. If the bend is not right, move the bend location mark and put the metal back into the bender. It is better to have to move the mark toward the starting end.

7. Figure 20-5 shows how you can bend angle iron with a sharp right-angle corner. Notice how you must cut a 90-degree notch before forming.

Other Types of Bending Equipment

Many different bending machines can be found in the school shop. Some are commercial models, and others are homemade. The two machines in Figures 20-6 and 20-7 have a series of holes to hold pins for different metal thicknesses, and sizes and styles of bends. A vise-mounted scroll and angle bender is shown in Figure 20-8. A roller handle can be attached to this machine to form scrolls.

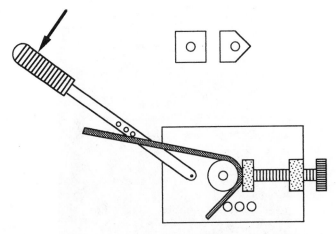

Fig. 20-6 This machine is being used to form a scroll. Note the two extra tools shown for bending angles and squares.

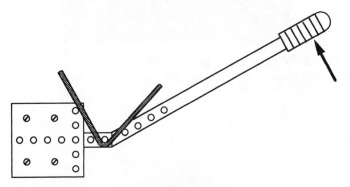

Fig. 20-7 This versatile tool can make angles of many sizes by changing the positions of the bending pins.

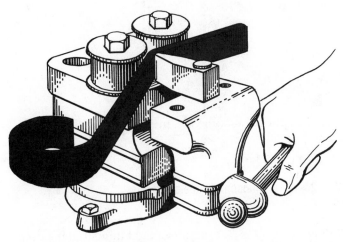

Fig. 20-8 This vise-mounted tool is being used to form a sharp bend in a piece of scroll work.

Industrial Machine Bending

There are four common bending techniques used in industry: draw, or rotary; stationary-die, or compression; roll; and stretch. Many others are used for special bending operations.

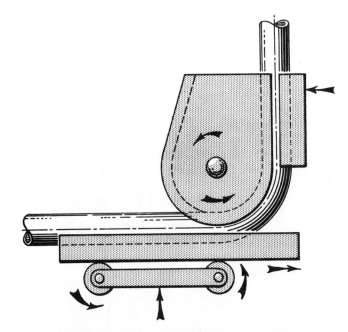

Fig. 20-9 Draw, or rotary, bending.

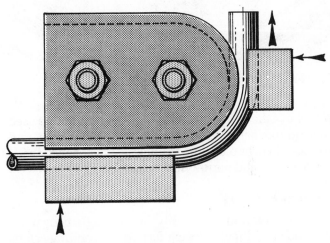

Fig. 20-10 Stationary-die, or compression, bending.

Draw Bending

Draw, or **rotary, bending** is done on a machine having a rotating bending die and stationary wiper block or die clamp. The tube is clamped to the die and bent as the die rotates (Figure 20-9).

Compression Bending

Stationary-die, or **compression, bending** is done on a machine having a fixed bending die and a movable wiper block. The tube is wrapped around the die by the wiper block (Figure 20-10).

Roll Bending

Roll bending is used to produce circular bends. The equipment consists of three or four rolls, positioned

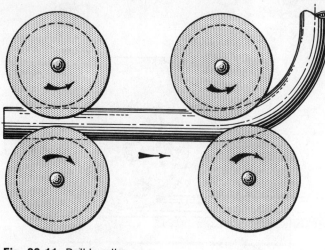

Fig. 20-11 Roll bending.

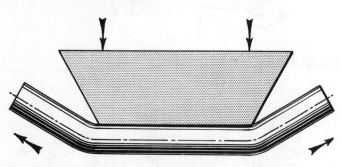

Fig. 20-12 Stretch forming or bending.

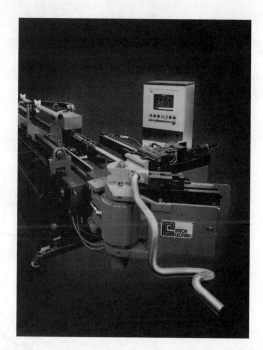

Fig. 20-13 An industrial tube-bending machine with programmable controls. (Courtesy of Eagle Technology Group)

Fig. 20-14 A handsome globe stand.

one above the other (Figure 20-11). This machine is used to bend rod, bar, or tubing into full or partial circles and to bend sheet or plate into cylinders. Wheel-shaped rolls are used to bend narrow sections.

Stretch Forming

When forming smoothly contoured parts by other methods is difficult, **stretch forming** or **bending** can be done. In this method, the material is stretched over a tool, or stretch, die (Figure 20-12). The entire piece is stretched by tension. Unlike the other types of bending, stretch forming uses no compression. This process can be used with strip, sheet, and plate metal. It is economical because only one die is required.

Heavy-Duty Bending

Powerful, hydraulic, computer-controlled tube and pipe benders are used in industry. The tube is clamped in position, and a movable die wipes it into the desired shape. The machines can be programmed to bend complex fuel and exhaust piping for automotive and industrial use (Figure 20-13).

Light Industrial Bending

In contrast to the heavy-duty bending described above, the interesting geographical globe structure shown

in Figure 20-14 is an example of light industrial bending. The globe stand is made of 3/8-inch mild steel, and its round and linear members are welded together. Its satin black powder-coat finish is attractive and durable.

KNOWLEDGE REVIEW

1. What is the thickest metal plate or bar that can be worked on school shop bender machines?

2. Write and use the formula for finding the length of material needed to make a bend in stock of a given radius.

3. Why is the bending form on the Di-Acro bender made with a smaller radius than that required for the bend itself?

4. Sketch two other types of bending machines.

5. List and describe the four common bending techniques used in industry.

6. Write a report on industrial metal bending.

7. Select a product illustration, such as the globe stand in Figure 20-14, and show the parts made by bending. Explain the possible bending procedures to the class.

Unit 21

Grinding Metal

OBJECTIVES

After studying this unit, you should be able to:

- Define *abrading*.
- Name three kinds of abrading.
- Describe the two main kinds of abrasives.
- Name two natural abrasives.
- Tell when artificial abrasives were invented.
- Name the two kinds of artificial abrasives.
- Explain how abrasives are graded into various sizes.
- List the number of sizes of flour abrasives.
- Name the material used to bond finer grains of abrasive together.
- Identify what grinding of tools is called.
- List the safety rules for grinding.
- Describe how to dress and true a grinding wheel.

KEY TERMS

abrading	dressing	grains	polishing
abrasive	emery	grinding	silicon carbide
aluminum oxide	flour sizes	grit	truing
buffing	glazed wheel	loaded wheel	vitrified
corundum			

Abrading is cutting or grinding away metal with abrasives. An **abrasive** is a material that will wear away something softer than itself. Abrasive grains are used to make abrasive paper, abrasive cloth, and grinding wheels. These can then be used on metal to remove small pieces and make it smooth (Figure 21-1).

Fig. 21-1 Metal grinding is a form of abrading. (Courtesy of Norton Company)

Fig. 21-2 Abrasives come in many forms and shapes. (Courtesy of Norton Company)

Kinds of Abrading

In metal finishing, there are three main kinds of abrading. **Grinding** is using abrasives to remove relatively large amounts of metal quickly. **Polishing,** which follows grinding, is used to smooth the surface and remove the scratches from grinding. **Buffing,** the final smoothing operation, gives a bright, mirror-like finish or a softer, satin-like appearance. Abrading is done either by hand or by machine. This unit deals with grinding. (Unit 23 presents information on polishing metals, and buffing is discussed in unit 44.)

Kinds of Abrasives

There are two main kinds of abrasive materials, although products utilizing these materials may come in many forms (Figure 21-2).

Natural Abrasives

Natural abrasives are found in nature. **Emery** and **corundum** are the two most common natural abrasives. Emery is about 60 percent aluminum oxide. Corundum is about 75 to 95 percent aluminum oxide.

Manufactured Abrasives

These abrasives were developed toward the end of the nineteenth century by scientists and engineers. **Silicon carbide** is made from coke, sand, salt, and sawdust. This mixture is heated to a high temperature in an electric furnace until an iridescent, or rainbow-like, blue-black crystalline material is produced. Silicon carbide is used to grind materials such as cast iron and bronze.

Aluminum oxide, although found naturally in emery and corundum, is also produced artificially from bauxite. This abrasive, like silicon carbide, is made in an electric furnace. The crystals are tougher than silicon carbide but not so hard. Aluminum oxide is used to grind such materials as steel.

Grain Size

Abrasives come from a furnace or a mine in chunks. These chunks are then crushed into **grains,** or fine particles. The grains are sorted by being passed through screens of different sizes. For example, one screen may have thirty-six openings per inch. The next smaller screen may have forty openings per inch. Grains that pass through the screen with the thirty-six openings but not through the next are numbered 36. This identifies the grain size, or coarseness. The higher the number, the smaller the grains. As you see in Table 21-1, the sizes range from coarse to fine. For sizes 280 and above, the grains are sorted in a different way and are called **flour sizes.**

Manufactured abrasives can be graded very accurately. The uniform size, shape, and hardness of each grain make them better than the natural abrasives.

Grinding Wheels

Wheels for grinding metal are available in many shapes, sizes, and **grits,** or grades. Thus, there is a wheel

Table 21-1 Grain sizes of abrasives

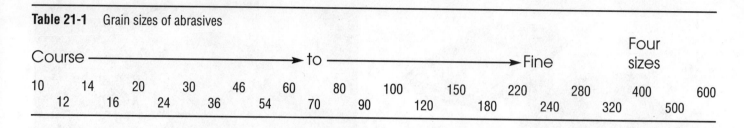

Course							to				Fine			Four sizes
10	14	20	30	46	60	80	100	150	220	280	400		600	
12	16	24	36	54	70	90	120	180	240	320	500			

that is right for each grinding operation. Grinding wheels are made by pressing and bonding or gluing the abrasive grains into the desired size and shape. The finer grains are bonded together with resin, rubber, or shellac. **Vitrified,** or ceramic-bond, wheels are also made for precision work. Soft metals are best ground with coarse grit wheels, such as numbers 36 and 46. For hard and brittle metals, use grit numbers 60 or 80.

Metal Grinding

Every metal shop has one or more grinders for sharpening tools and for general grinding. This kind of grinding is called *offhand* because the workpiece is applied to the grinder freehand. The *bench grinder* is small and is mounted on a bench. It has grinding wheels at both ends of a shaft that extends out from an electric motor. The *pedestal grinder* is similar except that it stands on the floor and is usually much larger.

Grinder Safety

Be sure to observe the following safety guidelines when using grinders.

- Always wear goggles or an eye shield. Do this even when the grinder is equipped with safety eye shields.

- Make sure the tool rest just clears the grinding wheel by about 1/16 to 1/8 inch. Most accidents happen when the workpiece becomes lodged between the revolving wheel and the tool rest. This can break the wheel and seriously injure the operator.

- Always choose the right kind of grinding wheel for the material. The most common for general grinding is an aluminum oxide wheel. When you replace a grinding wheel, use one that has the same thickness, diameter, and size of arbor hole as the original wheel.

- Use only the face of the wheel, never the sides.

- Make sure the wheel is dressed, or sharpened, properly.

- When a shop has two grinders, use one for general grinding and the other for sharpening tools.

Fig. 21-3 Using the wheel dresser. Wear safety glasses during this operation.

Dressing and Truing a Wheel

As a grinding wheel is used, several things happen: (1) the wheel becomes clogged with small bits of metal, resulting in a **loaded wheel;** (2) the abrasive grains on the wheel face wear smooth, and the wheel loses its grinding action—called a **glazed wheel;** and (3) the wheel wears irregularly with grooves or high spots.

When any of these things happen, the wheel must be dressed and trued. **Dressing** is sharpening the edge of the wheel by exposing new abrasive grains. **Truing** is straightening and balancing the wheel. Both can be done at the same time using the following procedure:

1. Select a *wheel dresser* that has a very hard, star-shaped steel wheel at one end of a holder (Figure 21-3). Put on safety goggles.

2. Adjust the tool rest so that the dresser will make contact with the grinding wheel on the wheel's centerline.

3. Start the wheel turning, then place the dresser on the tool rest with its handle tilted upward at an angle.

4. Slowly press the dresser against the face of the wheel until it "bites." Then move the dresser from side to side across the wheel to make the wheel surface straight.

 CAUTION: Hold the dresser rigidly on the tool rest so that it does not vibrate.

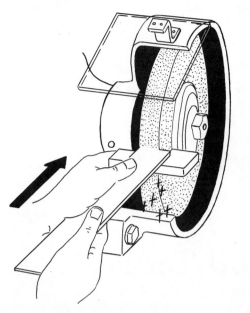

Fig. 21-4 Grinding a piece of metal. Hold the metal firmly on the tool rest and push it gently into the wheel. When grinding, use adequate lighting, keep the eye shield in place, and wear goggles.

Abrasive-stick and *diamond-tip dressers* are also used in the shop, much the same way as mechanical dressers. You smooth a wheel by passing the dresser back and forth over the face of the wheel. Use very light pressure.

Grinding Metal

Use the following guidelines when grinding metal.

1. Check the safety rules.
2. Hold the workpiece firmly on the rest and guide it back and forth to grind a straight edge (Figure 21-4). Keep the metal cool by dipping it in water.
3. To grind a curve or a semicircle, swing the workpiece in the proper arc.
4. Whenever an edge is ground, a small burr forms on the lower side. Dress this off with a file or grinder.

KNOWLEDGE REVIEW

1. Name the two main types of abrasive materials.
2. What manufactured abrasive is used on cast iron and bronze? On steel?
3. What determines how abrasives are graded?
4. How are grinding wheels made?
5. For safety, how much clearance should there be between the wheel and the tool rest?
6. What part of the wheel should you always use for grinding?
7. Define *dressing* and *truing*.
8. Report to the class on how grinding wheels are made.

Unit 22

Sharpening Hand Tools

OBJECTIVES

After studying this unit, you should be able to:

- List the instructions for sharpening tools.
- Describe how to sharpen a slotted-head screwdriver.
- Describe how to dress or sharpen a Phillips-head screwdriver.
- Explain the angle used to grind a flat cold chisel.
- Explain why a mushroom head on a chisel is dangerous.
- Describe the grinding angle for a center punch.
- Describe the grinding angle for a prick punch.
- Explain how a hollow punch should be sharpened.
- Identify the angle for sharpening a twist drill.
- Describe the device used to check the angle of a twist drill.

KEY TERMS

drill-grinding gage   relief

All hand tools used for metal cutting must be carefully maintained. Dull tools can be both dangerous and frustrating. Be sure you have learned the procedures for sharpening them properly. Otherwise, you may ruin the tool.

General Instructions

The following general instructions apply to the sharpening of all hand tools.

- Make certain that you select the right wheel for sharpening. Use a fine wheel of 60 to 80 grit.

- Check to see that the wheel and grinder are in good condition. The wheel should not be cracked or loose. The tool rest should be tight and the eye shield in place.

- Be sure that the water pot is filled. Keep the tool cool by dipping it into water often. This will prevent overheating and softening of the tool.

- Wear goggles. Be sure you have no loose clothing to get caught in the machine.

- Wipe all grease or oil from your hands and the tools. Oily tools are difficult to hold while grinding and may slip.

Sharpening Screwdrivers

Screwdrivers used for *slotted-head* screws should have a flat, blunt point with tapered sides. Grind or file either side of the point a little at a time. Then hold the screwdriver flat and grind or file off the end (Figure 22-1). Never sharpen a screwdriver to a sharp edge. Also, do not overheat the tool as this will soften it.

Phillips-head screwdrivers should be dressed or sharpened lightly with a sharp file.

Sharpening Cold Chisels

The cutting edge of a *flat cold chisel* is ground to an angle of 60 to 70 degrees with the edge at a slight arc (Figure 22-2). Hold one side of the cutting edge lightly against the face of the wheel and move it back and forth in a slight arc. Grind first one side and then the other to form a sharp edge. Cool the chisel often. Never allow it to become overheated and lose hardness.

The body of the chisel is softer than the cutting edge. The head will therefore mushroom, or flatten, after it has been used for a while. This mushroom is dangerous. It could cut your hand, or a piece of it could break off and hit your eye. Always grind off the mushroom when it forms.

Cape, diamond-point, and *roundnose chisels* should be ground in a similar way. Grind them carefully to retain their normal shapes. They can easily be ruined.

Sharpening Punches

Because punches are round or cone-shaped, they should be turned slowly against the grinding wheel. This will allow them to keep their shape. Cool the punches often in water.

Grind a *center punch* to an included angle of 90 degrees (Figure 22-3). *Prick punches* should be ground to angles of 30 degrees. Be careful not to burn the fragile points. Adjust the tool rest so that the punch meets the wheel face at the desired angle. Turn the punch during grinding to make the point symmetrical (the same shape all around). Dip the punch in water often to avoid overheating it. Do not grind away more metal than is necessary.

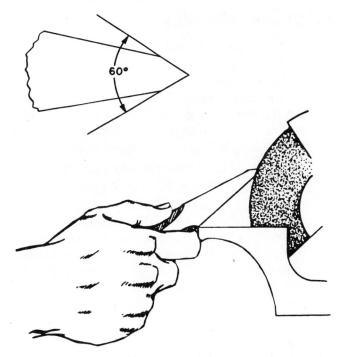

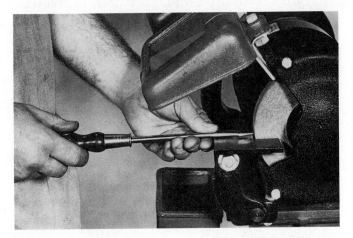

Fig. 22-1 Sharpening a screwdriver. Grind each side so that it tapers a little, and then grind the tip square. Dip the tool in water often to keep it cool.

Fig. 22-2 Grinding a cold chisel. The point should be ground at an angle of about 60 to 70 degrees. Never let the chisel get so hot that you cannot hold it comfortably in your bare hands while grinding. Too much heat can remove the controlled temper in the metal.

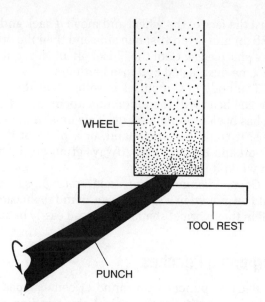

Fig. 22-3 Grinding a center punch. Turn the tool slowly to form a nicely rounded shape, ending in a sharp point.

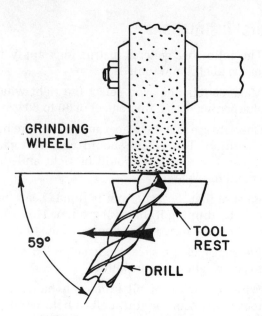

Figure 22-4 Note that the drill is held at an angle of 59 degrees to the grinding wheel.

Hollow punches should be ground on the outside only. Never attempt to grind the inside because this will change the size. The bevel should be ground slowly to an angle of about 45 degrees. *Solid punches* are ground only at the cutting tip. Do not change the diameter by grinding the sides.

Sharpening Twist Drills

To sharpen a twist drill, follow the procedure described below.

1. Check the condition of the twist drill. Usually, only a small amount of grinding is necessary to put the drill in good condition. Get a new, larger-size drill as a sample or guide. Make sure that the face of the grinding wheel is trued and dressed.

2. Grasp the drill near the point in your right hand, with your left hand holding the shank.

3. Hold the lip of the drill at an angle of 59 degrees to the grinding wheel. Grind the cutting edge slightly (Figure 22-4). Then turn the drill in a clockwise direction, at the same time swinging the shank down in an arc of about 12 to 15 degrees. Practice this before you grind the drill.

4. Grind a little off each cutting edge.

5. Check with a **drill-grinding gage** to make sure that the cutting edges are the same length and at the same angle with the axis—59 degrees. Check also to see that there is **relief,** or clearance, of 12 to 15 degrees from the cutting lips to the heel (Figure 22-5). Specifically, watch for the following:

    a. Make sure there is enough, but not too much, lip relief.

Fig. 22-5 Checking the point with a drill-grinding gage. Make sure it is ground so that the lips are of the same length and at the same angle. (Courtesy of The L. S. Starrett Company)

    b. Make sure both lips are the same length.

    c. See that the lips are ground at the same angle.

Do not overheat the drill. The point of carbon-steel drills may be cooled in water.

KNOWLEDGE REVIEW

1. What grit, or grade, of wheel should you use for sharpening tools?

2. Why should tools be kept cool while being sharpened?

3. To what angle should the cutting edge of a cold chisel be ground?

4. Why should you turn punches slowly against the grinding wheel?

5. At what angle should you grind drills?

6. Prepare a wall chart showing the proper grinding angles for some selected cutting tools.

Unit 23

Polishing with Abrasives

OBJECTIVES

After studying this unit, you should be able to:

- Name the major safety rule for polishing.
- Describe the backing for sheets of silicon carbide and aluminum oxide.
- Describe crocus cloth.
- Name the material applied to soft cloth wheels for polishing.
- Name the material used to apply abrasives to hard wheels.
- Describe the grain numbers of abrasives used for medium polishing.
- Name the forms in which coated abrasives are available.

KEY TERMS

abrasive cloth
abrasive wheels

coated abrasives

crocus cloth

polishing compound

In unit 21, you learned that polishing follows grinding in finishing metal. Polishing makes the metal smoother and removes grinding scratches. This is done by holding the surface of a workpiece against an abrasive-coated wheel or belt. Both hand and machine methods can be used.

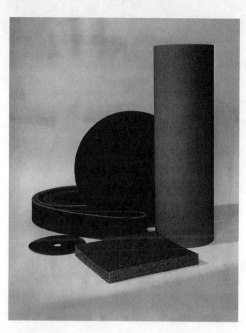

Fig. 23-1 Common forms of coated abrasive materials. (Courtesy of Norton Company)

Fig. 23-2 Polishing with a plastic fiber wheel. (Courtesy of 3M Abrasive Systems Division)

 Safety Rule: Always wear safety goggles when machine-polishing metal.

Three main methods of machine-polishing metal are used in the school shop: (1) polishing compounds and cloth wheels; (2) abrasive-covered wheels; and (3) abrasive-coated belts, disks, sheets, and drums.

Using Abrasive Cloth for Hand Polishing

Abrasive cloth or paper has abrasive material attached to it with glue or some other adhesive. It comes in 9- by 11-inch sheets or in rolls 1/2 inch to 3 inches wide. Abrasive cloth used in the shop comes in the common grain sizes: 60 for medium-coarse work, 80 to 90 for medium-fine, and 120 to 180 for very fine (Figure 23-1).

Flexible abrasive sheets made of a soft nylon web filled with abrasive grains and resins are also used. These sheets are made with either silicon carbide or aluminum oxide. They come in grades of very fine, fine, medium, and coarse.

Crocus cloth is a very fine abrasive cloth made with a red iron oxide coating. It is used to produce a very fine finish in final buffing operations (see unit 44). The procedure for hand polishing is as follows:

1. Cut a strip of abrasive cloth from a roll or sheet.

2. Wrap it around a flat stick or file.

3. Apply a few drops of oil to the metal surface.

4. Rub the cloth back and forth as if you were sanding. Do not rock the tool; keep it flat.

5. After all scratches have been removed, the abrasive grains will float in oil on the surface.

6. Reverse the cloth, exposing the back. Rub back and forth to get a high polish.

Polishing with Compounds and Wheels

Attach a clean, soft cloth wheel to the head of the polishing machine. Then select a stick of greaseless **polishing compound.** This is an abrasive mixed with glue in stick form. Turn on the machine and hold the abrasive stick against the turning wheel until the face is coated. This coating will dry quickly. Then, holding the workpiece firmly in your hands, move it back and forth across the wheel until the scratches have been removed. Keep the workpiece below the centerline of the wheel for safety.

Polishing with Abrasive Wheels

Polishing is often done with **abrasive wheels—** firm felt or leather wheels covered with abrasive grains. In some operations, the wheel is coated with an adhesive and then rolled in abrasive grains. A more efficient type of wheel is made of tough, compressed plastic fibers, such as nylon, and can remove workpiece scratches quickly (Figure 23-2). Some of these wheels are also impregnated with abrasive grains. Flexible, fibrous *flap wheels* are used for contour polishing (Figure 23-3). Some flap wheels are made of strips of coated abrasive cloth fastened to an arbor.

Fig. 23-3 Contour polishing with a flap wheel. (Courtesy of 3M Abrasive Systems Division)

Polishing with Coated Abrasives

As stated earlier, common **coated abrasives** include *belts, drums,* and *disks* for use on appropriate machines (Figure 23-4). In these polishing operations, the workpiece should be rested on the table, held firmly in your hands, and moved gently into the abrasive tool. Apply even pressure as you work the piece back and forth as needed. Be sure to wear goggles and a face mask for safety.

Special equipment is used for industrial polishing operations. For example, the machine in Figure 23-5 has a continuous, through-feed conveyor which holds the workpieces and carries them into the abrasive belt polishing heads. Flat products such as ice-skate blades, wrenches, and knife blades are processed on such machines.

KNOWLEDGE REVIEW

1. What safety rule should always be followed when machine polishing?

2. How is abrasive material attached to abrasive cloth?

3. What is a polishing compound?

4. In what forms are coated abrasives available?

5. Make a report on the method of manufacturing one of the two artificial abrasives. The manufacturers will send you bulletins giving you this information.

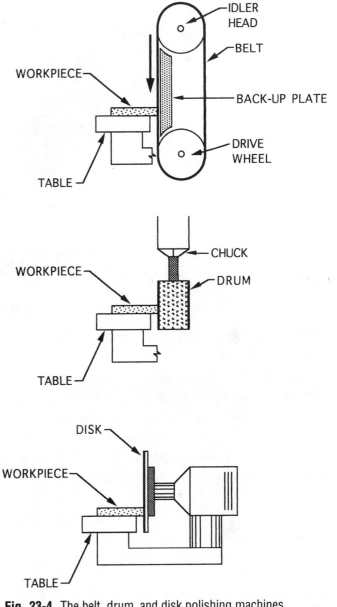

Fig. 23-4 The belt, drum, and disk polishing machines.

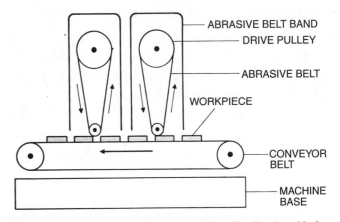

Fig. 23-5 An industrial polishing machine. The first head is for rough polishing and the second for fine polishing.

Unit 24

Cutting Threads

OBJECTIVES

After studying this unit, you should be able to:

- Define a *thread*.
- Name the major terms used to describe a screw thread.
- Name the two forms of the American National Thread System.
- Identify the meaning of 8-32 for an internal thread.
- Identify the meaning of 1/4-28 for a threaded rod.
- Tell how the American National System is similar to the Unified System.
- Describe an Acme thread and tell where it is used.
- Name the number of series of American Standard pipe threads.
- Tell how metric threads differ from the Unified System threads.
- Explain the meaning of M5 × 0.8-C.
- Name two classes of fits for metric threads.
- Define *tapping*.
- Identify what a die is used for.
- Describe a screw plate.
- Name three styles of taps.
- Name two methods of holding a tap.
- Identify what device is used to hold a die.
- Describe what to do when cutting internal threads if numbers are not available.
- Describe how to cut an internal thread.
- Explain how to cut external threads.
- Identify the uses of pipe threads.
- Describe how pipe taps and dies are marked for size.
- Name four industrial methods of threading and tapping.

KEY TERMS

adjustable die	external thread	pitch	screw plate
axis	internal thread	pitch diameter	tap
crest	lead	root	thread
depth	major diameter	screw-pitch gage	thread angle
die	minor diameter		

Threaded fasteners are used to join parts semipermanently. This joining system is used in many products. For example, the axle on a motorcycle is held to the fork with nuts. This feature makes it possible to remove the wheel to repair a tire (Figure 24-1).

A **thread** is a helical, or spiral-like, cut on the inside of a hole or the outside of a rod or pipe (Figure 24-2). Threads are used to assemble parts, transmit and pass along motion, and make adjustments.

Fig. 24-1 How many different threaded fasteners can you find on this motorcycle? (Courtesy of Buell Motorcycle Company)

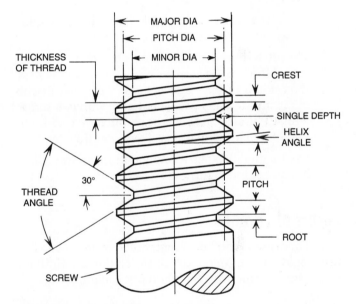

Fig. 24-2 Parts of a customary thread.

Screw-Thread Terms and Definitions

Refer to the following definitions as you encounter screw-thread terminology in this unit.

- An **internal thread** is a screw thread cut on the inside of a hole, as in a nut.

- An **external thread** is a screw thread cut around the outside surface of a rod or pipe, such as a bolt.

- The **major diameter,** sometimes called the *outside diameter,* is the largest diameter of a screw thread.

- The **minor diameter,** sometimes called the *root diameter,* is the smallest diameter of a screw thread.

- The **pitch diameter** is the diameter of an imaginary cylinder. This cylinder is of such size that its surface would pass through the screw-thread

forms at the level where the width of the forms equals the width of the spaces between the forms. This definition holds true for straight threads. The pitch diameter determines thread size in the customary system. For metric threads, size is found by gaging.

- The **crest** is the top intersection of the two sides of a screw thread.

- The **root** is the bottom intersection of the sides of two adjacent screw-thread forms.

- The **axis** of a thread is a line running lengthwise through the center of the material on which the thread is formed.

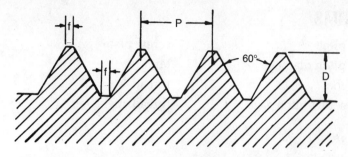

Fig. 24-3 American National thread form; *P* is the pitch, *f* is the width of the root and crest, and *D* is the single depth of the thread. The ISO metric thread has the same thread form, but it is not interchangeable with customary threads.

- The **depth** of a thread is the distance between the crest and the root on a line that is at right angles to the axis of the thread.

- The **thread angle** is the angle formed by the sides of two adjacent screw-thread forms.

- The **pitch** is the distance from a point on one screw-thread form to the corresponding point on the next form. Pitch in inches is equal to 1 divided by the number of threads per inch. For example, a screw with 10 threads per inch has a pitch of 1/10 inch. Pitch for metric threads is given in millimeters.

- **Lead** is the distance a screw will move into a nut in one complete turn.

Forms of Threads

There are several different thread form systems in use today, including: American National, Unified, Acme screw, American Standard pipe, and metric.

American National

The *American National Thread System,* based on customary units, is the most common one in the United States (Figure 24-3). In this system, the sides are at an angle of 60 degrees, and the crest and root are flat. The two common series are as follows:

1. *National Coarse* (NC), which has threads in sizes from 1 to 12 and from 1/4 to 4 inches, is for general-purpose work.

2. *National Fine* (NF), which has threads in sizes from 0 to 12 and 1/4 to 1 1/4 inches, is used for precision assemblies like automotive and aircraft engines.

Taps are tools used to cut *internal* threads. Dies are used to cut *external* threads. Below 1/4 inch, taps and dies are marked by a gage number corresponding to the machine screw size (Table 24-1). For example, the next size smaller than 1/4-inch National Coarse is

Table 24-1 Thread Series

American National Fine (NF)			American National Coarse (NC)		
Size of tap	Threads per in.	Tap drill	Size of tap	Threads per in.	Tap drill
#4	48	43	#4	40	43
#5	44	37	#5	40	38
#6	40	33	#6	32	36
#8	36	29	#8	32	29
#10	32	21	#10	24	25
#12	28	14	#12	24	16
1/4	28	3	1/4	20	7
5/16	24	1	5/16	18	F
3/8	24	Q	3/8	16	5/16
7/16	20	25/64	7/16	14	U
1/2	20	29/64	1/2	13	27/64
9/16	18	33/64	9/16	12	31/64
5/8	18	37/64	5/8	11	17/32
3/4	16	11/16	3/4	10	21/32
7/8	14	13/16	7/8	9	49/64
1	14	15/16	1	8	7/8

Tap drill based on 75 percent full thread.

12-24. This does not mean 12/24 inch. It means that the tap or die is for a machine screw made from stock with a No. 12 gage diameter and that there are 24 threads per inch.

Taps and dies are specified as in the examples that follow:

EXAMPLE 1: 8-32 NC for a hole.

This means that you need a tap made for a No. 8 machine screw that has 32 threads per inch, which is the National Coarse series.

EXAMPLE 2: 1/4-28 NF for a rod.

This means that you must use a 1/4-inch die that will cut 28 threads per inch, which is the National Fine series.

Unified

In 1948, Great Britain, Canada, and the United States agreed on a thread form that is very similar to the American National System. It is called the *Unified System.* The two most common thread series are United Fine (UNF) and United Coarse (UNC) (Figure 24-4).

These series also have a 60-degree angle thread. The crest of the external thread may be flat or rounded. The root of the external thread is rounded either on purpose or as a result of a worn tool. The internal thread has a flat crest and rounded root. This makes it

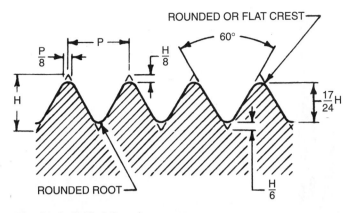

Fig. 24-4 Unified thread.

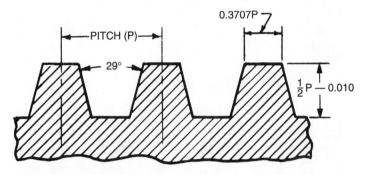

Fig. 24-5 Acme thread.

possible to interchange these series with the American National series.

Acme Screw

The *Acme screw thread* is a heavy-duty thread with a thread angle of 29 degrees. It is used for producing movement on machine tools and the like (Figure 24-5).

American Standard Pipe

American Standard pipe threads come in three series. The most common is the taper pipe thread (NPT), which is 3/4 inch per foot. You will find that a 1/8-inch pipe tap is about the same size as a 3/8-inch National Fine or National Coarse tap. This tap requires a letter *R* tap drill. Most electrical connections are made to fit this size.

Metric

Metric threads are designated much like the American National and Unified Systems. They are similar in appearance, and similar gages are used to check them. However, metric threads are not interchangeable with Unified threads, as their outside diameters and pitches are sized in millimeters. A typical designation for a metric thread is M5 x 0.8-C. In this example, M = the thread symbol for the International Organization for

Table 24-2 ISO Metric threads

Diameter	Pitch	Tap drill	Diameter	Pitch	Tap drill
M 1.6	0.35	1.25	M 20	2.5	17.50
M 2	0.40	1.60	M 24	3.0	21.00
M 2.5	0.45	2.05	M 30	3.5	26.50
M 3	0.50	2.50	M 36	4.0	32.00
M 3.5	0.60	2.90	M 42	4.5	37.50
M 4	0.70	3.30	M 48	5.0	43.00
M 5	0.80	4.20	M 56	5.5	50.50
M 6	1.00	5.00	M 64	6.0	58.00
M 8	1.25	6.75	M 72	6.0	66.00
M 10	1.50	8.50	M 80	6.0	74.00
M 12	1.75	10.25	M 90	6.0	84.00
M 14	2.00	12.00	M 100	6.0	94.00
M 16	2.00	14.00			

Standardization (ISO); 5 = the diameter in millimeters; 0.8 = the thread pitch in millimeters; and C = the class of fit, or tightness.

There are two classes of fit for metric threads: *general purpose* and *close fit*. The grade designation for general purpose is 6g for external threads and 6H for internal threads. These grades are comparable to Unified thread classes 2A and 2B, respectively. The external thread grade for close fit is 5g–6g. The internal thread grade for close fit is 6H, the same as for general purpose. If a thread symbol does not have any designation after the pitch, it is a general-purpose thread. For close fit, the capital letter *C* is used after the pitch, as in the example above.

The complete ISO metric thread standards include three pitch series: *coarse, fine,* and *constant*. Metric countries around the world use fifty-seven different combinations of pitches and diameters in the fine and coarse series. The United States, however, has decided to use only twenty-five diameter-pitch combinations. The diameters in these combinations range in size from 1.6 mm (about 1/16 inch) to 100 mm (about 4 inches). Spark plugs have always had ISO metric threads in the constant-pitch series.

Thread cutting is identical for both the customary and metric systems. Tap-drill sizes can be found on the ISO metric thread chart in Table 24-2.

Measuring the Diameter and Pitch of Threads

The diameter of a thread is measured with a *hold gage* or a *micrometer*. There are two common ways of measuring pitch. For customary threads, place a customary rule along the threads. Count the number of grooves in one inch (Figure 24-6). This will be the pitch.

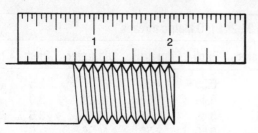

Fig. 24-6 A simple method for finding the number of threads per inch with a steel rule. Notice that there are eight threads per inch.

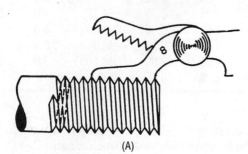

(A)

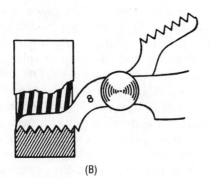

(B)

Fig. 24-7 The screw-pitch gage can be used for gaging both external (A) and internal (B) threads.

Cutting Threads

Threads are cut at the bench with taps and dies. A **tap** is a hardened steel tool with a threaded portion. It is used for cutting internal threads. This is called *tap-*

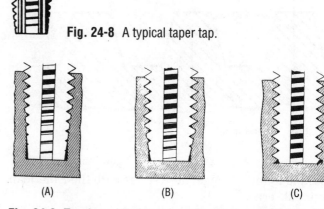

Fig. 24-8 A typical taper tap.

(A)   (B)   (C)

Fig. 24-9 Tapping a blind hole by using the taper (A), plug (B), and bottoming (C) taps.

ping. A **die** is a tool used to cut external threads. This is called *die cutting.* A **screw plate** is a set of taps and dies of the most common sizes.

Taps

A tap (Figure 24-8) is turned with a special wrench that is placed on a square at the end of the shank. Hand taps come in three styles: *taper, plug,* and *bottoming.*

When cutting threads completely through an open hole, only a taper tap is needed. When cutting threads to the bottom of a blind, or closed hole, the taper tap is used first, then the plug tap, and finally the bottoming tap (Figure 24-9).

On a customary tap, the diameter and the number of threads per inch are stamped on the shank. For example, if it is stamped 1/2-13 NC, it means that the thread is 1/2 inch in diameter, there are 13 threads per inch, and it is an American National Coarse thread. A metric tap shows diameter and pitch in millimeters, as in, for example, the legend M6X1.

For metric threads, place a metric rule along the threads and check the pitch in millimeters. This method is difficult, however. A simpler way to find the pitch for metric threads is to use a **screw-pitch gage** of the proper type. This gage consists of a group of thin blades. Each blade has saw-like teeth that equal standard threads cut along one edge. These blades are stamped with the pitch in millimeters. Try several blades against the screw thread until one fits exactly (Figure 24-7).

There are also customary screw-pitch gages stamped in threads per inch.

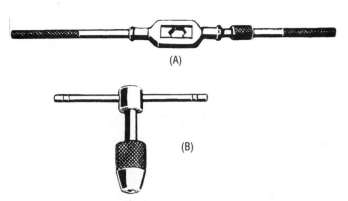

Fig. 24-10 The bar (A) and T-handle (B) tap wrenches.

Small taps are held in a T-handle tap wrench. Larger sizes are held in a bar tap wrench (Figure 24-10).

Cutting Internal Threads

Follow the guidelines below for cutting internal threads.

1. Determine the thread size and pitch needed. Select the correct tap drill. For customary-size threads, look at a tap-drill chart (Table 24-1). Suppose, for example, that you need a tap drill for an 8-32 NC tap. The table calls for a No. 29 drill. If number or letter drills are not available, use the closest fractional drill. In this case, it is 9/64. Another way of finding what customary-sized tap drill you need is to subtract the pitch of one thread from the tap's diameter. For example, to find the tap-drill size for a 3/4-16 NF, subtract 1/16 from 3/4 (or 12/16):

    $$12/16 - 1/16 = 11/16 \text{ inch}$$

    The tap drill cuts a hole slightly larger than the minor diameter of the thread. The tap-drill sizes given in Table 24-1 produce a thread 75 percent of full-thread depth. It is only 5 percent weaker than a 100 percent thread.

    To find a tap drill for a metric thread, simply subtract the pitch from the diameter. For example, for a tap labeled M6X1, the right drill would be 6 minus 1, or a 5-mm drill (Table 24-2).

2. Lay out the location for the hole and drill it carefully.

3. Choose the correct tap and fasten it in the tap wrench.

4. Clamp the workpiece firmly in a vise, with the hole in a vertical position, if possible.

5. Insert the tap in the hole. Grasp the tap wrench with one hand directly over the tap. Apply some pressure to it to turn it.

6. Check to see that the tap is at right angles to the workpiece. Once the tap is started, you need not apply pressure. It feeds itself into the hole (Figure 24-11).

Fig. 24-11 Cutting internal threads with a hand tap.

7. Apply a little cutting oil to the tap, and turn it clockwise. Every turn or so, reverse the direction a quarter turn or less to free the chips. Never force a tap—it can easily be broken. If it sticks, back it out a little.

8. Remove the tap by backing it out. Sometimes, it may stick. If it does, work it back and forth before removing it.

9. Check the threaded hole. Is it a good, clean-cut, sharp thread? Are the threads cut to the correct depth?

10. In cutting a thread in a blind hole, you must be especially careful as the tap nears the bottom. Remove the tap quite often to clean out the chips. If you fail to do this, the tap will not reach the bottom. After using a taper tap, follow with a plug tap and then a bottoming tap.

Threading Dies

The solid threading die is nonadjustable and is generally used to cut external threads on a rod. It is designed to produce the same type of running fit you would find on an ordinary nut and bolt. If you wish to change the fit of a rod thread to make it tighter or looser, an **adjustable die** must be used.

The *adjustable split die* has a special adjusting screw built into the side of the die. By tightening this screw, shown as *A* in Figure 24-12, the die will spread open slightly, will cut less deeply into the rod, and the fit

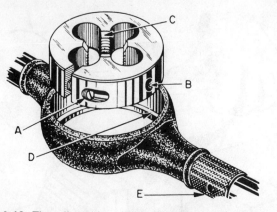

Fig. 24-12 The adjustable split die. The letters refer to the description in the text.

Fig. 24-13 Note that pipe taps produce tapered threads.

in the tapped hole will be tighter. The shallow hole at *B* is positioned opposite the adjustable handle, *E* of the diestock. When the handle is tightened, it holds the split die against the adjusting screw to maintain the setting while the die is cutting. The cutting teeth of the die are chamfered, or beveled, at *C* to help start the die squarely on the rod to be threaded. The die is placed in the diestock with the unchamfered teeth against the shoulder, *D.*

Cutting External Threads

Follow the guidelines below for cutting external threads.

1. Grind a slight bevel on the end of the rod or pipe to get the die started more easily.

2. Fasten the die in the diestock, with the side with the size to the top or away from the guide. The die has tapered teeth that should start the threading.

3. If necessary, put in a guide sleeve, or adjust the guide so that it just clears the workpiece.

4. Clamp the workpiece in a vise in either a vertical or a horizontal position.

5. Place the die over the end.

6. Cup your hand over the die, and turn to start it.

7. Apply some cutting oil when cutting steel.

8. Turn the die clockwise about two turns and then back about one turn.

9. After the thread is cut to the right length, remove the die.

10. Try the threaded rod or pipe in the threaded hole or nut. If the thread binds, close the die slightly by turning the setscrew. Recut the thread.

Pipe Threads

Pipe threads are used for joining pipes that carry liquids or gases. The thread most used is the American

Table 24-3 Pipe threads

Pipe diameters			Threads per inch	Tap-drill size
Nominal size	Actual inside	Actual outside		
1/8	0.270	0.405	27	11/32
1/4	0.364	0.540	18	7/16
3/8	0.494	0.675	18	19/32
1/2	0.623	0.840	14	23/32
3/4	0.824	1.050	14	15/16
1	1.048	1.315	11-1/2	1-5/32
1-1/4	1.380	1.660	11-1/2	1-1/2
1-1/2	1.610	1.900	11-1/2	1-23/32
2	2.067	2.375	11-1/2	2-3/16
2-1/2	2.468	2.875	8	2-5/8

Standard taper pipe thread (NPT). It has the same form as the American National thread, but it tapers by 3/4 inch per foot (Figure 24-13).

Pipe taps and dies are marked to correspond to pipe sizes, which are measured by their *inside* diameters (Table 24-3). For example, a 3/8-inch pipe tap is larger than 3/8 inch, and is roughly equal to a 5/8-inch UNC tap.

For most pipe work, only the external thread needs to be cut since the fittings are already threaded. Pipe is first cut to length with a hacksaw or pipe cutter. Then a tapered reamer is used to remove the burr on the inside of the pipe. Pipe dies are made so that one size can cut a range of pipe threads. For example, one size can be used to cut threads from 1/2 inch through 3/4 inch. On most die heads, the dies can be removed and reversed for cutting threads close to a wall.

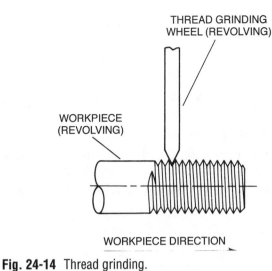

Fig. 24-14 Thread grinding.

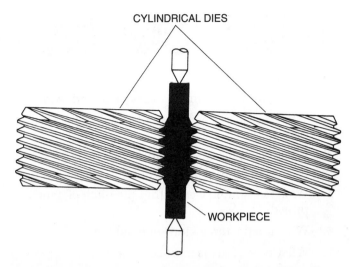

Fig. 24-15 Thread rolling between cylindrical dies. Three dies are used; one has been removed to make this illustration clearer.

Industrial Threading and Tapping

In industry, machines are used to produce screw threads on cylindrical surfaces. Threading is performed on external surfaces and tapping on internal.

Threading

Threading is done with machines by four general methods: cutting on a lathe, die chasing, grinding, and rolling. All can produce various external thread shapes.

Cutting on a Lathe

Cutting on a lathe is done by using a single-point cutting tool shaped to match the thread to be produced. (Refer to unit 57.) Automatic controls feed the tool along the turning workpiece at the proper rate to make a uniform spiral cut. Several cuts along the same path are needed to finish the thread.

Die Chasing

Die chasing is done using a hollow die with internal cutters, similar to that used for hand threading. The end of the workpiece is placed against the opening in the die head. Either the die or the workpiece is rotated as the other remains stationary. The process is the same as screwing a nut onto a bolt except that the die cuts threads as it or the workpiece turns. The die has a series of cutting edges. Each edge "chases" the preceding edge along the same path so that the thread is cut deeper.

Grinding

The grinding of threads is done with a grinding wheel shaped to match the thread contour desired on the workpiece. Otherwise, it is like surface and cylindrical grinding. Thread grinding can be done by using table or wheel motion to produce a shaped path along the surface of the workpiece (Figure 24-14).

Rolling

Bolts and screws made by cold heading and extrusion are usually threaded by rolling. The rolling of threads is entirely a forming process. No metal is cut from the piece.

Thread rolling is most often performed by rolling the workpiece between two flat dies. The faces of the two dies are cut with the pattern of the thread to be formed. The space between them is less than the diameter of the unthreaded bolt or screw. One die does not move while the other one moves over it. As the fastener is rolled, metal is forced into the grooves of the dies to form the threads.

Threads are also rolled by using three rotating, cylindrical dies with threads cut in their surfaces (Figure 24-15). They are mounted so that each almost touches the other two, a triangular arrangement similar to a three-roll bender. The fastener is inserted into the space between the three dies. The rotating dies form the threads.

Tapping

Tapping is the most common method of machining internal threads. It is done with a cylindrical tap that has cutting edges shaped like threads around its surface. Either the tap or the workpiece is rotated as the tap is fed into the hole or into the end of a tubular part. Machines used for tapping include drill presses, lathes, and special machines similar to drill presses.

KNOWLEDGE REVIEW

1. What is a thread?

2. Define the screw-thread terms *pitch* and *lead*.

3. What thread system is usually used in the United States?

4. What is the measuring unit for the diameters and pitches of metric threads?

5. What is a simple way to find the pitch of a metric thread?

6. What are the three kinds of hand taps?

7. What type of die is most common?

8. Should you apply pressure when starting a tap? After starting a tap?

9. To get a die started easily, what should you do to the end of a rod or pipe?

10. What is the most commonly used kind of pipe thread?

11. Name four methods of industrial threading.

12. Write a report on the different uses for these threads: National Fine, Acme, and NPT.

13. Write a report on industrial methods of threading.

Unit 25

Assembling with Metal Fasteners

OBJECTIVES

After studying this unit, you should be able to:

- List the uses of rivets.
- Name four kinds of rivet heads.
- Describe the device used to protect a round-head rivet.
- Describe the proper length of a shank of a rivet.
- Describe a bolt.
- Name six kinds of threaded fasteners.
- Identify the purpose of nuts and washers.
- Name four kinds of nuts.
- Describe two kinds of screwdrivers.
- Name five kinds of pliers.
- List six kinds of wrenches.
- Tell how metric wrenches are marked for size.
- Describe six hints for using assembly tools.
- Describe how to assemble a product with screws.

KEY TERMS

adjustable wrench	diagonal cutter	pipe wrench	slip-joint pliers
bolt	groove-joint pliers	rivet	socket wrench
box wrench	long-nose pliers	rivet set	stove bolt
capscrew	machine bolt	setscrew	stud
carriage bolt	machine screw	side cutter	thread-forming screw
combination wrench	open-end wrench		

There are many fasteners you can use to assemble metal projects. Rivets are used to put a project together permanently. Threaded fasteners such as bolts or screws are used when the project may need to be taken apart or adjusted.

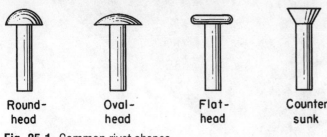

Fig. 25-1 Common rivet shapes.

Table 25-1 Common rivet sizes (shank diameter)

Customary (in.)	Nearest ISO metric equivalent (mm)
1/16	1.6
1/8	3
5/32	4
3/16	5
1/4	6
5/16	8
3/8	10

Rivets and Riveting Tools

A **rivet** is a permanent fastener with a head and a shank. Rivets for assembling mild-steel or wrought-metal projects are made of soft steel, sometimes called *soft iron.* They are available with flat, round, oval, or countersunk heads, in diameters from 1/16 to 3/8 inch, and in lengths from 1/14 to 3 inches (Figure 25-1, Table 25-1). The most common are the 1/8- and 3/16-inch round-head or flat-head 1-inch rivets. If the project is made of copper, brass, or aluminum, choose nonferrous rivets of the same metal. The most common size for small art metal projects is the 1/8-inch round-head rivet.

A **rivet set** or a *riveting plate* protects the round-head rivet as it is set. Rivet sets come in various sizes with cone-shaped holes that fit the rivet heads tightly.

Assembling with Rivets

Use the following procedure and guidelines when assembling with rivets.

1. Select the rivets. If the rivet is to be part of the design, a round-head may be chosen. A flat-head should be selected if the assembly is not meant to be noticed. If there are to be round-heads on both sides of the project, the rivets must be long enough to go through both pieces and extend beyond by one and one-half times the length of the diameter (Figure 25-2). If the back is to be flush, the rivet must extend by only a small amount, just enough to fill the countersunk holes.

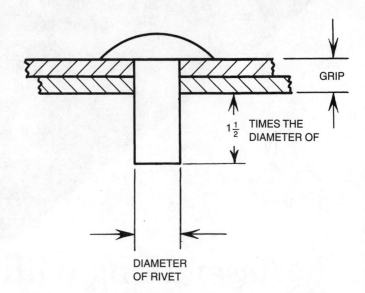

Fig. 25-2 A rivet that extends through both workpieces. About one and one-half times the diameter of the rivet is needed when forming a round-head.

2. Lay out the location of the rivet and drill the hole.

3. Countersink, if necessary. On most projects, at least the back will be flush; but when countersunk rivets are used, both sides must be countersunk.

4. Check the rivet for length. Cut off any excess with a saw or with bolt clippers.

5. Insert the rivet in the hole. Be sure that the two pieces are pressed firmly together with no burr separating them. If the rivets are round-head, protect the head by holding it in a rivet set or riveting plate.

6. To round off the shank, strike it first with the flat of the hammer to fill up the hole. Then round it off with glancing blows. If the back is to be flat, strike the rivet with the peen of the hammer to fill in the countersunk hole. Then finish by striking it with the flat of the hammer.

7. To rivet together curved parts or scrolls that cannot be held against a flat surface, proceed as follows:

   a. Cut off a piece of scrap rod of a size that will slip into the curve under the rivet hole.

   b. Drill a cone-shaped hole about the size of the rivet head toward the center of the rod.

   c. Place the rivet head in the hole in the rod, and do the riveting as before (Figure 25-3). It may be necessary to open the scroll slightly to complete the riveting.

8. Check the riveting job. Is the head damaged? Did you bend your project in riveting it? Is the second

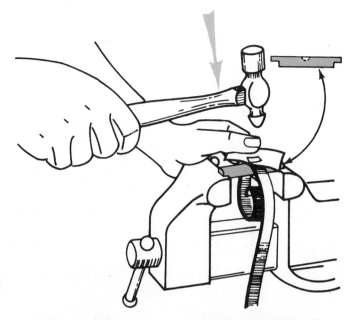

Fig. 25-3 Riveting a base to a scroll. A cone-shaped hole has been drilled in the small metal rod that supports the rivet.

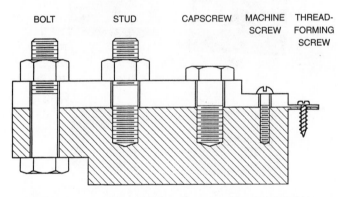

Fig. 25-4 Types of threaded fasteners used in metalworking operations.

head misshapen? If the back is flush, is the surface rough? Are the two parts too loosely fastened together?

Threaded Fasteners

Threaded fasteners include all those cylindrical shapes whose surfaces have threads cut or rolled into them. The basic forms include: bolts, studs, capscrews, machine screws, and thread-forming screws (Figure 25-4).

While the terms are often used interchangeably, there is a difference between a screw and a bolt. A *screw* is generally fitted into prethreaded holes, and holds parts without the use of a fitted nut. A *bolt* fits through predrilled holes in parts to be joined, and requires a matching nut to secure it. Descriptions of the common types of these fasteners are given below.

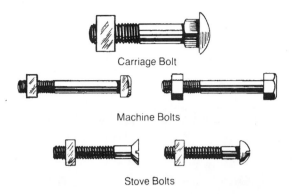

Fig. 25-5 Common types of bolts.

Thread-forming screws are small, fully threaded fasteners that cut their own threads in the parts to be joined. An example of this type is the sheet metal screw. **Machine screws** are fully threaded, made of brass or steel, range in gage sizes from 4 to 12, and fit into tapped holes for small-part assembly. They are occasionally used with washers and nuts. **Capscrews** are larger and tougher than machine screws, and also fit into tapped holes. They are partially threaded and their sizes range up to one inch in diameter and six inches in length. **Studs** are tough steel fasteners which are force-fitted into tapped holes for permanent assembly. The protruding threaded portion of the stud accepts a part to be secured with a nut. The engine block of an automobile uses studs. **Bolts,** as stated earlier, require nuts to complete part assemblies. The common forms are shown in Figure 25-5.

Types of Bolts

Carriage bolts are used to assemble wooden parts such as workbenches and farm equipment. Note that they have a square section below the head which is drawn into a wooden part to secure it. Their diameters range from 1/4 to 3/4 inch, with lengths from one to six inches. **Machine bolts** are partially threaded, with a size range similar to the carriage bolt. They are used for heavier assembly operations. **Stove bolts** are fully threaded and used to assemble cabinets and machinery. They come in diameters of 1/8 to 5/16 inch, and lengths of 3/8 to 4 inches.

Head Shapes and Styles

A variety of common and special threaded fastener *head shapes* is shown in Figure 25-6. The square and hexagon bolt heads are designed for heavy tightening with wrenches. The slotted heads are tightened with a standard screwdriver, but with less force than a wrench. Too much force will strip a slotted head.

The recessed *head styles* shown in Figure 25-7 permit greater tightening power than the slotted heads.

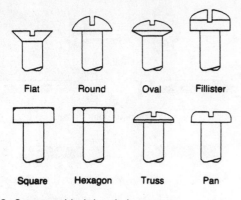

Fig. 25-6 Screw and bolt head shapes.

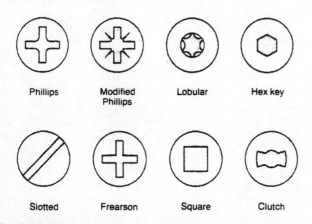

Fig. 25-7 Different styles of recessed screw and bolt heads. Each of these requires a special driving tool.

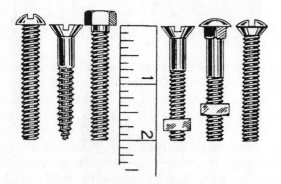

Fig. 25-8 Measuring the lengths of bolts and screws.

These recessed heads also lend themselves to the use of automatic driving tools, especially the Phillips and modified Phillips which are self-centering. Refer to Figure 25-8 to learn how to measure the lengths of threaded fasteners.

Other Threaded Fasteners and Assembly Hardware

Setscrews are made in a range of head and point shapes and are used to fasten pulleys and collars to

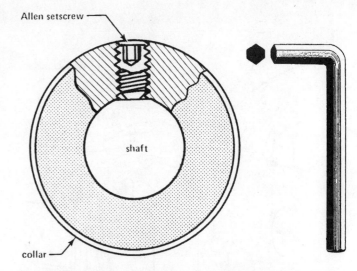

Fig. 25-9 A typical setscrew installation. The hex key or Allen wrench is used to tighten the setscrew.

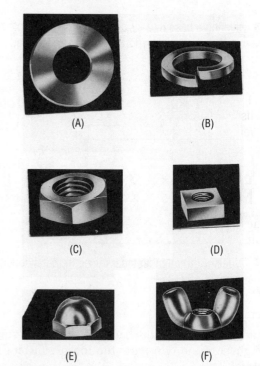

Fig. 25-10 Common washers and nuts: (A) plain or flat washer, (B) lock washer, (C) hexagon nut, (D) square nut, (E) cap nut, and (F) wing nut.

shafts. They come in a variety of sizes according to need. A typical setscrew installation is shown in Figure 25-9.

Many other threaded fasteners are also available. *Nuts* and *washers* are often used with screws and bolts in fastening operations. *Cap nuts* enclose the thread for protection and appearance. *Square* and *hexagon nuts* are used for general holding jobs. *Wing nuts* are finger-tightened and are used where quick and easy assembly is needed. *Flat washers* prevent nuts from becoming loose. All these devices come in a wide range of kinds, materials, and sizes (Figure 25-10).

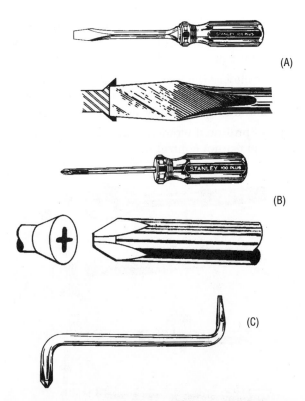

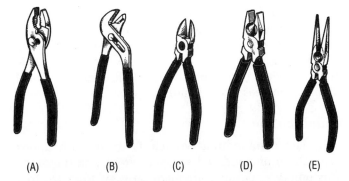

Fig. 25-12 Common types of pliers: (A) slip-joint, (B) groove-joint, (C) diagonal cutter, (D) side cutter, and (E) long-nose. (Courtesy of Stanley Tools, Division of the Stanley Works, New Britain, CT)

Fig. 25-11 Screwdrivers: (A) slotted head, (B) Phillips head, and (C) offset. (Courtesy of Stanley Tools, Division of the Stanley Works, New Britain, CT)

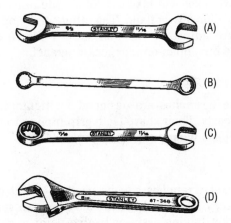

Fig. 25-13 Common types of wrenches: (A) open-end, (B) box, (C) combination, and (D) adjustable. (Courtesy of Stanley Tools, Division of the Stanley Works, New Britain, CT)

Tools for Assembling with Bolts, Nuts, and Screws

The most common tools used for assembling parts with bolts, nuts, or screws are screwdrivers, pliers, and wrenches.

Screwdrivers

Screwdrivers are made in many diameters and lengths. The larger sizes have a square shank. This makes it possible to get extra leverage by using a wrench with the screwdriver. The blade is made for either slotted heads or for recessed (Phillips) heads (Figure 25-11).

In choosing a screwdriver, make sure that the blade fits the slot in the screw to be driven. The *offset screwdriver* is used to reach screws in out-of-the-way places.

Pliers

Pliers are used to hold and grip small articles and to make adjustments (Figure 25-12). They are not substitutes for wrenches, however, and should never be used as such. There are five common types of pliers. **Slip-joint** and **groove-joint** pliers are used for holding round bars and pipe and for doing general work. **Side cutters** are used for cutting and gripping wire to do electrical work. **Long-nose,** or *chain,* pliers are used for get-

ting into out-of-the-way places and for holding small parts. And **diagonal cutters** are used for cutting wire.

Wrenches

A variety of wrenches are used in metalworking (Figure 25-13). These include: open-end, box, combination, adjustable, pipe, and socket wrenches.

Open-End

Open-end wrenches are made with an opening in one or both ends. Customary-sized ones are available with openings from 5/16 to 1 inch. The head and opening are set at an angle, usually 15 or 22 1/2 degrees, to the body. This is so the nut can be tightened in small areas. The smaller the openings, the shorter the lengths.

Box

Box wrenches are very popular because they can fit into small places for tightening nuts. The modern box wrench is a twelve-point wrench that has twelve notches around the circle for gripping the nut. The nut can be moved a short distance and then the wrench changed to another position. The box wrench is made with a straight end or with the end offset at an angle of 15 or 45 degrees to the handle.

Combination

The **combination wrench** is made with a box wrench on one end and an open wrench on the other. Customary-sized ones are made in sizes from 1/4 to 1 1/4 inches.

Adjustable

Adjustable wrenches are also very popular, since one or two sizes can take care of a wide range of bolts and nuts. One jaw is fixed. The other can be opened or closed. An adjustable wrench should not be used in place of a box, open-end, or socket wrench.

Pipe

Pipe wrenches are designed for tightening pipe and rod parts. Their sharp teeth grip pipe or rod to prevent slippage when turning (Figure 25-14).

Socket

Socket wrenches can be used to get at hard-to-reach places. Most socket wrenches come in sets, with each socket having a six-or twelve-point grip (Figure 25-15). The set includes an extension bar, a ratchet handle, and other components.

Metric Wrenches

All styles of fixed-opening wrenches are available in metric sizes. The size in millimeters is stamped on the wrench. This size is the width of the bolt head across the flats, *not* the diameter of the bolt's threaded part. For example, an M16 bolt requires a 24-mm wrench (Table 25-2).

Hints for Using Assembly Tools

The following guidelines for using tools will help you with your assembling operations.

- Never use a screwdriver as a punch or chisel.

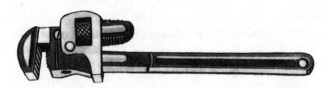

Fig. 25-14 A typical pipe wrench.

- Never use a pliers with a round-shank screwdriver. Some square-shank screwdrivers are designed for use with a wrench.
- Always choose the wrench that fits the nut or bolt. A loose-fitting wrench will round off the corners of the head. It may also cause serious injury if it slips.
- Always pull on a wrench. Never push it, because then you cannot control it.

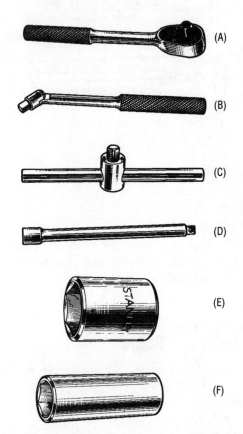

Fig. 25-15 Socket wrench tools: (A) ratchet handle, (B) flex handle, (C) sliding T-handle, (D) extension bar, (E) standard socket, and (F) deep socket. (Courtesy of Stanley Tools, Division of the Stanley Works, New Britain, CT)

Table 25-2 Metric bolts and nuts and matching metric wrench sizes

Bolt and nut	Wrench size (mm)
M6	10
M8	13
M10	17
M12	19
M14	22
M16	24
M20	30
M22	32
M24	36

- Never use a piece of pipe for added leverage. It is better to get a heavier wrench.
- Never strike a wrench with a hammer. Sometimes, it may be absolutely necessary to tap it if a nut or bolt sticks. In these cases, be sure to use a soft-faced hammer.

Assembling with Screws

Use the following procedure and guidelines when assembling with screws.

1. Select the machine screw and a nut, if needed. The screw must be long enough to fasten the two pieces. If the clearance hole is drilled through, the screw should extend beyond the second piece so that a nut can be fastened to it. Usually, however, a clearance hole is drilled in the first piece, and the second hole is tapped.

2. Lay out the location of the hole and drill the clearance hole in the first piece. This hole should be the same size as the outside diameter of the screw threads. If a flat-head screw is used, countersink the hole.

3. Drill and tap the hole in the second piece (see unit 24). Sometimes, it is a good idea to clamp the pieces together and drill them with a tap drill. The first hole can then be enlarged to the clearance-hole size.

4. Choose the right size screwdriver so that you will not mar the screw head or the workpiece. Thread the screw into the tapped hole or nut.

KNOWLEDGE REVIEW

1. What different shapes can rivet heads have?

2. How are screws and bolts measured?

3. What is a machine screw used for? A setscrew?

4. How can you tell the difference between a metric hexagon bolt and a customary one?

5. What is a box wrench used for? A pipe wrench?

6. How can you tell the size of a metric wrench?

7. What will a loose-fitting wrench do to a bolt?

8. Make a sample display board showing some of the common metal fasteners. You can obtain samples from your school shop or from your hardware dealer.

Unit 26

Introduction to Sheet Metal

OBJECTIVES

After studying this unit, you should be able to:

- Describe the industries that employ many sheet metal workers.
- List some of the hand tools used by sheet metal workers.
- Name four kinds of metals used in sheet metal products.

KEY TERMS

sheet metal operations

We use hundreds of sheet metal products in our lives. Motorcycle and automobile bodies, tools and equipment, modern buildings, and household appliances are but a few examples (Figure 26-1).

Historically, people have made a variety of sheet metal items, such as the interesting body armor seen in Figure 26-2, which is similar to that worn by knights and warriors of times past. These pieces were made by hammer-beating metal sheets into wooden or sandbag forms to fit the bodies of their wearers. Hand skills such as cutting, forming, riveting, and forge welding were commonly used.

Today, thousands of workers are employed in sheet metal and metal-related industries using more modern tools and equipment, and requiring different skills. Construction workers prepare and install metal roofing, siding, panels, storefronts, and heating and ventilating ductwork in large buildings such as the one in Figure 26-3.

In the automotive industries, highly-trained assemblers, welders, and polishers build truck cabs and farm equipment (Figure 26-4). In the following units on sheet metal you will learn some of these basic sheet metal skills and the related equipment.

172

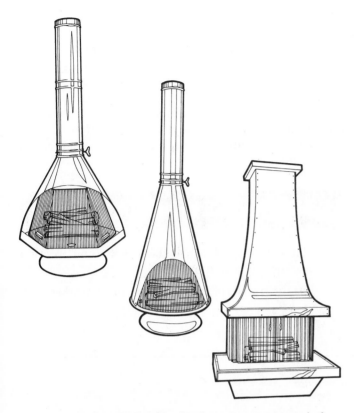

Fig. 26-1 These free-standing fireplaces are constructed of sheet metal.

Fig. 26-2 Jörg T. Sorg the Younger (possibly), German, c.1522–1603, Elements of armor for field and tournament from a garniture, etched, gilded and blackened steel; iron; brass; leather; cord, c.1570–80, ht.: 185.4 cm, George F. Harding Collection, 1982.2411a–r. (Photograph © 1996, The Art Institute of Chicago, all rights reserved)

Fig. 26-3 Attractive, durable anodized aluminum panels cover this modern building. (Courtesy of Aluminum Company of America [ALCOA])

Fig. 26-4 This truck cab is made of tough, lightweight aluminum sheets. (Courtesy of Aluminum Company of America [ALCOA])

Tools

To work with sheet metal, you must learn to use a number of hand tools and machines. You must also learn tool and machine safety.

In this part of the book you will be learning to use hand tools such as hand snips and punches; layout tools; special equipment for bending, cutting, and seaming; and equipment for soldering and welding to make joints. The skills you learn will also help you at home. You can use them, for example, for the many small repair jobs to be done on home appliances, fixtures, and automobiles.

Materials

Sheet metal ranging from a few thousandths of an inch to 1/8 inch in thickness is made from both ferrous and nonferrous metals. In the school shop, most of the metals you use will be between 20 and 25 gage. The most common sheet metals are: brass, aluminum, mild steel, and copper. Some of these will be embossed or have etched patterns in them. Some sheet steels may have coatings of zinc (galvanized sheet) or black iron oxide.

Sheet metal work calls for extra caution to prevent accidents. Each piece of sheet metal as it is being cut and handled is a potentially dangerous cutting tool. The razor-sharp burrs that appear on metal as it is cut can easily slice into your finger and produce a painful scratch or cut. Be very careful in working with sheet metal. Remove dangerous burrs or snags with a hand file.

Operations

Some of the more important **sheet metal operations** you will be learning about in this unit are shown in Table 26-1. These operations must be mastered in order to produce satisfactory work. They are widely used in manufacturing. You can also use them to make a number of interesting projects, such as mailboxes, storage cabinets, toolboxes, bookends, and containers of many kinds.

KNOWLEDGE REVIEW

1. List some of the things sheet metal construction workers make and install.
2. List three kinds of tools that sheet metal workers use.

Table 26-1   Sheet metal operations

Cutting operations	Purpose	Tools used
shearing	cut to size, separate metal	tin snips, hand and power shears, squaring shears, notcher, circle shears
punching	make holes	hollow punch, solid punch

Forming operations	Purpose	Tools used
bending	create angular, cylindrical, and conical shapes	forming stakes, hand seamer, bar folder, cornice brake, box-and-pan brake, press brake, forming rolls
turning	produce functional shapes on circular pieces	rotary machine with beading, crimping, wiring, burring, and turning rolls

Fastening operations	Purpose	Tools used
riveting	produce a permanent joint	rivet set, pop-rivet tool
threaded fasteners	produce a semipermanent joint	sheet-metal screws, self-tapping screws
seaming	produce a permanent mechanical joint, such as grooved, standing, folded, or double seams	hand groover, forming stakes, rotary machine
soft soldering	produce a permanent watertight joint	soldering copper, torch, gun, furnace, Bunsen burner
gluing	produce a permanent watertight joint	contact cement, epoxy cement, clamps

3. Name four metals commonly used in sheet metal.

4. Name three main groups of sheet metal operations.

5. Find out all you can about the sheet metal worker and give a report to the class. Use the *Occupational Outlook Handbook* (U.S. Government Printing Office, Washington, D.C.) and other sources to get this information.

Unit 27

Developing Patterns

OBJECTIVES

After studying this unit, you should be able to:

- Describe a pattern or stretchout.
- Identify the use of a hem.
- Name two kinds of hems.
- Describe a wired edge.
- Identify some uses for seams.
- Name three kinds of patterns.
- Describe how to make a layout for a box.
- Identify the purposes of notches.
- Describe the shape of a development for a cylinder.

KEY TERMS

hem	pattern	seam	triangulation
parallel-line	radial-line	stretchout	wired edge

Before a sheet metal project can be made, a pattern must be developed. A **pattern,** or **stretchout,** is a flat shape laid out on a piece of sheet metal prior to cutting. If patterning is done correctly, the metal can then be formed into a three-dimensional object (Figure 27-1).

A stretchout can be drawn directly on the metal. Most often, however, the pattern is first developed on paper, then transferred to the metal. This can be done by placing the paper pattern over the metal, with a piece of carbon paper between them, and tracing. However, in sheet metal, the usual practice is to place the pattern over the metal and prick-punch around the outline. If many pieces of the same kind are to be made, a *template,* or heavy pattern, is made of metal. Then this can be held firmly in place and you can trace around it.

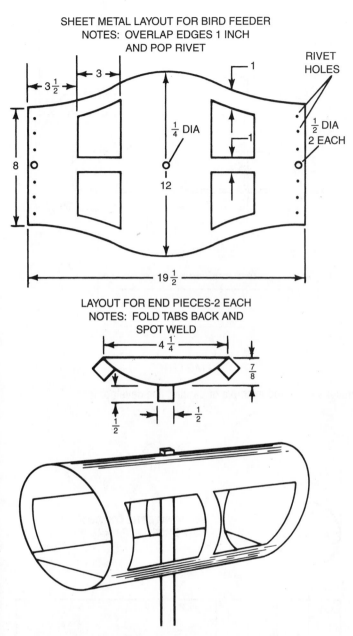

SHEET METAL LAYOUT FOR BIRD FEEDER
NOTES: OVERLAP EDGES 1 INCH
AND POP RIVET

RIVET HOLES

½ DIA
2 EACH

¼ DIA

19½

LAYOUT FOR END PIECES-2 EACH
NOTES: FOLD TABS BACK AND
SPOT WELD

4¼

⅞

½

½

Fig. 27-1 Typical stretchout for a bird feeder. The parallel-line method for developing this is described in this unit.

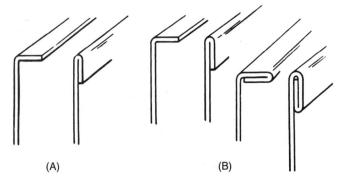

Fig. 27-2 (A) Single hem. (B) Double hem.

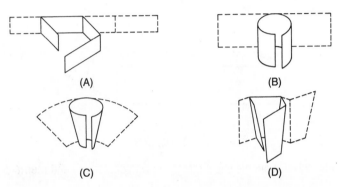

Fig. 27-3 Three kinds of pattern developments: (A and B) parallel-line development to make a box or a cylinder, (C) radial-line development to make a cone, and (D) triangulation to make a two-transition piece.

Hems, Edges, and Seams

In making a pattern, you must also make allowances for hems, wired edges, and seams. A **hem** is a folded edge used to improve the appearance of the work and to strengthen it. A *single hem* may be any width. In general, the heavier the metal, the wider the hem. A *double hem* is used when additional strength is needed (Figure 27-2). Hems are made in standard fractions such as 1/4 inch, 3/8 inch, 1/2 inch, and so on.

A **wired edge** is a sheet metal edge folded around a piece of wire. The edges of such items as trays, funnels, and pails are often wired for added strength. The amount allowed for a wired edge is two and one-half times the diameter of the wire itself. For example, if the wire has a diameter of 1/8 inch, then 5/16 inch is allowed for the edge.

Seams are mechanical joints between pieces of sheet metal. The common kinds of *lap seams* are plain, offset or countersunk, and inside or outside corner (see unit 32). Lap seams are usually finished by riveting, soldering, or a combination of both. *Grooved seams* are used most often in joining two parts of a cylindrical object. There are three thicknesses of metal for the seam above the place where the two pieces join. For example, on a 1/4-inch seam, an allowance of 3/4 inch of extra metal must be made. On the layout, half this amount, or 3/8 inch, is added to either side or end.

Kinds of Patterns

Three kinds of pattern developments are used in sheet metal work (Figure 27-3).

1. **Parallel-line,** for making rectangular objects, such as boxes and trays, and cylindrical objects, such as funnels, tubes, pipes, scoops, and the like.

2. **Radial-line,** for making cone-shaped objects, such as funnels, buckets, or tapered lampshades.

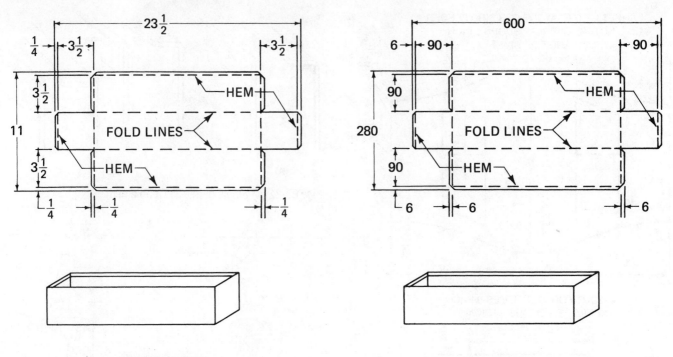

CUSTOMARY   METRIC

Fig. 27-4 A simple sheet metal box with customary and metric dimensions is a good example of parallel-line development.

3. **Triangulation,** for making transition pieces, such as the pipes that join square and round shapes. These patterns are commonly used in making fittings for air-conditioning and heating systems.

Making a Simple Layout for a Box

To make a layout for a sheet metal box, follow these steps:

1. Draw the square or rectangle equal to the size of the box bottom (Figure 27-4).

2. Draw the side and ends so that they are of equal height.

3. Add material for the corner lap seam.

4. Add material for the hem.

The seams may be joined by soldering, spot welding, or riveting.

Laying Out Notches

Notches are laid out on a pattern to help in forming, assembling, and completing the sheet metal object. A slant notch is cut at 45 degrees across the corner when you use a single hem. A V notch is often used at the corner of a box.

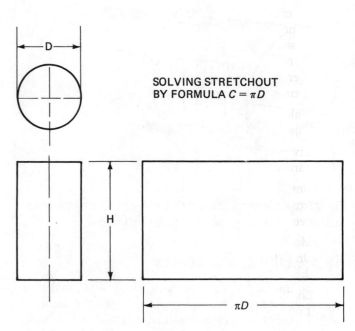

SOLVING STRETCHOUT
BY FORMULA $C = \pi D$

Fig. 27-5 Developing a cylinder by finding the circumference.

Parallel-Line Development of a Cylinder

The stretchout of a simple can-shaped object is rectangular (Figure 27-5). One dimension of the rectangle will equal the height of the cylinder, or can. The other dimension will be the circumference of the cylin-

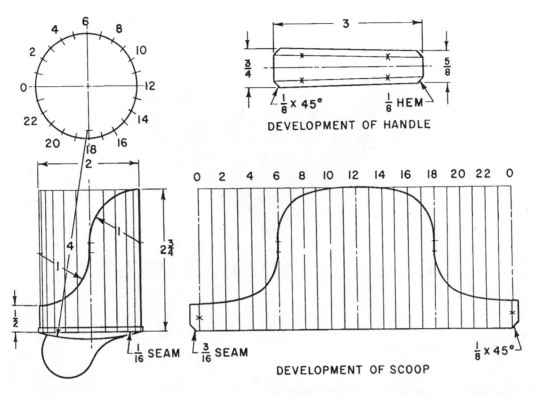

Fig. 27-6 The development for a scoop.

der. The circumference rule can be used to find the circumference of any cylinder for which you know the diameter (see unit 9).

To make the pattern for the layout of a scoop or any other object that has one end shaped, follow these directions:

1. Make a top-view and a front-view drawing of the object (Figure 27-6).

2. Divide the top into several equal parts. The more parts, the more accurate the stretchout.

3. Draw light vertical lines from the top view to the front view. Number the lines. (For clarity, only even numbers are shown in the figure.)

4. Make a stretchout to the right of the front view. Draw light, vertical, equally spaced lines. Number them in the same way as those on the top view.

5. Locate line 1 on the front view and project it to line 1 on the stretchout. Do this for all the other lines.

6. Join the points with a French curve.

7. Add a certain amount on the ends for a grooved seam.

KNOWLEDGE REVIEW

1. How is a pattern on paper usually transferred to sheet metal?

2. What is a hem? A seam?

3. How much extra metal is needed for a wired edge?

4. What type of pattern development is used for a box? A scoop?

5. What is triangulation used for?

6. Develop a pattern for a tool box.

Unit
28

Cutting Sheet Metal

OBJECTIVES

After studying this unit, you should be able to:

- Tell what cutting sheet metal is called.
- Name the tool used to cut large sheets into smaller pieces.
- Describe punching.
- Name five kinds of hand snips.
- Explain how to do straight cutting with snips.
- Describe how to cut outside curves.
- Explain notching.
- Name the tool used to punch large holes.
- Describe the use of electric shears.
- Explain how to cut sheet metal on squaring shears.
- Describe the use of a tab notcher.
- Describe the machine for cutting rings.

KEY TERMS

aviation snips	punching	shearing	straight snips
duck bill snips	ring-and-circle shears	squaring shears	tab notchers
hollow punches			

The sheet metal you use in the school shop usually requires some kind of cutting. Cutting metal sheets is called **shearing.** *Squaring shears* are used to cut large sheets into pieces that can be more easily handled. *Hand snips* and *shears* are used to cut the workpiece to the size and shape of the pattern. In the sheet metal industry, large machines do this shearing (Figure 28-1).

Another kind of sheet metal cutting is called **punching.** This is making holes in the metal. These holes are usually round, but other shapes can be cut with special dies.

In this unit, you will learn how to cut sheet metal with shears and punches.

Fig. 28-1 Industrial shearing using a power shear. (Courtesy of Clearing Niagara)

Fig. 28-3 Cutting with the straight snips. Cut at the right side of the material whenever possible. (Cut at the left side if you are left-handed.)

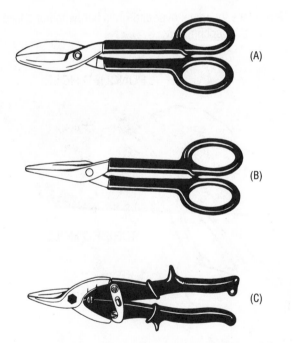

Fig. 28-2 Common sheet metal cutting snips: (A) straight, (B) duck bill, and (C) aviation. (Courtesy of CHANNELLOCK, Inc.)

Hand Snips and Shears

There are many types of hand snips and shears used in sheet metal work. Three are shown in Figure 28-2. **Straight snips** are used for cutting straight lines in sheet metal that is 22 gage or thinner. They are made in different sizes with cutting jaws from 2 to 4 1/2 inches long. Straight snips are also used to make outside cuts on large-diameter circles. **Duck bill snips** are used for in-

side cutting of intricate work. These snips have narrow curved blades that allow you to make sharp turns without bending the metal. **Aviation snips** can be used for all kinds of cutting. They are made with left, right, or universal cutting blades.

Making Straight Cuts

To make a straight cut in sheet metal, first select the straight snips or the aviation snips. Then follow the procedure below.

1. Hold the sheet metal in your left hand and the snips in your right hand. Be careful of sharp edges.

2. Open the snips as far as possible and insert the metal. Hold the straight side of the blade at right angles to the sheet.

3. Squeeze the handle firmly and cut to about 1/4 to 1/2 inch from the point of the blade. Reopen the snips and complete the cut. The edge should be even and clean. Avoid jagged, slivered cuts. Whenever possible, cut to the right of the layout line (Figure 28-3).

Cutting Outside Curves

Aviation snips may be used for cutting an outside curve in sheet metal. Follow the steps below.

1. Hold the metal in your left hand and rough-cut to within about 1/8 to 1/4 inch of the layout line.

2. Carefully cut up to the layout line and around it, making a continuous cut. The scrap metal will tend to curl out of the way during the cutting (Figure 28-4).

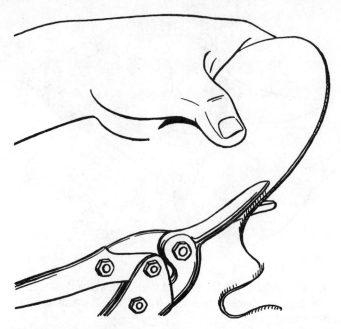

Fig. 28-4 Cutting an outside curve with aviation snips. Notice how the thin edge of metal curls away.

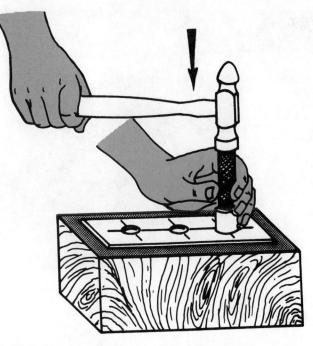

Fig. 28-5 Place the metal over end-grain hardwood or a lead block. Strike the punch firmly.

Notching

To cut notches or tabs in sheet metal, open the snips only part way, and use the portion near the point of the blade for cutting. This will prevent any cutting past the layout line.

Punches

Hollow punches are used to cut holes 1/4 to 3 inches in diameter, in sizes varying by 1/8 inch. You must punch or drill a hole before cutting an internal opening in sheet metal.

Punching Large Holes

To punch a hole in sheet metal, follow the procedure below.

1. Locate the hole on the sheet metal and draw its exact size with dividers.

2. Choose a punch of the correct size.

3. Place the metal over end-grain hardwood or a soft lead block (Figure 28-5).

4. Hold the punch firmly over the layout and strike it solidly with a heavy ball-peen hammer. Try to punch the hole with one or two blows. Never tap lightly, as this makes ragged edges.

Cutting Inside Curves

Duck bill snips may be used to cut an inside curve in sheet metal following the punching operation. The steps are listed below.

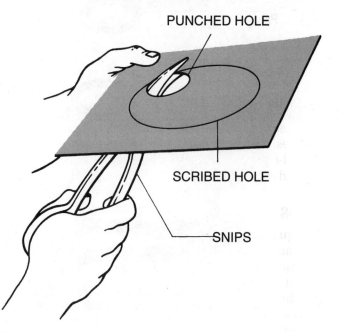

PUNCHED HOLE

SCRIBED HOLE

SNIPS

Fig. 28-6 The punched hole permits you to start the cut.

1. Punch or drill a hole in the waste stock. You can also make this opening by cutting two slits at right angles with a cold chisel and a hammer. This hole should be large enough to allow duck bill snips to get started (Figure 28-6).

2. Insert the snips from the underside. Rough out the inside opening to about 1/4 inch of the layout line.

3. Trim the hole to size.

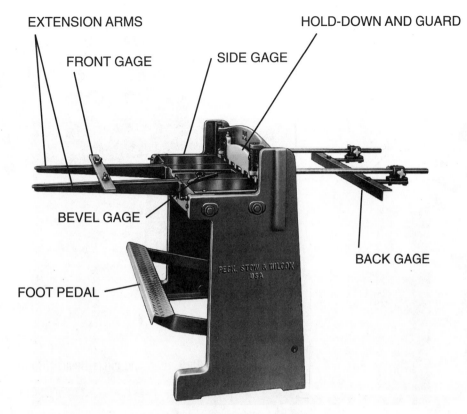

EXTENSION ARMS

FRONT GAGE

SIDE GAGE

HOLD-DOWN AND GUARD

BEVEL GAGE

BACK GAGE

FOOT PEDAL

Fig. 28-7 Parts of the squaring shears. (Courtesy of Roper Whitney Company)

Machine Cutters

Sheet metal can be cut with a number of simple machines, including squaring shears, tab notchers, and ring-and-circle shears. These machines can handle both large and small cutting jobs.

Cutting with Squaring Shears

Squaring shears, operated by foot, are used to square and trim large pieces of sheet metal (Figure 28-7). The size of the machine is determined by the width of sheets it will cut. The common sizes are thirty or thirty-six inches. A back gage controls the length of cut when the metal is inserted from the front. A front gage does the same for metal inserted from the back. The side gage is adjustable and is kept at right angles to the cutting blade. Sheet metal 18 gage or lighter can usually be cut on the squaring shears. Note the following guidelines:

1. To cut long sheets, insert them from the back. To cut several pieces to the same length, set the front gage to this length. Use the graduated scale on the top of the bed for this purpose.

2. The left edge of the sheet should be firmly pressed against the left gage and the end of the sheet against the front gage.

3. Hold the sheet down on the bed with both hands, then press the foot pedals with your foot.

 Caution: Keep your fingers away from the cutting blade at all times.

Also, never let anyone near the front of the cutting blade when you are working there.

4. When cutting smaller pieces, feed the metal in from the front of the shears against the back gage. This gage can also be set easily to the right length. *Never attempt to cut band iron, wire, or any heavy metal on the squaring shears.* This will nick the blade, which will then make a notch in every edge you cut.

Cutting with Tab Notchers

Tab notchers will handle small cutting jobs. They will cut a six-inch by six-inch corner notch in one stroke. They will also cut tabs and notches for box corners (Figure 28-8).

To use, just slide the workpiece up tightly against the guides, hold firmly, and pull the handle down.

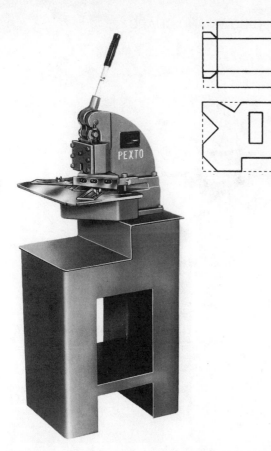

Fig. 28-8 Using a notcher for cutting tabbed corners. The insert shows the kinds of cuts that can be made on this machine. (Courtesy of Roper Whitney Company)

Cutting with Ring-and-Circle Shears

Ring-and-circle shears are used to cut rings or circles of sheet metal up to 20 gage in thickness (Figure 28-9). Industrial models are used for heavier stock.

To use, set the machine to the desired diameter of cut. Then place the workpiece between the holding jaws and clamp tightly. Turn the handle slowly, and use your hand to ease the metal through the slitters, or cutting wheels. Be careful of sharp edges.

KNOWLEDGE REVIEW

1. What is cutting metal sheets called?

2. Straight lines and outside curves are cut with what hand tool? Inside curves and intricate work?

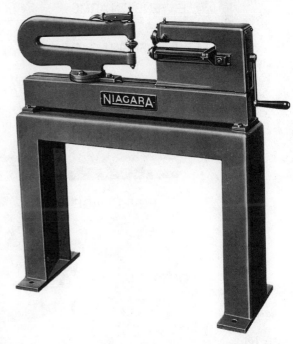

Fig. 28-9 Ring-and-circle shears. (Courtesy of Clearing Niagara)

3. Can aviation snips perform all kinds of sheet-metal cutting?

4. What hand tool is used for cutting around cylindrical objects?

5. When punching, should you strike the punch solidly or lightly?

6. What machine is designed to cut large pieces of sheet metal?

7. What kinds of material should you never cut on the squaring shears?

8. The hand snips and squaring shears both apply the principles of a lever and inclined plane in cutting sheet metal. Explain how these principles aid in cutting.

Unit 29

Bending Sheet Metal

OBJECTIVES

After studying this unit, you should be able to:

- Name four kinds of stakes.
- Describe the device used to hold stakes.
- Explain why a metal hammer should not be used when bending sheet metal.
- Explain the use of a setting-down hammer.
- Describe how to make angular bends.
- Describe how to make a hem using hand tools.
- Describe how to form a cylinder by hand bending.
- Name the stake used for forming cone-shaped objects.

KEY TERMS

blowhorn stake

conductor stake

bench plate

hand seamer

hatchet stake

hollow mandrel stake

setting-down hammer

Every sheet metal project you make will have to be bent into some shape. Sheet metal can be formed or bent in many ways. Some of the common methods are described in this unit.

Bending Equipment

A **stake** is a device for supporting or giving shape to metal that is being formed. Metal stakes are made in many sizes and shapes for bending all kinds of work (Figure 29-1). Some of the most common ones include: the **conductor stake,** for forming round sheet metal objects; the **hollow mandrel stake,** for forming, seaming, and riveting; the **hatchet stake,** for making sharp-angle bends; and the **blowhorn stake,** for forming large cone-shaped objects.

Stakes are held in a *stakeholder* or *bench plate* that is fastened to the bench. If stakes are not available, many other common metal pieces, such as round, square, and flat bar stock or pieces of railroad rail or angle iron, can be used. Also, the **hand seamer,** or *handy seamer,* can be used for bending an edge or folding a hem (Figure 29-2).

A *wooden* or *rawhide mallet* with a smooth face is needed for bending, since a metal hammer would dent the metal workpiece. A **setting-down hammer** is used for setting down flanges and hems and for making certain kinds of seams. A *riveting hammer* is used for sheet metal riveting (Figure 29-3).

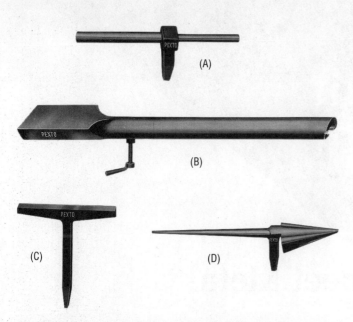

Fig. 29-1 Common kinds of stakes: (A) conductor, (B) hollow mandrel, (C) hatchet, and (D) blowhorn. The stakes are placed in the openings of the bench plate when in use. (Courtesy of Roper Whitney Company)

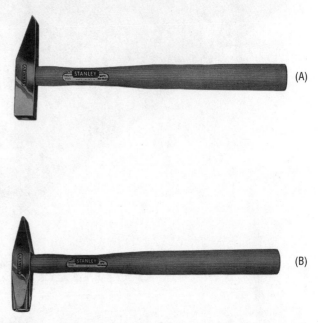

Fig. 29-3 Hammers for sheet metal work: (A) setting-down and (B) riveting. (Courtesy of Stanley Tools, Division of the Stanley Works, New Britain, CT)

Fig. 29-2 A hand seamer. Note the two adjusting screws for setting the depth of the bend. (Courtesy of Roper Whitney Company)

Fig. 29-4 Squaring off a sharp bend over a hatchet stake with a rawhide mallet.

Making Angular Bends

Follow the procedures below for making different types of angular bends in sheet metal.

1. To make a sharp angle bend, use a hatchet stake. Hold the work over the stake, with the bend line over the sharp edge. Now, press down with your hands on either side of the stake to start the bend. A mallet is used to square off the bend (Figure 29-4).

2. To bend a wide piece of metal by hand, place the metal on the bench with the bend line over the edge. Clamp a piece of angle iron over the bend line with one or two C clamps. Press the metal down with your left hand, and strike it with the side of a mallet. Strike the metal with the face of the mallet to square off the bend (Figure 29-5).

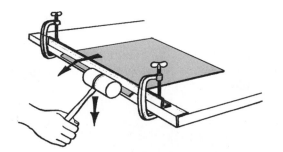

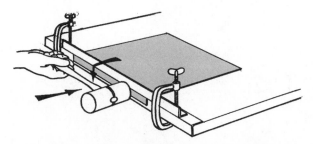

Fig. 29-5 Start the bend by striking the metal with the side of the mallet, forming the metal a little at a time. Square off the bend with the face of the mallet. If a hem must be bent, reverse the metal and flatten the edge.

Fig. 29-6 The hand seamer can be used to make sharp-angle bends or to form hems by hand.

Bend slowly. If you bend one part too much, a kink may form.

3. To make a hem by hand, first bend the hem at a right angle over the edge of a bench. Then reverse the workpiece and reclamp it on the bench. Close the hem by striking it with a mallet.

4. If a hem must be bent on a small piece of metal, use a hand seamer (Figure 29-6). Grasp the metal, with the edges of the jaws at the bend line, and turn the edge to the right angle. If you are making

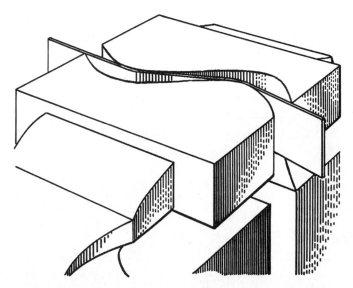

Fig. 29-7 Bending an irregularly shaped piece by squeezing it between two wooden forms.

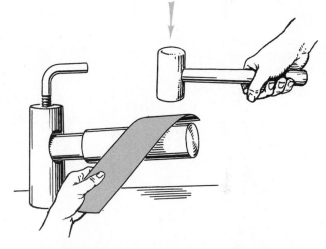

Fig. 29-8 Using a mallet to form heavier metal into a cylindrical shape. Strike the metal as you push it over the form.

a hem, bend it as far as it will go. Then open the seamer and squeeze the metal to close the hem.

5. For irregular bends, cut two pieces of hardwood to shape. Squeeze the metal between them (Figure 29-7).

Bending Cylinders

Choose a stake, pipe, or rod with a diameter equal to, or smaller than, the diameter of the curve to be bent. Place the metal over the stake, with one edge extending slightly beyond the center of the bending device.

If the metal is thin, force the sheet around the stake with your right hand to form it. If the metal is heavy, use a mallet to strike the sheet with glancing blows as you feed it across the stake (Figure 29-8).

KNOWLEDGE REVIEW

1. List four kinds of sheet metal stakes.

2. If metal stakes are not available, what are some substitutes you can use?

3. What device is used for bending an edge or folding a hem?

4. Why should you use a wooden or rawhide mallet for bending?

5. Suggest methods of bending sheet metal using common objects found around the home. Sketch the items you could use.

Unit 30

Bar Folder and Brake Bending

OBJECTIVES

After studying this unit, you should be able to:

- Describe a bar folder.
- Explain how to make a hem on a bar folder.
- Tell the difference between a fold and a hem.
- Describe how to make a rounded fold for a wired edge.
- Name three kinds of metal brakes.
- Describe how to use a cornice brake.
- Calculate how to set the gage for a 1/4-inch wired edge.

KEY TERMS

bar folder   box-and-pan brake   cornice brake   hand-operated press brake

Metal sheets can be bent better and faster on a bar folder or brake than by hand. Bar folders and brakes are especially useful in industry for precision bending (Figure 30-1).

The Bar Folder

The **bar folder,** or folding machine, comes in various sizes designated by the maximum length of bend (Figure 30-2). The most common size is thirty inches. The bar folder will form open and closed bends in widths 1/8 to 1 inch on metal as heavy as 22 gage. It is used to fold edges and to prepare folds for various seams and wired edges. It is adjusted in two ways:

1. To regulate the *depth* of fold, turn the gage-adjusting screw knob in or out.

2. To regulate the *sharpness* of the fold, loosen the nuts or lever on the wing, and adjust the wing.

The wing is lowered when the bar folder is used to make a rounded fold for a wired edge. There are stops at the left end of the bar folder that will limit the fold to 45 or 90 degrees. An adjustable collar can be set for any angle. Figure 30-3 shows some common bends that can be made on a bar folder.

Making a Fold or Hem

A fold must be made before making a folded or grooved seam or a hem. The procedure is as follows:

1. Adjust for the depth of fold by turning the gage-adjusting knob in or out until the depth is shown on the depth gage. Adjust the back wing for the

189

Fig. 30-1 This 100-ton capacity programmable hydraulic press brake can produce bends in 10-gage mild. Both rigid and flexible dies can be used. (Courtesy of Clearing Niagara)

correct sharpness of bend. If it is a 45- or 90-degree bend, set the stop. Make a trial fold on a piece of scrap sheet metal.

2. Insert the metal in the holder and hold it firmly in place with your left hand.

3. Pull up on the handle with your right hand until the correct amount of fold is obtained.

4. If a hem is being made, remove the metal from the folder. Place it on the flat bed with the hem upward, then flatten the hem (Figure 30-4).

Making a Rounded Fold for a Wired Edge

To prepare a metal sheet for a wired edge, follow the steps below.

1. Hold the wing vertically and loosen the wedge locknut. Then turn the wedge-adjusting screw until the distance between the wing and the edge of the blade is about equal to the wire diameter. Tighten the wedge locknut.

2. Set the gage to a distance equal to one and one-half times the wire diameter. Insert the metal in the bar folder and make the bend.

Try this on a piece of scrap metal before attempting it on your project. With practice, you can make a variety of other bends on the bar folder.

Brakes

The **cornice brake** is used for making bends and folded edges on large pieces of stock (Figure 30-5). The size is determined by the maximum length of bend it can make. The 5- and 7-foot lengths are the most common. This machine is quite easy to operate:

1. Lift the clamping bar levers and insert the metal in the brake. Tighten the bar levers, with the layout line directly under the front edge of the upper jaw. Check one edge of the metal and the front edge of the upper jaw with a square.

2. Lift up the bending wing until you have the angle of bend you want. Go a few degrees past this angle, because the metal tends to spring back.

The cornice brake works well for most jobs. It is not possible, however, to bend all four sides of a box on it. A **box-and-pan brake** is used for this purpose. The upper jaw is made up of various widths of removable fingers. You use only the number of fingers necessary for the length of bend you want. Figure 30-6 shows the steps in bending all four sides of a box.

The **hand-operated press brake** will perform not only simple angle and radius bends, but also flanging,

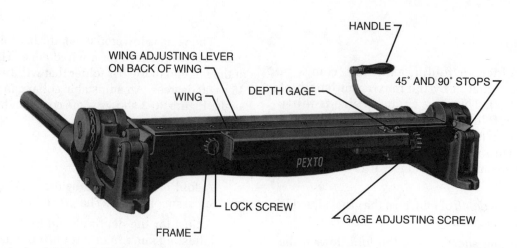

Fig. 30-2 Parts of a bar folder, or folding machine. (Courtesy of Roper Whitney Company)

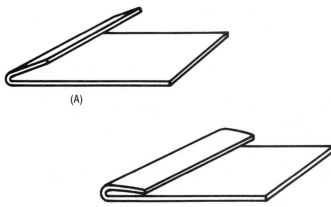

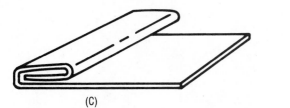

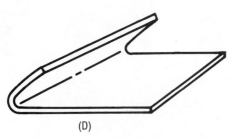

Fig. 30-3 Some common bends that can be made on a bar folder: (A) sharp bend for seam or hem, (B) single hem, (C) double hem, (D) open bend for wire-reinforced edge.

Fig. 30-5 Parts of a cornice brake. (Courtesy of Roper Whitney Company)

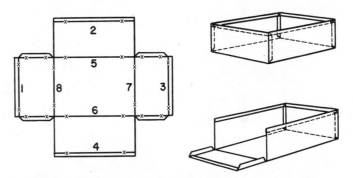

Fig. 30-6 Steps in completing a box on a box-and-pan brake: (1) Make the first hem. (2) Make the second hem. (3) Make the third hem. (4) Make the fourth hem. (5) Make a 90-degree bend on one side. (6) Make a 90-degree bend on the second side. (These first six steps can be completed with all the fingers in the bar folder.) (7) Fit just enough fingers together to equal the width of the box. Bend the first end. (8) Reverse the metal and bend the second end. (When you complete steps 7 and 8, the sides may flatten a little as you bend the two ends.)

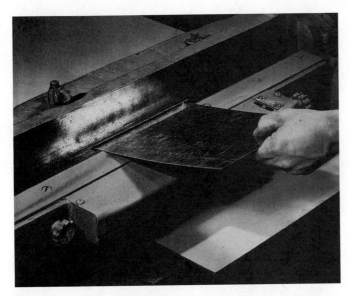

Fig. 30-4 Place the metal with the folded edge up on the bed, then pull down on the handle with some force to flatten the hem.

hemming, seaming, flattening, punching, blanking, and drawing operations.

KNOWLEDGE REVIEW

1. On the bar folder, how do you regulate the depth of fold? The sharpness of fold?

2. Can a bar folder make any degree of bend?

3. What kind of brake is needed to bend all four sides of a box?

4. Name some special operations that can be performed on the hand-operated press brake.

5. Report on a trip you make through a sheet metal shop.

Unit 31

Roll Forming Cylindrical Parts

OBJECTIVES

After studying this unit, you should be able to:

- Give another name for forming rolls.
- Name the number of cylinders on forming rolls.
- Identify the maximum thickness of metal that can be shaped on forming rolls.
- Describe the method of forming a cylinder.
- Tell why there are grooves in the cylinders of forming rolls.
- Explain how cone-shaped objects can be formed.

KEY TERMS

forming rolls   slip-roll forming machine

The quickest and easiest way to form cylinders and cones is to use **forming rolls,** also called **slip-roll forming machines.** These machines are used in both school shops and industry (Figure 31-1).

Forming Rolls

The most common forming rolls have two-inch rolls and are thirty or thirty-six inches wide. They will form metal as heavy as 22 gage. The machine in Figure 31-1 has three rolls. The front two are gear-operated by turning the handle. The back one is the idler roll; it does the actual forming. You can move it up or down to form larger or smaller cylinders. You can also move the lower front roll up and down by the two front adjusting screws to take different thicknesses of metal. To remove the cylinder after it has been formed, you slip the end of the upper, or slip roll, out of place. The lower and back rolls have grooves cut along their right sides. These grooves are for forming wire or for forming cylinders when the wired edge is already installed on the workpiece.

Forming a Cylinder

To form a cylinder with forming rolls, follow the procedure below.

1. Lock the upper roll in position. Adjust the lower roll to be parallel to it. Leave just enough clearance between the rolls for the metal to slip in under slight pressure (Figure 31-2).

2. Adjust the back roll to about the position needed to form the cylinder. Make sure that the back roll is parallel to the other rolls. You can test these settings on scrap sheet.

3. Insert the sheet between the front rolls and turn the handle. Just as the metal enters, raise it slightly to start the forming. Then lower it to catch the back roll (Figure 31-3).

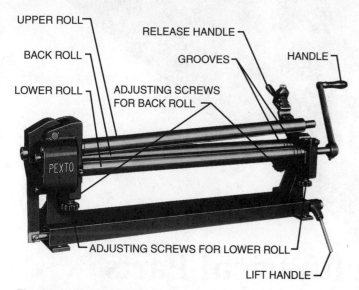

Fig. 31-1 Parts of forming rolls, or slip-roll forming machines. (Courtesy of Roper Whitney Company)

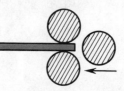

Fig. 31-3 Forming a cylinder: (A) Insert the sheet in the front rolls. Turn the handle forward. (B) Just as the metal enters, raise it slightly with your left hand to start the cylinder.

Fig. 31-4 The cylinder is beginning to form here. (Courtesy of Clearing Niagara)

Fig. 31-2 Adjust the lower roll screws until the metal workpiece is snug between the upper and lower rolls. (Courtesy of Roper Whitney Company)

Fig. 31-5 When the cylinder is formed, open the release handle and raise the upper roll to remove the cylinder.

4. Continue turning the handle to shape the cylinder (Figure 31-4). If the cylinder is too big, bring the sheet back to the starting position and readjust the back roll. After the cylinder is formed, release the upper roll to remove the metal (Figure 31-5).

Rolling a Cylinder with a Wired Edge

If the cylinder to be formed has a wired edge, change the procedure for rolling as follows:

1. Adjust the front rolls with a slightly wider space at the right side than at the left.

2. Place the sheet between the front rolls, with the wired edge down and in the groove of the correct size (Figure 31-6). Now proceed as before.

Forming Cone shapes

Use the following guidelines for forming a cone.

1. Adjust the front rolls as above. Set the rear roll at an angle that is about the same as the taper of the cone, with the left end of the roll near the front rolls.

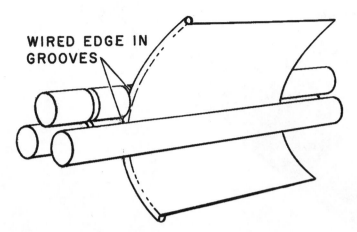

Fig. 31-6 Forming a cylinder with a wired edge.

2. Insert the sheet with the short side to the left and the long side to the right. Hold the short side as you turn the handle. Allow the short side to go through the rolls more slowly than the long side. (Figure 31-7). This forms the cone.

KNOWLEDGE REVIEW

1. How many rolls do forming rolls have?
2. Which roll does the forming?
3. Which roll do you adjust so that the machine can take more than one thickness of metal?

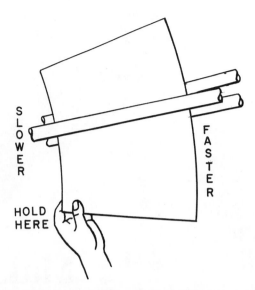

Fig. 31-7 Forming a cone-shaped object. Hold one corner so that one side goes through the rolls faster than the other side.

4. After a cylinder is formed, how is it removed from the machine?
5. How do you set the rear roll for forming a cone?
6. Which of the basic machines discussed in unit 15 are applied in forming rolls? Make a sketch of each.
7. Prepare a research report on roll forming in industry.

Unit 32

Making Seams

OBJECTIVES

After studying this unit, you should be able to:

- Name and describe four kinds of seams.
- Name the hand tool used to lock a grooved seam.
- Calculate the amount of extra material needed for a folded seam.
- Explain how to fold the ends when making a cylinder or rectangular object.
- Describe the width of the groove on a number 5 hand groover.
- Name the tool used to strike a hand groover.

KEY TERMS

butt seam

folded seam

grooved seam

hand groover

lap seam

seaming

Seaming is joining sheet metal parts with mechanical or soldered joints. Seams are sometimes soldered to make them watertight. In this unit, you will learn to make some simple seams using the hand tools found in the shop.

Types of Seams

A number of different seams are used in sheet metal work (Figure 32-1). The **butt seam** has two pieces of metal butted together and hard soldered. The **lap seam** is made by lapping the edge of one piece of metal over the edge of the other and riveting or soldering them together. The most common kind of lap seam is the *plain lap*. If the two metal surfaces must align, an *offset lap* is made. An *inside* or *outside corner lap* may be used on the corners of sheet metal projects. **Folded** and **grooved seams** have two folded edges hooked together. In a grooved seam, the edges are locked, with three thicknesses of metal above the joined sheets.

Hand Groovers

The **hand groover** is used to lock a grooved seam (Figure 32-2). The groover should have a groove about 1/16 inch wider than the width of the finished seam. For example, hand groover No. 2, which has a 5/16-inch groove, is made to lock a 1/4-inch seam. The common sizes are No. 0, No. 2, and No. 4. (See the chart in Figure 32-3.)

Making a Folded or Grooved Seam

To make a folded or grooved seam between two pieces of sheet metal, follow the steps and guidelines listed on the following page.

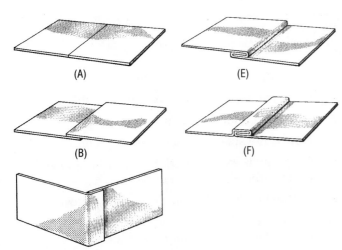

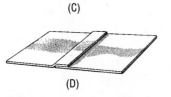

Fig. 32-3 For a grooved seam, an allowance of three times the width of the seam must be made. For heavier sheet stock, a small additional allowance is made for the height of the seam.

Fig. 32-1 Common seams for sheet metal joining include: (A) butt, (B) lap, (C) corner lap, (D) offset lap, (E) folded, and (F) grooved.

1. Determine the width of the seam, and allow extra material equal to three times the seam width. For example, if it is a 1/8-inch seam, allow 3/8 inch. Sometimes, a small additional amount is allowed for the rise in the seam, especially if the metal is heavier than 22 gage. This amount should be about one and one-half times the thickness of the metal. On most projects, add half the allowance to either end of the metal (Figure 32-3). Sometimes, it is a good idea to notch the corners. This makes a neater seam.

2. Fold the edges by hand or in a bar folder. The metal should be bent as for an open hem. When using a bar folder, adjust the machine to the seam width and fold. If the seam is to be made on one continuous piece, such as a cylinder, remember to fold the ends in *opposite directions.*

3. If the project is to be of some particular shape—cylindrical or rectangular, for example—form it at this point.

4. Place the metal over a solid backing such as a stake, if it is circular, or a metal table, if it is flat. Hook the folded edges together. To make a folded seam, gently strike the seam along its length to close it, using a wooden mallet. Avoid nicking or denting.

5. To make a grooved seam, select a hand groover that is 1/16 inch wider than the seam. Hold the groover over the seam, with one edge of the groover over one edge of the seam. Strike the groover solidly with a metal hammer to close one end of the seam. Slide the groover along as you strike it, to complete the seam (Figure 32-4). You can further lock the seam with prick-punch marks about 1/2 inch from either end of the seam.

6. Check the seam after it is locked. Are the joined sheets of metal level? Is the seam well-formed, without nicks, and smooth?

The small number stamped on the hand groover indicates the size.

Number	Size of Groove	
	inch	mm
6	1/8	3
5	5/32	4
4	7/32	5.5
3	9/32	7
2	5/16	8
1	11/32	9
0	3/8	10

Fig. 32-2 A hand groover and groover chart. (Courtesy of Roper Whitney Company)

Fig. 32-4 To complete the seam, slide the groover along as you strike it.

1. Name the main kinds of seams.
2. How do a folded and a grooved seam differ?
3. When is a double seam needed?
4. How is the size of the hand groover chosen?
5. How much extra material should be allowed to make a folded or grooved seam?
6. With which tool do you close a folded seam?
7. Make a display board of the common types of seams. Mount each sample and identify.

Unit 33

Wiring, Burring, Beading, and Crimping

OBJECTIVES

After studying this unit, you should be able to:

- Describe turning.
- Name the machine used for turning operations.
- Name the operations a combination rotary machine can perform.
- Describe how the size of a wire is measured.
- Explain how to turn an edge.
- Describe how to close a wired edge by machine.
- Describe when to use burring rolls.
- Explain how to turn a flange.
- Explain the use of beading.
- Describe the process that allows one object to slip into a mating part.

KEY TERMS

beading

burring

crimping

rotary machine

turning

wiring

Turning in sheet metal work is forming metal on disk or cylindrical shapes. A **rotary machine** (Figure 33-1) is used for these operations. A single *combination rotary machine* will turn, wire, or burr an edge and do beading and crimping (Figure 33-2). Larger shops have a separate machine for each operation. These operations serve to decorate, stiffen, or prepare a workpiece for wiring or seaming.

Wiring an Edge

Wiring an edge stiffens the article and eliminates sharp edges. The wire used is made of mild steel, with a galvanized or copper coating to protect it from rusting. The size is measured by the American Wire Gage System. The most common sizes are 10 (about 1/8 inch in diameter), 12, 14, and 18.

How an edge is wired depends on whether the wire is put on before or after the article is formed. If the project is a simple cylinder, the wire is put on the flat sheet before rolling. In this case, use the bar folder to turn the edge. You will need only the wiring rolls of the rotary machine to finish the job. If the project is a cone, such as a funnel, the wiring is done after it is shaped. Both turning and wiring rolls are needed. In making a

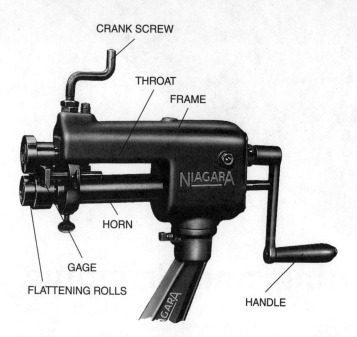

Fig. 33-1 The rotary machine is used for turning operations in sheet metal work. (Courtesy of Clearing Niagara)

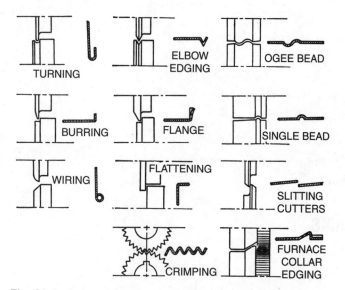

Fig. 33-2 A rotary machine can perform these operations.

layout for a wired edge, remember to allow two and one-half times the diameter of the wire for the edge.

Turning an Edge

The first step in forming a wired edge is to turn the edge using the turning rolls of the rotary machine.

1. Install the turning rolls on the rotary machine.

2. Set the gage at a distance equal to two and one-half times the diameter of the wire from the center of the groove.

3. Slip the metal between the rolls and against the gate. Tighten the upper roll until it just grips the metal.

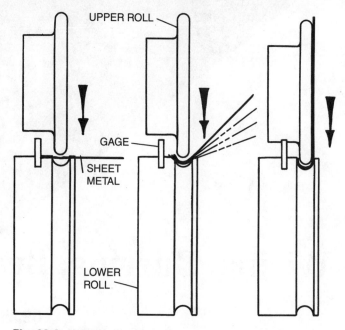

Fig. 33-3 Using the turning rolls to form the edge for a wired edge. Notice how the edge is formed a little at a time.

4. Holding the metal against the gage with your left hand, turn the handle with your right hand. Make one complete turn. The metal must track the first time. Never let the roll run off the edge of the metal.

5. Tighten the upper roll a little more—about one-eighth of a turn. At the same time, raise the outside edge of the container a little higher. Make another complete turn.

6. Continue tightening the upper rolls after each turn. Raise the cylinder until the edge is U-shaped to receive the wire (Figure 33-3).

7. Loosen the upper roll to remove the metal from the machine.

Closing the Wired Edge by Machine

To close a wired edge, complete the following procedure.

1. Place the wiring rolls on the rotary machine. Adjust the gage. Its distance from the sharp edge of the upper roll should equal the diameter of the wire plus twice the thickness of the metal.

2. Place the project between the rolls with the wired edge up and against the gage. Tighten the rolls at the point where the edge is already turned until they grip the metal.

3. Turn the handle as you feed in the metal to set the wired edge (Figure 33-4).

4. Loosen the upper rolls and remove the work.

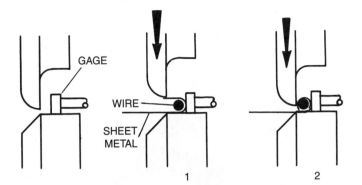

Fig. 33-4 Closing a wired edge, using the wiring rolls.

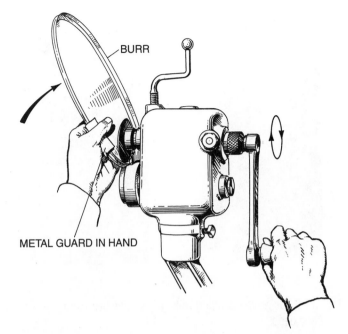

Fig. 33-6 The burr has begun to form. The disk is being raised to help form it. The metal guard protects the hand.

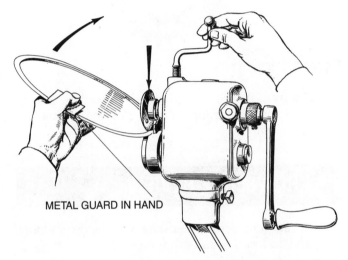

Fig. 33-5 Using the burring rolls to turn a burred edge on the bottom of a container. Notice that the disk is inserted and that slight pressure is applied to the metal between the rolls.

Turning a Burr or Flange

The burring rolls of the rotary machine are used mostly to turn a flange, or rim, on a cylinder and to turn a burr on a round bottom in making a double seam to attach it to a cylinder. The second is a tricky operation. It must always be practiced on scrap stock. The following is the procedure for **burring,** or turning a burr, on a round bottom:

1. Place the burring rolls on the rotary machine. Adjust the upper and lower rolls. The distance between the sharp edge of the upper roll and the shoulder of the lower roll should be equal to the thickness of the metal.

2. Set the gage away from the shoulder of the lower roll by a distance slightly less than the width of the burr.

3. Bend a little piece of scrap metal into a U shape to protect your hand from the sharp burr. Place this in the round of your hand between thumb and forefinger.

4. Grasp the circular end piece with your thumb on top and forefinger below toward the center of the disk.

5. Holding the edge of the disk firmly against the gage, turn the upper roll down until there is slight pressure on the metal (Figure 33-5).

6. Turn the handle slowly, carefully tracking the burr. This must be done the first time around. The beginner usually allows the disk to escape the rolls, making tracking difficult.

7. Apply a little more pressure by tightening the upper roll. Then turn the handle a little faster as you slowly raise the disk from a horizontal to an almost vertical position (Figure 33-6). Continue tightening the upper roll as the burr is formed.

8. Loosen the upper roll to remove the disk.

9. If you have done a good job of turning the disk, the end piece will slip over the flange on the cylinder with a little snap to form the bottom of the container.

Turning a flange on a cylinder is a similar operation, except that the edge is turned only 90 degrees.

Beading

Beading is strengthening and decorating containers with either a *simple* bead or an *ogee,* or S-shaped, bead. This stiffens the workpiece much as sheet metal is made stronger by corrugation. This operation is quite simple:

1. Place the beading rolls on the rotary machine (Figure 33-7). Set the gage according to where you want the bead.

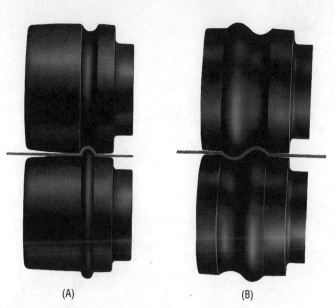

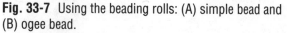

Fig. 33-7 Using the beading rolls: (A) simple bead and (B) ogee bead.

2. Tighten the upper roll enough to form the metal lightly.

3. Turn the handle, tracking the bead the first time. Continue tightening the upper roll until the bead is completely formed.

Crimping

Crimping is drawing in the edge on the end of a cylindrical object, particularly a heating pipe or stack. It is done so that the object will slip easily into a mating part. The procedure is as follows:

1. Place the crimping rolls on the rotary machine (Figure 33-8). Adjust the gage to the proper length of crimp.

2. Slip the cylinder between the rolls, with the edge against the gage.

3. Applying moderate pressure with the upper roll, turn the handle to form the first impression.

Fig. 33-8 Crimping rolls.

4. Apply more pressure to deepen the crimp, making sure that the crimping rolls follow the first impression.

KNOWLEDGE REVIEW

1. What one machine can wire an edge, turn a burr, and do beading and crimping?

2. What kind of wire is used for wiring an edge?

3. How much allowance should be made for a wired edge?

4. When turning a burr on a round bottom, what special equipment do you put on the rotary machine?

5. What is *beading*?

6. What is *crimping*? Why is it done?

7. Make a drawing of each sheet metal operation done on a rotary machine.

8. Prepare a research report on turning machines used in industry.

Unit 34

Rivet, Screw, and Adhesive Fastening

OBJECTIVES

After studying this unit, you should be able to:

- Describe the fasteners used to fasten parts together permanently.
- Name the fasteners used for semipermanent joints.
- Name two kinds of tinner's rivets.
- Describe how the size of tinner's rivets is measured.
- Describe the shape of the head of tinner's rivets.
- Explain how to make a hole for installing tinner's rivets.
- Describe a rivet set.
- Describe a riveting hammer.
- Tell how far from the edge rivets should be located.
- Describe pop rivets.
- Describe two kinds of self-tapping sheet metal screws.
- Tell what size solid punch should be used for 28-gage sheet metal.
- Name some of the advantages of adhesive bonds.
- Describe how to apply contact cement.
- Explain the process for using epoxy cement.

KEY TERMS

adhesive	hand lever punch	self-tapping sheet metal
adhesive films	pop rivet	screws
contact cement	riveting hammer	solid punch
epoxy cement	rivet set	tinner's rivets

Metal sheets are often fastened together permanently with rivets. Sometimes, a joint is also soldered or glued to make it leakproof. Adhesives can give a very strong, permanent joint. Sheet-metal screws provide a semipermanent joint.

Aircraft structures are made of metal frames and ribs covered with strong, thin, lightweight aluminum alloy sheathing, or skin. Computerized riveting machines are generally used to fasten these sections together. A huge device clamps the skin to a section of a wing frame, for example. An automatic riveter is programmed to locate a rivet position, drill a hole, insert an aluminum rivet, head the rivet, and then mill it flush with the surface of the skin. The smooth wing structure then has no rivet "dimples" to cause an uneven airflow and turbulence (Figure 34-1). This is a fast and efficient way to install the over 750,000 flush rivets found on the body, wings, and stabilizers of a modern jetliner.

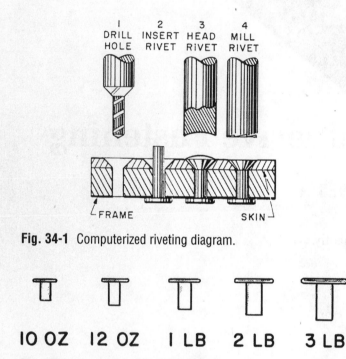

Fig. 34-1 Computerized riveting diagram.

Fig. 34-2 The actual size of sheet metal rivets.

Rivets and Riveting Tools

Sheet metal rivets are usually made of mild steel. *Black-iron rivets* have a black iron oxide coating. *Galvanized rivets* have a zinc coating. Both are also known as **tinner's rivets.** They make a strong joint. The size of the rivet is indicated by the weight per thousand. For example, a "1-pound" rivet indicates that one thousand rivets weigh one pound (Figure 34-2). All tinner's rivets are flat-head. The length of the shank depends on the size. Sizes are shown in Table 34-1. Tinner's rivets may also be made of copper.

A **solid punch** is usually used to make the holes for tinner's rivets. These punches are numbered according to size from 6 to 10. Some companies letter the punches from B (3/32 inch) to I (3/8 inch). Each number fits a certain size of rivet. For example, a No. 8 punch is made for 2-pound rivets (see Table 34-1). Holes can also be made in sheet metal with the **hand,** or **hand lever punch.**

A **rivet set** is used to set and head the rivets. It is made in various sizes to match the rivets. Choose the

Table 34-1 Guide for selecting punch, rivet, and rivet set.

Gage of metal	Solid punch		Size of rivet (weight per 1000)	Rivet set	
	Letter	Number		Number	Size of hole
30	B	10	10 to 12 oz	8	0.110
28	C	9	14 oz to 1 lb	7	0.128
26	C	9	1 lb	7	0.128
24	D	7 or 8	2 lb	5	0.149
22	D	7 or 8	2 1/2 lb	5	0.149
20	E	6	3 lb	4	0.166

rivet set by number or by matching the rivet shank to the hole of the rivet set.

A **riveting hammer** is the best one to use for riveting operations. It has a flat face on one side and a beveled cross peen on the other. However, a ball-peen hammer may also be used.

Pop rivets are available with either a hollow or a solid core. These blind rivets can be installed quickly and easily. They are inserted and set from the same side of the work with a hand tool (Figure 34-3). This is particularly useful in assembling hollow sheet metal objects, such as boxes or ducts.

Riveting

Follow the order of operations and guidelines below for fastening metal sheets with rivets.

1. *Obtain the proper materials.*
   Select the correct size of rivets and rivet set and a solid punch or drill.

2. *Lay out the location of the rivets.*
   If several are to be set along a lap seam, they should be placed one and one-half to two times their diameter from the edge. They should be equally spaced along the seam.

3. *Punch or drill the holes.*
   If the metal is thin, the holes are usually punched. Place the sheets over end-grain hardwood or over a lead block. Hold the solid punch over the place where the hole is to be. Strike the punch solidly with a hammer to form the hole (Figure 34-4).

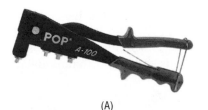

(A)

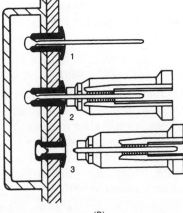

(B)

Fig. 34-3 (A) The pop riveting tool. (B) The installation diagram shows: (1) rivet is installed; (2) tool jaws are set in position; and (3) tool jaws pull rivet stem, setting the rivet. (Courtesy of POP, Emhart Fastening Teknologies, Industrial Division)

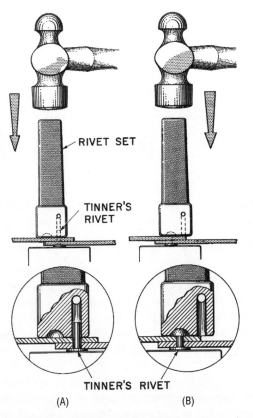

Fig. 34-5 (A) The hole of the rivet set is placed over the shank of the rivet, and the sheets of metal are drawn together. (B) The cone-shaped hole is used to form the shank or to head the rivet.

Fig. 34-4 Using a solid punch to form holes for riveting. Place the sheet metal over end-grain hardwood or a lead block.

4. *Set the rivet.*
    Place the rivet with the head down on a flat surface or, if cylindrical objects are being riveted together, on the crown of a stake. Slip the metal over it so that the shank comes through the hole. Place the hole in the rivet set over the shank. Then strike the rivet set with a hammer once or twice (Figure 34-5). This will flatten out the sheet metal around the hole and draw the two sheets together. Do not strike the rivet set too hard. Also, be sure to keep it square with the workpiece because sheet metal dents easily.

5. *Head the rivet.*
    Use the flat face of the riveting hammer to strike the shank squarely with several blows. The shank will expand, filling the hole tightly. The top of the shank will flatten a little. Now, place the cone-shaped depression of the rivet set over the shank, and strike the set two or three times to round off the head. You may also use a ball-peen hammer to set and head rivets in sheet metal work.

6. *Determine the order of riveting.*
    If you are setting several rivets in a row, it is a good idea to punch and rivet the center hole first. Then punch and rivet the other holes from the center outward.

7. *Check your work.*
    Have you dented the sheet metal around the rivet? Is the head well-shaped? Do you have the rivets too close to the edge?

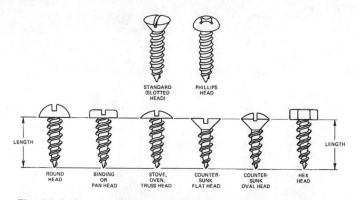

Fig. 34-6 Common head shapes for type A sheet metal screws.

Self-Tapping Sheet Metal Screws

Self-tapping sheet metal screws are made to cut their own threads in mild-steel sheet and softer aluminum alloys. These screws are a quick, sure means of joining sheet metal. They are used often in making ducts for ventilation, air-conditioning, and heating. They are also very useful in home maintenance and repair jobs.

There are two kinds of self-tapping sheet metal screws. *Type A* is pointed and is made for joining sheet metal no thicker than 18-gage galvanized iron or soft aluminum (Figure 34-6). *Type B* has a blunt point and is used for joining both lighter and heavier stock. Both kinds come in several head shapes with either slotted or Phillips heads. Some common sizes of type A are No. 6 diameter by 3/8-inch length, No. 6 diameter by 1/2-inch length, and No. 8 diameter by 3/4-inch length.

Installing Sheet Metal Screws

To fasten with sheet metal screws, follow the steps below.

1. Locate and prick-punch the site of the holes.

2. Choose a drill size that equals the root diameter of the screw. For example, use a 7/64-inch drill for a No. 6 screw and a 1/8-inch drill for a No. 8 screw. You can also judge the drill size by holding the drill in back of the screw thread. Drill the hole.

3. Line up the hole with a punch. Make sure that the two pieces of metal are held firmly together as the screw is started.

Adhesive Bonding

Adhesives are sticky substances such as cements and glues that are used to join surfaces. Adhesive bonds offer a number of advantages over other fastening methods. First, an adhesive bond acts as many tiny fasteners. Thus, the load is distributed uniformly over the entire

area of contact. Second, adhesive bonds are inconspicuous. Third, adhesive bonds form a shield between the pieces they join. This reduces the corrosion that can occur when the pieces are of different metals.

Using Adhesives

In sheet metal work, adhesives are used mainly for joining thin sheet or strip parts that are too low in bearing strength for threaded fasteners and too thin to withstand welding heat. In the school shop, you can use adhesives to provide a fast, permanent, watertight joint for small metal parts. You can also use them to fasten metal to other materials such as wood and plastic.

Fastening with Contact Cement

Contact cement is a type of adhesive that is applied to *both* of the sheets to be joined. It is especially useful for adhering one sheet of metal to another or to a piece of wood. The solvent for contact cement is lacquer thinner.

Design your project so that you have as wide a joint as possible. Clean the joint surfaces thoroughly. Apply the contact cement to both workpieces and allow them to dry for about fifteen minutes. When the workpieces are dry, carefully place them together. Then put a block of wood over the joint and rap sharply with a hammer or mallet. *Be very careful: Once you touch the glue surfaces together, they cannot be moved.*

Fastening with Epoxy Cement

Epoxy cement is special plastic adhesive that is very tough. It can be used similarly to contact cements except that the joint surface can be smaller. The solvent for epoxy cement is acetone or a special epoxy cleaner. Use epoxy cement as follows:

1. Clean the workpieces carefully.

2. Squeeze out equal parts of the hardener and resin; mix thoroughly.

3. Apply to *one* workpiece (Figure 34-7). *Use gloves.* Do not get any adhesive on your hands or face.

4. Press the joint together, remove excess glue, and allow it to dry for about eight hours.

Industrial Adhesives

The two basic types of adhesives used in industry are structural and nonstructural. High-strength *structural* adhesives are those which will join almost any material—metal, wood, plastic, or ceramic—over extended periods of time, and can bear a considerable load. These adhesives will result in a bond which is as strong as the materials being joined. A list of typical applications includes:

Fig. 34-7 Using epoxy cement to join wood and metal.

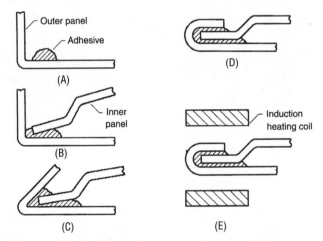

Fig. 34-8 Vehicle body panel bonding system employing a sheet metal lock joint and an adhesive. A ribbon of adhesive is flowed into the outer panel joint (A). The inner panel is pressed into place (B). The outer panel joint is partially bent to secure the pieces (C). The bending process is completed (D). The assembly is placed under heater coils to cure the bond (E).

1. *Aircraft:* wing and stabilizer panels and interior bulkheads.
2. *Automotive:* gas tanks, headlights, brake linings, body panels, and fuel filters.
3. *Electronics:* speakers, relays and controls, coils, and cabinets.
4. *Hardware:* water faucets, lawn sprinklers, and tools.
5. *Building construction:* metal window frames, building panels, and roofing systems.
6. *Packaging:* shipping and food containers and containers for scientific and electronic equipment.

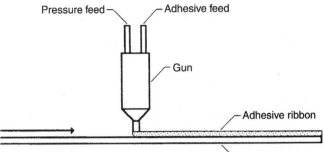

Fig. 34-9 Robotic adhesive guns are programmed to lay ribbons of paste or powder adhesives on workpieces.

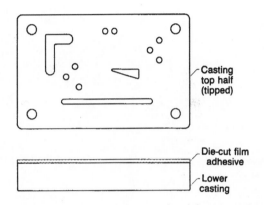

Fig. 34-10 Adhesive film is being used to join two cast manifold sections.

7. *Sports:* golf clubs, bows and arrows, and telescopic rifle sights.

An example of structural adhesive application appears in Figure 34-8. Here the parts of the front hood of an automobile are permanently assembled with a combination sheet metal joint and an engineering adhesive. A robotic gun is often used to lay the adhesive ribbon on the hood panel (Figure 34-9). This method is also used in the aircraft industry to assemble interior panels.

Non-structural adhesives are used in non-load-bearing applications such as cushions, gaskets, decorative trim, molding, weather stripping, and bottle and package labels. These are essentially bonds used to hold lightweight materials in place, where strength is less important.

Adhesive films are die-cut to size and shape, and applied dry to the mating parts of an assembly. A typical metal-casting example is shown in Figure 34-10. This is a fast and efficient method for producing strong permanent bonds between parts.

Engineering adhesive (double-sided) tapes also can be used. These tapes are commonly used in the home for hanging posters and pictures on walls and for project assembly in the home workshop.

KNOWLEDGE REVIEW

1. Write a description of the materials and coatings used in the production of sheet metal rivets.

2. Sketch the tools used to punch riveting holes, and to set and head the rivets.

3. Explain how to lay out the location of rivets along a lap seam.

4. Describe some advantages of using pop rivets.

5. Name the two kinds of self-tapping sheet metal screws.

6. How do you select the drill size for a sheet metal screw?

7. List some advantages of adhesives over other sheet metal fasteners.

8. Describe the differences between the methods of using epoxy and contact adhesives.

9. Explain the difference between structural and non-structural adhesives. List some examples of each.

10. Collect some samples of products assembled with mechanical and adhesive fasteners. Analyze these and prepare a class report on your findings.

Unit 35

Soldering

OBJECTIVES

After studying this unit, you should be able to:

- Name two kinds of solder.
- Identify what metals are used in standard solder.
- Explain the conditions that must be met before soldering.
- Name three heating devices used for soldering.
- Identify the metals in lead-free solder.
- Explain why lead-free solder is used for pipes that carry drinking water.
- Name three forms of solder.
- Describe fluxes.
- Name four kinds of fluxes.
- Explain the use of noncorrosive flux.
- Describe how to tin a soldering copper.
- Tell why cleanliness is so important when soldering.
- Describe how to solder a seam or joint.
- Explain sweat soldering.

KEY TERMS

butane torch	hard soldering	solder
electric soldering copper	propane torch	soldering copper
flux	soft soldering	soldering furnace

Soldering is joining two metal parts with a **solder,** a third metal that has a lower melting point. In **soft soldering,** a tin-lead solder and relatively low temperatures are used. In **hard soldering,** a silver solder and high temperatures produce neater, stronger joints (see unit 43).

Before you can solder, the following conditions must be met: (1) the metal must be clean, (2) the correct soldering device must be used, and it must be in good condition, (3) the correct solder and flux, or soldering agent, must be chosen, and (4) the proper amount of heat must be applied. If you follow these rules, you will get a good solder joint.

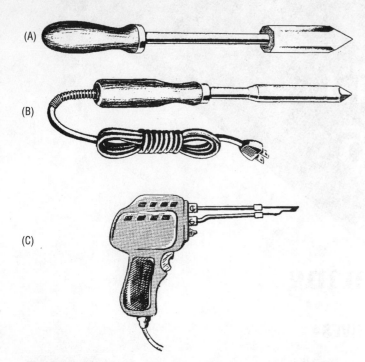

Fig. 35-1 Common heating devices for soldering: (A) a regular soldering copper, (B) an electric soldering iron, and (C) an electric soldering gun.

Soldering Devices

For most soldering, you will need some type of device to heat the soldering copper. The most common kinds are: soldering coppers, electric soldering coppers, the soldering furnace, the propane torch, and the butane torch.

Soldering coppers are purchased in pairs (Figure 35-1A). The coppers of a two-pound pair weigh one pound apiece. This is an average size for regular sheet metal work. They are available in smaller or larger sizes, from three ounces to three pounds per pair. They are sold in pairs so that one can be used while the other is being heated.

An **electric soldering copper** is much more convenient because it will maintain uniform heat and can be used wherever there is an electric outlet (Figure 35-1B). Electric coppers are specified by their wattage. They range from 50 to 300 watts. A 150- to 200-watt soldering copper is a good size for most sheet metal soldering. One as small as 25 watts might be used for electronic work. One as large as 300 watts might be used for rugged work.

Another type of electric soldering copper is the *soldering gun* (Figure 35-1C). This gives instant heat when the trigger is pulled. It is used primarily for electrical work.

Both nonelectric and electric soldering coppers are commonly called *soldering irons*.

The **soldering furnace** (Figure 35-2), **propane torch,** (Figure 35-3), and **butane torch** (Figure 35-4) are also heating devices used for soldering.

Fig. 35-2 Combination soldering and melting furnace. (Courtesy of Johnson Gas Appliance Company)

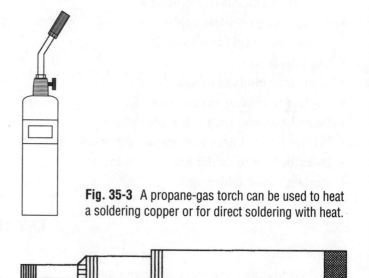

Fig. 35-3 A propane-gas torch can be used to heat a soldering copper or for direct soldering with heat.

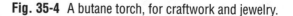

Fig. 35-4 A butane torch, for craftwork and jewelry.

Solders

In the past, *soft solders* were alloys of 50 percent tin and 50 percent lead, and some of this is still used in industrial applications. However, because of health and environmental reasons, solders now are lead-free alloys of tin, silver, antimony, and copper, in various combinations. Solders for home and school shop use commonly contain 96 percent tin and 4 percent silver, and melt at about 430°F. Lead-free solder must be used for joining copper water pipe or for joining any project that will hold water or food. This solder is a combination of copper and silver. Lead is a health hazard.

Hard solders are low-temperature brazing alloys and are discussed in unit 43.

Solder is available in bar, solid wire, and powder forms. More convenient are the acid-core and rosin-core wire types. These have flux in the hollow center of the wire. Soldering paste, containing both solder and flux in paste form, is also available. Solder in preformed shapes is also used for convenience and neatness.

Fluxes

When metal is exposed to the air, a film of oxide, or rust and tarnish, forms on it. The process of oxidation increases greatly when metal is heated. This must be prevented because the oxide tends to keep the metal from reaching soldering temperature and from uniting with the solder. A material called a **flux,** therefore, is used to: (1) remove the oxide from the metal, (2) prevent the formation of new oxide, (3) reduce surface tension so that the molten solder will flow easily by capillary action, and (4) assist alloying action of the solder with the workpiece.

There are two types of flux, corrosive and noncorrosive. The first is more effective but must never be used on electrical connections. It must also be removed from any metal after soldering by washing in hot water. The noncorrosive type is used for all electrical and electronic work. The better commercial kinds can be used on tin plate, copper, brass, and other alloys of copper. Fluxes are available in paste and liquid forms.

Tinning a Soldering Copper

After a soldering copper has been used for some time, or if it gets overheated, the point becomes covered with oxide. Then the heat cannot flow to the metal. To correct this, the point must be cleaned and covered with solder. The process is called *tinning a soldering copper,* and is outlined below.

1. File the point of the soldering copper with a mill file until the clean exposed copper appears.

2. Heat the soldering copper until it turns to yellow or light brown. Then do one of the following:

   a. Apply acid-core or rosin-core solder to the point (Figure 35-5).
   b. Rub the point on a bar of sal ammoniac, and apply a few drops of solder.

Fig. 35-5 One way to tin a soldering copper is to apply an acid-core or rosin-core solder to the point and rub the point on a smooth surface.

   c. Dip the point in liquid flux, and rub with a bar of solder.

3. Wipe the point with a clean cloth to remove the excess molten solder.

Cleanliness

Solder will never stick to a dirty, oily, or oxide-coated surface. Beginners often ignore this simple point. If the metal is dirty, clean it with a liquid cleaner. If it is black annealed sheet, remove the oxide with abrasive cloth, and clean it until the surface is bright. A bright metal, such as copper, can be coated with oxide even though you cannot see it. This oxide can be removed with any fine abrasive.

Soldering a Seam or Joint

Follow the procedure below when soldering a seam between two pieces of metal.

1. Place the two pieces on a soldering table that has a non-heat-conducting top, such as firebrick.

2. Clean the area of the seam.

3. Make sure that the seam is held together properly. If necessary, place small weights on it.

4. Apply a coat of flux with a swab or brush. Use rosin flux on most metals. A raw-acid flux is needed for galvanized iron. If wire-core solder is used, the area will be fluxed as it is soldered.

5. Heat the soldering copper just until the point can melt solder quickly. *Never allow the soldering copper to become red-hot.*

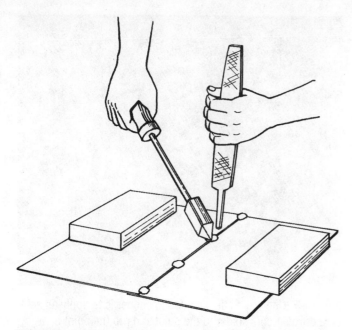

Fig. 35-6 Hold the seam or joint together firmly as you tack it with solder at several points.

WIRE CLIPS

OVERLAY

LONG-NOSE PLIERS

TORCH

Fig. 35-7 Make several wire clips, as shown, to hold two or more pieces together for sweat soldering, and hold the pieces over a torch until the solder begins to ooze out of the edges.

6. Tack the seam in several places (Figure 35-6). Put the point of the soldering copper on the seam. Leave it there until the flux sizzles. Then immediately put a small amount of solder directly in front of the point. Never apply the solder directly to the copper, since this merely makes the solder run without joining the two pieces.

7. Start at one end of the seam. Hold the copper with the tapered side flat on the metal until the solder melts.

8. Move the soldering copper along slowly in one direction only, never back and forth. Put more solder in front of the point as needed. If necessary, press the freshly soldered part of the seam together with a tool such as a file until the solder hardens.

9. Clean the seam with warm running water if acid flux has been used. Baking soda mixed in the water will neutralize the chemical action.

10. Check your work. Is the soldered seam smooth? Are the metals really soldered together? Did you use too much solder?

Soldering a Right-Angle Joint

A right-angle joint is soldered using the same basic principles as above, with the following guidelines:

1. Block up the pieces so that the corner seam is in a horizontal position. Use bricks or charcoal blocks.

2. Apply flux at several points along the joint. Tack the two pieces together.

3. Flow the solder on along the seam.

Soldering an Appendage

Handles, feet, clips, and other similar appendages are fastened in place as follows:

1. Apply flux and a little solder at the point where the appendage is to go.

2. Apply a little flux again. Hold the appendage in place. Heat the area until it is sweat-soldered (see below). Hold the appendage with pliers, since it usually becomes too hot to handle before the solder melts.

Sweat Soldering

Sweat soldering is soldering two or more pieces, one on top of the other. You cannot see any of the solder afterwards. This type of soldering is often used to fasten together several thicknesses of metal to make a heavier part. The handle of a letter opener is a good example. Follow these steps when sweat soldering:

1. Flux one surface and apply a thin coat of solder. Flux the opposite surface. Clamp the pieces together with paper clips or small wire clips.

2. Hold the pieces over a torch until the solder oozes out at the edge (Figure 35-7). For a neat job, use a small amount of solder.

3. Sweat soldering can be used to make a joint. Cover the surfaces to be joined with a thin coat of solder. Holding the two parts together, apply heat with a soldering copper until they are fastened.

In Focus: *Early Metal Can Forming Technology*

Every day, millions of Americans buy millions of cans of food and drinks from grocery stores. These tasty and satisfying products are packaged in containers made by modern canning technology processes. The instances of people getting food poisoning from these tins are rare, for food packaging is a safe and well-regulated industry. But such was not always the case.

During the years from 1845 to 1848, a British team of explorers set out to discover a northwest passage between the Atlantic and Pacific Oceans. This team, called the Franklin Expedition, traveled by boat and dog team through the Arctic region of North America in search of this route.

Much of the food they carried with them was packed in tin cans which had just been developed for food preservation. The can body (Figure 35-8) was made by bending a thin sheet of tin-plated steel to form a small cylinder. The body ends lay over one another to make a lap seam. A tinsmith then ran a bead

of hot solder along the inner and outer edges of the lap to form an airtight and watertight joint.

Next, a flanged or lipped top piece was formed and fitted to the body and soldered in place, inside and out. The bottom end was then soldered from the outside. The inner edge of the bottom piece was soldered through the filler hole in the can top. The food was pushed in through the filler hole, and the cans were placed in boiling water to cook the food in the can. With the cooking complete, the can was removed from the boiling water and the filler cap was placed in position and soldered (Figure 35-9). The solder used in these fastening operations contained about 90 percent lead and 10 percent tin.

This type of canned food was carried and eaten by the explorers, and 129 people died from food poisoning. The reasons for this tragedy were faulty soldered joints and a very high lead content solder which caused the tinned food to spoil.

The solder used in modern can technology contains little or no lead. This safe solder is an alloy of tin, silver, antimony, and copper. Many modern cans are deep-drawn to form a one-piece container, to which a lid is attached by roll-seaming after filling with a food or beverage. Today, the millions of people who consume the food contained in such cans have little need to worry about food poisoning.

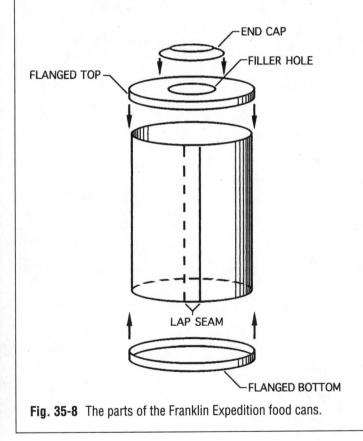

Fig. 35-8 The parts of the Franklin Expedition food cans.

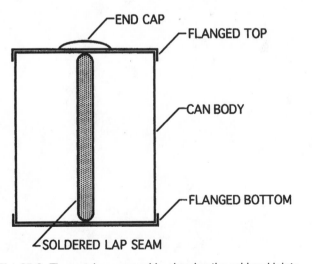

Fig. 35-9 The metal can assembly, showing the soldered joints.

Soldering a Box

The sides of a simple box are sometimes soldered to the bottom. Often, the sides are hard-soldered together before the bottom is soft-soldered to the sides. Solder the bottom as follows:

1. Clean the surface with steel wool. Pay special attention to the area of the bottom to be soldered.

2. Hold the bottom to the sides with black-iron wire, making sure that the joint fits tightly.

3. Add a little flux to the joint.

4. Hold the joint over the flame of the torch until the flux boils. Move the box back and forth to heat the joint evenly. Reapply a little flux.

5. Touch the joint with wire solder until it flows along the edge. Work quickly to keep oxide from forming.

KNOWLEDGE REVIEW

1. What is *soldering?*
2. Name five kinds of heating devices commonly used in soldering.
3. What is soft solder made of?
4. How does metal oxidation hinder soldering?
5. Which kind of flux can be used on electrical connections? Which cannot?
6. What must you do when the point on a soldering copper becomes coated with oxide?
7. How can you remove oxide from a bright metal, such as copper?
8. How much should the soldering copper be heated?
9. When soldering a seam with a soldering copper, where do you put the solder?
10. What is *sweat soldering?*
11. Study the chemistry and physics of soft soldering, and report on (1) applied heat, (2) capillary action, and (3) chemical change of fluxes.

Unit 36

Spot Welding

OBJECTIVES

After studying this unit, you should be able to:

- Define *spot welding.*
- Tell how spot welding compares with rivets for holding metal parts together.
- Name two kinds of spot welders.
- List the parts of a spot welder.
- Describe the process of spot welding.
- Name two kinds of spot welders used in industry.
- Describe the adjustments that can be made on industrial spot welders.

KEY TERMS

seam welding   spot welding

Spot welding is a resistance-welding process in which the heat is generated by resistance to an electric current. Spot welding is used frequently in sheet metal fabrication. Two pieces of ferrous sheet metal are held together under slight pressure between two copper electrodes. A heavy electric current is then passed through the metal. The resistance of the metal to the current heats those spots on both sheets where the current passes to a plastic, or soft, state. The two pieces are thus welded together (Figure 36-1). If a spot weld is made correctly, it is stronger than a rivet of the same diameter. Nonferrous metals cannot be spot welded.

Spot-Welding Machines

Spot-welding machines include portable models that are operated by hand (Figure 36-2) and floor models that are operated by foot (Figure 36-3). Spot welders have controls for regulating pressure, heat, and time. Many floor models are water-cooled to protect the electrodes.

Spot Welding

Follow the steps below to complete a spot weld.

1. Be sure the two pieces of metal are clean. A small spot welder will weld clean pieces of mild steel up to 1/8 inch in combined thickness or two pieces of 20-gage galvanized sheet.
2. Lap the two pieces to be joined about 1/4 to 1/2 inch.

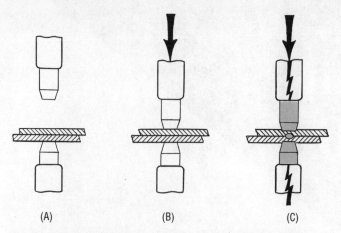

Fig. 36-1 How spot welding is done: (A) Sheets are placed between tips. (B) Pressure is applied to the sheets. (C) Electric current causes sheets to weld by fusion.

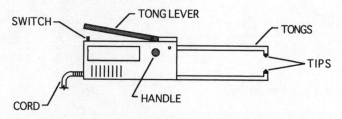

Fig. 36-2 Parts of a hand-operated portable spot welder.

3. Turn on the electric power.

4. Holding the two pieces between the copper electrodes, apply pressure to complete the weld.

Other Applications

Spot welding can also be used to join metal bar and rod stock of many sizes. You must check the capacity of your spot welder before doing some of this heavier work.

Industrial Spot-Welding Processes

Large, computer-controlled, robotic machines are commonly used to spot weld automobile body parts together. This type of equipment is described in unit 84.

Another industrial operation called **seam welding** is often used to join metal sheets. This is a type of spaced spot welding where two wheel electrodes roll along the lapped sheet metal seam to produce a series of spot welds (Figure 36-4). The machine can be programmed to lay the spots at precise intervals, close or far apart, according to the design needs of the product.

When the spots touch or overlap each other, a continuous, strong, and leakproof joint can be made. Special equipment is available to seam weld coated metals, such as galvanized (zinc-coated) sheets, by a combination of resistance heat and roller pressure. This process utilizes a

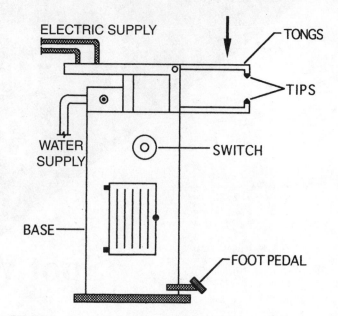

Fig. 36-3 Parts of a foot-operated pedestal spot welder.

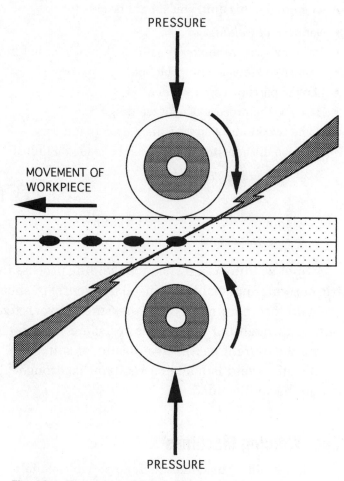
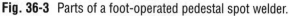

Fig. 36-4 The seam welding process.

continually renewable copper electrode wire. The wire carries the melted coating away from the weld areas and provides a clean area for reliable resistance welds (Figure 36-5). In operation, the electrode starts as an ordinary

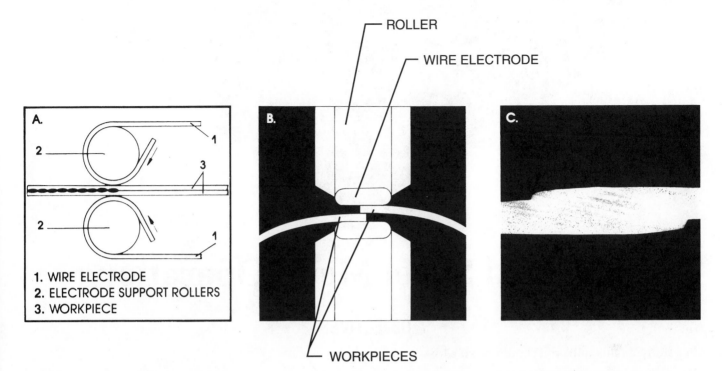

Fig. 36-5 (A) The seam welder diagram. (B) The seam welding process. (C) The completed weld. (Courtesy of Soudronic Ltd.)

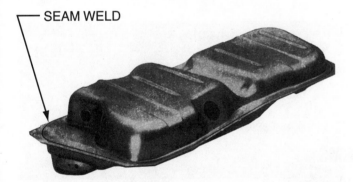

Fig. 36-6 A seam-welded automotive fuel tank. The seam weld follows the perimeter of the tank. (Courtesy of Soudronic Ltd.)

round copper wire. Passing through the support rollers on the welder, it is work-hardened, elongated, and reshaped to match the particular welding conditions. After the welding, the wire is salvaged as copper scrap.

Round, rectangular, flat, and irregularly shaped products can be processed by this method. The auto-

motive fuel tank in Figure 36-6 is a good example. It was produced on a computer-controlled machine with a robotic arm which followed the contours of the tank to complete the weld.

KNOWLEDGE REVIEW

1. What is *spot welding?*
2. What generates the heat in spot welding?
3. How does the strength of a spot weld compare with that of a rivet?
4. What other metal pieces can be spot welded besides metal sheets?
5. Study the metal products in your home. See if you can identify parts that were assembled by spot welding.

Unit 37

Industrial Sheet Metal Operations

OBJECTIVES

After studying this unit, you should be able to:

- Describe the theory of cold working.
- Identify and describe the twelve basic industrial metal-shearing operations.
- Explain the parts and functions of metalworking die-sets.
- Explain the difference between progressive and transfer dies.
- Explain the difference between metal bending and metal drawing.
- Describe the several kinds of bending dies.
- Identify and describe the common shearing and forming machines.
- Explain metal spinning.
- Explain the roll forming process.

KEY TERMS

annealing	drawing	press-brake	shearing
bending	forming	progressive die-sets	slitting
blanking	lancing	punching	slotting
cold working	nibbling	roll forming	spinning
cutting off	notching	shallow drawing	transfer die
deep drawing	parting	shaving	trimming
die-set	perforating		

Most of the skills you learned in part 2 are known in industry as **cold working** methods of changing the sizes and shapes of metal sheets. These changes occur when the metal is at room temperature, or below the point at which it softens. The two basic cold working methods are **shearing** and **forming**. Industry also uses these processes to convert metal sheets into usable products. In industry, *blanking* is one type of shearing, and *deep drawing* is one type of forming. Hot working process examples include flame cutting and forging.

In product manufacture, cold working is preferred because it is less expensive, easier to do because you are working with cool workpieces, sizes are more accurate, and it produces a better finish. Also, the metal grain structure follows the contour of the product shape, resulting in a stronger piece. However, cold-formed metal can become brittle, or *work-hardened.* At this point, it must be **annealed,** or softened with heat, before the forming can continue. The major disadvantage of cold working is that it can distort the grain of the metal and weaken it or cause it to crack.

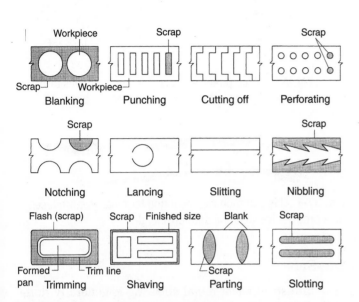

Fig. 37-1 Chart of shearing operations.

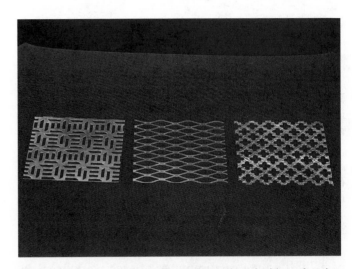

Fig. 37-2 These sheet samples were produced with perforating die-sets. The center example is made with slitting dies.

Shearing Operations

A chart of industrial metal shearing processes is shown in Figure 37-1. In each example, note the difference between workpiece and scrap. Wood, plastic, paper, leather, and cloth sheets can also be sheared by the same or similar methods.

Blanking

Blanking is a shearing operation used to produce workpieces (blanks). Blanks are usually cut from a larger piece of material on a press machine using a punch having the shape of the blank, and a die which mates with the punch. This punch and die arrangement (properly called *die-sets,* commonly called *dies*) is also used for most other shearing operations.

Punching

When holes of various shapes and sizes are cut into a workpiece, the process is called **punching.** Punching is also called *piercing.*

Cutting Off

Workpiece blanks are also produced by **cutting off,** in which shearing occurs along a line extending across the width of the stock. As shown in Figure 37-1, blanks are generally nested for greater material economy, and require only simple and inexpensive dies.

Perforating

Perforating is a method of producing holes of various shapes in a sheet for use as the material for screens, sieves, or decorative grills (Figure 37-2). In addition, selected areas are perforated to permit fastening of the material with bolts or rivets. Most perforating operations generate several holes in the workpiece with one stroke of the press. (When only one hole is made with each stroke, the operation is called *punching.*) To reduce the load on the equipment, the multiple punches are of different lengths so that they do not pass through the material at the same time.

Notching

Shearing metal slugs from the edges of a workpiece is called **notching.** This process is often used to produce shapes which may be difficult using ordinary blanking methods. Notches also serve to relieve stresses in compound bends and thereby eliminate wrinkles.

Lancing

Lancing operations produce an incomplete shearing cut in workpieces. The example in Figure 37-3 shows several lanced knockout holes in an electrical outlet box. Because no material is removed, no scrap is produced.

Slitting

Slitting consists of cutting coiled or sheet stock lengthwise by passing the material between circular

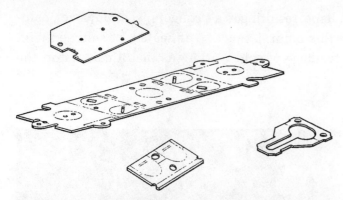

Fig. 37-3 Sheet metal blanks were processed by several shearing operations to produce the parts for an electrical box. How many shearing operations can you identify?

cutters. This provides a fast, efficient method for producing sheet stock for additional operations.

Nibbling

Nibbling is used to produce irregular blanks, especially if only a small number of such workpieces is needed. Nibbling is done by punching a series of overlapping, small holes around the outline of the blank. The edges of nibbled blanks may be rough and usually must be smoothed by trimming. This process is often substituted for blanking when the tooling costs cannot be justified. Nibbling also is used to remove material from the interior of a workpiece.

Trimming

Trimming is the process of separating excess metal from a formed part, such as an automotive oil pan. The excess metal appears as a deformed or uneven flash surrounding a pressed metal part; a waste area is used to grip the workpiece while it was being formed.

Shaving

Shaving also is a method of improving the edge quality of a blanked part or punched hole. Shearing operations seldom leave a perfectly straight edge due to the nature of force cutting. The shaving operation employs dies with extremely close tolerances so that the cut edge is straight and accurate. Shaving also is used to obtain accurate part dimensions.

Parting

Parting is a method of producing blanks by cutting away a piece of unwanted material lying between them.

Slotting

The punching of elongated holes is called **slotting**, to produce long openings in workpieces.

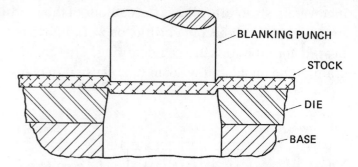

Fig. 37-4 A blanking punch and die. The punch moves through the stock or workpiece into the die to produce the blank.

Fig. 37-5 This common electrical ring terminal was produced on a progressive shearing and forming die.

Die-Sets

Industrial sheet metal shearing is generally done with **die-sets** (Figure 37-4). The upper part of this tool is called the *punch,* and the lower part is the *die.* In operation, the metal sheet is placed on the die plate and the punch is forced through the work and the die to complete the cutting. Most of the shearing processes in Figure 37-1 are done with such die-sets.

Strips of metal (workpieces) produced by slitting can be processed with **progressive die-sets.** Here the workpiece is moved (or progressed) through a series of connected die-cutting operations. Piercing (or punching) and notching are done in the first stage; notching and forming in the second stage; final forming in the third stage; and cutting off or blanking in the fourth stage. The automotive electrical ring-terminal in Figure 37-5 was made with this type of die setup. Note that this terminal was produced with a progressive die which used both cutting and forming die-sets. Progressive dies can have any number of stages, according to the requirements of the product design.

Some products, such as automobile hubcaps and metal cans, cannot be made with progressive dies. Instead, **transfer dies** are used, where the workpiece is moved (or transferred) from one die station to the next, very often with a robotic arm. All types of die-sets can be designed for metal forming as well as cutting operations.

Forming Operations

Forming is giving shape to a workpiece, generally without adding or removing any of the workpiece material, except for perhaps trimming or shaving the

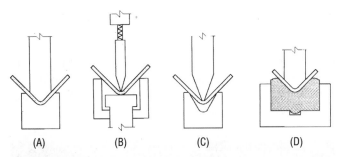

Fig. 37-6 Examples of press-brake die-sets used to bend sheet metal: (A) bottom, (B) three point, (C) air, and (D) flexible.

edges. This is accomplished through bending and drawing operations.

Bending

Bending is the process of forming sheet metal in one direction only, around a straight axis. Earlier, you learned about the cornice brake and the bar folder, both of which are types of sheet metal bending machines. Industry also uses large mechanical and hydraulic **press-brakes** to bend workpieces of thicker metal and greater size.

Some of the kinds of press-brake die-sets are shown in Figure 37-6. Note that the bottom bending type (A) requires that a special punch and die be made for each separate bend, while the three-point and air types (B and C) can produce controlled bends of different angles. The flexible bending die-set (D) requires that only the punch be made for each size and shape of bend. The die is always a tough, flexible rubber pad. It is easier to make a punch than it is a die.

Drawing

Drawing processes shape metal in many directions, such as forming a pan-like container. A simple drawing die-set is shown in Figure 37-7.

Drawing includes both shallow and deep product applications. **Shallow drawing** produces items such as trays, dishes, and automobile hubcaps and trim parts. Because these products have a gentle depth, there is only a moderate movement of material and little distortion. In **deep drawing,** the metal is stretched more severely and requires more powerful equipment and special dies.

A simpler, *single-action press* is needed for shallow drawing. Deep drawing requires a *double-action press,* where the blank is placed in the drawing die; the hold-downs are lowered to secure it; and the punch forced down to draw the blank into the die, thereby forming the round container. Aluminum soft-drink cans are drawn in *triple-action presses.* Here a secondary punch rises up into the die cavity to give a concave shape to the bottom of the can, for strength and stability.

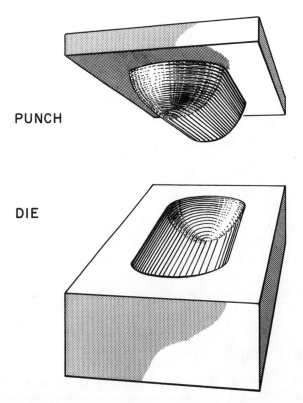

PUNCH

DIE

Fig. 37-7 A simple type of drawing die.

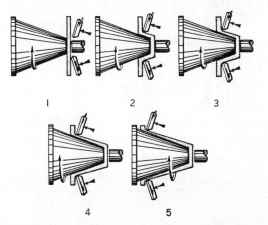

Fig. 37-8 This missile nose-cone section is being formed by spinning. The numbers show the different steps of the process. The operation is very similar to metal spinning, described in unit 45.

Metal Spinning

The **spinning** process is used to produce hollow shapes from sheet metal disks by forcing the metal around a rotating hard steel mandrel (or form). The workpiece moves with the rapidly-turning mandrel. As it does, it is shaped by pressure tools or rollers that squeeze and stretch it around the mandrel. The rollers exert a pressure to draw and thin out the blank (Figure 37-8). Note that the blank is heavier at the base than it is at the top. Many rocket

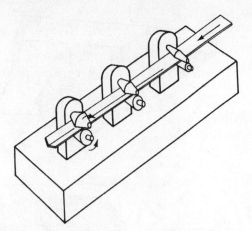

Fig. 37-9 Roll forming is a cold-forming method used to produce angles, channels, and tubes.

and missile nose cones are made this way, as well as industrial parts.

Roll Forming

Long, flat, thin strips of metal are shaped into straight lengths of various cross sections by **roll** **forming.** A roll forming machine can be compared to a series of small rolling mills arranged in a straight line (Figure 37-9). Typical shapes are angles, channels, and tubes. Copper tubing used as water pipe is made by passing the strip through pairs of contoured rollers until its edges meet, and then welding the edges together.

KNOWLEDGE REVIEW

1. List some advantages of cold-working of metals. List a disadvantage.

2. Explain the twelve basic industrial metal-shearing operations.

3. What are die-sets? Describe their two major parts.

4. What is the difference between progressive and transfer dies?

5. Describe the theories of metal shearing and metal forming.

6. Write a research report on shearing and forming machines.

IN FOCUS: *Making Aluminum Baseball Bats*

The familiar metal baseball bats are made from round seamless aluminum tubes. Such seamless tubes prevent any weak areas from developing; welded tubes may crack at the welded area and fail. High-strength aluminum alloys of zinc and magnesium are used to resist denting, bending, and breaking. Aluminum bat manufacture begins with a tube about 2.5 inches in diameter and

from 50 to 60 inches long, depending upon the type of bat to be made. Two bats are made from each tube. The methods used to form the tube into a bat are as follows (Figure 37-10):

1. The tube ends are tapered between two rapidly rotating impacting dies to form the bat handles. This process is called *swaging*. The tube is then cut in two.

2. A special metal spinning process is next used to give the bat its final shape and taper, and to close the head end of the bat. Part of the handle end is cut off to give the bat its proper length.

3. Two heat-treating processes give the bat its maximum hardness and strength.

4. The bat is polished and then either coloranodized or given a baked-on plastic finish.

5. The bat is then filled with a lightweight, flexible plastic urethane foam which hardens to deaden the sound of the bat hitting the ball, and to further strengthen it.

6. The rubber grip is then slipped on the bat handle, and the end knob is laser-welded in place.

7. The bat is decorated with colorful trademarks and tabs to identify its model and size. The bat is now ready for use.

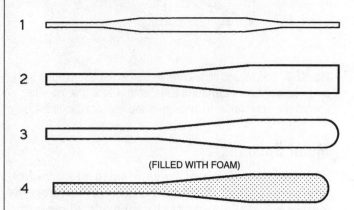

(FILLED WITH FOAM)

Fig. 37-10 The primary metal-forming steps required to manufacture an aluminum baseball bat: (1) Swage ends to proper taper. (2) Cut swaged tube in half. (3) Spin to final shape and round the end. (4) Fill the bat with foam. (5) Fit rubber hand grip and weld end knob in place.

Unit 38

Introduction to Art Metalwork

OBJECTIVES

After studying this unit, you should be able to:

- Describe the art metalworking activity.
- Describe the purposes and activities of metalworking guilds in history.
- Write lists of art metal processes and materials.
- Engage in the designing and making of a quality art metal craft piece.

KEY TERMS

art metalwork   guild   materials   processes

The designing and making of creative craft pieces is known as **art metalwork.** These are generally hand-crafted, one-of-a-kind projects. While machines and equipment are used in making them, there is no notion of mass production in this creative activity.

Historically, metalworking began with the craft guilds of the Middle Ages in Europe, and with similar organizations in Asia. A **guild** is an association of people with a common interest. The craft guilds were therefore composed of people with common craft skills and interests. They were known as silversmiths, goldsmiths, tinsmiths, and gunsmiths, among others, according to their specific skills. These master crafters created fine metal pieces such as the flintlock gun in Figure 38-1. The delicate designs were chiseled and tooled and etched. The guilds were the forerunners of the modern craft and trade unions.

Creating Art Metal Pieces

Art metalworking begins with design, a topic which was described and illustrated in unit 11. It may be helpful to review this information. The sketching of ideas and experiments with materials can lead to some interesting craft ideas. Experiment with structures to create forms for candleholders, vases, or even for a table (Figure 38-2).

Just "playing around" with materials can lead to many exciting results. For example, the clever dog structure in Figure 38-3 was created from thin brass strips wrapped around wooden dowels for temporary support, and then soldered and riveted. To make the animal sculptures in Figure 38-4, shapes were cut from slabs of 1/4-inch aluminum, and joined with adhesives and rivets.

In Figure 38-5, pieces of brass tubing were sawed at angles, and their bottoms plugged with hard soldered discs. They were then buffed and clustered together with adhesives to form an attractive flower vase. The insides of the tubes were treated with a gun-bluing liquid

Fig. 38-1 French, Paris, Presentation flintlock gun, probably made for a Turkish firm Hubert, Jean Le Clerk and Fatou (firm dated 1780–1825), chiseled and gilded steel; silver; wood, 1810 or 1816, 1.: 149 cm, George F. Harding Collection, 1982.2306 side 1. (Photograph © 1996, The Art Institute of Chicago, all rights reserved)

Fig. 38-2 Experimental wire structures.

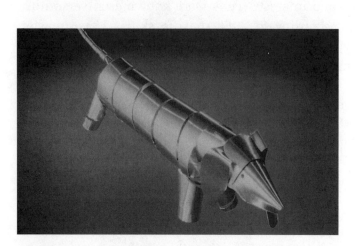

Fig. 38-3 A clever sheet metal sculpture. (Courtesy of Aluminum Company of America [ALCOA])

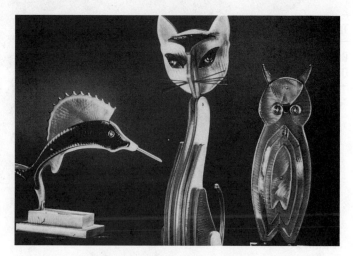

Fig. 38-4 Creative metal sculpturing examples. (Courtesy of Aluminum Company of America [ALCOA])

Fig. 38-5 Many brass-tubing vases are possible as art metal activities.

for contrast, and the entire assembly was then coated with clear satin spray lacquer.

The candy dish in Figure 38-6 was cut to shape from sheet brass and then beaten into a sandbag to give it form. When polished and buffed, three tapered feet were hard soldered in place. A spray lacquer completed the project. These are but a few examples of exciting art metal projects you can create.

Processes and Materials

Besides the tool operations used in bench and sheet metalwork, many special **processes** are used in art metalwork (Table 38-1). In the next seven units, you

Table 38-1   Art metal operations

Cutting Operations	Equipment used
sawing and piercing	hacksaw, band saw, jeweler's saw
shearing	tin snips, scissors, bench shears
abrading	grinder and abrasives
punching	hollow and solid punches
etching	etching liquids and resists

Forming operations	Equipment used
bending	sheet-metal bending tools
spinning	spinning tools and machine
sinking	hammers and forms
raising	hammers and forms
tapping	hammer and tapping tools
chasing	hammer and chasing tools
stamping	hammer and stamping tools
tooling	molding or modeling tools
fluting	hammer and fluting tools
doming	hammer and dapping set
scalloping	scalloping tools
flaring	hammer and flaring tools

Fastening operations	Equipment used
soldering	soldering supplies
brazing	brazing supplies
riveting	riveting tools
adhering	cement and adhesives

Finishing operations	Equipment used
enameling	enameling supplies and kiln
planishing	planishing hammers and stakes
buffing	buffer and compounds
coloring	coloring chemicals and heating devices
coating	lacquers, paints, and plastic sprays

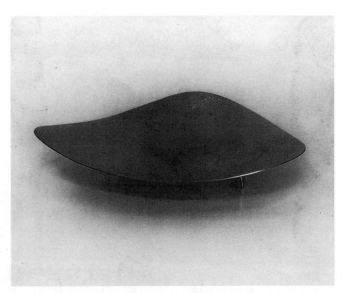

Fig. 38-6 This copper dish is an example of art metal-forming.

range in thickness from 36 to 18 gage, and are popular because they are attractive and easy to work. In addition, metal rods, bars, and wires can become parts of art structures and supports.

The art metalwork experiences and skills which you gain in these units can also lead to career opportunities in jewelry design and repair, craftwork design and manufacture, and fine silverware and tableware industries. Art metalwork also is a satisfying hobby activity because it requires little workspace in the home, and few tools and machines are required.

KNOWLEDGE REVIEW

1. Write a research paper on the art metalworking field and its career and leisure-time opportunities.

2. List some processes related to art metalwork.

3. List some of the common metals used in art metalwork. Why are nonferrous metals generally used?

4. Write a research paper on the history of metalworking guilds in foreign countries. Consider specific regions, such as Africa, Asia, or Europe.

will learn about hand shaping, forming, and spinning. You will become familiar with surface decorating operations such as piercing, planishing, fluting, flaring, coloring, and etching. These are the skills required to produce quality artwork.

The **materials** generally used in art metalwork are copper, brass, aluminum, and pewter sheets. These

Unit 39

Sawing and Piercing

OBJECTIVES

After studying this unit, you should be able to:

- Describe sawing.
- Identify the process for making interior cuts.
- Name the frame used to hold saw blades.
- Tell which direction the teeth should be pointing when installed in the frame.
- Describe the blade that should be used for average cutting.
- Give another name for a bench pin.

KEY TERMS

piercing   sawing

In art metalwork, **sawing** involves cutting intricate outline shapes out of metal. Making interior cutout designs is called **piercing.**

Saw Frames and Blades

The *jeweler's saw* frame is similar to a hacksaw frame. However, it is lighter and takes smaller blades. These blades are from 3 to 6 1/2 inches long (Figure 39-1). The U-shaped frames commonly come in 2 1/2-inch and 5-inch depths. Jeweler's blades are sized from No. 6/0, smaller than a thread, to No. 14, over 1/16 inch wide. For average light work, use a No. 2/0 or No. 1/0. For slightly heavier work, use a No. 1 or No. 2 (Figure 39-2).

If you do not have a jeweler's saw frame, you can use blades with pinned or bent ends installed in a *coping saw* frame.

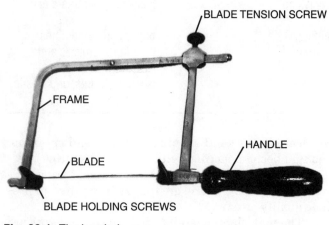

Fig. 39-1 The jeweler's saw.

SIZES   SAW BLADES

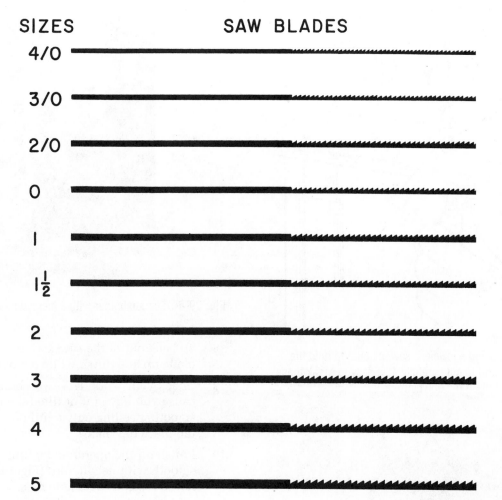

Fig. 39-2 Relative sizes of jeweler's saw blades.

Cutting with a Jeweler's Saw

To get the best results when cutting with a jeweler's saw, follow the guidelines below.

1. Adjust the frame so that it is about one-half inch longer than the blade. Fasten the blade in one end with the teeth pointing *toward the handle.*

2. Hold the frame against the edge of the bench, pressing the handle with your body, and fasten the handle end. If piercing is to be done, a hole must be drilled in the waste stock and the work slipped over the blade before the second end is fastened (Figure 39-3).

3. Fasten a bench pin, or V block, in a vise or on the end of a table. It should be at about elbow height when you are seated so that you can sit down and relax as you cut.

4. Hold the metal over the V block with your left hand. The cutting should be done in the V part of the block. Hold the saw vertically in your right hand.

5. Start sawing in the waste stock with up-and-down movement, using the entire length of the blade.

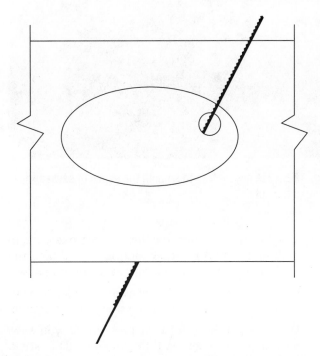

Fig. 39-3 A typical setup for metal piercing.

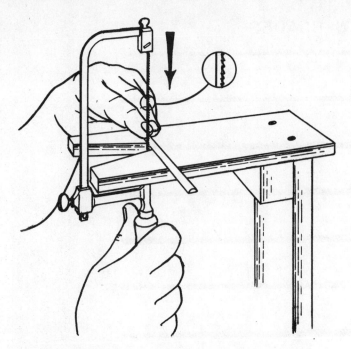

Fig. 39-4 When using a jeweler's saw, be sure to hold the workpiece firmly and to cut carefully. It is easy to break these thin saw blades. Cut on the downstroke.

Fig. 39-6 A candleholder with a pierced base.

Fig. 39-5 Use needle files to smooth the edges of sawn and pierced projects.

The cutting is done on the downstroke. Try to maintain a uniform, easy motion. Avoid bending the blade, for it can break very easily (Figure 39-4).

6. Come up to the layout line at a slight angle, then cut along just outside it. Move the saw a little at a time.

7. Do not apply any forward pressure as you saw, even at a sharp corner. If the blade tends to stick, apply a little soap or wax.

8. To back out of the saw kerf, continue moving the blade up and down while removing it.

9. Check your work. Have you broken the blade because you forced it or tilted the frame? Is the line smooth, needing only a little filing? Have you cut the piece too small?

10. Finish off the operation by filing the sawed edges smooth with needle files (Figure 39-5). Be careful not to break these fragile files.

With practice, you will be able to create many different art metal pieces using these basic methods. The candleholder in Figure 39-6 is a good example of such art metalwork.

KNOWLEDGE REVIEW

1. What is *piercing?*

2. How is a jeweler's saw frame different from a hacksaw frame?

3. How do the blades for the jeweler's saw frames range in size?

4. When the blade is in the frame, which way should the teeth point?

5. How much of the blade do you use for sawing?

6. How can you keep the blade from sticking?

7. Make a tour of a jewelry repair shop or jewelry store. Find out what tools are used by the jeweler to make and repair such things as brooches, pins, earrings, and rings. Report to the class.

Unit 40

Sinking and Raising

OBJECTIVES

After studying this unit, you should be able to:

- Identify another name for sinking.
- Define *raising.*
- Describe how to sink a rectangular plate.
- Describe how to form a tray by beating it down in a form.
- Calculate the amount of metal needed to form a bowl.
- Name the material used in a pickling solution.
- Identify the safety rule when mixing acid and water.
- Describe how to anneal copper.

KEY TERMS

annealing   pickling   raising hammer   sinking
beating down   raising

Shallow trays, plates, and bowls are formed by **sinking** or **beating down** the metal. Bowls, irregularly shaped dishes, and spoons can be formed by **raising.**

Sinking

The process of sinking is done in three common ways: sinking rectangular plates in a vise, sinking or beating down in a wooden or metal form, and beating down over a wood or metal block.

Sinking Rectangular Plates

The process of sinking a rectangular plate is described below.

1. Lay out and cut a piece of metal to a rectangular shape slightly larger than the finished size of the project.

2. Mark a line that shows the area to be sunk.

3. Cut two pieces of hardwood or angle iron equal in length to the longest side of the metal.

4. Get a rounded hardwood stick or metal punch and a mallet or hammer.

5. Fasten the metal in a vise between the wood or metal jaws, with the layout line just above the edge of the jaws.

6. Hold the punch just above the vise jaws and strike it firmly with the hammer (Figure 40-1). Work along from one side to the other, sinking the metal

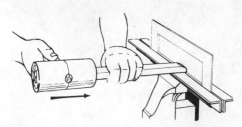

Fig. 40-1 Sinking a rectangular plate.

Fig. 40-2 Forms for beating down. You can make your own by cutting the shape in a piece of plywood and mounting it on another piece of wood.

a little at a time. Rotate the metal a quarter turn and repeat. Anneal as needed. (Annealing is discussed later in this unit.)

7. When the forming is complete, the edge will be stretched out of shape. Trim it with snips and file it. Then decorate the metal. Either the edge or the recessed portion or both can be planished (see unit 41).

Beating Down a Tray into a Form

Follow the steps below for beating down metal into a recessed form.

1. Select a metal or wooden form with a recessed section of the desired size (Figure 40-2). For simple, small trays, the top of an electric-outlet box is good. You can make a round, oval, or irregular form by cutting out the center of a 3/4-inch piece of plywood to the right shape and fastening the outside to another piece of plywood.

2. Cut a piece of metal slightly larger than the size of the tray. Small trays can be made of 24- or 22-gage metal. Larger ones require 20- or 18-gage metal. The metal can be rectangular or roughly the outside shape of the finished project.

Fig. 40-3 Shaping a dish with a wooden hammer.

3. Lay out a line showing the section to be beaten down.

4. Hold the metal over the form or, if you wish, clamp it in place. This can be done by fastening strips of wood around the edge of the metal with C clamps. The metal can also be nailed to the form at the corners. These are waste materials that will be cut away later.

5. Use a metal or wood forming hammer or a round-pointed hardwood stick and a wooden mallet. Begin just inside the layout line. Strike the metal with glancing blows to outline the area to be sunk (Figure 40-3).

6. After you have gone completely around the form once or twice, the metal will require annealing.

7. Continue forming, working in toward the center in rows until the metal is stretched to the bottom of the form. Then strike the metal with angular blows to form the edge of the recess clearly.

8. If the metal is not clamped, remember to keep the edge flat by striking it with the flat of a wooden mallet (Figure 40-4).

9. Finish the project by cutting, filing, and decorating the edge as desired.

Raising

Raising is forming a shaped piece of flat metal by stretching it with a raising hammer over simple forms (Figure 40-5). These forms can be used for any size of bowl. **Raising hammers** are made of steel and come in a variety of shapes and sizes (Figure 40-6).

Raising a Bowl

The procedure and guidelines for raising a bowl are outlined below.

1. Find the amount of material needed by bending a piece of wire to the cross-sectional shape of the

Fig. 40-4 Flatten the edge of the dish with a mallet several times during the forming operation.

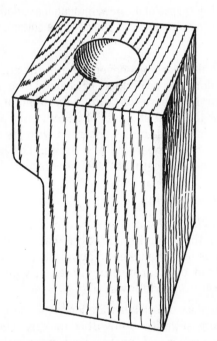

Fig. 40-5 A form for raising a bowl. You can make your own by gouging out an opening in a piece of hardwood.

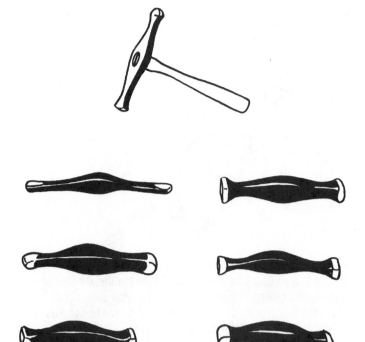

Fig. 40-6 Steel raising hammers.

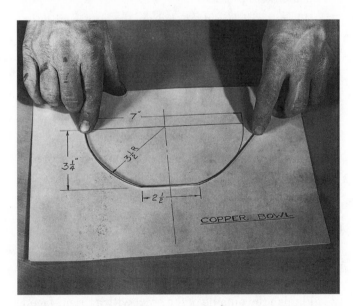

Fig. 40-7 To find the diameter of the metal disk, bend soft wire to the cross-sectional shape of the bowl. You can then measure the length of the wire. Another method is to step off the length with dividers.

bowl (Figure 40-7). By straightening the wire, you can determine the diameter of the disk.

2. Select a piece of metal. Use a compass to lay out the required disk shape. Cut the disk out with tin snips.

3. With a pencil compass, draw several circles, one inside the other, on the disk. The center of the disk should be the center of these circles, which should be spaced equal distances apart. These serve as guidelines for the forming.

4. Fasten the block in a vise with the end grain up. Pound or gouge out a depression in the wood.

5. Place the disk on the block. Then strike it near the edge with the raising hammer. Turn the disk after each blow (Figure 40-8). Make the blows uniform, and overlap them. Never allow the metal to wrinkle so badly that it folds over, as this would ruin the piece. Continue hammering around the circles until you reach the center of the disk.

Fig. 40-8 Raising a bowl. Place the metal disk over the form. Hold it at a slight angle to the form. Strike the metal near the edge with a raising hammer. Turn the disk after each blow. Make the blows uniform and overlapping. As the bowl takes shape, lower the angle at which you hold it.

Fig. 40-9 The shapes of the bowl through the various steps of raising. The completed bowl is shown at the bottom.

6. By this time, the disk will be hard and must be annealed and pickled.

7. When the disk is clean, redraw the circles with the pencil compass.

8. Continue to stretch the metal by hammering it as you turn it (Figure 40-9). As the forming proceeds, lower the disk over the depression. Do not be discouraged because the bowl looks very crude at this time. Anneal and pickle it often. Continue forming the metal until it is about the shape you want.

9. Select a metal stake that has about the curve of the bowl. Place the bowl upside down over the stake. With a wooden or rawhide mallet, work out the dents and irregularities.

10. With a surface-height gage, mark the edge true. Trim it with tin snips, and file the edges smooth.

11. Anneal and pickle the bowl, and clean and polish it in preparation for planishing.

12. Planish the outside of the bowl with a round-faced hammer (see unit 41).

13. Decorate the edge, make a base, or add appendages.

The early forming of the bowl can also be done over a sandbag. Fill a heavy canvas bag about two-thirds full with fine sand. Form a depression in the bag, and proceed as with a wooden form.

Raising skill requires many hours of practice.

Annealing and Pickling

As metal is shaped, it becomes work-hardened. It must then be **annealed,** or softened by a heat-treating process. It also becomes covered with dirt and oxide. It must therefore also be **pickled,** or cleaned in an acid bath. During forming, these two things are done at the same time. After the final shape is obtained, only pickling is needed. This discussion is concerned with the annealing and pickling of copper and aluminum. The hardening, tempering, and annealing of steel are discussed later.

Pickling Solutions

A *pickling solution* is a mixture of acid and water. Remember that pickling solutions can be dangerous. Proper steps must be taken to avoid burning your skin, injuring your eyes, or burning holes in your clothing. Mix the solution as follows:

1. Get a glass or earthenware jar large enough to hold the largest project you will be making. The jar must have a cover and must be kept in a well-ventilated place, since acid fumes rust steel tools rapidly.

2. Pour a gallon or multiple of a gallon of water into the jar to fill it half to three-fourths full.

3. Slowly add ten ounces of sulfuric acid for each gallon of water, stirring the solution with a glass or wooden rod to keep it from overheating.

 CAUTION: Always pour the acid into the water, never the reverse.

Annealing and Pickling Copper and Its Alloys

For a project made of copper or a copper alloy, the procedure for annealing and pickling is as follows:

1. Heat the metal to a dull red over a Bunsen burner, with a torch, or in a soldering furnace. If you use a soldering furnace, be careful that the hot copper does not touch any old solder. This would pit the metal.

2. Heat slowly and evenly to bring the entire piece up to annealing temperature.

3. Pick up the heated article with copper tweezers or tongs, and *slide* it into the pickling solution. *Never drop it,* making a splash. The rapid cooling will anneal the metal and clean off the dirt and oxide.

4. Remove the article from the solution with the tweezers or tongs. Clean it under warm running water.

Annealing can also be done with water instead of pickling acid, but this does not remove the oxide.

Pickling the Final Shape

When the forming is over and you do not want to soften the project again, place it in the solution without heating. Leave it there five to ten minutes. Take it out with tweezers, wash it under warm running water, and dry it in clean sawdust or with paper towels. *Never touch the project with your fingers after it has been cleaned for the last time in preparation for adding a finish.*

KNOWLEDGE REVIEW

1. In what two ways can metal be shaped into trays, plates, and bowls?

2. In sinking a rectangular plate, how do you position the metal in the vise?

3. What thickness of metal do you use for beating down a small tray into a form? A large tray?

4. What can you use to find the diameter of the piece from which the bowl is to be formed?

5. What is the purpose of drawing circles on the workpiece?

6. What kind of hammer is used for the forming?

7. What tool is used to work out surface irregularities after the forming?

8. What must be done to work-hardened and dirty art metals?

9. What is a pickling solution?

10. What must you always remember about mixing acid and water?

11. What color is copper at annealing temperature?

12. In annealing, should the article be heated quickly or slowly?

13. How should an article be placed in the pickling solution? How should it be removed?

14. For pickling only, how long should an article remain in the pickling solution?

15. Find out how formed metal parts such as body parts for cars are manufactured, and make a report to the class.

16. Make a report on the methods of forming used in industry.

17. Study the chemistry and physics of annealing and pickling metal. Tell what happens and why.

Unit 41

Planishing and Decorating

OBJECTIVES

After studying this unit, you should be able to:

- Describe planishing.
- Explain how to planish a deep bowl.
- Describe the method used to planish a tray.
- Explain how to calculate the number and position of edge decorations.
- Describe doming.

KEY TERMS

doming   planishing

The surface or edge of a tray, plate, bowl, dish, or bottle can be beautified in many ways. A number of these are described in this unit.

Planishing

Planishing is making the surface of metal smooth by hammering it (Figure 41-1). It is similar to peening except that planishing hammers and stakes are used. Planishing removes blemishes, bruises, and irregularities that have developed during the forming. On deep dishes or bowls, the outside surface is planished. On shallow trays or plates, however, the upper surface and sometimes only the edge are planished. To do a good job of planishing, you must have: (1) a smooth hammer (Figure 41-2), (2) a smooth stake (Figure 41-3), and (3) clean metal.

Planishing a Deep Bowl or Dish

Follow these guidelines for planishing a deep bowl:

1. Choose a good planishing hammer and a round-faced stake with about the same curvature as the bowl. Both should be free of nicks and scratches; use a buffer and polishing compound to remove any that are present. Fasten the stake in a vise.

2. Draw a few circular pencil lines around the outside of the bowl as guides.

3. Hold the bowl over the stake with your fingers underneath and your thumb on top. This will let you hold and guide it securely. Keeping your elbow close to your body, pound with a regular rhythm and uniform wrist movement. Beginners often strike the metal unevenly and too hard. When you pound the right way, you hear a clear ring as the hammer strikes the metal. When you do it the wrong way, you hear a dull thud.

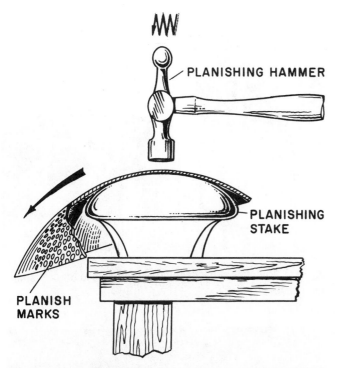

Fig. 41-1 Planishing a bowl. Begin at the center of the project and work outward. Have the blows overlap slightly. Keep the hammer strokes uniform.

Fig. 41-2 Common shapes of planishing hammers. Each hammer produces a different hammer mark.

4. Begin to planish at the center. Do one area at a time, moving the article a little after each blow. Work from the center outward, following the circles. Have the blows overlap slightly. Rub the surface with fine abrasive cloth to highlight the areas that have been planished.

Fig. 41-3 Planishing stakes.

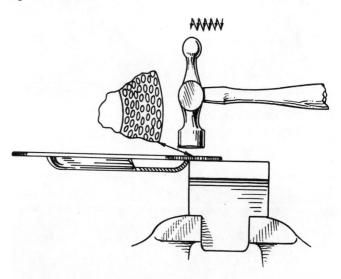

Fig. 41-4 Planishing the edge of a tray.

Planishing Trays or Dishes

The procedure for planishing trays or dishes is the same as described above, except that you must support the surface to be planished over solid metal or a lead block (Figure 41-4).

Layout for Edge Decoration

To lay out the position for decorations that are to be repeated around an edge, do the following:

1. Cut a disk of paper of the same size as the project. Fold the paper into the number of parts in the design.

2. Clip the corners of the folded paper, open it up, and lay it on the metal. Mark the places for the decorations with a pencil or with chalk (Figure 41-5). If necessary, draw a line from the edge to the center and locate the exact position of the area to be decorated. This can also be done by trial and error with dividers.

Shaping an Edge

After the edge has been shaped and planished, it should be cut to finished size, filed, and smoothed with

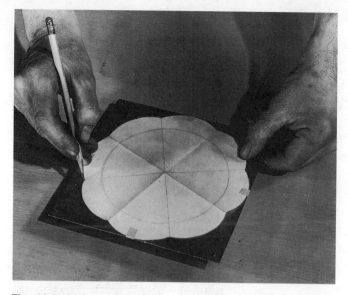

Fig. 41-5 Divide an object into equal parts by folding a piece of paper, clipping the corners, and using it as a pattern.

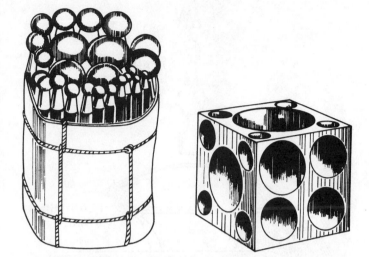

Fig. 41-7 Punches and a dapping block are used to form small round shapes in metal.

Fig. 41-6 To make a dome, hold the metal over a small conical depression in a wooden form. Strike this spot with a ball-peen hammer or a small round-faced punch.

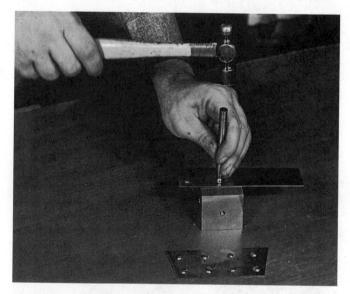

Fig. 41-8 Using a dapping block to form half-spheres.

abrasive cloth. In addition, you may further decorate the edge by filing recesses in it at regular intervals. Strike the edge at right angles with a dull chisel to raise or shape it or to add other decorations.

Doming

A **dome** is a raised, rounded shape that can be placed at equal intervals or in groups around the edge of an object. Domes can also serve as feet on the bottom of a dish or tray. There are three methods for making domes:

1. Cut or gouge a small conical shape in a wood block. Then hold the face of the object over the hole. Strike the back of the metal at the layout position with a ball-peen hammer or with a round-faced punch (Figure 41-6).

2. Cut a piece of pipe that has an inside opening equal to the diameter of the dome. Round the inner edge of the pipe. Choose a ball-peen hammer that will fit part way into the pipe. Fasten the pipe in a vise. Hold the item over the pipe, place the ball-peen hammer, and strike it with another hammer.

3. *Dapping blocks* and *punches* are also used to make domes (Figure 41-7). They are also used for making half-spheres, which can be soldered together to make a ball (Figure 41-8).

KNOWLEDGE REVIEW

1. How is planishing different from peening?

2. How can you tell whether you are handling the planishing hammer in the right way?

3. What can you use to lay out a decoration that is to be repeated around an edge?

4. What is a dome?

5. Make a chart that shows the different kinds of surface decorations and how they can be used. Sketch or make a sample of each kind.

6. Visit a body shop to find out in what ways "bumping" (pounding dents out of fenders) and planishing resemble the forming operations in art metal. Make a report to the class.

Unit 42

Etching

OBJECTIVES

After studying this unit, you should be able to:

- Describe etching.
- Identify what industrial process is similar to etching.
- Describe a resist.
- Name four materials that can be used as a resist.
- Explain the safety rule when mixing water and acid.
- Explain the formula to use when etching with an acid solution.
- Describe the masking tape method of etching.
- Explain the asphaltum varnish etching process.
- Explain how beeswax can be used as a resist.
- Tell what kind of etching solution should be used for copper, brass, and pewter.

KEY TERMS

acid-resist enamel   beeswax   hydrochloric acid   nitric acid

asphaltum varnish   etching   muriatic acid   resist

Etching is a process of surface decoration in which a chemical eats away part of the metal surface. The similar industrial process is called *chemical machining* (see unit 61).

Materials

Resists are materials used to cover the areas of metals that are not to be etched. For simple etching in which only lines or narrow shapes are cut, **beeswax** makes a good resist. For more complicated patterns, **asphaltum varnish** and **acid-resist enamel** applied with a good camel's-hair brush are excellent. For straight-pattern work, the metal can be covered with *masking tape* or *plastic shelving paper*. The area to be eaten by acid can be cut away with a sharp knife (Figure 42-1).

For etching copper, brass, and pewter, use a solution of one part **nitric acid** to one part water. Remember the rule about mixing acid and water.

 SAFETY RULE: Pour the acid into the water. Use a glass or earthenware jar to mix and store the acid.

Fig. 42-1 This bookend plate was covered with plastic sticky paper before it was immersed in the etching solution.

Fig. 42-2 Trace a design from the printed pattern using carbon paper. Hold the pattern and carbon paper in place with masking tape.

To etch aluminum, combine one part **muriatic,** or **hydrochloric, acid** and one part water. A nonacid solution can also be used. It comes in powder form and is added to boiling hot water. Use six tablespoons of powder to one quart of water (75 ml per liter). With this solution, the etching can be done in about thirty-five minutes.

The Etching Process

Follow the procedure and guidelines below for etching metals.

1. The surface of the metal must be buffed and perfectly clean. Clean it either with fine abrasives or by pickling it in an acid bath. Then do not touch it. Your fingerprints can ruin the etching because they leave an oily film that acts as a resist.

2. Apply the resist in one of the following ways:

   a. Masking tape method.

   Cover the article with masking tape and lay out the design on, or transfer it to, this surface. Cut around the area to be exposed with a sharp knife. Carefully lift off the tape.

   b. *Asphaltum varnish* or *acid-resist enamel method.*

   Carefully transfer the pattern to the metal by tracing with carbon paper (Figure 42-2). Paint the varnish or enamel on the areas to be protected from the acid (Figure 42-3). If the piece is flat or if the entire article must be put into the acid bath, both surfaces must be covered. The edges, in this case, should be given two or three coats. If you are etching a tray or bowl, however, you only have to treat the top surface. This is because you can pour the acid into the object. Be sure to apply enough resist. The area should look black. Allow it to dry for twenty-four hours.

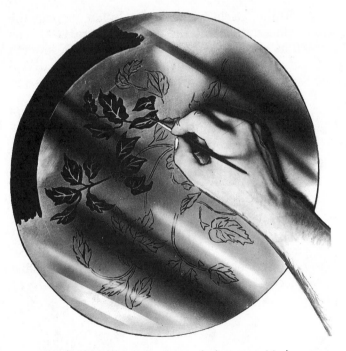

Fig. 42-3 Decide what parts of your design are not to be etched. Coat those parts with asphaltum varnish or acid-resist enamel. Apply the resist carefully with a camel's-hair brush.

   c. *Beeswax method.*

   Melt the beeswax, coat the surface of your project with a layer about 1/16 inch (1.5 mm) thick. Use an awl or scriber to trace the outline of your design through the beeswax coating. Be sure to press hard enough to reach the metal. Scrape the beeswax off any large areas that are to be etched.

Fig. 42-4 Apply an acid or a nonacid solution by pouring it into the tray.

Fig. 42-5 The completed etched plate.

3. Put the article in the acid solution, or pour the acid solution into the article. (Figure 42-4). The time needed for etching depends on the strength of the acid and the depth of etching wanted. Approximate times are: one hour for copper, one and a half hours for brass, fifteen minutes for pewter, and thirty minutes for aluminum. Stir the acid during the etching with a narrow wooden stick wrapped in cotton.

4. Take the article from the solution, rinse it in water, and then remove the resist. If asphaltum varnish has been used, clean the article with paint thinner.

5. Check your work. Is the design clearly outlined? Is the edge ragged? This would be caused by too rapid action or by a poorly applied resist. The result of good and careful work is a fine finished piece (Figure 42-5).

KNOWLEDGE REVIEW

1. In etching, what happens to a metal surface?

2. What are the materials called that protect the areas not to be etched? List three of these materials.

3. What acid is used to etch copper, brass, and pewter?

4. Name two methods of applying the resist.

5. What does etching time depend on?

6. Describe the chemical action that takes place in etching. Find your answer in an encyclopedia.

Unit 43

Soldering Art Metal

OBJECTIVES

After studying this unit, you should be able to:

- Define *hard soldering.*
- Identify another name for the hard soldering process.
- Describe the kinds of metals that are hard soldered.
- Name five kinds of heating devices for hard soldering.
- Describe the grades of silver solder.
- Identify the temperature of the colors "just visible red," "dull red," and "cherry red."

KEY TERMS

blowpipe   MAAP gas kit   silver solder   sparklighter

hard soldering

A metal-joining process that utilizes silver-alloy solders is called **hard soldering** (also known as *silver soldering* or *silver brazing*). This is a brazing process used to join copper, brass, silver, gold, and some steels. Hard soldering produces a neater and tougher joint than soft soldering, and is generally used in art metalwork.

Tools and Materials

Hard soldering is done at higher temperatures than soft soldering. It therefore requires the special equipment shown in Figure 43-1. The **blowpipe** is a simple torch connected to air and natural gas sources found in many shops. The valves can be adjusted to produce the correct flame pattern. This system is especially desirable for fragile workpieces where a low-pressure flame is needed.

The **MAAP gas kit** consists of a MAAP (methyl acetylene propadiene) tank, an oxygen tank, hoses, and a nozzle (Figure 43-2). It is convenient to use and can produce temperatures up to 5,000°F.

The familiar propane torch (unit 35) also can be used for small jobs; and the common gas welding outfit described in unit 66 for larger pieces.

All of the above gas torches should be lit with a **sparklighter.** Never use a match because you may suffer a serious burn. Check with your instructor before using any of the hard soldering torches.

Silver solder comes in many different compositions or formulas for commercial and industrial use. One commonly used in the school shop is made of 70 percent silver, 20 percent copper, and 10 percent zinc, and can be worked at about 1,300°F. There are other compositions designed for specific materials, and for different

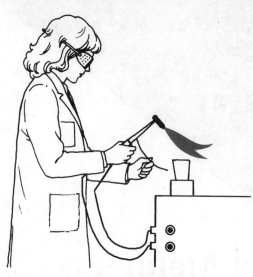

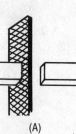

Fig. 43-3 The hard-solder brazing process: (A) filing and (B) applying flux.

Fig. 43-1 A blowpipe with natural gas and air from a bench source.

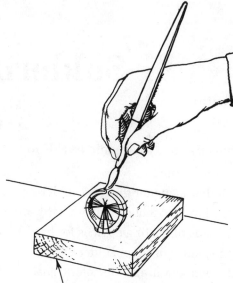

Fig. 43-4 Brush flux onto the wired joint.

temperatures. Special *fluxes* must be used when hard soldering.

When two or more hard soldered joints must be made on the same project, solder the first joint and let it cool. Then coat the joint with a mixture of *buffing rouge* and *paint thinner* and let it dry. (Special heat-sink pastes are also available.) This will keep the first joint from remelting when the second joint is soldered.

The Hard Soldering Process

Using a finger ring project as an example, study the following procedure to do the soldering:

1. Clean and true the joint with a file. Then file a slight V notch at the joint to hold the solder (Figure 43-3).

2. Bend the joint ends together until they touch. Bind the ring with wire to keep it from springing apart. Place the ring on a brick or a charcoal block so that it does not move. Apply flux (Figure 43-4).

3. Flatten the solder if necessary. Cut it into tiny pieces on a sheet of clean paper. Flux the joint with a small brush.

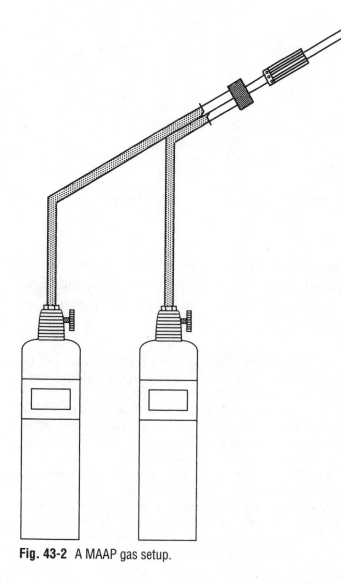

Fig. 43-2 A MAAP gas setup.

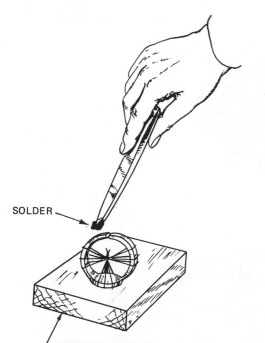

Fig. 43-5 Applying small bits of solder to the joint with tweezers.

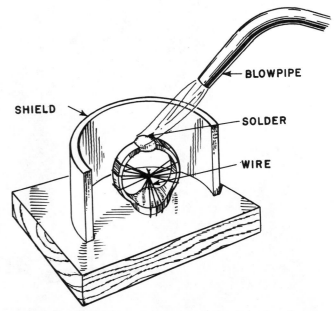

Fig. 43-6 A shield placed behind the object helps concentrate the heat. This makes it easier to solder.

4. Carefully apply the small bits of solder along the joint with tweezers (Figure 43-5). Use slightly more than is necessary to fill the joint. The camel's-hair brush is handy for picking up the bits of solder.

5. With the blue part of the flame, preheat the joint until the flux dries out. This will hold the solder in place.

6. With the inner point of the flame, heat the joint to the correct temperature. Follow the heat colors as a guide. If possible, heat the joint from the back (Figure 43-6). Avoid putting the heat directly on the solder. Do the soldering quickly. Do not hold the flame on the article too long. When the solder melts, it will flow or run fast. Remove the heat right away. If pieces of different size or thickness are being joined, apply most of the heat to the thicker piece.

7. Put the article in a pickling solution, then rinse it with water.

8. Check your work. If the solder rolls into balls and does not join, either the work is not clean or there is not enough heat (the solder is being heated and not the joint). If the joint melts away or the silver burns and smokes, the flame is too hot.

Other Fastening Methods

For very small workpieces and some jewelry findings, soft solder or adhesives are sometimes used. These are described in units 34 and 35.

KNOWLEDGE REVIEW

1. Which metals usually require hard soldering?

2. How does a hard-soldered joint differ from a soft-soldered one?

3. What are some good sources of heat for hard soldering?

4. What is silver solder?

5. How can you keep one hard-soldered joint from melting while you solder another one nearby?

6. Why is it necessary to bind the joint together before soldering?

7. Is the flame applied to the joint or to the solder?

8. How do you know when to remove the heat?

9. What is wrong if the solder rolls into balls and does not run?

10. Compare hard soldering and brazing. Describe the differences in materials and procedure.

Unit 44

Buffing and Coloring

OBJECTIVES

After studying this unit, you should be able to:

- Describe the processes of buffing and coloring.
- Describe the types of compounds and their uses.
- Use buffing equipment properly and safely.
- Use coloring chemicals properly and safely.
- Use cold conversion finishes properly and safely.

KEY TERMS

buffing   cold conversion   coloring   heat coloring

buffing compound

Two common methods used to improve the appearance of an art metal project are buffing and coloring. **Buffing** gives a piece a smooth, bright luster. **Coloring** changes the tone of the project metal, giving it a different look.

Buffing Tools and Machines

All small scratches and imperfections must first be removed by polishing (see unit 23). A *buffing machine* (similar to a polishing machine) produces a high shine or luster. A *buffing wheel* (or buff) can also be fastened to a lathe or drill press. These wheels are made of cotton, flannel, or felt. The sides are sewn together, leaving a soft outer edge. To use them, you must first coat the wheel edge with an abrasive material called a **buffing compound**. Use a different wheel for each kind of compound.

The most common compounds are pumice, tripoli, rouge, rottenstone, and whiting. Their descrip-

tions and uses are found in Table 44-1. Other greaseless compounds are available which are somewhat cleaner to use. They are available in stick or cake form.

The Buffing Process

The basic procedure and guidelines below will help you with buffing a project.

1. Mount the wheel on the machine. These generally screw onto a tapered, threaded shaft. Press the stick compound lightly against the outer surface of the wheel as it turns (Figure 44-1). Wear goggles and tuck in loose clothing when buffing.

Table 44-1 Materials for buffing metal

Name	Use	Source	Color
pumice	for scrubbing and cleaning; also for cutting down and polishing	powdered lava	white
tripoli	for polishing brass, copper, aluminum, silver, gold, platinum	decomposed limestone	yellowish brown
rouge	to burnish or produce a high color or luster	red iron oxide	red
rottonstone	for polishing precious stones, soft metals	decomposed shale	reddish brown or grayish black
whiting	for final polishing	calcium carbonate (pulverized chalk)	white

Fig. 44-1 Apply a stick or cake of buffing compound by holding it lightly against the edge of a soft wheel. The buffer should operate at a speed of 1,750 revolutions per minute.

Fig. 44-2 Hold the project lightly against the face of the wheel, a little below center.

2. Move the project back and forth against the wheel below its center (Figure 44-2). Put more compound on the wheel as needed. Wipe the project clean.

3. Change to another wheel and a finer compound to finish the buffing.

4. Wash the project in hot water and dry it with a clean cloth. Hold the clean piece in a gloved hand to prevent fingerprints.

5. To preserve the high luster, coat the project at once with a brush or spray lacquer.

Satin Finish

A low-luster (not shiny) satin finish can be applied by lightly holding the project against a soft wire wheel (Figure 44-3). This gives an attractive, softly scratched appearance. Some compounds can also produce such a finish.

Fig. 44-3 For a satin finish, a wire brush can be fitted on the grinder. Always wear safety glasses; fine bits of wire from the wheel can fly off and injure your eyes.

Table 44-2   Formulas for chemical finishes for nonferrous metals

Brown

4 ounces (120 g) iron nitrate
4 ounces (120 g) sodium hyposulfite
1 quart (1 L) water

Dull matte

1 pint (0.5 L) hydrochloric acid
1 pint (0.5 L) 40% ferric chloride solution

Green antique

3 quarts (3 L) water
1 ounce (30 g) ammonium chloride
2 ounces (60 g) salt

Red

4 ounces (120 g) copper sulfate
2 pounds (1 kg) salt
1 gallon (4 L) water

Pickling and cleaning

1 pint (0.5 L) sulfuric acid
1 pint (0.5 L) nitric acid
4 pints (2 L) water

Satin dip

1 pint (0.5 L) hydrofluoric acid
3 pints (1.5 L) water

Coloring Metals

To produce light and dark tones on a metal surface, you can color it with chemicals. Remember that these are dangerous.

 SAFETY RULE: Wear an apron, gloves, and eye and face protection when using coloring chemicals.

Always check with your instructor before using chemicals. A chart of these is shown in Table 44-2.

The Coloring Process

The chemical often used on brass, copper, and silver is liver of sulfur, or potassium sulfide. Apply this as follows:

1. Dissolve a piece of liver of sulfur about the size of a small marble in a gallon of hot water.

2. Be sure that the article is clean. If necessary, scrub with pumice and water. Hold the workpiece with tongs, and wear gloves.

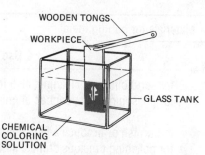

Fig. 44-4 Coloring with chemicals. Hold the article with wooden tongs or with pliers whose jaws are covered with masking tape.

3. Dip the article until it turns the color you want. Copper and brass colors vary from brown to black. Silver turns black (Figure 44-4).

4. If you wish, hand-polish the article with steel wool to highlight certain areas. In some work, the crevices are usually left black and the rest highlighted.

Cold Conversion

Cold conversion is an oxide finishing method used on many art metal pieces. *Gun-bluing liquid* chemicals are a good example. The workpiece can be dipped into it, or it can be rubbed on with a clean cloth. These chemicals penetrate the metal surface to produce an attractive and durable finish, which can be further protected with lacquer spray. Separate cold conversion finishes are available for ferrous and nonferrous metals. Some come in kit form, containing a cleaner, cold finish, and coating. Follow the directions on the instruction sheet. Remember to wear protection, and apply as follows:

1. Clean the workpiece with soap and warm water or a cleaner, and rinse. You generally do not have to wipe dry.

2. Most cold finishes work better on a slightly warmed workpiece. Use a heat gun for this, or place it in an oven.

3. Hold the piece in a clean glove or tongs, dip a clean cloth into the liquid, and rub the moist cloth back and forth on the workpiece (Figure 44-5). You can control the depth of the color by rubbing harder and longer. You may wish to use a test piece to become familiar with the process.

Coloring with Heat

Attractive, interesting colors can be given to metals through the process of **heat coloring.** First, thoroughly clean and buff the article. Then heat it with a gas torch or over a flame. Watch the colors as they appear, and remove the workpiece when the desired colors are

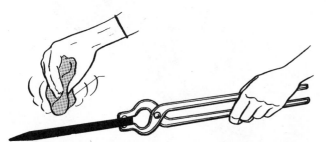

Fig. 44-5 Applying gun-bluing solution to a steel workpiece.

present. Quench immediately in water or oil, and rub gently with steel wool to highlight the metal.

KNOWLEDGE REVIEW

1. Describe the difference between buffing and coloring.
2. List some of the materials used for buffing.
3. How can you give a project a satin finish?
4. What chemical is most often used to color brass, copper, and silver?
5. How can you color metals without using chemicals?
6. List some safety precautions to follow when using chemicals.
7. What are cold conversion finishes?
8. Write a research report on gun-bluing operations used by companies that manufacture sport rifles.

Unit 45

Metal Spinning

OBJECTIVES

After studying this unit, you should be able to:

- Describe metal spinning.
- Identify what kind of chucks are used for spinning.
- Describe the major parts of a spinning lathe.
- Name six kinds of spinning tools.
- Describe the materials used for making chucks.
- Name two types of chucks.
- Identify what metals are best suited for spinning.
- Describe how to spin a simple bowl.
- Describe how to polish a product.

KEY TERMS

all-purpose flat tool	chuck	metal spinning	solid chuck
back stick	cutoff tool	pointed tool	spinning lathe
ball tool	follow block	roundnose tool	trimming tool
beading tool	forming tool	sectional chuck	

Metal spinning is a machine shaping process in which a metal disk is revolved on a lathe. As the disk turns, a spinning tool presses it over a form called a **chuck.** Gradually, the disk takes the shape of the chuck. Plates, bowls, and many other objects are made in this way. Some industrial spinning is done by hand in much the same way as in the school shop. Production machines are also used. Spin forming is frequently less expensive than other machining.

Equipment, Tools, and Materials

Various machines, hand tools, and materials are used in metal-spinning operations in industry and the shop. A **spinning lathe,** for example, is a solidly built machine which operates at speeds from 300 to 2,000 rpm (revolutions per minute) (Figure 45-1). Most spinning is done at 1,800 rpm. A *tool rest* is needed that has a crossbar with holes into which fulcrum pins are fitted. A *live,* or *ball-bearing tailstock, center* is

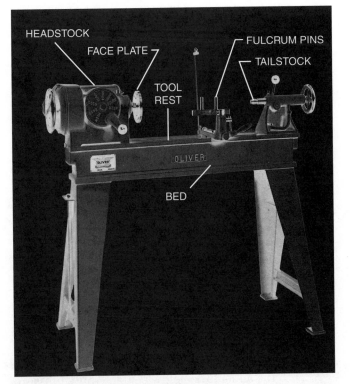

Fig. 45-1 Parts of a spinning lathe.

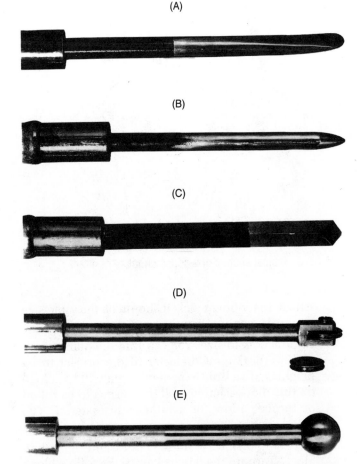

Fig. 45-2 Spinning tools: (A) all-purpose flat, or forming, tool; (B) roundnose, or pointed, tool; (C) cutoff, or trimming, tool; (D) beading tool; (E) ball tool.

used. The pressure against this center holds the metal against the chuck.

Spinning tools come in many forms and shapes (Figure 45-2). They can be purchased or can be made from old wood-turning tools. Common ones include:

1. The **all-purpose flat,** or **forming, tool,** used during most of the spinning. One part of the tip is flat for smoothing purposes. The other side is rounded for "spinning to the chuck." The place where the flat and round sections join is also rounded, but with a smaller radius.

2. The **roundnose,** or **pointed, tool,** used for hooking the disk to the chuck at the start of the spinning and for shaping small curves.

3. The **cutoff,** or **trimming, tool,** used for trimming the extra metal from the edge of the spun object.

4. The **beading tool,** used for turning the edge of a spun project for a beaded lip.

5. The **ball tool,** used for the first steps in spinning hard metals such as brass and steel. It is used to bring the metal near the chuck but not to finish shaping it over the chuck.

6. The **back stick,** which supports the back of the metal as it is spun. This should be about one inch in diameter and about fifteen to twenty-four inches long. One end should have a flat wedge shape. The back stick can also be used to roll the edge. An old broom handle makes a good back stick.

Solid chucks used for shop spinning can be made of wood (Figure 45-3), such as oak, maple, or cherry, or of metal. The **sectional chuck,** which can be taken apart, is used for more complicated work. The chuck can be turned to the desired shape after it has been put on the lathe. A **follow block** fits over the live tailstock center. This is a piece of wood slightly smaller in diameter than the smallest part of the chuck.

The metals best suited to spinning are copper, aluminum, pewter, brass, silver, and certain kinds of steel. Copper and soft aluminum are most commonly used. When spun, copper and brass must be kept soft by annealing.

Lubricants are also used in spinning operations. The lubricant can be yellow laundry soap or a tallow candle.

Spinning a Simple Bowl

The process of spinning a simple bowl is outlined below.

1. Choose a chuck, or turn one in the shape you want. Fasten the chuck to the headstock spindle.

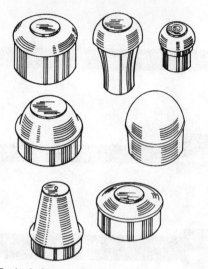

Fig. 45-3 Typical shapes of wooden chucks.

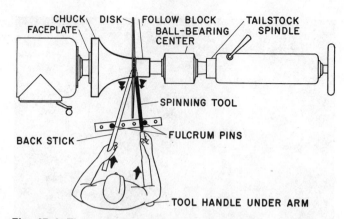

Fig. 45-4 The metal has been hooked over the chuck. Now the waves and wrinkles are being smoothed out with a spinning tool and back stick. Always keep the metal in a cone shape.

Attach the follow block; it should be the same size or a little smaller than the base of the chuck.

2. Cut the disk to size. You can find the diameter by adding the largest diameter of the project to its height. With a little experience, you will be able to do this more accurately. If the edge is to be rolled or turned, allow for a little extra material. The metal should be soft or annealed before you start.

3. Insert the disk between the chuck and the follow block. Turn up the tailstock until the follow block fits snugly against the disk. There are two ways of centering the disk in the chuck:

   a. Turn the lathe over by hand, tapping the disk first on one side and then on the other until it runs true.

   b. Adjust the lathe to the slowest speed. Turn on the power. Hold a back stick between the edge of the disk and the tool rest. Start with the follow block firmly against the disk. Loosen the tailstock hand-wheel a little as you press lightly with the back stick against the revolving disk. When it runs true, quickly tighten the tailstock handwheel.

CAUTION: Centering the disk in the chuck is rather dangerous, so be sure you never stand directly in line with the edge of the disk.

4. Stop the lathe. Adjust the tool rest about 1 1/2 to 2 inches away from the edge of the disk. Adjust the spindle speed to about 900 to 1,200 rpm. Put a fulcrum pin about one inch ahead of the disk.

5. Turn on the power. Never stand directly in line with the disk. It could fly out and hit you. Apply a little tallow candle or yellow laundry soap to both sides of the disk to lubricate it.

6. Hook the disk to the base. Choose a flat, or forming, tool. Hold the handle under your right arm, grasping the tool toward the front of the handle with your right hand. Hold the blade of the tool firmly against the fulcrum pin with your left hand. Place the rounded side of the tool against the disk, with the end a little below center. Apply pressure, moving the point of the tool down and to the left while moving the handle up and to the right to seat the metal against the chuck. Once you do this, the metal cannot fly out of the lathe. Work from the center outward. As you do this, the metal sometimes tends to "dish out," causing it to wrinkle. Anneal as needed.

7. Insert a second fulcrum pin ahead of the disk. Hold the point of the back stick against the front of the disk opposite the flat tool that is against the back of the workpiece. Press equally on both tools. Now move them slowly toward the outside to straighten the disk until it is cone-shaped. Always keep the unformed disk to this shape (Figure 45-4).

8. Stop the machine. Move the tool rest within about one-half inch of the outside of the disk. Start the lathe. Trim the edge. Holding the trimming tool firmly on the rest, press in slowly until the tool cuts all the way around the outside (Figure 45-5). File the trimmed edge lightly to remove the burr.

9. Reset the toolholder about two to three inches from the disk. Put one fulcrum pin a little ahead of the disk and the other behind it. Hold the flat tool against the tool rest, with the handle under your right arm. Move the handle to the right and up to shape the metal against the chuck. Always keep the point of the tool moving back and forth on the metal in an arc of about two inches. If you force the metal in one direction only, it thins out and cracks. Also, never let the point of the tool stop and rest on the metal, as this makes ridges. Always be care-

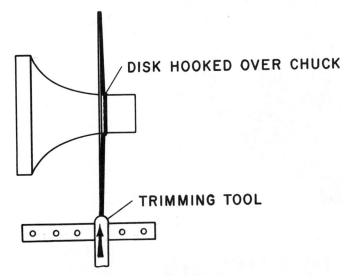

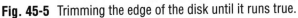

Fig. 45-5 Trimming the edge of the disk until it runs true.

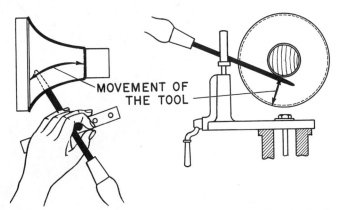

Fig. 45-7 Another way of holding the tool is shown here. Apply pressure on the metal and move it back and forth in an arc, as shown. Notice how the spinning tool moves to follow the arc. Remember that the spinning tool must be guided with the entire body. The pressure of the tool causes the metal to form over the chuck.

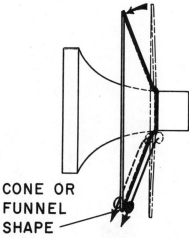

Fig. 45-6 Here is the metal disk before and after it has been shaped. The waves and wrinkles are removed and it is kept in a cone or funnel shape.

ful to keep the edge from "dishing out" and becoming wrinkled (Figure 45-6). On shallow dishes, the spinning can be done quickly and can be finished before the metal becomes hard. If a deeper dish is being formed, anneal the metal at this time.

10. Replace the disk and lubricate both surfaces again. Move the fulcrum pin nearer the headstock. Al-

ways place the pin such that you are forming the metal about one inch on either side of it. Use the back stick as needed to straighten the metal to a cone shape. Work the metal in the curves and around the corners (Figure 45-7).

KNOWLEDGE REVIEW

1. What is the advantage of metal spinning over other machining?

2. On what kind of machine is metal spinning done?

3. What metals are best suited for spinning?

4. When preparing to spin a project, how do you find the diameter of the metal disk?

5. What safety precaution should be taken when spinning metal?

6. What shape should the unformed disk always have while spinning?

7. List some industrial products made by the spinning process.

8. Prepare a bulletin board display of some very large spun products.

Unit 46

Introduction to Forging

OBJECTIVES

After studying this unit, you should be able to:

- Identify the forging processes used in industry.
- Describe what the forging process does to metal.
- Describe the age of the forging trade.
- Describe the equipment needed to do industrial forging.
- Explain how forgings are usually processed after they are cool.
- List the equipment needed to do forging in the school lab.
- Name the metal used in most forgings.

KEY TERMS

blacksmith   forging

Forging is a method of forming metal by hammer blows or pressure. It is used to shape parts where great strength is needed. Forging improves the quality of the metal, refines the grain structure, and increases strength and toughness (Figure 46-1).

Forging by hand was probably the earliest method of forming iron. The trade is as old as civilization. In hand forging, such as that done by the **blacksmith,** metal is heated in the forge and shaped over an anvil with various kinds of hammers and tools. In manufacturing, forging is done to shape metal objects that must withstand great stress, such as crankshafts, gears, and axles. It is very much like the work of the old-time blacksmith. However, machine power is used instead of the blacksmith's arm. Moreover, dies, or molds, replace the hammer and anvil.

Forging Processes

Forging is a hot-metal forming method related to other processes like extrusion and hot rolling.

Basic equipment in production forge shops includes various kinds of power hammers and presses. To make a connecting rod, for example, a bar of metal is heated to white-hot in a furnace. Next to the furnace is the power hammer. The hammer face is shaped something like the outside of the connecting rod. The white-hot metal bar is squeezed between the upper and lower dies. Sharp hammer blows squeeze the hot metal into the dies. Forgings must usually be machined in the machine shop after they are forged.

Most forgings are made of steel. However, brass, bronze, copper, and aluminum are also shaped in this manner. To be really skillful in a hobby involving ornamental-metal or wrought-metal work, you must be able to do hand forging (Figure 46-2).

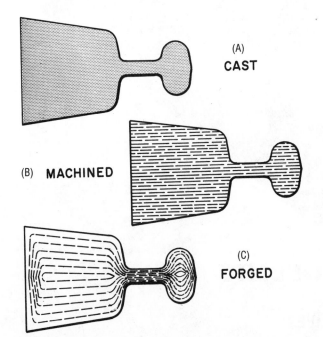

(A)
CAST

(B) **MACHINED**

(C)
FORGED

Fig. 46-1 Grain structure of metal produced by different methods of forming: (A) In casting, there is uniform grain structure throughout the part. (B) In machining, the part is weakened at the thin section. (C) In forging, the grain structure is such that added strength is obtained at the thinner sections.

Forge Work in the School Lab

You will learn to use many different pieces of forge equipment and tools in the school lab. Hammers, tongs, anvils, and anvil tools are commonly used. You must use the forge safely and efficiently. The skills you learn will aid you in making tools such as chisels.

KNOWLEDGE REVIEW

1. What effects does forging have on metal?
2. What is the difference between the work of the blacksmith and modern production forging?

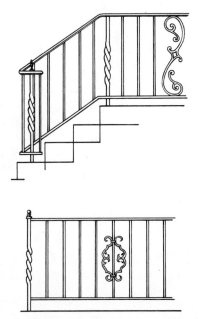

Fig. 46-2 Many forging operations are used in making these attractive ornamental-iron railings.

3. Name two kinds of equipment used in production forging.
4. What are the chief forging metals?
5. Make a list of some of the parts of an automobile that are made by forging. If possible, visit a forge shop.

Unit 47

Hand Forging

OBJECTIVES

After studying this unit, you should be able to:

- Describe hand forging.
- List the equipment needed for hand forging.
- List the steps in lighting a gas or oil furnace.
- Describe tapering and drawing out.
- Describe the process of upsetting.
- Describe how to do bending.
- Name the anvil tools used for special forging operations.
- Identify the safety precautions to follow when lighting a fire.

KEY TERMS

anvil	drawing out	hardy	tapering
bending	flatter	neutral flame	tongs
blacksmith's hammer	forge	punch	twisting
chisel	fuller	swage	upsetting

Hand forging is useful in making repairs on metal parts around the home, farm, or shop. It is also a good way of making tools and ornamental ironwork.

Equipment

For large ornamental-metal projects, a variety of forging equipment is needed. The **forge** may be a small gas furnace to supply the heat for forging (Figure 47-1). **Anvils** (Figure 47-2) are made of cast iron or steel, or with a cast iron base and a welded steel *face,* or top. The steel anvil is best. The *horn* of the anvil is used for shaping circular metal parts. Various anvil tools fit into the *hardy hole.* The anvil must be solidly mounted for forming the metal.

The hot metal from the forge is held with **tongs.** There are many shapes, but the most common are shown in Figure 47-3. For working the metal, there should be at least two sizes of **blacksmith's hammers.** One of 1 1/2 or 2 pounds and another of 3 or 3 1/2 pounds will do all the necessary work (Figure 47-4). Also, various *anvil tools* are needed if much forging is to be done (Figure 47-5). These include top and bottom fullers, top and bottom swages, punches, chisels, hardies, and flatters.

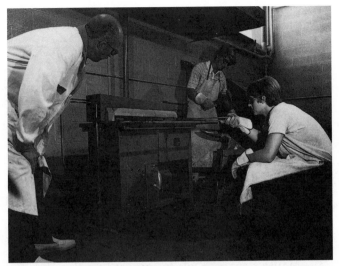

Fig. 47-1 A typical gas forging furnace. Always wear protective clothing when forging metal. (Courtesy of Johnson Gas Appliance Company)

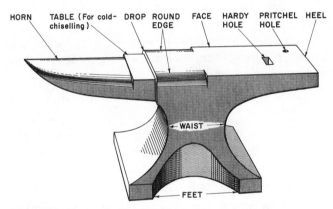

HORN TABLE (For cold- DROP ROUND FACE HARDY PRITCHEL HEEL
chiselling) EDGE HOLE HOLE

WAIST

FEET

Fig. 47-2 Parts of an anvil.

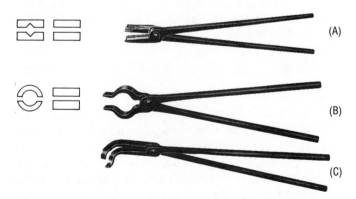

(A)

(B)

(C)

Fig. 47-3 Tongs: (A) flat lip, showing regular and grooved jaws; (B) pickup, showing curved and flat lips; and (C) offset.

Building a Forge Fire

The steps in lighting a typical *gas* or *oil furnace* are as follow:

1. Light a piece of paper and place it in the opening.

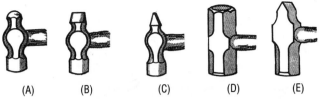

(A) (B) (C) (D) (E)

Figure 47-4 Forging hammer heads: (A) ball-peen, (B) straight-peen, (C) cross-peen, (D) double-face sledge, and (E) cross-peen sledge.

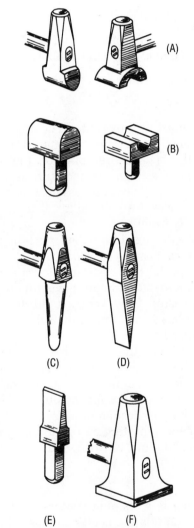

(A)

(B)

(C) (D)

(E) (F)

Fig. 47-5 Anvil tools: (A) top and bottom fullers, (B) top and bottom swages, (C) punch, (D) chisel, (E) hardy, and (F) flatter.

2. Turn on a small supply of air, then turn the fuel valve a little until the furnace lights up.

 CAUTION: Never look into the opening as you turn on the fuel. Always have your instructor present when lighting the forge.

3. Let the furnace heat up.

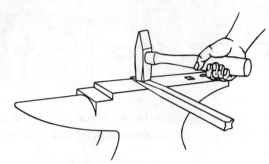

Fig. 47-6 Tapering a cold chisel.

4. Turn on more fuel and air until the fire burns with a **neutral flame.** This flame is blue and nonoxidizing.

Forging

Metal must be brought to the correct temperature before shaping. For example, most mild steel should be heated until it is cherry-red in color. Tool steel should not be heated to as high a temperature as mild steel. Of course, thinner parts require less heat than heavier and thicker ones. Never allow the metal to become so hot that sparks fly from it.

Remove the metal from the forge from time to time to see how it is. It is very easy to burn it. Wear a face protector and special heat-resistant gloves and an apron. Note that these latter items have traditionally been made of asbestos. However, this substance has been found to be hazardous to health. Substitutes for it are now being used.

Always select the forging hammer best suited to the size of the workpiece. Use a small hammer on thin metal and a larger hammer on heavier stock. Then use your shoulder and wrist for medium blows and your whole shoulder for heavier blows.

The main hand-forging operations are: tapering, drawing out, bending, twisting, and upsetting.

Tapering

Tapering is a shaping process for forming such tools as cold chisels and center punches. It is also done to shape the ends of many wrought-metal projects. The process is as follows:

1. Heat the end to be shaped to the right temperature.

2. Grasp the opposite end firmly with tongs. Place the heated section on the face of the anvil. Hold it at an angle equal to about half the amount of taper you want (Figure 47-6).

3. Strike the heated portion with the flat of the hammer, turning it to keep the tapering even. If the metal is rectangular, make a quarter turn after every few blows. If it is octagonal, turn it to the op-

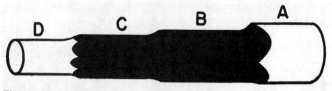

Figure 47-7 The steps in drawing out a round bar into a smaller diameter: (A) original shape, (B) hammering to square, (C) hammering to octagon shape, and (D) hammering again to round.

posite side. For round stock, turn the workpiece a little after each hammer blow. Make the blows firm and sharp.

4. Always keep the metal at forging temperature. Reheat the metal to a dull red, then use light hammer blows to smooth out the tapered portion. If a rectangular shape is being tapered, another person can hold a flatter against the tapered portion as you strike the tool to smooth out the taper.

Drawing Out

Drawing out is done to make a piece longer and thinner. Figure 47-7 shows a piece of round stock in various stages of being drawn out.

1. Heat the area to forging temperature; to a bright red color.

2. Hammer out the section to a square shape. Use heavy blows that come straight down and are square to the work.

3. Heat the metal again. Form it into an octagon.

4. Round off the section by striking the workpiece firmly while turning it on the anvil face. If you are drawing out square stock, round it off first. If the stock remains square while being drawn out, the metal may crack or distort badly.

Bending

Bending is done in various ways:

1. To make a sharp square bend, heat the area at the bend line. Place the workpiece over the anvil face at the point where the corner of the anvil is rounded (Figure 47-8). Strike the extended portion with glancing blows. Then square off the bend over the corner of the anvil.

2. Another way of bending hot metal is to place it in the hardy hole and pull or bend with tongs.

3. You can also make a right-angle bend by putting the hot metal in a vise, with the bend line at the upper edge of the jaws.

4. Curved shapes are formed over the horn of the anvil (Figure 47-9). To do this, first heat the area.

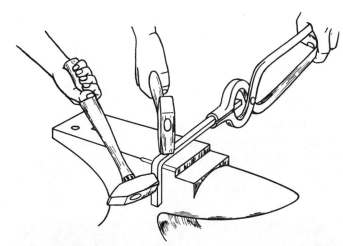

Fig. 47-8 Starting a square bend over the face of an anvil.

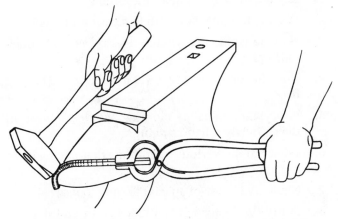

Fig. 47-9 Making a circular bend over the horn of an anvil.

Then place the stock over the horn at about the right curvature. Then strike the workpiece with overlapping blows as you move the metal forward to form the curve.

Twisting

Twisting can best be done by heating the area and then bending the stock in a vise as with cold bending.

Upsetting

Upsetting is the opposite of drawing out. It is increasing the area of the metal by decreasing its length. Follow this procedure:

1. Heat the section to the proper temperature.
2. If the piece is short, hold it over the end of the anvil and strike the end to flatten it out (Figure 47-10). If the metal tends to bend, straighten it out by striking it as it lies flat on the anvil.
3. If the end of a long piece is to be enlarged, heat this area. Fasten the metal in a vise, with the heated

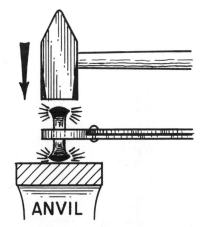

Fig. 47-10 Upsetting both ends of a short piece of stock.

end extending above the jaws. Strike the end with a hammer to increase the size.

Using Anvil Tools

There are a number of anvil tools that have been designed for special forging operations, as shown in Figure 47-5. They are held by a second person while the hammering is being done.

Fullers

Fullers are used to shape round inside corners and angles and to stretch metal. When the forging is worked between the top and bottom fullers, the top fuller is struck with a sledge. The top fuller is often used alone to make a depression on the upper side of a forging lying flat on an anvil. It has a handle like a hammer and is held on the work while a helper strikes it.

Swages

Swages are used for smoothing and finishing. They are made in many sizes, depending on the work for which they are intended. They are used in pairs, each consisting of a bottom swage and a top swage. The bottom swage is inserted in the hardy hole of the anvil. The workpiece is put in the groove of the bottom swage. The top swage, also grooved, is placed over the workpiece and struck to smooth and straighten it (Figure 47-11).

Punches

Punches are placed over a heated workpiece and struck with a hammer. In finishing a hole, the punch is held on the work over the pritchel hole of the anvil. The slug of the stock drops through this hole.

Chisels

Chisels used in forging are thinner than cold chisels. This lets them penetrate the workpiece more

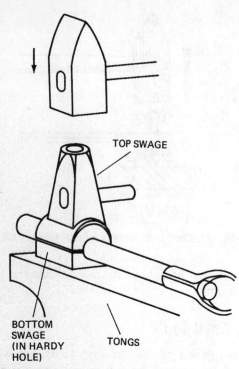

Fig. 47-11 Using the swage to smooth and straighten a forged rod.

quickly. Their edges are also thin because a tool used to cut hot metal does not have to be as strong. They are usually ground to an angle of about 30 degrees. They are held over the workpiece and struck with a hammer. Chisels are also used to make grooves or shoulders on the metal.

Hardy

The **hardy** is a hot-and-cold chisel made to fit into the hardy hole of the anvil. It is used mainly as a bottom-cutting tool. Metal is cut by placing it on the hardy and striking it with a hand hammer. The hardy is used for cutting metal bars, wires, and rods. It may be used on both hot and cold steels.

Flatters

Flatters are like swages, except that they are used with flat workpieces. They are used for smoothing work and for producing a finished appearance by taking out the uneven surface left by the hammer or other tools. Their use requires a helper.

KNOWLEDGE REVIEW

1. List the main equipment for forging.
2. What is the safe way to light a gas or oil forge?
3. How do you know when mild steel is hot enough for forging?
4. How do you insert a piece of metal into the forging fire?
5. What kinds of safety clothing should be worn in the shop when hand forging?
6. What are the various operations that can be done in hand forging?
7. Name five kinds of special forging tools.
8. Study the history of the blacksmith and explain the importance of this occupation in the early development of our country.
9. Write a research report on the industrial production of some hand forged tools.

Machine Forging and Other Hot-Forming Processes

OBJECTIVES

After studying this unit, you should be able to:

- Describe hot forming.
- Explain recrystallization.
- Describe the wrought structure of metal.
- Define *inclusions.*
- Tell how to identify flow lines.
- Describe hot rolling.
- Identify some of the uses of forging in industry.
- Describe two kinds of forging presses.
- Name two kinds of forging operations.
- Name three other kinds of forging methods.
- Explain the process of extruding.

KEY TERMS

cold forging	flow lines	hot rolling	open-die forging
cold heading	forging hammer	hot working	ram
extruding	forging press	impression-die forging	recrystallization point
fiber	hot forming	inclusion	swaging

As metal is heated, it becomes plastic and, therefore, easy to form. Because of this, industry does much of the mechanical working of metal when it is hot. Many metal products begin in the form of castings and are then reworked by hot forming, cold working, or machining.

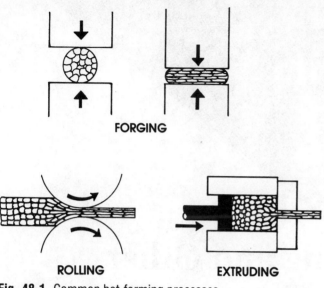

Fig. 48-1 Common hot-forming processes.

Those methods used to create shapes from heated metals are called **hot forming** or **hot working** (Figure 48-1). Hot forming is deforming metal by hammering or pressing it at a temperature higher than its recrystallization point. The **recrystallization point** is the temperature below which a metal's internal structure can no longer be rearranged with ease. Thus, above this point, a metal can be shaped with less force than is needed to shape it when it is cold. Also, because plastic flow *rearranges* the metal at those temperatures instead of *distorting* it, the metal does not work-harden.

Hot forming breaks up the large grains in a casting and gives the material a *wrought,* or worked, structure. Wrought metal has more uniform properties and is better suited for many uses than cast metal.

Hot forming also stretches out the **inclusions**—the impurities or small particles of nonmetallic elements that are present in all but the purest of metals. These inclusions are stretched out in the directions in which plastic flow occurs during forming. The stretched inclusions are the **flow lines,** or **fiber,** that can be seen on the surface of a cut through a hot-formed part. The flow lines affect the properties of hot-formed parts. A metal has greater load-carrying ability in the direction of its flow lines than it does across the flow lines. For that reason, a hot-formed part is usually designed so that the flow lines run in the direction of greatest load during service.

There are a number of hot-forming processes used in industry, such as rolling, forging, and extruding, the latter of which is also done as a cold-forging operation. These processes are described in this unit.

Hot Rolling

Hot rolling is an important process in steel, aluminum, and copper mills for forming industrial raw materials. In the hot-rolling process, hot metal ingots go through rolling mills to produce finished products such as sheet, pipe, bar, rod, plate, and others. There are many kinds of rolling mills, but all of them operate on the same principle: The metal ingot is squeezed between two rotating rolls. These rolls are closer together than the thickness of the ingot. The slabs from this first rolling are trimmed, reheated, and rolled again. A final operation is coiling the sheets for easy handling.

Forging

Forging is widely used in industry for hot-forming finished shapes. Rods and bars made by hot rolling are often formed into final shape by forging. Metals that can be forged commercially include wrought iron, steel, copper, and brass. Aluminum and zinc can be forged at low temperatures.

Industrial forging is forming hot metal by hammering, pressing, or intermittent rolling. The metal is first heated to a high temperature. It is then shaped by the pressure applied by one surface, the hammer, as the metal rests on another surface, the anvil. Forged products are remarkably free of concealed or internal defects. Therefore, they have great strength and toughness. For example, automobile parts requiring these qualities, such as the crankshaft connecting rods and wheels, are forged. Hand tools such as wrenches and pliers, hoist hooks, and vehicle parts are made by forging.

When metal has been forged, it has a fibrous structure resembling the grain structure of wood. Like wood, the metal is stronger and tougher in the direction of the grain.

Forging Machines

All forging equipment shapes the metal between two dies. One die holds the workpiece, the other delivers the pressure. The dies may be flat surfaces or cut to correspond to the finished shape of the piece. There are two general kinds of forging equipment: forging hammers and forging presses.

Forging Hammers

Forging hammers deliver a high-speed impact. This is done by a **ram,** or heavy weight, that is raised and then dropped or driven downward into the workpiece. Forging hammers are usually named by the method used to raise the ram. *Board-drop hammers* have a ram mounted on the end of a vertical hardwood board (Figure 48-2). *Air-lift hammers* use air to raise the ram. *Steam hammers* are the largest and most versatile forging hammers. They are constructed like air-lift hammers but operate differently. The ram is raised and driven downward by steam pressure on a piston (Figure 48-3).

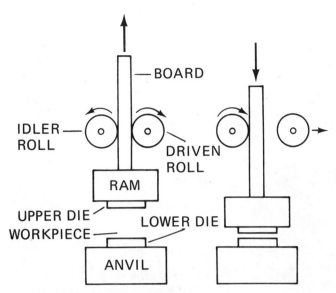

Fig. 48-2 A board-drop hammer.

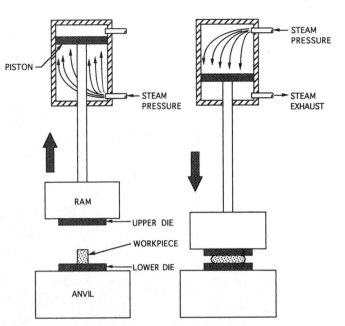

Fig. 48-3 A steam hammer used in industry.

Forging Presses

Forging presses are often preferred to forging hammers for many operations. For example, since nickel alloys retain much of their toughness at forging temperatures, the slow movement of presses gives the alloys more time to flow into the contours of dies. Also, the squeezing action of presses has a kneading effect on the metal. This causes deeper deformation in the workpiece than is normally achieved with forging hammers. In general, larger forgings can be produced on presses than with hammers. There are two kinds of forging presses: mechanical and hydraulic.

The *hydraulic forging press* has a large piston attached to the top of the ram, similar to the steam hammer. Hydraulic pressure moves the piston down for the work stroke and up for the return stroke. These presses are sometimes equipped to automatically vary the forging speed and pressure. Hydraulic presses are made in a wide range of sizes.

Forging Operations

There are two kinds of forging operations: open-die and impression-die.

Open-Die

In **open-die forging,** the workpiece is shaped between dies that do not completely confine the metal. The working surfaces of the dies are either flat or contain only simple shapes. In most open-die forging, the workpiece does not take the shape of the dies. The shape is obtained by manipulation of the workpiece. For example, a round rod can be forged into a square bar by hammering along its length with open, flat dies.

Impression-Die

In **impression-die forging,** the workpiece is shaped between dies that completely enclose the metal. The upper and lower dies are cut out so that when they come together, the hollow space between them makes a mold shaped like the forged piece. The workpiece is not manipulated during forging. The entire piece is placed between the dies and hammered or pressed until the metal fills the die cavity (Figure 48-4).

Before either kind of forging operation can be done, the metal must be heated to working temperature in a large gas or oil furnace. Forging is then done with hammers or presses between the dies. After the forging is complete, the workpiece must be cleaned, heat-treated, straightened, inspected, and often machined.

Cold Forging

The process of **cold forging** is used to make small metal parts, and includes such methods as cold heading, extrusion, and swaging. Parts are worked cold for the same reasons as those listed for cold forming in unit 37.

Cold Heading

Cold heading is an upsetting operation. The workpiece (or blank) is gripped in a stationary die, and its end is struck by a heading ram. The blow spreads out the end of the workpiece to produce the head. Fasteners such as screws, nails, bolts, and rivets are so formed from metal rod or wire. The resulting grain structure makes a part which is very much tougher than a similar machined part.

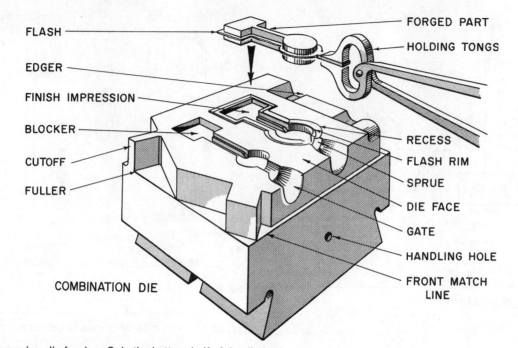

Fig. 48-4 Impression-die forging. Only the bottom half of the die is shown.

Cold heading is done on automatic equipment that cuts the blank from a coil or stock, forms the head and shank of the fastener, and ejects the formed part. Two dies are needed to upset the end of the blank: one holds the blank; the other delivers the single forming blow. Fasteners with smooth undersides, such as rivet heads, are usually formed within the ram or punch. Heads with shaped undersides but flat tops, such as some wood screws, are usually formed in the holding die. Parts shaped on both sides of the head, such as carriage bolts, are formed partly in the punch and partly in the die.

Large or complex heads require more than one forming blow. The two-blow header upsets the blank in two steps. The blank is held in the same die for both blows. However, it is formed by two different punches. The first punch preforms the head into a cone. A cone shape is used because it results in less bending of the blank, especially when large blank extensions are required. The second punch finishes the head. Two-blow heading is illustrated in Figure 48-5.

Extruding (Cold)

While most extrusion work is done with heated workpieces or billets, tin and lead and some other very soft alloys can be extruded cold. This process is explained later in this unit.

Swaging

Swaging is a way of tapering or decreasing the diameters of round metal bars and tubes (Figure 48-6). The swaging machine has a ring with small die hammers mounted at the inside edge. As the ring spins, a cam sys-

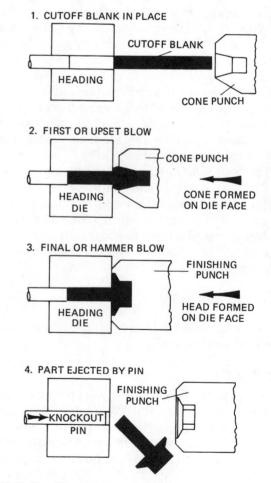

Fig. 48-5 Two-blow cold-heading diagram.

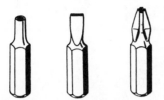

Fig. 48-6 A typical swaging die, showing the hammers in the closed and open positions. Standard and Phillips screwdriver tips are made on this machine from the rounded blank shown. Further forging operations are necessary to complete them.

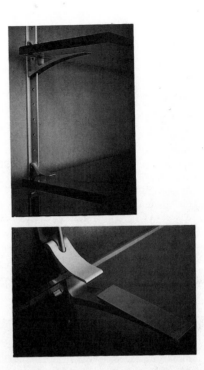

Fig. 48-7 Extruded aluminum shelf brackets. (Courtesy of Parallel Design Partnership Limited; design by M. Ali Tayar, photograph by David Sundberg.)

tem intermittently forces the hammers against the workpiece. This impact and pressure action gives shape to the workpiece. The continual, very rapid blows of the whirling hammers quickly reduce the metal to a smooth round shape. Needles, bicycle spokes, fishing rods, screwdriver tips, and furniture legs are swaged products.

Extruding

Extruding (or extrusion) is a process by which a warm metal billet is forced through an opening in a die. The extruded part has a cross section of the same shape as the die opening. The basic principle of extruding is the same as that of squeezing toothpaste from a tube. The paste ribbon on your toothbrush has the exact shape of the tube opening (see Figure 48-1).

In this process, a die is placed in an extrusion press. A heated billet is then put into the machine cylinder behind the die. A hydraulic ram applies pressure to the billet, forcing the softened metal through the die opening. A major advantage of extrusion is that long pieces of uniform and complex shapes can be produced. For example, study the attractive and functional aluminum shelf bracket shown in Figure 48-7. It was made from a long extrusion cut into sections 3/4 inch wide.

A similar process is *impact extruding,* where a part is formed by one high-speed blow. The metal blank can be worked cold or warm. One type of this impact work is the *reverse* method, shown in Figure 48-8. Here, a metal blank (or slug) of the exact size needed to form the part is placed in a die. When a high-speed blow is

applied to the punch, the metal literally mushes around the punch to form the part. In other words, instead of forcing the metal through a die as shown in Figure 48-1, a *reverse* flow of metal occurs. Here the die controls the bottom and outside wall shapes of the part, and the punch shapes the inside. This method is used to make tubes for items such as marking pencils, like the one shown in Figure 48-9.

KNOWLEDGE REVIEW

1. Define the term *hot forming*. List some of its advantages.

2. List three hot forming processes.

3. Describe the action of a forging hammer.

4. Describe the open-die and impression-die forging operations.

5. Explain the advantages of cold forging.

6. Explain the differences between the cold heading, extrusion, and swaging processes.

7. Describe the difference between forward billet extrusion and reverse impact extrusion.

8. Visit a local forging industry and give an oral report on the types of equipment used, the products made, and the safety practices you observed.

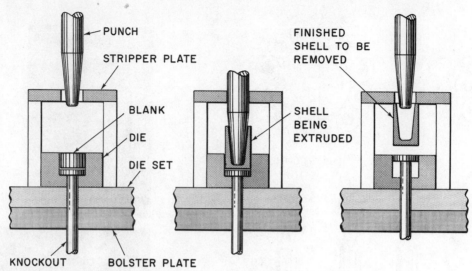

Fig. 48-8 Steps in producing a simple cylindrical part by reverse impact extruding.

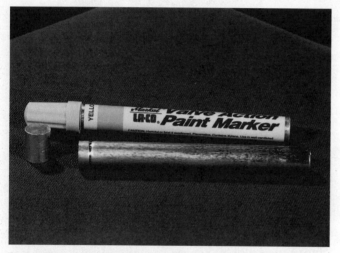

Fig. 48-9 This marker was made by the impact extrusion process. The 1/2-ounce slug, the extruded barrel, and the finished product are shown.

Unit 49

Introduction to Casting

OBJECTIVES

After studying this unit, you should be able to:

- Describe the general casting process.
- Name several specific casting processes.
- Describe the process of sand casting.
- Name the most common ferrous metal used in casting.
- Name five common foundry occupations.
- Describe the two types of industrial foundries.

KEY TERMS

chipper	foundry	melting and pouring	patternmaker
coremaker	grinder	metal casting	pourer
coremaking	jobbing foundry	molding	production foundry
finisher	melter	pattern	

Very simply, **metal casting** is the process of pouring molten metal into a cavity (or mold), where it cools and hardens to form a part (Figure 49-1). It is also called *founding,* and a **foundry** is therefore a factory which has all of the equipment needed to produce metal castings. Many other materials besides metal are also cast: water is poured into a tray and frozen to produce ice cubes; liquid ceramic is poured into a mold to make the insulator for a spark plug; and plastic is injection molded to make a tool handle. In this unit we shall study metal castings and how they are made.

Metal casting is an ancient process, over 5,000 years old. It probably began when primitive people accidentally surrounded a campfire with copper-bearing rocks, and the heat melted out the copper, which has a low melting point. Much later, they made a simple bellows from animal skins to direct more air to the fire, making the process more efficient. Eventually, these early people found they could place bits of copper ore into a stone container (a very early crucible) to control the melting, and add tin to the mix to make bronze alloys. With this technology they learned to make spearheads and cutting tools, and to beat the copper into sheets to form pots. Since then, casting has become both an art and a science, which has resulted in hundreds of

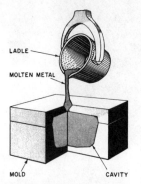

Fig. 49-1 The casting process. When hardened, the molten metal takes the shape of the mold cavity. This is how cast products are made.

Fig. 49-2 This bronze chesspiece was cast as a student project.

Fig. 49-3 A typical flat pattern.

Fig. 49-4 Shown are the two wood halves of a split pattern for the movable jaw of a machinist's vise.

products from jewelry to bicycles to automotive engine parts (Figure 49-2).

Modern foundry methods include: sand casting, shell molding, investment casting, die casting, and many others. Many high-production foundries have computer-controlled production lines. In one plant that produces automotive engine castings, a sign reads, "The sand used in this foundry is never touched by human hands, except out of curiosity." This is good, for foundries used to be hot, dusty, dirty, and foul smelling; not very pleasant places in which to work. Modern plants are safer and more pleasant, and most dirty jobs are left to the robots.

Patterns

Like all castings, the chesspiece shown in Figure 49-2 was made from a pattern. A **pattern** is a replica (or copy) of the part to be cast, and from this pattern a mold is made. Into this mold is poured molten metal to make

the casting. *Loose patterns* are models of the pieces to be cast, and may be of the flat or split pattern types. *Flat patterns* are simpler to make and easier to use (Figure 49-3). *Split patterns* are used for irregular or round parts. They are made in two pieces so that they can be removed from the two halves of a rammed sand mold. They are harder to make and require more complicated molding processes. The large vise jaw in Figure 49-4 is an example of a split pattern.

When many castings of the same product are needed, a *match plate* is used. The patterns are split, and their halves are permanently fixed to both faces of a plate. With this method, mold-making is much simpler and more efficient. These match plates can also be used for single castings. An advantage of this type of pattern is that the gating system, which allows for the flow of the molten metal to the mold cavity, is a part of the plate. A typical match plate for a paper punch body,

Fig. 49-5 A match plate for the bodies of paper punches. Note that both sides of the plate are shown.

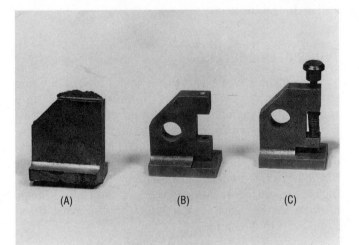

Fig. 49-6 The raw casting (A); the machined casting (B); and the assembled paper punch (C). This piece was produced from the match plate pattern in Figure 49-5.

produced in a school shop, appears in Figure 49-5. In Figure 49-6 you can see the rough casting of the punch body, the completely machined body, and the punch final assembly. This is how castings are processed into finished products.

Patterns are made of wood, plaster, metal, or plastic, with wood being one of the more common materials. Foamed polystyrene plastic and wax patterns also are used. Here the pattern is imbedded in the sand or plaster molding material. The patterns are not removed. They are left to melt or vaporize as the molten metal fills the mold. A separate pattern obviously must be used for

Fig. 49-7 The steel part on the left has been produced from the foam pattern on the right.

each mold, as it is "lost" in the casting process. A foam pattern and final casting are shown in Figure 49-7.

Patterns are an important part of the molding process, and skilled patternmakers with woodworking, plastics, and metal experience are employed to prepare them.

Foundry Practice

Besides patternmaking, founding involves three other basic skills: molding, coremaking, and the melting and pouring of molten metals.

Molding

Molding is preparing the mold, or casting cavity, and it is the heart of the process because all other foundry activities are grouped around it. The purpose of a mold is to form a cavity that will hold molten metal until it becomes solid. A mold is made of sand or plaster, and for special foundry work, metal. An entry port, or sprue, is provided for pouring the molten metal into the mold cavity (left when the pattern is removed), as shown in Figure 49-1.

Coremaking

Coremaking is closely related to molding, because the core becomes a part of the mold. Many split patterns have openings or holes in them, and a core is needed to produce these as part of the casting process. This is more efficient, for otherwise the holes would have to be cut into the finished casting. An example of a cast part, and the core pattern used to produce it, is shown in Figure 49-8. The core prints shown in the drawing are part of the pattern and are used to hold the core within the mold.

A core is made by packing a sand and binder mixture into a core box of the desired shape. The core is dried, removed from the box, and baked in an oven

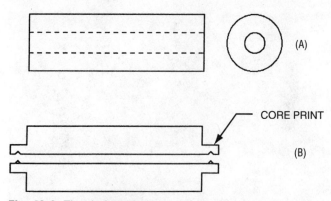

Fig. 49-8 The shaft guide (A) and the split pattern (B) used to produce the casting. Note the core prints at each end of the pattern to support the sand core.

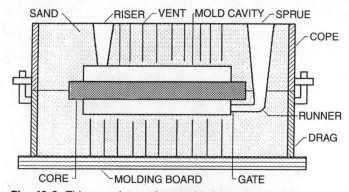

Fig. 49-9 This complete and assembled rammed flask shows the core in place ready for casting the shaft guide in Figure 49-8.

where it becomes strong enough to handle. The core is then placed in the mold cavity on the core prints. Later, the core can be broken out to leave a hole in the cast piece. An assembled mold, with the core in place, appears in Figure 49-9.

Melting and Pouring

Melting and pouring is a science that requires knowledge and skill if high-quality castings are to be produced. Melting is done in oil, gas, coke, or electric furnaces. In large industrial foundries, raw metals and scrap are placed in furnace crucibles and heated. Temperature control is important, because metal which is overheated will result in poor castings. When the metal is ready for pouring, it is tapped from the furnace and poured into the mold sprue. Automatic pour systems are commonly used. In smaller foundries, the melting occurs in smaller crucibles and smaller furnaces, and the pouring is done by hand.

Metals Used in Casting

Most castings are made from ferrous metals, principally gray iron, steel, and malleable iron. About three-fourths of all foundry workers are employed in ferrous foundries. The others work in the nonferrous foundries. There, copper alloys, aluminum, magnesium, lead, and zinc alloys are poured.

Foundry Occupations

Foundries employ engineers, scientists, technicians, skilled workers, semiskilled workers, and laborers. The engineer and the metallurgist control the quality of the metal. An engineer also checks the castings and molds and supervises many foundry technicians. The patterns are made by highly skilled woodworkers and metalworkers called **patternmakers.** The molds in

the foundry are made by skilled and semiskilled workers called **coremakers.** The furnace operator is a **melter.** The worker who pours the castings is a **pourer.** The clean castings are finished by workers called **chippers, grinders,** and **finishers.** Altogether, about 450,000 people work in foundries throughout the United States. Of this number, about 80,000 are employed in professional, office, managerial, and sales jobs. There are over 8,000 engineers, chemists, metallurgists, and other scientists in the industry.

Types of Foundries

There are two basic types of foundries: jobbing and production. The **jobbing foundry** is often very small. There, many different kinds of castings of all sizes and weights are made in limited numbers. Also, one casting is often made to fill a single order. The **production foundry** is usually larger. The work is generally limited to a few patterns from which thousands of castings are made. Most foundry work in the United States is done in the Great Lakes states, in the West Coast states, and in Alabama.

Foundry as a Hobby

For people who like to make models, pouring small castings can be a lot of fun and a fascinating hobby. Small parts or fittings for model airplanes, trains, and boats can be made in sand, plaster, or metal molds. Those interested in jewelry making can build molds from plaster of paris to make brooches, rings, and other pieces. Many other projects can also be made.

KNOWLEDGE REVIEW

1. List some typical cast products.
2. List the methods of making castings.

3. Name at least five special occupations involved in foundry.

4. What are the two basic kinds of foundries?

5. What kinds of hobbies are related to foundry work?

6. Find out all you can about one of the major occupations in foundry. Write a report or present it orally in class.

Unit 50

Making a Metal Casting

OBJECTIVES

After studying this unit, you should be able to:

- Describe patternmaking.
- Name two kinds of patterns.
- Compare a shrink rule with a standard rule.
- Explain why a pattern must have a draft.
- Explain why fillets and rounds are used.
- Describe how to make a sand core.
- Describe the different kinds of molding sands.
- Name several pieces of foundry equipment.
- Explain why a parting compound is used.
- Name some of the metals used in casting.
- Describe a good gating system.
- Explain how to make a simple mold.
- List the safety rules involved with pouring a casting.
- Describe how to finish a casting.

KEY TERMS

cope	gating system	riddle	slag
crucible furnace	natural molding sand	riser	sprue
draft	parting compound	rounds	synthetic molding sand
drag	parting line	runner	tempering
fillets	permeability	shrink rule	type metal
gate			

There are many interesting cast projects which can be made in the school shop. Some examples are: model brass cannons, bookends, paperweights, tool handles, and furniture parts. The method generally used in the school shop is sand casting, and you will learn how to do this in this unit.

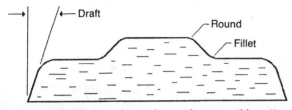
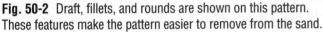

Fig. 50-1 A shrink rule.
(Courtesy of The L. S. Starrett Company)

Fig. 50-2 Draft, fillets, and rounds are shown on this pattern. These features make the pattern easier to remove from the sand.

Fig. 50-3 This anvil paperweight casting was made from a split pattern.

Patternmaking

As you learned in the previous unit, you must first design and make a *pattern* of your casting project. These are usually made of clear pine or mahogany, which are easy to saw and carve. An original, flat cast article can often be used as a pattern (see Figure 49-3).

Since molten metal shrinks as it cools, some patterns must be made larger than the finished article to account for this shrinkage. You must use a **shrink rule** (Figure 50-1) to do this. The inches on this special rule are actually longer than regular inches. This shrinkage problem is especially important if you are making a project with closely fitting parts, such as a drill press vise. In most one-piece craft projects, it is not necessary to use the shrink rule.

A pattern also must have a **draft**, which is a slight taper on the pattern sides and ends; **rounds**, which means that sharp outside corners are broken or rounded; and **fillets**, which is the filling of inside corners with wax or wood paste (Figure 50-2). All of these features will make it easier to remove the pattern after it has been rammed in sand.

Flat Pattern

Flat patterns can be made by drawing the project on a piece of wood. If the object is to be 3/8 inch thick, the wood should be that thick. Cut the pattern on a band saw or scroll saw. Wear safety glasses, roll up your sleeves, remove your necktie and your wristwatch, and tuck in your long hair—do everything necessary to prevent serious accidents when using these saws. Next, use a wood rasp or belt sander to round the corners, and give the edges a bit of draft. Any carving of the surface should be done very carefully at this time. Finish with a final hand-sanding. The pattern must be perfectly smooth without any slivers or tears. Apply a coat of varnish or lacquer to preserve it.

Split Pattern

If the project you have designed has a round or irregular shape, such as the small anvil paperweight in Figure 50-3, you must prepare a split pattern. An example of this type of pattern appears in Figure 49-4.

The procedure for making a split pattern for the anvil casting is described below.

1. Prepare accurate, full-size, top and front view drawings of the anvil (Figure 50-4). Make the drawings on heavy paper so that they can be cut out and used as tracing templates.

2. Prepare two wooden workpieces, slightly larger than the front view of the anvil, and half as thick. Locate the alignment holes on the front view and mark their centers with a scriber.

3. Place the two pieces together and lock them in a drill press vise. Drill one 3/16-inch hole through the workpieces.

4. Remove the workpieces from the vise and separate them. Place a drop of wood glue around one end of a 3/16-inch wooden dowel, and use a hammer to gently tap it all the way through the hole in one of the workpieces. Wipe off any excess glue.

5. Round the end of the dowel and reassemble the workpieces. Lock the workpiece in the vise once again and drill the second hole.

6. Separate the pieces and install the second dowel. Round and slightly taper both dowels, so the split pattern will fit snugly together.

7. Cut a piece of newspaper to fit between the split halves of the pattern block. Punch holes so it fits over the dowels. Wipe a thin coat of glue on the paper and press it in place. Next, cover the exposed side of the paper with glue. Wipe excess glue from

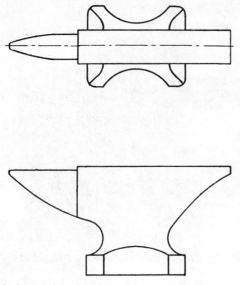

Fig. 50-4 A two-view drawing of the anvil. The centerline shows the "split" of the pattern.

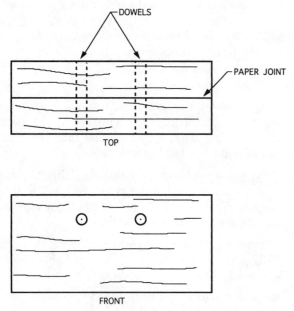

Fig. 50-5 The wood workpieces for the anvil split pattern.

the dowels and reassemble the pattern halves. Lock them in a vise (gently, with not too much pressure) to hold them until the glue has dried. You have now prepared a paper joint (Figure 50-5).

8. Remove the workpiece and trace the front view of the anvil on it. Carefully cut the shape with a scroll saw or band saw, and file and sand it smooth.

9. Next, trace the top view template on the piece, and file and sand the anvil to shape. Continue this work until the pattern is smooth and accurate. Make sure the pattern has sufficient draft, that all sharp corners are rounded, and that fillets have

been added where necessary. Apply a coat of varnish or lacquer.

10. The finished pattern should now be clamped in a vise, and the paper joint opened with a knife or chisel. This can be tricky and dangerous, so have your instructor help you with this operation.

The above instructions describe the preparation of a typical split pattern. It is now ready to be rammed in a flask.

Molding Sands

The sand used for making molds is of two basic kinds: synthetic and natural. **Synthetic molding sand** contains certain proportions of many different substances. For example, it may include a clear, washed, pure silica sand mixed with clay, carbon materials, and water. Whatever the kind of sand used, it must be heat-resistant, must stick together, and must retain its shape well. It must also have **permeability.** This characteristic lets the gases from the hot metal poured into the mold escape through the sand. Synthetic sands are used mostly in iron and steel foundries, which require very high temperatures.

Natural molding sand is used just as it is dug out of the ground, with the possible addition of a little water. Natural sands are found principally in the eastern Great Lakes region, in Kentucky, and in Missouri. Natural sands are most widely used for nonferrous metals, which require less heat. New sand is light brown, but it becomes dark with use. The sand dries out when used or stored, so it must be tempered by moisture. **Tempering** is done by sprinkling water on the sand and mixing it in with a shovel or riddle. The sand is tempered properly when a clump picked up and squeezed retains the sharp impression of your fingers. It should feel moist to your hand. It should break off sharply when you tap it. If you have added too much water, large amounts of sand will stick to your hand.

Parting compound, a dry powder, is dusted between the **cope,** or top, and **drag,** or bottom, sections of the mold. This keeps them from sticking together. You can use a fine beach sand for this purpose, but the commercial parting compounds are better.

Foundry Equipment

Some special pieces of equipment are needed for simple foundry work. These include: a flat molding board, a flat bottom board, flask weight or clamps, bench rammer, sprue pin, riser pin, draw spike or screw, rapping bar, strike-off bar, riddle, bellows, wet bulb, slick and oval, spoon and gate cutter, trowel, sprinkling can, shovel, and a few other molder's tools (Figure 50-6).

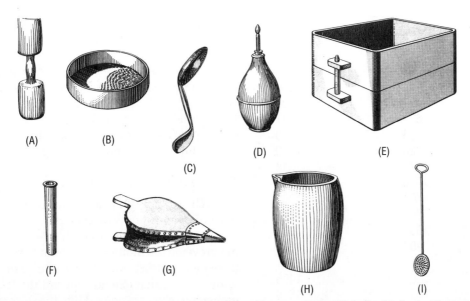

Fig. 50-6 Common casting tools include: (A) rammer, (B) riddle, (C) spoon slick, (D) bulb, (E) flask, (F) sprue pin, (G) bellows, (H) crucible, and (I) skimmer.

Fig. 50-7 A small crucible furnace. (Courtesy of Johnson Gas Appliance Company)

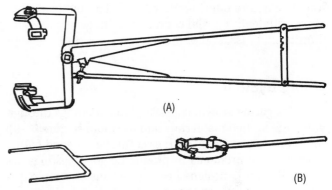

Fig. 50-8 (A) Crucible tongs. (B) Crucible shank.

ing (Figure 50-8). For protection, the pourer must wear clear goggles, a face shield, and special heat-resistant gauntlet-type gloves, leggings, and apron.

Foundry Metals

Practically every metal can be melted and poured into castings. However, many metals have very high melting points (Table 50-1). These must be heated with special equipment. Nonetheless, many of the nonferrous metals can be easily heated with the equipment available in the average shop. Pure lead or **type metal,** an alloy of lead and antimony, is adequate for simple projects. However, lead and lead alloys should never be used for making objects that come in contact with food or are likely to be used by small children. This is because lead is poisonous. Aluminum or a die-casting metal alloy are best for the small shop. To help purify the metal and eliminate gas, a *flux* is usually added. This is sometimes done during the melting process. But

A *flask* holds the sand mold. It is very important that the molding board be perfectly flat. This lets you get a clean parting between the cope and drag sections of the flask.

A furnace is needed to melt the metal. The *soldering furnace* can be used for light metals such as tin, lead, and pewter. Small amounts of these metals also can be melted with a welding torch. The **crucible furnace** in Figure 50-7 can be used to melt aluminum, brass, copper, bronze, and gray iron, and is the safest and most convenient to use. The crucible is a light ceramic container used to hold the metal as it is melted. Often, the crucible also becomes the pouring ladle. The crucible is lifted with tongs and placed in a crucible shank for pour-

Table 50-1 Melting Temperature of Selected Metals

Metal	°F	°C
tin	449	232
babbitt metal	462	239
lead	621	327
zinc	788	420
aluminum	1220	660
bronze	1675	913
brass	1706	930
silver	1761	960
gold	1945	1063
copper	1981	1083
cast iron	2200	1204
carbon steel	2500	1371
nickel	2651	1455
wrought iron	2750	1510

with aluminum and aluminum alloys, it is generally done just before pouring. Powdered charcoal is used as a flux with lead. Special commercial fluxes are used for aluminum, brass, and other alloys.

Gating System

The **gating system** consists of the openings through which the molten metal runs into the mold (Figure 50-9). The liquid metal flows with ease through a good gating system. Many problems in producing good castings are caused by poorly designed gating systems. A simple gating system includes the sprue, runner, riser, and gate.

Sprue

The **sprue** is a vertical opening through which the molten metal enters the gating system. The sprue is

formed by a *sprue pin.* This is often a tapered round wood pin. In industry, however, it has been found that a tapered rectangular pin gives better results. The *pouring basin* is a cup-shaped opening at the top of the sprue. The molten metal is poured into this basin, from which it enters the sprue proper. The *sprue base,* or *sprue well,* is a rectangular opening at the base of the sprue. Molten metal fills this base before it enters the runner. The sprue base makes the flow of liquid metal more even by reducing its turbulence.

Runner

The **runner** is the horizontal opening through which the molten metal flows to the gates. If the runner is long, it should decrease in size as each gate is passed. This tapering prevents the molten metal from flowing too quickly to the far end of the runner. The *runner extension* is the part of the runner that extends beyond the last gate. The first molten metal that is poured into the system contains many impurities. These impurities become trapped in the runner extension. Therefore, they cannot ruin the casting.

Riser

The **riser,** which may or may not be needed, provides a reserve of molten metal that helps control the flow of metal to the mold. When aluminum is used for a heavy or thick casting, a riser is not needed. A riser is usually needed for a thin casting with a large area. The volume of the riser must always equal the volume of the mold cavity itself. The riser can be placed directly over the mold cavity or between the mold cavity and the sprue hole. *When casting aluminum,* never place the riser on the side of the mold cavity opposite the sprue.

Gate

A **gate** is a horizontal opening from the runner to the mold cavity. The number of gates needed depends on the size of the mold.

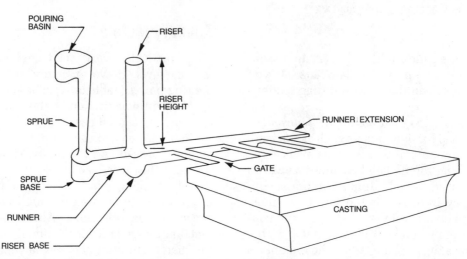

Fig. 50-9 This machine table has been cast in aluminum. The waste material, including the sprue, riser, and gate, must be cut off.

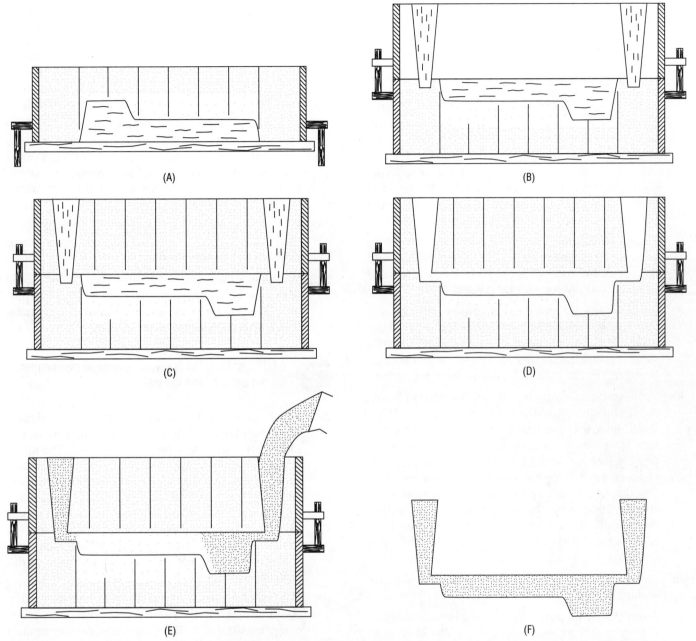

Fig. 50-10 (A) Ram pattern in drag; add vents. (B) Assemble flask; install sprue and riser. (C) Ram cope; add vents. (D) Separate cope and drag; remove pattern, sprue, and riser; cut gates; reassemble cope and drag. (E) Pour molten metal to fill mold cavity. (F) Remove cooled casting from flask, with sprue, riser, and gate runner attached.

Making a Simple Mold

These are the steps to follow in making a simple mold (Figure 50-10):

1. Select or make a pattern of the article or piece to be cast.

2. Place the pattern on the molding board with the draft, or smaller side, up.

3. Make a chalk mark on one side of the cope and drag of the flask. Place the drag face down on the molding board with the pins pointing toward the floor. The flask should be large enough for the pattern, sprue, and riser. The molding board should just cover the flask. It should not be so large that it is hard to handle.

4. Dust the pattern with parting compound.

5. Shovel some molding sand into the **riddle,** or sand sieve. Shake the riddle back and forth over the pattern until the pattern is covered with at least one inch of sand. Then fill the drag about half full with

riddle sand. With your fingers, tuck the sand around the pattern and into the corners of the drag. When this is done, shovel sand in until it is a little higher than the top of the drag.

6. Using the peen end of the rammer, tuck the sand around the outside edges of the mold first. Then, with the butt end, ram the sand firmly around the pattern (Figure 50-11). This step is very important. In order to get a good casting, you must have a sharp impression of the pattern in the sand. However, if the sand is rammed too hard, it loses its ability to "breathe" and thereby allow hot gases to escape. Ram from the edges, and work toward the pattern.

7. Again, riddle the sand into the drag until it is heaping. Tuck the sand around the edges first and then back and forth over the pattern until the sand is firmly but not too solidly packed.

8. Strike off the excess sand by using a metal or wood bar. This bar should be completely square. Pull the sand off the top of the drag. Then sprinkle a little sand on the surface to form a bed for the bottom board. Place the bottom board on the drag.

9. Holding the molding board and the bottom board tightly together, turn the drag over. Now lift off the molding board. You will see the bottom of the pattern in the even surface of the sand. This is the **parting line.** Carefully blow off any extra particles of loose sand from the surface with the bellows. Smooth any rough spots with a slick, trowel, or spoon. Then dust the parting line with parting compound. Be careful not to use too much of the compound. An excess would dry out the surface of the sand.

10. Put the cope in place. Make sure that it slips on and off easily.

11. Insert the sprue pin. If the casting is quite thick, a riser pin should also be inserted. The shrinkage that occurs when the metal cools will thus take place in the riser rather than in the casting. Risers are generally larger in diameter than the sprue pins and have no taper. If a split pattern is being used, put the cope part of the pattern in place and dust it with parting compound.

12. Repeat steps 5 through 8 for the cope section of the flask. Sometimes, the sprue and riser holes are cut with a hollow metal tube *after* the cope has been rammed.

13. To help hot gases escape, it is necessary to vent the mold by punching small holes in the cope or drag section of the mold. These vent holes should go down to within 1/8 inch of the pattern. An old bicycle spoke or a 1/16-inch welding rod make good vent rods. You must not damage the pattern. This is the best method for punching the holes: Slowly push the vent rod down until you touch the pattern. Then pull it back out about 1/8 inch. Take hold of the vent rod at the level of the sand. Now punch several holes in the sand. Be careful to stop each time your fingers touch the sand (Figure 50-12).

14. Wiggle the sprue pin around until it becomes loose. Carefully pull it out, then pack the sand into a funnel shape around the sprue hole. This pouring basin can also be cut with a trowel. Remove the riser pin as you did the sprue pin. It is not necessary to make a funnel here.

15. Lift the cope section of the flask straight up. Set it on its edge out of the way. Make sure the sprue and riser holes are open and clean.

16. With a wet bulb or a camel's-hair brush and a *little* water, carefully moisten the sand around the pattern. Wetting the sand keeps the mold from crumbling when the pattern is removed.

 CAUTION: Water is the worst enemy of molten metal. Do not get the sand too wet.

17. Tap a spike into the wooden pattern, then lightly rap the spike on all sides until you can see that the pattern is completely loose in the sand. The spike or a large wood screw can be used to pull the pattern out of the mold. Now *carefully* lift up the pattern. If you do this right, the pattern should leave a clean-cut imprint in the sand. If the sand crumbles a little, you can repair the mold with a molder's tool. If you are using a metal pattern, screw a threaded rod or bolt into the back of the pattern to remove it from the sand.

18. Bend a small piece of sheet metal into a U shape about one-half inch wide. You can use a commercial gate cutter also. Carefully cut a gate, or a small groove, into the drag section of the mold. The size of this gate will depend on the size of the casting. Start with a gate about one inch wide and about one-half inch deep. Do not try to do this in one stroke. First, cut a small stretch. Then throw the sand away. When a riser is used, the gate must be cut from the sprue to the riser and then from the riser to the casting cavity. The gate from the sprue to the riser is a little smaller than the gate from the riser to the casting cavity. After cutting these gates, pack down all the loose sand. Then round out the edges where the gates meet the parting line. This is very important. If all the loose sand is not removed from the gates, the molten metal will wash it into the casting cavity. The sand will leave pinholes in the casting.

Fig. 50-11 Tuck the sand around the edges with the peen end of the bench rammer, then ram the drag with the butt end until the sand is firmly packed.

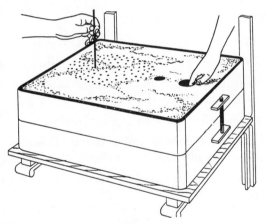

Fig. 50-12 Venting and preparing the sprue hole.

19. Repair the mold as needed by adding small bits of sand and smoothing them down with the right molding tools. Blow off any loose sand with the bellows.

20. Close the mold by replacing the cope section. Carefully guide the cope by putting your fingers along the sides. The pins of the drag should line up with the holes of the cope. Be careful to put the cope back on exactly as you took it off. Do not turn it around. Make sure that the chalk marks are on the same side. Put the mold on the floor and scatter some loose sand around it. The mold should now be clamped or weighted. If you are not going to pour the casting right away, cover the sprue and riser holes. This will keep dirt from falling into them.

Pouring the Casting

Once the mold is completed you can then proceed to pour the casting. Note the safety precautions and follow the steps listed.

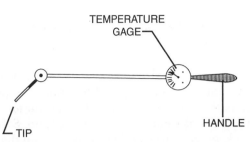

Fig. 50-13 The tip of the pyrometer is carefully placed into the molten metal to check the temperature.

Fig. 50-14 Using bent-handle crucible tongs to remove the crucible.
(Courtesy of Johnson Gas Appliance Company)

 CAUTION: You must wear safety clothing when pouring the casting. Put on clear goggles, a face shield, and special heat-resistant gauntlet-type gloves, leggings, and apron.

1. Select a crucible of the right size. It should be large enough to hold enough molten metal to fill the mold cavity, sprue, and riser. Place the crucible in the furnace. Put in the pieces of metal to be melted. Light the furnace.

2. Heat until all the metal is at pouring temperature. Check often when it is nearing this temperature. Use a lance-type pyrometer for this (Figure 50-13). Overheating will produce a defective casting. Turn off the furnace, add the flux, and stir. The **slag**, or impurities, will float to the top of the crucible. This should be skimmed off.

3. Take the crucible out of the furnace with crucible tongs (Figure 50-14). Put it in a crucible shank. Pick up the crucible shank so that your body is parallel to it. Your arm and safety clothing will help shield you from the heat.

SAFETY RULE: Stand to one side of the mold, never over the top of it.

Pour the metal as rapidly as possible. Once you have started to pour, keep the sprue hole full. Stop pouring as soon as the sprue hole remains full. The metal will start to solidify as soon as it touches the sand. Safe practice is to keep your face and body away from the mold. This is because steam rises from it. Also, metal may spurt up and out of the mold at this time. This is another reason for doing a good job of venting the mold.

4. Let the metal cool. The time will vary depending on the thickness and size of the casting. Separate the cope and drag.

Shake out or break up the mold. Remove the casting. Be careful, the casting is still hot. You should handle it with tongs.

Finishing the Casting

These final steps will complete the casting:

1. A casting is ready to be finished when it is cool enough to be handled with bare hands. Cut off the gates, sprue, and riser with a hacksaw or a metal-cutting band saw.

2. You will find that the parting line shows on the casting. This may be filed or ground off and the edges completely smoothed. Castings can be finished in many different ways. Decorative castings can be scratch-brushed, sandblasted, painted, or enameled. In many cases, felt bases are added for such objects as lamps and bookends. Most commercial casting must be machined on the drill press, lathe, shaper, milling machine, or grinder.

KNOWLEDGE REVIEW

1. What are most patterns for sand casting made of?

2. Define *draft* and *fillet.*

3. What is used to provide an opening in a casting?

4. What are the characteristics of good molding sand?

5. What special clothing must be worn when the casting is poured?

6. What metals can be cast?

7. Name the main parts of a gating system.

8. In sand casting, what is the parting line?

9. What do you call the container in which metals are melted for pouring?

10. In what ways can castings be finished?

11. Make a pattern for a project that will become a part of the school shop equipment.

12. Make a bulletin board display of the sand-casting process.

Unit 51

Industrial Casting Methods

OBJECTIVES

After studying this unit, you should be able to:

- Name the six major casting methods.
- Explain why sand casting is the most common casting method.
- Name the second most important method of casting.
- Identify two kinds of die-casting machines.
- Explain how to do permanent-mold casting.
- Describe shell-mold casting.
- Identify two other names for investment casting.
- Describe another casting method that is similar to sand casting.

KEY TERMS

centrifugal casting   lost-wax process   plaster-mold casting   sand casting

die casting   permanent-mold casting   precision casting   shell-mold casting

investment casting

Until recently, the casting process was more an art than a science. Nonetheless, many new casting methods have been developed in industry to meet the ever-increasing demand for metal products. The average five-room home, for example, contains over two tons of metal castings. The average automobile has over six hundred pounds of castings. In fact, nine out of every ten industries make use of castings either in their products or in the machines that make the products (Figure 51-1).

Six major casting methods are in general use: sand; die; permanent mold; shell mold; precision, or investment; and centrifugal. There are also other methods for specialized purposes, such as plaster-mold casting. There are certain advantages and disadvantages to all casting methods. Generally, industry decides on the method to be used mostly on the basis of the quantity, size, and complexity of the parts to be made.

Figure 51-2 Steelmakers tap furnaces and run the molten metal into molds to produce steel products. (Courtesy of American Iron & Steel Institute)

Figure 51-1 This 18-speed heavy-duty truck transmission has a cast steel outer shell and many interior gears. The shifting pattern is shown in the inset. (Courtesy of Eaton Corporation)

Sand Casting

Sand casting is the most important method used in industry. It is low in cost and requires very little complicated equipment. It also allows great flexibility in product design. Design changes can be made even after production has begun. Large castings, such as machine bases and transportation equipment, are usually made by sand casting. The major disadvantage of sand casting is that products are less uniform in size. Consequently, products must be made a little larger and later machined down.

In school foundries and many small commercial job shops, sand casting is largely a hand process done with very simple tools and equipment. By contrast, the large modern foundry has huge, complex machines that are often controlled by computers. These machines are fed by conveyor systems that eliminate much of the heavy work of the old-style foundry.

The modern industrial foundry uses many operations in casting. First, the melting department provides a continuous flow of molten metal to the molding de-

partment. In automotive foundries, the metal is melted in cupolas, or furnaces fired by coke and high-velocity air (Figure 51-2). The molten metal runs out through a trough to a holding ladle. There, the molten metal remains until it is needed. Then a transfer ladle is filled with molten iron. This ladle is conveyed to the molding machines by an electrically driven monorail system or a special lift truck.

The sand for cores and molds is stored outside the foundry. Since its level of moisture must be uniform before it can be used, it must usually be dried. This is done in a continuous rotary gas-fired sand drier. The sand is then moved through a series of large machines called *mullers*. These mix the sand with oil and cereal grain in exact amounts. From there, the sand is taken to the core-making machines and to the molding department. Many castings have hollow interiors. Special machines are used to make the cores needed to form these hollow sections. Cores are also needed for castings of irregular shape.

At the molding machines, the flask is positioned over the pattern. A controlled amount of sand is dropped on the pattern from an overhead sand hopper. The pattern and flask are jolted, and the sand is leveled and squeezed (Figure 51-3). Then the pattern is taken out of the mold. If cores are needed, they are put in place at this time. The flask with the mold in it is closed and put on a conveyor. Then the conveyor moves the flask to the pouring station. There, the molten metal is poured into the mold to form the casting.

The continuous conveyor takes the mold through a cooling tunnel. This lets the casting cool and harden enough to be removed from the mold at the shake-out area. Then the casting is conveyed to a cleaning room. There, all sprues, gates, and other unnecessary parts and irregularities are removed. Last, the casting is cleaned and inspected.

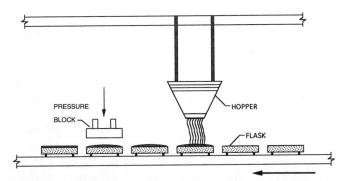

Figure 51-3 A hopper drops measures of sand into the flasks, which move beneath it on a conveyor. The pressure block moves down to compact the sand.

It is then shipped to the machining area. There, all necessary machining is done. A final inspection ensures good quality. Each casting must meet certain standards and must be free of defects. One way to check this is to use an X-ray machine to examine its structure.

Die Casting

Die casting is an important method of making castings. It is used when a large number of quality parts are needed. In the automobile industry, die casting has been used to make everything from door handles to pistons. Parts as large as engine blocks can be produced by this method. However, there are limits to the size and complexity of the parts that can be made. In addition, only nonferrous metals and alloys are suitable for die casting. Most die castings are made from aluminum or alloys based on zinc. Copper and magnesium are also used.

In die casting, an expensive set of dies must first be machined from alloy steel. These dies are mounted in a die-casting machine. This machine has a rugged frame that supports and opens the die halves in perfect alignment.

In production, the molten metal is forced into the closed dies under pressure. As the metal goes in, the air is forced out. When the metal has solidified, the dies open and the casting is ejected. For complex castings, a core must be inserted in the dies.

There are two general kinds of die-casting machines: hot-chamber and cold-chamber (Figure 51-4). The *hot-chamber die-casting machine* is used for metals with low melting points, primarily zinc alloys. The melting pot is built into the machine. The cylinder leading to the die chamber always contains molten metal. The hot-chamber method can produce from three hundred to seven hundred fifty castings per hour.

The *cold-chamber die-casting machine* is similar except that the cylinder is not in the liquid metal. Instead, the metal is poured in the "cold chamber" through a port from a ladle that holds only enough

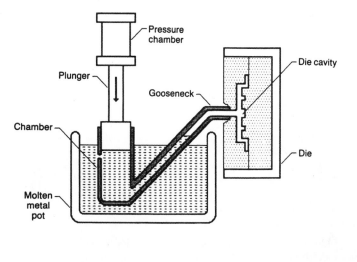

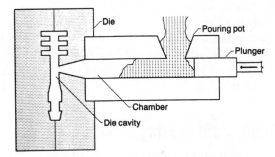

Figure 51-4 Hot chamber (above) and cold chamber (below) die casting.

metal for one casting. The cold-chamber method can be used for casting copper alloys and other metals with a high melting point. The cold-chamber method can produce between sixty and four hundred castings per hour.

Permanent-Mold Casting

Permanent-mold casting involves pouring molten metal into a permanent metal mold or die. It is used only to make relatively simple parts. However, the castings made with this process have a better surface finish and grain structure and are more accurate than sand castings. Thus, less machining is needed. The molds consist of two or more parts and must be very accurate. In mass production, they are held in a large machine that rotates through a cycle. These molds are expensive. Thus, this method is used only if a large number of identical castings are needed.

To make the casting, the mold is heated. Then a coating is sprayed into the casting cavity. This provides the insulation needed to prevent direct contact between

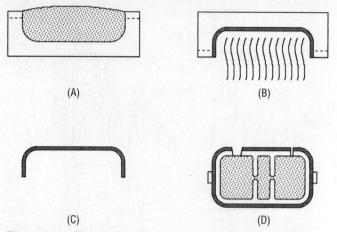

(A)   (B)

(C)   (D)

Figure 51-5 The shell-molding process: (A) A heated metal pattern is filled with the sand mix; (B) the pattern is cooled, and excess sand poured out; (C) the shell is cured; and (D) two shell halves are glued together, with the core in place. The closed mold is now ready for casting.

the metal mold and the molten metal. The mold is filled using the force of gravity. After the metal becomes solid, the mold is opened and the casting is taken out. If a casting requires an opening, a core is inserted between the two parts of the permanent mold before it is closed for pouring. As in die casting, molds can be reused.

Shell-Mold Casting

Shell-mold casting uses a mold in the shape of a thin, two-part shell. The shell mold is made from dry sand and resin. It produces an accurate casting with a very smooth surface that needs little cleaning and finishing.

A metal pattern must first be produced. Then the pattern is heated. A mixture of sand and resin is applied to its surface to form the thin shell. Any extra sand is removed in a rollover operation that dumps the sand and

resin back into the dump box. The shell is removed in halves from the pattern. It is baked until the resin binder is cured. The resulting mold is very light and thin. The mold halves are then put back together with clamps, adhesive, or some other device (Figure 51-5). The mold must be supported as the molten metal is poured into it. The mold is broken to remove the casting.

Shell-mold castings are very accurate and have smooth surfaces. Though shell-mold castings are of high quality, the metal patterns and the equipment for making and pouring the molds are expensive. Generally, only small castings weighing less than twenty-five pounds are made by this method.

Investment Casting

Investment casting, also called **precision casting** and the **lost-wax process,** is used to make very accurate castings that are too complex for other methods (Figure 51-6). This technique is used by dentists for making small gold castings for tooth repair. It is also used for jewelry.

The pattern, often very complex, is made by injecting wax into a die. A pattern can also be made by hand. The pattern is covered with a fine coating of ceramic investment material, similar to plaster of paris. More investment material is mixed. The mounted pattern is placed in the flask, and the investment is poured in. The filled pattern is then placed on a vibrator to remove all the bubbles. When the investment has set, the flask is placed in a kiln to burn out all the wax. While the flask is still warm, it is locked in a casting machine.

The casting metal is placed in the crucible and heated with a torch until it melts. The lock button on the casting machine is pushed to release the casting arm. This then spins rapidly, forcing the metal from the crucible into the mold cavity. Next, the flask is removed with tongs and soaked in water to soften the investment. The casting is first dipped in a pickling solution

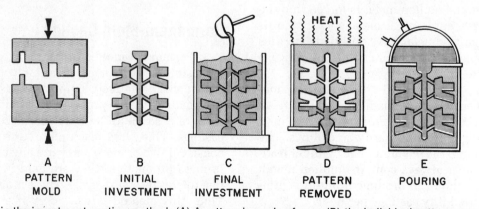

| A | B | C | D | E |
| PATTERN MOLD | INITIAL INVESTMENT | FINAL INVESTMENT | PATTERN REMOVED | POURING |

Figure 51-6 Steps in the investment-casting method: (A) A pattern is made of wax; (B) the individual patterns are fastened together to form a "tree" and placed in a flask; (C) ceramic plaster material is used to cover the pattern; (D) the flask with the patterns in it is turned upside down and then baked to remove the wax; and (E) the molds are turned upright and molten metal is poured into the die cavity.

and then cleaned with a stiff brush under running water. Defects are touched up with a hand grinder, and sprues, gates, and risers are cut off.

In industrial investment casting, clusters of patterns are joined before investment. The flask is then placed on a vibrating table to pack the material solidly around the pattern. The flask and pattern are turned over and then baked to melt out all the wax. A perfect mold remains for the final casting. The mold is turned over again, and the casting is poured. The casting has the exact shape of the original wax pattern. The mold is broken up to remove the casting. Sprues, gates, and risers are then ground off.

Other Casting Methods

In **centrifugal casting,** a rotating mold is filled with molten metal. The rotation forces the molten metal to flow into the detailed surfaces of the mold cavity. The spinning continues until all the metal has solidified. In industry, centrifugal casting is used primarily for making large hollow products, such as pipes.

Plaster-mold casting is very similar to sand casting, except that plaster is used in place of sand. This gives the final casting a smoother finish and greater accuracy. It is possible to produce castings of very intricate design with the plaster-mold method. In plaster-mold casting, wet plaster is poured around the pattern. The plaster hardens with a high degree of detail. The mold is sometimes made in sections. It is baked in an oven for several hours to remove the moisture. All the remaining steps in making a plaster mold—the use of cores, the pouring, and the casting—are the same as for sand casting. The process is used primarily for low-melting-temperature metals, such as aluminum, bronze, and brass.

KNOWLEDGE REVIEW

1. Name six basic industrial casting methods.

2. Which casting method is the most important and is low in cost?

3. What two kinds of machines are used in die casting?

4. In permanent-mold casting and in shell-mold casting, what are the molds made of?

5. What are the other two names for investment casting?

6. What kinds of products are made by centrifugal casting?

7. Plaster-mold casting is similar to what other method?

8. Prepare a research report on one of the industrial casting methods.

9. Collect several castings made by different casting methods. Prepare a display of these and identify each.

Unit 52

Basic Machining Methods

OBJECTIVES

After studying this unit, you should be able to:

- Identify the four major metal manufacturing methods.
- Describe the meaning of the term *machining.*
- Understand what machine tools are, and how they are used in the school and in industry.
- Understand the basic elements of machining.
- Describe the process of turning, and the machines used in it.
- Identify the major parts of a lathe.
- Describe the process of milling, and the machines used in it.
- Identify the major parts of a milling machine.
- Describe the process of drilling, and the machines used in it.
- Identify the major parts of the drill press.
- Describe other machining operations, such as boring, shaping, broaching, grinding, bandsawing, and reaming.

KEY TERMS

bandsawing	fixed head drill	milling	turning
boring	grinding	multiple-spindle drill	turning center
broaching	horizontal band saw	radial drill	turret drill
drilling	horizontal mill	screw machine	turret lathe
drill press	internal grinder	shaping	vertical band saw
engine lathe	machine tool	surface grinder	vertical mill
external cylindrical	machining		
grinder			

Metal machining is one of the more interesting and demanding activities in the school shop. Besides making some attractive and useful projects, you will learn many new skills. But just what is machining?

Machining is one of the four major metal-manufacturing methods. The others are forging, cold forming, and casting. All are used in making metal products. Industry generally uses machining processes when very accurate and smooth workpiece surfaces are needed. However, machining is often more expensive than other methods.

In **machining,** machine tools are used to remove chips from a metal workpiece to produce the desired form and size. A **machine tool** is a piece of power-driven equipment used to perform these cutting operations. The most common of these tools are lathes, milling machines, drilling machines, shapers, and grinders. There are also many special-purpose machines adapted from these basic machine tools.

Basic Elements of Machining

The cutting tool on a machine can be of two kinds: *single-point,* as on the engine lathe or shaper; or *multiple-point,* as on the milling machine, drill, or reamer. Cutting tools are made from high-speed steel, carbide, diamond, ceramics, or silicon carbide and aluminum oxide abrasives. Work devices called *fixtures* and *jigs* are used to hold workpieces and control the cutting tools in machining.

The materials from which parts are machined are standard bar stock, forgings, or castings. Bar stock may be carbon steel, tool steel, stainless steel, aluminum, copper-based alloys, or special metals such as magnesium and titanium.

Many industrial machines are similar to those found in the school shop, except that they are larger, more complex, and generally have automatic controls for speed and accuracy.

Basic Types of Metal Cutting

An introduction to the primary metal-cutting operations performed with machine tools is given below. These machining methods include: turning, milling, drilling, boring, shaping, broaching, grinding, and bandsawing.

Turning

Turning is cutting a workpiece as it revolves on a lathe. A single-point cutting tool moves against the revolving stock (workpiece) to produce cylindrical shapes, as shown in Figure 52-1. Turning can also produce screw threads, tapered parts, and perform drilling, reaming, counterboring, and chamfering operations. In the school shop, the **engine lathe,** the most basic of all machine tools, is used for these operations (Figure 52-2).

Two common types of lathe-turning operations are: turning between centers and chuck turning. When *turning between centers,* the workpiece is held by its ends between the *headstock* and the *tailstock* of the machine. In *chuck turning,* the workpiece is placed in a *chuck* which has adjustable jaws to hold it securely. The chuck is fastened to a *headstock spindle* which rotates the chuck, and no tailstock support is required. This chuck arrangement is common for turning short workpieces, for shaping the ends of workpieces, and for drilling, reaming, and tapping operations.

Many special kinds of lathes are used in both industry and the school shop. The **turret lathe** is a machine used to make identical parts. It differs from the engine lathe in that a six-sided turret replaces the tailstock (Figure 52-3). All six faces of this turret can hold tools of various kinds that can be used on the workpiece, one after another. For example, one face may hold a drill; the next, a reamer; the next, a tap; and so on. As the turret is turned or indexed, it automatically

Figure 52-2 The engine lathe is the basic turning machine used in school shops and in industry. (Courtesy of Clausing Industrial, Inc.)

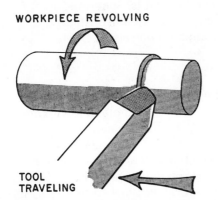

WORKPIECE REVOLVING

TOOL TRAVELING

Figure 52-1 Lathe turning is done when a cutting tool travels into a revolving workpiece.

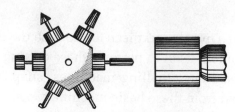

Figure 52-3 Typical setup for a turret lathe.

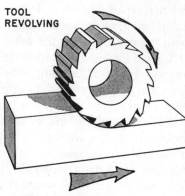

TOOL REVOLVING

WORKPIECE TRAVELING

Figure 52-4 Milling is done when a workpiece travels into a revolving cutting tool.

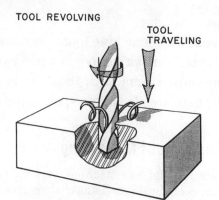

TOOL REVOLVING

TOOL TRAVELING

Figure 52-5 In drilling, a revolving tool is fed into a stationary workpiece.

Figure 52-6 The drill press in use. Note that the operator is wearing safety goggles and that the workpiece is securely held in the drill press vise.

rotates to the next tool. There are also toolholders on the front and rear cross slides to hold special shaping and cutoff tools. Common operations performed in sequence on a turret lathe include drilling, boring, reaming, threading, turning, and cutting off. The order of operations needed determines the arrangement of tools in the turret.

There are many other special-purpose lathes used in schools, small job shops, and in the manufacturing industries. **Screw machines** are high-production turret lathes designed to automatically make screws, bolts, and threaded shafts. Multiple spindle, multiple tool vertical and slant bed **turning centers** are employed for numerous industrial mass production applications. Some of these more advanced, computer-controlled machines will be described later in section V of this book.

Milling

Milling is feeding a workpiece into a rotating tool that has many cutting edges (Figure 52-4). Milling is the opposite of turning. In turning, the workpiece rotates while a single-point cutting tool is fed into it. In milling, the multitoothed rotating tool is stationary and the workpiece moves into it. The milling machine is one of the most important machine tools. The work can be accurately controlled so that interchangeable parts can be produced.

There are many kinds of milling machines used in industry. The two most common are the horizontal mill and the vertical mill. On the **horizontal mill,** the arbor shaft is horizontal, or parallel to a movable table. The

vertical mill has a tool-holding spindle which is perpendicular, or vertical, to the table, and works much like a drill press.

Modern industry uses many variations of the horizontal and vertical mills. Some are computer-controlled and capable of very sophisticated cutting operations. These are called *milling centers.*

Drilling

Drilling is using a rotating cutter shaft, or drill bit, to produce a cylindrical hole in a workpiece, as shown in Figure 52-5. Drilling machines make such holes in metal, wood, plastic, and other materials. For simple drilling, a small bench or floor **drill press,** as shown in Figure 52-6, is generally used. The parts include a table, to support a workpiece locked in a vise, and a vertical powered spindle directly above the table. The drill bit is attached to this spindle by a three-jawed chuck, which is lowered into the work to provide the cutting action.

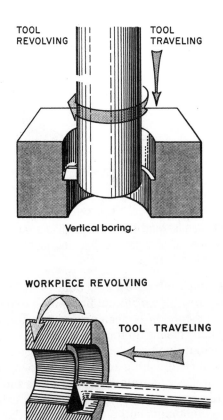

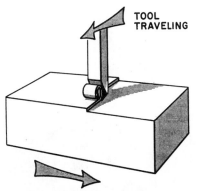

Figure 52-8 Shaping is done by moving a cutting tool into and across the surface of a workpiece. The workpiece moves, or indexes, after each cut.

Figure 52-7 Vertical boring is done by moving a revolving cutter down into a fixed workpiece. Horizontal boring is done by moving a fixed tool into a revolving workpiece; an operation that can be performed on a lathe.

Industry also uses many other kinds of drilling machines. The **fixed head drill** is designed for heavy work requiring large drill bits up to 3 inches (75 mm) in diameter. The worktable and the feed mechanism are power driven. A **radial drill** is designed for large pieces that cannot be moved easily for multiple drilling operations. The entire drilling *head* can be moved to different locations for machining holes. This drill has an *arm* that swings around a *column,* and a precision drilling head that moves along this arm. The **multiple-spindle drill** has several spindles that can drill holes at the same time. A **turret drill** uses six or more drilling devices located around a turret, much like those of a turret lathe. This machine can do drilling, tapping, reaming, and other operations, often under computer control.

Boring

Boring is enlarging or finishing the inner surfaces of an existing hole. The usual method is to use a single-point tool that feeds into the hole while the work rotates. However, on some very large workpieces the tool turns while the work is fixed. There are two kinds of boring machines: *horizontal* and *vertical*. Their operation is shown in Figure 52-7.

Shaping

Shaping is machining metal by making straight-line, back-and-forth cuts across it (Figure 52-8). This machine uses a single-point cutting tool similar to a lathe tool. The tool is clamped in a tool post, and a ram pushes it across the work on the cutting stroke. The ram then raises and withdraws the tool on the return stroke. The work is clamped on a horizontal table which moves a distance of one cut-width after each stroke. Shapers are seldom used in mass production, but they are often found in small job shops. They have been largely replaced by milling machines.

Broaching

Broaching is a cutting operation which is related to shaping (Figure 52-9). It is used to create internal shapes by pulling a tool called a *broach* through an existing hole to change the geometry of that hole. This tool has a number of cutting edges or teeth similar to a file. Each tooth is slightly larger than the preceding one, so that a shape is "machined" by forcing the tool through the hole. In the illustration shown, a keyway, or slot, is being cut by this force action. A broach with external teeth can enlarge holes, give them a smooth finish, or change their shapes. A hollow broach with internal teeth can finish or change the shape of the outside of a workpiece. Broaching is a fast and inexpensive machining operation.

Grinding

Grinding is removing excess material from a workpiece by using rotating abrasive wheels. There are three basic types of these machines. The **external cylindrical grinder** is used to finish the outer surface of stock that is cylindrical or conical in shape. **Surface grinders** produce finished surfaces on flat workpieces held in a vise or on a magnetic table (Figure 52-10). **Internal grinding** is exactly what the term implies; the inside surface of a cylinder or pipe is ground smooth. Both internal and

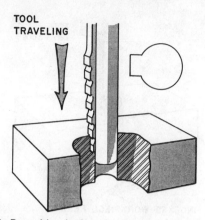

Figure 52-9 Broaching is done by moving a taper-tooth cutting tool into a hole in a fixed workpiece.

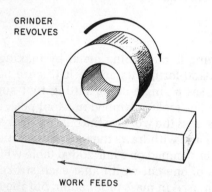

Figure 52-10 Surface grinding uses a revolving tool on a flat workpiece.

cylindrical grinding can be of the *centerless* type, where regulating rolls and/or workpiece rest blades support the work, instead of mounting it between centers.

Bandsawing

Bandsawing is cutting stock to size and/or shape on a special saw, as shown in Figure 52-11. The blade of this saw is a metal band with teeth cut into one edge. The band moves on upper and lower wheels to produce both internal and external shapes. Both straight-line and contour cutting can be done on **vertical band saws. Horizontal band saws** are used to straight-line cut metal workpieces to length for additional machining operations on lathes or mills. If a complicated part must be made, the large masses of material can be removed quickly by internal bandsawing.

Other Machining Processes

Several additional machining operations are closely related to the ones described above. These can be performed on drilling, milling, and boring machines, either manually or automatically. *Drills* produce finished

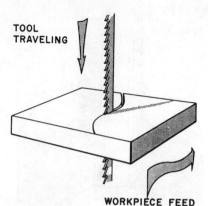

Figure 52-11 Bandsawing is done by feeding a workpiece into a moving metal band with teeth cut into one edge.

holes in castings, enlarge existing holes, and finish holes to accurate sizes. *Step drills* cut holes with two different diameters, and *core drills* clean up and true holes in castings. *Spotfacing* cleans up the area around a cast hole so that a washer and bolt can be used in an assembly operation. *Reaming* is finishing an existing hole to a smooth, very accurate size. A *counterbore* produces a shoulder on the inside of a hole to accept a bearing or a special screw fastener. A *countersink* cuts a cone shape at the end of a hole. This allows a flat-head screw to be used when fastening two parts together. Flute or single point *boring* is truing or enlarging a hole, as described above.

KNOWLEDGE REVIEW

1. Name the four major metal-manufacturing methods.

2. What materials are machine cutting tools generally made of?

3. Describe the differences between chuck turning and turning between centers.

4. Describe the differences between the horizontal and vertical milling machines.

5. Describe several kinds of drilling machines used in the school and in industry.

6. How does boring differ from drilling?

7. Describe the different kinds of grinding processes.

8. Describe the relationship between broaching and shaping.

9. Explain the band machining process and its special uses.

10. With your instructor, arrange a field trip to a local machining industry. Write a report on the different kinds of machine tools found there.

11. Write a research report on modern machining methods in industry.

Unit 53

Precision Measurement

OBJECTIVES

After studying this unit, you should be able to:

- Understand the importance and use of precision measuring tools.
- Read the slide caliper, micrometer, and vernier caliper.
- Recognize the difference between customary and metric precision measuring tools.

KEY TERMS

depth micrometer   micrometer   slide caliper   vernier caliper

Machined products require the use of precision measuring tools to ensure their quality. These tools differ from those described in unit 13 because they are more accurate. For example, the spring caliper and scale shown in Figure 53-1 can be used to provide a suitable measure for lathe-turning a piece of stock to a 1 1/8-inch rough size. A micrometer then would be needed to finish-turn the stock to a 1.000-inch diameter for thread-turning. The tools used in beginning machine shop courses include the slide caliper, the micrometer, and the vernier caliper. You must learn to use these in order to do quality, precision machining.

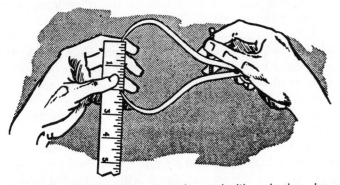

Figure 53-1 Spring calipers must be used with scales in order to make measurements.

Slide Calipers

The main disadvantage of using the common spring calipers is that they do not give a direct reading of the caliper setting. You must always measure this setting with a scale. To overcome this disadvantage, a **slide caliper** is used (Figure 53-2). This instrument can be used for outside, inside, and length measurements. The *back face* of the caliper is used as an ordinary rule. Stamped on the frame of the *front face* are the words "OUT" and "IN," for use when making outside and inside measures. For example, when measuring the inside diameter of a hole or tube, insert the *sliding* and *fixed*

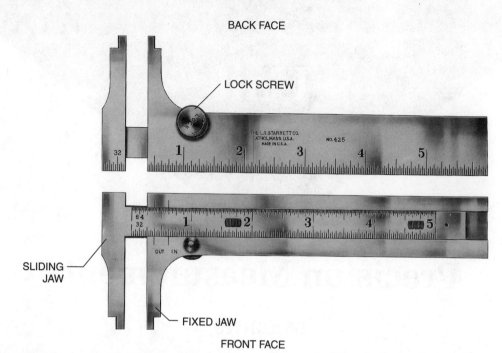

BACK FACE

LOCK SCREW

SLIDING JAW

FIXED JAW

FRONT FACE

Figure 53-2 Slide calipers are used like spring calipers, but they will make direct-reading and more accurate measurements. Note the placement of the scales on both the back and front faces. (Courtesy of The L. S. Starrett Company)

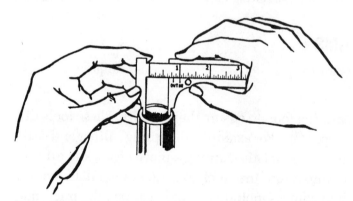

Figure 53-3 Using the slide caliper to determine the inside diameter of a piece of metal pipe.

jaw tips into that hole. Read the measurement at the reference line marked "IN" (Figure 53-3). A *lock screw* secures the jaws in position during use. Slide calipers are generally made in three-inch and five-inch sizes, and are available in both inch and millimeter models. These calipers are convenient, direct-reading tools, and are used when extreme precision is not required.

The Customary-Inch Micrometer

Micrometers are not only far more accurate than slide calipers, but they also give a direct reading without the use of a scale. Some of the most common micrometers are the outside, inside, and depth types. There are many others used for special precision measuring in the

shop and in industry—available in both customary-inch and metric models. The one used most often in the school shop is the two-inch outside micrometer (Figure 53-4), and it works in the following way.

There are forty threads to the inch on the micrometer screw. As it turns, it changes the space between the two measuring faces. The thimble turns with the screw. Each complete turn of the thimble equals 1/40 (0.025) inch. There are also forty lines to the inch on the sleeve. Each time you make a complete turn, you change the reading on the sleeve by one small division. Every four divisions on the sleeve are marked 1, 2, 3, and so on, representing 0.100 inch, 0.200 inch, 0.300 inch, and so on. The tapered end of the thimble is divided into twenty-five equal parts. One complete turn of the thimble equals 0.025 inch. A turn of the thimble of one twenty-fifth, one small division on the thimble, equals 0.001 inch. The scale on the thimble is marked 0, 5, 10, 15, 20, making it easy to read in thousandths (0.001) of an inch. Other micrometers are graduated to read in ten-thousandths (0.0001) of an inch for yet more accurate measurements.

When using the micrometer to measure round stock, hold the stock in one hand and the micrometer in the other hand so that the *thimble* rests between the thumb and the forefinger (Figure 53-5). Hold the *frame* against the palm of your hand with your third finger. Guide the stock to the *anvil*, and turn the thimble to bring the *spindle* against the stock to make the measurement. The thimble should be turned gently, "feel-

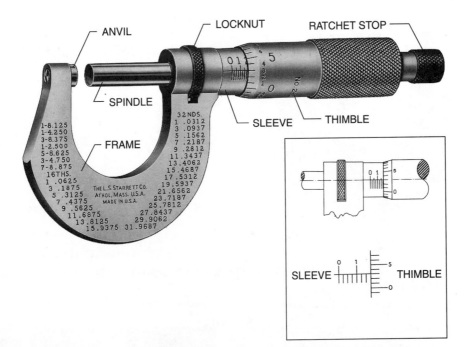

Figure 53-4 The parts of a customary micrometer. Note that the reading of this setting is 0.178 inch.
(Courtesy of The L. S. Starrett Company)

Figure 53-5 Note how the micrometer is held when measuring a piece of round stock.
(Courtesy of The L. S. Starrett Company)

ing" the contact pressure against the stock. You can also use the *ratchet stop,* which will automatically stop turning when the necessary contact pressure has been reached. Never overtighten the thimble, as this will damage the micrometer. Now read the markings on the sleeve and the thimble, and calculate your measurement of the metal bar. The *locknut* will secure the thimble so that you will not lose the measurement while using the micrometer.

Reading a Customary Micrometer

Follow the sample procedure below for reading a measurement on a customary micrometer.

1. Refer to Figure 53-4. Look at the number of divisions you can see on the sleeve. You see three divisions past the number 1, or seven full divisions. This indicates 0.175 inch (0.025 times 7).

2. If the end of the thimble is between two division lines on the sleeve, you must read the scale on the thimble. In Figure 53-4, you see three small divisions past the 0. This indicates 0.003 inch (0.001 times 3).

3. By adding the reading on the sleeve to the reading on the thimble, you get 0.175 plus 0.003, or 0.178 inch.

The Metric Micrometer

On a metric micrometer, the pitch of the spindle screw is one-half millimeter (0.5 mm). Thus, one full turn of the thimble moves the spindle toward or away from the anvil by 0.5 millimeter.

The reading line on the sleeve is marked in millimeters, with every fifth millimeter being numbered

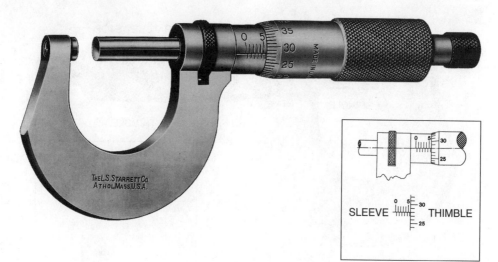

Figure 53-6 A metric micrometer is used like the customary instrument, but the readings are in millimeters. This setting reads 5.78 mm. (Courtesy of The L. S. Starrett Company)

from 0 to 25. Each millimeter is also divided in half (0.5 mm). Each time you give the thimble one full turn, you change the reading on the sleeve by 0.5 millimeter.

The tapered end of the thimble is divided into fifty equal parts, every fifth line being numbered from 0 to 50. Since one turn of the thimble moves the spindle 0.5 millimeter, each thimble division equals 1/50 of 0.5 millimeter, or 0.01 millimeter. Thus, two thimble divisions equals 0.02 millimeter, three equals 0.03 millimeter, and so on.

Reading a Metric Micrometer

Follow the sample procedure below for reading a measurement on a metric micrometer.

1. Refer to Figure 53-6. Count the number of divisions on the upper sleeve. There are five full divisions, indicating 5 mm (5 times 1.00).

2. Count the divisions on the lower sleeve which lie beyond the 5 mm mark. There is one, indicating 0.50 mm (1 times 0.50).

3. Count the divisions on the thimble, shown by the thimble number mark which lines up with the sleeve centerline. There are twenty-eight indicating 0.28 mm (28 times 0.01).

4. Add the sleeve divisions to the thimble divisions to get a total reading of 5.78 mm (5.00 plus 0.50 plus 0.28).

Other Micrometers

The many special kinds of micrometers include those for inside and depth measures, and for checking the pitch diameters of threaded parts. The use of the **depth micrometer** is shown in Figure 53-7. There also

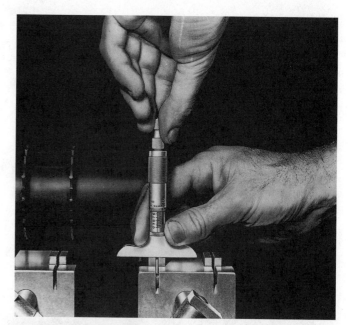

Figure 53-7 The depth micrometer is used to measure the depths of holes and slots. (Courtesy of The L. S. Starrett Company)

are dual-reading micrometers for customary-inch and metric measures; and direct-reading models which give a digital reading without having to add sleeve and thimble values.

The Vernier Caliper

The disadvantage of the micrometer is that because of its size it can be used only for measuring small workpieces or parts. Also, different models are needed to make inside and outside measurements. The **vernier caliper** is a tool which is able to provide accurate mea-

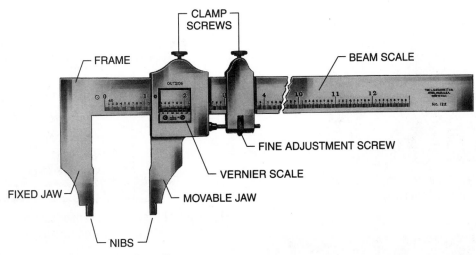

Figure 53-8 The parts of the customary vernier caliper. (Courtesy of The L. S. Starrett Company)

surements, both inside and out, over a large range of sizes. There are pocket models which usually measure up to three inches, but others are available in sizes up to four feet. Metric calipers also are used. The most common for shop use are the six-inch and twelve-inch sizes. This caliper looks and is used much as the slide caliper, but is more accurate.

The vernier caliper consists of an L-shaped *frame* with a *fixed jaw* that has an engraved *beam scale* (Figure 53-8). A *movable jaw* can slide on the beam to take measurements, and a *vernier scale* is attached to this movable jaw. The jaw is moved against the workpiece by pushing it with the fingers, and then the *fine adjustment screw* is used to complete the operation. The *clamp screws* lock the jaws while the instrument is in use.

Principles of the Vernier Scale

As stated above, the vernier caliper has two separate scales which must be read together in order to make a measurement. It is important to understand the principles and theory of the scales in order to be comfortable and confident when using the caliper. The *beam scale* of the tool is graduated, or divided, in 40ths (0.025) of an inch. Every fourth division is numbered, and represents a tenth (0.10) of an inch. The *vernier scale* is graduated into twenty-five divisions, numbered 0, 5, 10, 15, 20, and 25, with each division representing one-thousandth (0.001) of an inch.

The twenty-five divisions on the vernier occupy the same space as twenty-four divisions on the beam scale. Since one division on the beam scale equals 0.025 inch, twenty-four divisions equals 24 times 0.025, or 0.600 inch; and twenty-five divisions on the vernier also equals 0.600 inch. It follows, therefore, that one vernier division equals 1/25th of 0.600, or 0.024 inch; and the difference between one beam division (0.025")

and one vernier division (0.024") equals 0.025 inch minus 0.024 inch, or 0.001 (one-thousandth) inch. It may sound confusing, but once you have used the caliper a few times it will make sense to you and become clearer.

If the tool is set so that the 0 mark on the vernier lines up with the 0 mark on the beam scale, the line to the right of the 0 mark on the vernier will differ from the line to the right of the 0 mark on the beam scale by 0.001 inch. And it follows that the second line will differ by 0.002 inch, the third line by 0.003 inch, and so on. The difference will continue to increase by 0.001 inch for each division until the 25 mark on the vernier coincides (lines up) with the 24 mark on the beam scale. Take the tool in your hands and try it out; you will see that this is what happens.

Reading a Customary Vernier Caliper

To make a measurement with this tool, note how many inches, tenths (0.100), and fortieths (0.025) the 0 mark on the vernier is from the 0 mark on the beam scale. To this value add the number of thousandths (0.001) indicated by the mark on the vernier which *exactly* coincides with a line on the beam. For example, in the illustration in Figure 53-9 the vernier has been moved to the right 1.000 plus 0.400 plus 0.025, which equals 1.425 inches as shown on the beam. Note that the eleventh line on the vernier coincides with a line on the beam, as indicated by the stars on the illustration. Therefore, the vernier reading of 0.011 must be added to the beam reading of 1.425 to get a total caliper reading of 1.436 inches. The best way to become familiar with the vernier caliper is to use it to measure some sample pieces of stock or metal parts, and write down the various beam and vernier values so you can check your measurements carefully (Figure 53-10).

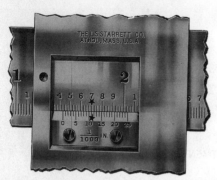

Figure 53-9 This vernier caliper reads 1.436 inches. (Courtesy of The L. S. Starrett Company)

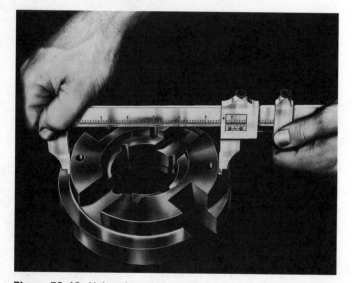

Figure 53-10 Using the vernier caliper to measure the outside diameter of a machine part. (Courtesy of The L. S. Starrett Company)

Figure 53-11 This metric vernier caliper reading is 41.68 mm. (Courtesy of The L. S. Starrett Company)

Figure 53-12 A direct-reading vernier caliper is much easier to use than other models. (Courtesy of The L. S. Starrett Company)

Reading a Metric Vernier Caliper

The metric caliper looks like, is used, and is read much like the customary. The beam scale is divided into 0.5-mm divisions, and every twentieth division is numbered 10, 20, 30, etc., meaning 10 mm, 20 mm, 30 mm, and so on (Figure 53-11). The vernier scale is graduated into two-hundredths (0.02) millimeter divisions, and every fifth division is numbered 0.10, 0.20, 0.30, 0.40 and 0.50, meaning 0.10 mm, 0.20 mm, and so on.

To read the caliper in Figure 53-11, note the number of half-millimeter divisions which lie between the 0 mark on the *beam* and the 0 mark on the *vernier*. There are 41.5, meaning 41.5 mm. Now find the mark on the vernier which lines up exactly with a mark on the beam (indicated by the stars in the figure). Count the number of divisions to the right of the vernier scale 0 mark. You will find there are nine, giving a vernier reading of 0.18 (9 times 0.02) mm. Add the two readings together to get a total of 41.68 (41.5 plus 0.18) mm.

Other Vernier Calipers

Dual-reading calipers are available which can be used for either customary-inch or millimeter measurements. These instruments can also be used to convert one measurement to the other. The *direct-reading* vernier caliper, shown in Figure 53-12, is used as you would the standard tool, but is much easier to read.

KNOWLEDGE REVIEW

1. What is the advantage of a slide caliper over a spring caliper?

2. On a customary micrometer, what distance does one turn of the thimble equal?

3. On a metric micrometer, how far does one full turn of the thimble move the spindle?

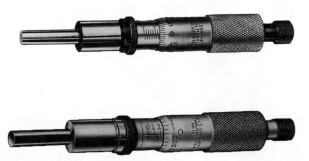

Figure 53-13
(Courtesy of The L. S. Starrett Company)

4. What is a depth micrometer used for?

5. Compare the operation of the micrometer and vernier caliper.

6. Explain to the class how the vernier caliper is used.

7. Select several pieces of metal bar stock and measure their thicknesses or diameters using the slide and vernier calipers and the micrometer. Record your readings and compare them with the readings of one of your classmates. Did you both get the same results?

8. Visit a local industry and write a report on the different kinds of precision measuring tools you saw being used.

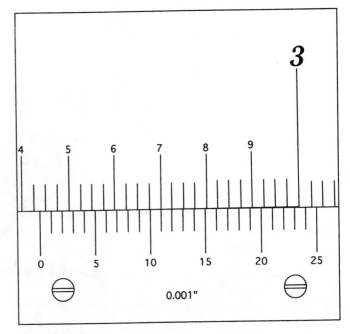

Figure 53-14

9. Study the customary-inch and metric micrometer settings in Figure 53-13 and write down your readings.

10. Study the vernier caliper setting in Figure 53-14 and write down your reading.

Unit 54

Bandsawing

OBJECTIVES

After studying this unit, you should be able to:

- Understand what bandsawing is, and list its advantages.
- Explain the differences between vertical and horizontal bandsawing.
- Recognize vertical and horizontal bandsawing machines.
- Describe the importance of band saw blade tooth shape and set.
- Install, track, and adjust a band saw blade.
- Select the band saw speed for a given cutting operation.
- Perform a simple bandsawing operation.

KEY TERMS

bandsawing   tooth set   tooth shape   vertical band saw

horizontal band saw

Bandsawing is a fast, efficient, and accurate method of making straight, contour, internal, and external cuts on almost any material. It is done with a thin, continuous blade, or band, made from high-speed or carbon steel with sharp teeth on one edge. The blade is stretched around two wheels, one above and one below the table that supports the workpiece. The lower wheel is power driven to move the blade, and the workpiece is cut as it is pushed into the moving blade. This arrangement is typical of the **vertical band saw,** which is used for most band cutting operations (Figure 54-1). The other common type is the **horizontal band saw,** which is designed primarily for cutting stock to length (Figure 54-2). A combination vertical and horizontal band saw also is available.

The advantages of bandsawing, also called *band machining,* are many. The band saw can cut material to any shape with the least amount of waste. Material is removed in sections instead of chips as in many other kinds of machining. The band saw cuts continuously and, as a result, chip removal is fast and accurate. Since the thin blade makes it possible to cut close to a layout line, material is saved. It cuts all shapes and designs with no limitation on angle, direction, or length of cut. Because it has many moving teeth, the contact time between the teeth and the workpiece is short and the saw cuts cooler. The band saw can be

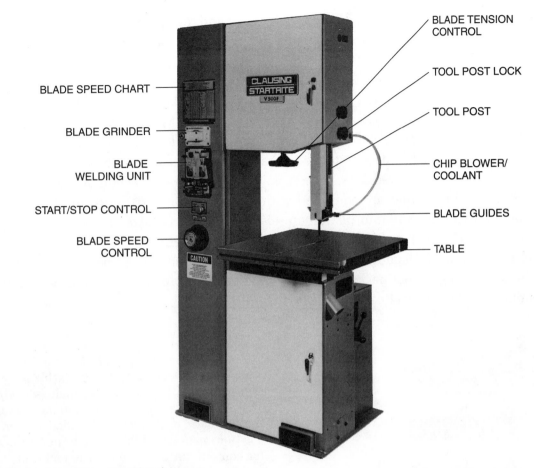

Figure 54-1 Basic parts of the vertical band saw.
(Courtesy of Clausing Industrial, Inc.)

Figure 54-2 Basic parts of the horizontal band saw.
(Courtesy of Clausing Industrial, Inc.)

used to make internal cuts because its blade can be easily cut and rewelded. Simple fixtures or holding devices are used in mass manufacturing products, but many operations can be done with the work held freehand on the table.

The size of the band saw is indicated by the diameter of its wheels and its throat capacity (the distance from the blade to the inside of the frame). Whereas most band saws have only an upper and lower wheel, some of the larger machines have a three-wheel arrangement. Many smaller machines are designed for wood, plastic, and metal cutting, while many larger ones are used exclusively for metal cutting.

Most saw tables can be tilted 45 degrees to the right and from 5 to 10 degrees to the left. On some large industrial models, the table can be moved forward or backward with hydraulic action by means of a foot pedal.

Variable-speed drive machines are equipped with a job selector chart built into the machine itself. This circular chart gives a variety of information on selecting the correct size of blade, cutting speed, feed, and other important information. By dialing the correct material on the chart, all other information is immediately visible. Some band machines are equipped with a tachometer or speed selector that tells the speed the blade is traveling in feet per minute. Many are also equipped with a butt welder for welding blades conveniently, under automatically controlled conditions.

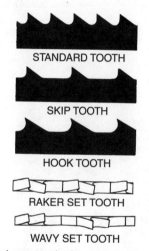

Figure 54-3 Band saw tooth shape and set patterns.

Band Saw Blades

There are many sizes and kinds of blades for use on different materials and jobs. The size or circumference of the blade is measured around the upper and lower wheels. Check this carefully in the operator's manual as the size will vary from machine to machine.

Tooth shape is important when selecting a saw blade (Figure 54-3). The *standard tooth* is preferred for cutting all hard brasses and ferrous metals. The *skip tooth* blade has widely spaced teeth to provide the added chip clearance needed for cutting softer materials such as wood, plastic, and nonferrous metals. The *hook tooth* blade has a 10-degree undercut which permits better feed and chip removal. This blade will cut the same materials as the skip tooth, but will do it faster and cleaner.

There are two types of **tooth set.** The *raker set* (or regular set) is generally found on blades which have two to twenty-four teeth per inch (see Figure 54-3). These blades have one tooth set to the left, one tooth to the right, and one unset tooth called a raker. This set is used on material which has a uniform thickness, and for contour cutting. *Wavy set* is found on saws which have eight to thirty-two teeth per inch. This set has groups of teeth bent alternately to the left and right, which greatly reduces the strain on individual teeth. Saws with wavy set are used where tooth breakage is a problem, such as in cutting thin stock, or where a variety of work is cut without changing blades.

Installing a Blade

Band saw blades do break accidentally during normal use, and a new blade must be installed. After installation, the blade must be tracked and adjusted before it can be used for cutting. It is wise to have your teacher watch you while this is being done, to make sure your procedure is correct. The proper procedure for installing a new blade is as follows:

1. Turn off the power to the machine.

2. Release the tension on the upper wheel.

3. Loosen the locking screws on all guides of the upper and lower guide assemblies and back off all the guides so they are out of the path of the new blade.

4. Remove the alignment pin or screw from the table slot and remove the table insert.

5. Open the wheel guards to expose the upper and lower wheels.

6. Uncoil the new blade. Be sure the teeth are pointed in the proper direction. Hold the blade with the teeth facing your body. If the teeth in your right hand point downward, the teeth are pointed in the proper direction. If they do not, try turning the blade inside out; they should then be in proper position.

7. Still holding the blade in proper position, carefully work it through the table slot and slip it onto the upper and lower wheels.

8. Replace the table alignment pin in the slot.

Tracking a Blade

To track a newly installed blade, follow the procedure below.

1. Tighten the tension on the upper wheel. Proper tension may be indicated by a dial indicator located on the tensioning mechanism. Often, however, proper tension is largely a matter of experience.

2. Turn the upper wheel slowly by hand, and at the same time adjust the tracking control until the blade runs on the center of the rubber tire. Rotate the upper wheel by hand several more times to make sure the blade stays on the wheels.

3. Check the tracking by clicking the power switch on and off. Use longer bursts of power-on until you are sure the blade runs true.

Adjusting the Blade Guides

The blade guide mechanism is located at the lower end of the saw guide post. Its purpose is to hold the blade in position for cutting, and to prevent blade breakage. To adjust the blade guides, do the following:

1. Loosen the locking screws of the upper guide. Bring the guide assembly forward until the front edges of the side supports are about 1/64 inch behind the tooth gullets. Then place slips of thin paper on either side between the guide and the blade. Push the guides against the paper and tighten the locking screws. The paper should provide a clearance of 0.003 or 0.004 inch on each side.

2. Repeat the above procedure on the lower guide assembly located below the table.

3. Move the backup guide wheel to a point 1/64 inch behind the back edge of the blade. The wheel should not touch the blade until a cutting load is applied. A simple way to set the backup wheel is to turn on the power and bring the wheel up until it begins to move with the blade. Then back the wheel off until it stops moving. Repeat on the lower guide.

4. Inspect the machine to make sure that none of the guide parts touch the running blade until a workpiece cutting load is applied.

Band Saw Safety

Observe the following guidelines and precautions when using a band saw.

- Check the workpiece to be sure that it is free of dirt and flaws which may damage the blade.

- Wear gloves when handling and repairing band saw blades.

- Wear safety glasses or a shield.

- Make sure that the saw blade has proper tension. Read the instructions or tension dial plate. Too much tension can cause the blade to break.

- Avoid backing out of a cut, as this may pull the blade away from the guides, off the wheels, and damage the blade and the workpiece.

- Cut round or irregular stock carefully, or use a holding fixture (Figure 54-4).

- Never cut a curve of small radius with a wide blade. See Figure 54-5 for guidance in selecting blades for making curved cuts.

- If you hear a rhythmic click as the material is being cut, this usually indicates a cracked blade or an improperly welded joint. Stop the machine and inspect.

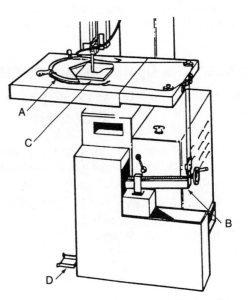

Figure 54-4 A special work holder (A) and feed weight (B) guide the workpiece (C) for safe contour cutting. The foot lever (D) controls the feed rate.
(Courtesy of Clausing Industrial, Inc.)

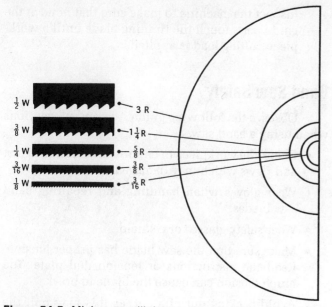

Figure 54-5 Minimum radii that can be cut with various widths of blades.

- If the blade breaks, shut off the power and stay away from the machine until it comes to a complete stop.
- Never try to free the blade while the wheels are still turning.
- Never place your fingers or arms in direct line with the blade.
- Make sure that all guards are in place before starting the machine.
- Never try to remove small pieces of material near the blade while the saw is operating. These scrap pieces are sharp and hot. Always stop the saw and use a small wooden stick, not your fingers.

Basic Band Saw Operating Techniques

The following list of band saw operating procedures and techniques will help you to make safe and accurate cuts in your workpieces.

1. Unless some type of cutting guide is used, draw guidelines on the workpiece to indicate the shape to be cut.

2. Check for proper size and kind of blade, speed, and feed. Most large machines have a built-in job selector chart on the cover of the upper wheel. This circular chart can be set for the kind of material to be cut. It indicates the kind of blade to use, speed, feed, and other information.

   Blade speed is an important factor in successful band saw use. If the speed is too fast for the material being cut, the teeth will not dig into the material. They will just ride over the surface of the

work, create friction, rapidly dull the cutting edge, and wear out the blade. The table in Figure 54-6 shows recommended saw speeds for different metal materials.

Under general conditions, an even feed pressure without forcing the work gives best results. A clean curl or chip indicates ideal feed pressure. Burned or discolored chips indicate excessive feed pressure which can cause excessive wear and tooth breakage.

3. Many machines have an air jet or cutting fluid tube to lubricate and cool the blade, and to blow away chips. They are necessary when cutting tough metal materials, or to blow away plastic and wood chips.

4. Lower the saw guide post so that it is about 1/4 inch above the stock. Feed the waste stock into the blade and come up to the layout line. Then follow the layout line, keeping the blade just to the outside of this line. When cutting, guide the stock with your left hand and apply forward pressure with your right (or the opposite, if you are left-handed). Move the stock into the work as rapidly as it will cut. Moving it too slowly will tend to dull the blade. Do not feed the work into the blade until the machine is at full operating speed. In all cutting, accuracy is important and can only be obtained by carefully following the layout line. A correct stance and position will help you to better control the work.

5. For straight sawing, a fence (Figure 54-7) or miter gage (Figure 54-8) should be used. A miter gage is used for cutting short lengths and angles, and a fence is used for ripping long pieces.

6. For cutting curves and other irregular shapes, come directly up to the layout line leaving only the small amount of material that might be necessary for finishing. A special guide is available for cutting circles (Figure 54-9).

7. Internal cuts can be made on any band saw which has a blade welding unit. To do internal cutting, first drill a hole in the waste stock. Then remove the blade from the machine and cut it apart with shears. Place the workpiece on the table and thread the blade through the hole. Weld the blade together, anneal it, and grind it. Move the workpiece and the blade into position and then install, track, and adjust the blade. Now make the internal cut. After the cut is completed, the blade must again be cut apart to remove the finished work from the machine.

Most of the procedures listed above refer to vertical band saws, but the information regarding blade selection and speeds also applies to horizontal saws. As stated at the beginning of this unit, horizontal band

TYPE OF METAL	Average Saw Speeds in fpm	
	Carbon-Steel Blades	High-Speed Steel Blades
Straight-carbon steels	150–175	300
Medium-carbon steels	100–150	200
Free-cutting low-carbon steels	150–175	300
Free-cutting medium-carbon steels	100–150	250
High-carbon steels	80–125	150
Manganese steels	70–125	200
Nickel steels	75–100	175
Nickel-chrome steels	50–100	200
Moly steels	75–135	200
Chrome-moly steels	50–100	225
Nickel-chrome moly steels	50–100	200
Nickel-moly steels	50–100	200
Aluminum bronze	700	
Beryllium copper	800–2000	
Manganese bronze	200–900	
Phosphor bronze	300–700	
Silicon bronze	200–900	
Aluminum castings and forgings	300–1500	
Aluminum extrusions	2000–3000	
Aluminum sheet, plate, rod, and tubing	3000–5000	
Copper, Brass	1500	
Babbitt, Lead	1500–3000	

Figure 54-6 Table of recommended cutting speeds.

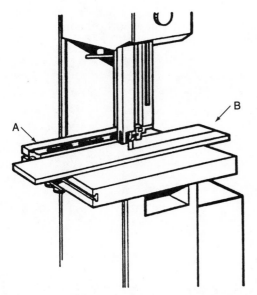

Figure 54-7 Use a fence (A) for making accurate long cuts on workpieces (B).
(Courtesy of Clausing Industrial, Inc.)

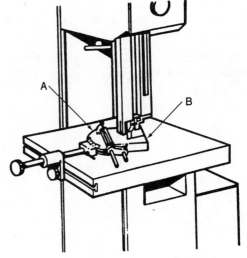

Figure 54-8 Use a miter gage (A) for cutting angles or short straight cuts on a workpiece (B).
(Courtesy of Clausing Industrial, Inc.)

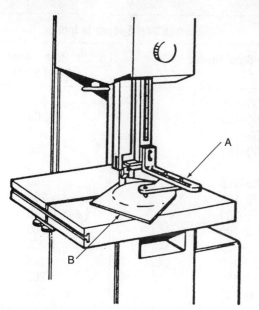

Figure 54-9 A special guide (A) controls the workpiece (B) when cutting circles.
(Courtesy of Clausing Industrial, Inc.)

saws are used mainly for cutting stock to length. The most important points to remember are to lock the stock firmly in the chuck (Figure 54-10), and to use plenty of coolant where required.

KNOWLEDGE REVIEW

1. Name the two common types of band saws and describe their uses.

2. List the advantages of bandsawing over other means of metal separating.

3. Describe the process of installing, tracking, and adjusting a band saw blade.

Figure 54-10 Be sure that the workpiece is securely fastened in the vise for horizontal cutoff work.
(Courtesy of Clausing Industrial, Inc.)

4. Explain the meaning of the terms *tooth shape* and *tooth set*.

5. Why are cutting fluids or air jets sometimes used?

6. Describe how you would cut a six-inch circle out of the inside of a ten-inch square sheet of 1/4-inch thick aluminum.

7. What should the speed in feet per minute be when cutting medium-carbon steel with a high-speed steel blade?

8. How can you tell when you are using the proper amount of feed?

9. Prepare a research report on computer-controlled industrial band saws.

Unit 55

The Engine Lathe

OBJECTIVES

After studying this unit, you should be able to:

- Understand the theory of lathe turning.
- Recognize the important parts of an engine lathe.
- Understand the functions of the headstock, tailstock, and carriage.
- Be familiar with the lathe controls.
- Recognize the various kinds of cutting tools.
- Grind a roundnose cutting tool.
- Select lathe cutting speeds.
- Recognize the common toolholders, and mount a tool in one.

KEY TERMS

bed	depth of cut	feed controls	tailstock
carriage	engine lathe	headstock	toolholder
cutting speed	feed		

The metal-cutting lathe found in the school shop is often called an **engine lathe,** and it cuts metal by revolving a workpiece against a sharp tool bit. It is an important machine, and the one the beginner usually learns to use first. There are many sizes and types of lathes used for a wide range of metal removal and separation operations (Figure 55-1).

The size of a lathe is determined by the swing and the bed length (Figure 55-2). The *swing* indicates the largest diameter of a workpiece that can be turned on the machine. As shown on the illustration, the swing is twice the distance from the headstock center to the nearest interference on the bed of the lathe. The *bed length* is the total length of the ways, or tracks, of the lathe. Another important lathe dimension is the *distance between centers,* which determines the longest workpiece that can be turned on that lathe. A common school shop has a 9-inch (228-mm) swing with a 3-foot (914-mm) bed.

Figure 55-1 Lathe turning is a basic machining operation. Here a taper is being cut on a chuck-mounted workpiece. (Courtesy of Clausing Industrial, Inc.)

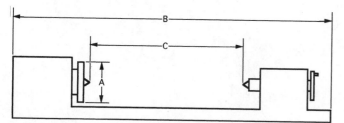

Figure 55-2 Important lathe size dimensions: (A) swing, (B) bed length, (C) distance between centers.

Parts of the Lathe

The main components common to all engine lathes are shown in Figure 55-3. Many companies make many different styles and sizes of lathes. Therefore, these parts may not all look alike, nor will the control levers be in exactly the same place on every machine. They do, however, all serve the same purposes in lathe operations.

Bed

The **bed** is the base or foundation of the lathe onto which all other main components or parts are attached. This is a heavy, sturdy, and durable steel casting. The top part of the bed has accurately machined and carefully aligned *ways,* which are the rails that support the headstock, carriage, and tailstock, and hold them in position.

Headstock

The **headstock** assembly is fastened to the left end of the bed (Figure 55-4). It consists of the headstock *spindle* and the mechanisms and controls for driving it. The spindle is the "live" end of the lathe; it drives the workpiece to be turned. It has either a long tapered nose, as shown, or a threaded nose onto which chucks and faceplates are screwed. The spindle has a tapered hole at the front end to hold the headstock, or *live center,* used in lathe turning.

The power for turning is provided by the electric motor located in the headstock. The power is delivered to the spindle by a series of belts and a gear train. On belt-driven lathes, power is delivered through belts to a

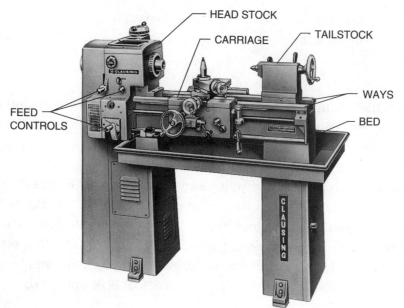

Figure 55-3 Important parts of the lathe. (Courtesy of Clausing Industrial, Inc.)

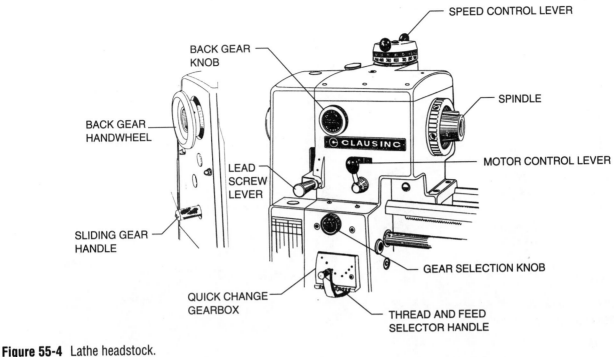

Figure 55-4 Lathe headstock.
(Courtesy of Clausing Industrial, Inc.)

step pulley that turns the spindle (Figure 55-5). The speed is changed by moving the belts to different positions. To obtain more torque (turning force) and slower speeds, a special back gear arrangement is used. This is operated by the *back gear knob,* the *back gear pin,* and the *back gear handwheel.* Modern lathes have variable speed control chuck mechanisms, operated by a *speed control lever,* eliminating the need for moving the belts on the pulleys. This is a safer, faster, and more convenient arrangement. The *motor control lever* has three

Figure 55-5 Step pulleys and gears power this engine lathe.
(Courtesy of Clausing Industrial, Inc.)

positions which control the rotational direction of the spindle. Moving the lever to off allows the spindle to stop. Moving the lever to forward or reverse directs the spindle to turn in either of those directions. You should always allow the spindle to stop before shifting to the reverse position. A number of other important tool feed and thread-cutting controls are described below.

Tailstock

The **tailstock** is the "dead" end of the lathe (Figure 55-6). It holds the tailstock, or *dead center,* for turning, and also holds various chucks and tools for operations such as reaming and drilling. The lathe bed and ways are clearly shown in the illustration. The tailstock assembly can be moved along these ways and locked in position with the *bed clamp lever.* The tailstock is made up of two castings; a *lower* which rests on the ways, and an *upper* which is fastened to the lower. The upper casting can be moved toward or away from the operator with a *setover screw* to offset the assembly for taper turning. A hollow *ram* moves in and out of the upper casting when the *ram handwheel* is turned. This ram has an inner taper into which the dead center and other tapered tools can fit.

Carriage

The **carriage** holds a variety of cutting tools, and has several major parts (Figure 55-7). The *saddle* is an H-shaped casting that fits over the bed and slides along the ways. The *apron* fastens to the saddle and hangs

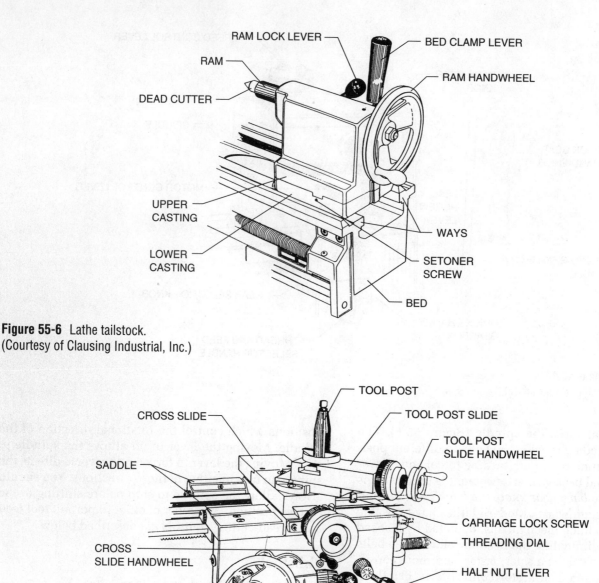

Figure 55-6 Lathe tailstock.
(Courtesy of Clausing Industrial, Inc.)

Figure 55-7 Lathe carriage.
(Courtesy of Clausing Industrial, Inc.)

over the front of the bed. It contains the gears, clutches, and levers for operating the carriage by hand or with power. The *carriage handwheel* is turned to move the carriage back and forth. A *carriage lock screw* can be tightened to secure the carriage. The *cross slide* is mounted on the saddle, and turning the *cross slide handwheel* moves it toward or away from the operator. The *tool post slide* (also called the compound rest) is fitted to the top of this cross slide. It can be adjusted to various angles for taper facing and locked in position. The upper part of the tool post slide can be moved in

and out with the *tool post slide handwheel*. The *tool post* with a ring collar and rocker base slides in a T slot on top of the tool post slide. Various cutting tools can be locked into it for cutting operations. A *clutch and brake handle* brings power to the carriage, or will stop it.

Feed Controls

The **feed controls** are used to direct cutting and threading operations. They consist of a headstock quick-change gearbox, lead screw and feed rod, as well as the

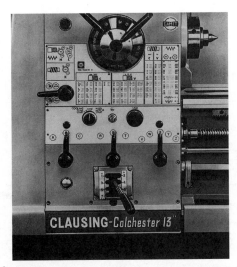

Figure 55-8 Lathe controls located on the headstock. This machine will cut both customary-inch and metric threads. (Courtesy of Clausing Industrial, Inc.)

gears and clutches in the carriage apron (refer to Figures 55-4 and 55-7). The *quick-change gearbox* is directly below the headstock assembly. Power from the left end of the spindle is fed through gears to it. The gearbox makes it possible to change the rotational rate of the lead screw per revolution of the spindle, and the subsequent movement of the carriage for thread cutting. The *lead screw lever* engages the lead screw, which controls the movement of the carriage. It has three positions. The *center* position is neutral; when the gear train is disengaged, the lead screw does not turn, and the carriage therefore does not move. The *lower* position causes the carriage to move toward the tailstock; and the *upper* position moves the carriage toward the headstock.

There are several control levers on or near the gearbox which relate to its operation. An *index chart* fastened to the gearbox tells you how to position these levers for any specific lathe operation. The *sliding gear handle* changes the lead screw/spindle ratio. There are two positions for it: in and out. The *gear selector knob* can be turned to positions A, B, or C. The *thread and feed selector handle* is a device for shifting gears in the quick-change box. Another arrangement of these headstock controls appears in Figure 55-8.

To get power for longitudinal (back-and-forth) feeding, the *power feed lever* on the carriage is moved to either the down or up position. Then the *clutch handle* is moved to engage the carriage. To get power for the cross slide, put the power feed lever in the opposite position. For thread cutting, the power feed lever is put in the center, or neutral, position to operate the *half nut lever*. This nut closes over the threads of the lead screw to move the carriage for thread-cutting operations.

Getting to Know the Lathe

All the control levers and knobs just described may be confusing to the beginner. You therefore should get to know the parts of the lathe, and learn how to operate the controls in order to become comfortable with the machine. The following list of suggestions will help you do this.

First, try the levers and handles with the power off.

- Move the carriage back and forth, and lock it in position.
- Move the cross slide in and out.
- Move the tool post slide in and out.
- Slide the tailstock back and forth, and lock it in position.
- Change the position of the belt on the pulley.
- Change the lead screw lever.
- Disengage the back gear pin, and pull the back gears into place.
- Adjust the selector handle on the quick-change gearbox.

Now try these adjustments after turning on the power and receiving the permission of your instructor.

- Operate the carriage with power feed.
- Operate the automatic cross-feed.
- Use the clutch and brake handle.
- Use the variable speed control lever.

Lathe Safety

When using the lathe, you must understand and obey the following safety rules:

- Do not attempt to operate the lathe until you have received instructions from your teacher and have passed a written safety test.
- Correct attire is important. Loose clothing and other items can get tangled in a moving lathe and cause a serious accident. Remove rings and wristwatches. Roll up your sleeves. Wear an apron. Put on your safety glasses. Do not wear loose sweaters or neckties. Tie long hair back or wear a cap.
- Do not try to lift heavy chucks and attachments alone. Always get help.
- Make certain your workpiece is set up securely and tightly when using chucks.
- When holding the workpiece between centers, make sure to use centers of the right size and with good points. Never use a soft center in the tailstock. Apply oil to the tailstock center and adjust it properly. If it is too tight, the point will heat up

and burn off. Some tailstock centers spin on a ball-bearing collar, so oil is not needed.

- Make sure the tool bits are sharp and ground to the correct shapes and angles. Set them at the proper height and angle to the work.

- Never remove the guards over belts and gears without permission from your teacher. Shut off the power at the switchboard before removing guards.

- After setting up the lathe, remove all wrenches, oil cans, and other tools from the work area. When the workpiece is held on a faceplate, give the faceplate one complete turn by hand to make sure the work will not strike any part of the lathe.

- Stop the lathe before making adjustments.

- Never measure the workpiece while it is turning.

- Never use a file without a handle.

- Never leave a chuck wrench in the chuck. This is dangerous; it can spin out and strike you.

- Keep rags, cotton waste, and brushes away from tools while turning.

- It is unsafe to have small-diameter work extend more than an inch or two (25 to 50 mm) from a chuck unless it is supported by the tailstock center.

- Always check the direction and speed of the carriage or cross-feed before you turn on the automatic feeds.

- Be careful not to run the carriage or cross slide into a turning chuck. This can cause an accident and can damage equipment.

- If you hear unusual noises coming from your machine, stop it, and find out what is causing them.

- Always start and stop your own machine. Never let other students do this for you.

- Do not allow others to stand around your machine. Accidents happen when your mind is not on your work.

Taking Care of the Lathe

Make sure that you clean and lubricate the lathe whenever you use it. Clean the machine by brushing off the chips and then wiping it with a clean cloth. Do not blow chips from the lathe with an air hose. This will cause them to lodge in gears and ways. Oil and grease the lathe, following the lubricating chart instructions that come with it.

Lathe Tools and Toolholders

The cutting tools, or tool bits, you will use are rectangular bars made of high-speed steel. The common tool shapes are shown in Figure 55-9. Note their uses for

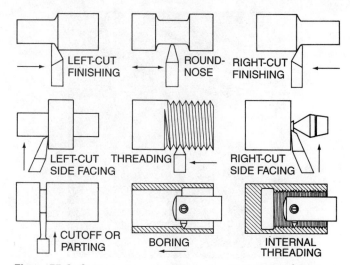

Figure 55-9 Some common cutting tools used in lathe operations.

various types of cutting. In addition, roughing tools are used to remove material quickly. The roughing operation is followed by various finishing tools to complete the turning.

Tools are ground to different shapes for different cutting tasks. The tool angles give functional shapes to the points and prevent the cutting edge from rubbing on the workpiece during machining. The basic angles are: side-relief, end-relief, back-rake, side-rake, end-cutting edge, and side-cutting edge.

There are recommended procedures to follow when grinding a tool bit. For example, to grind a roughing tool:

1. Make sure the face of a medium-coarse grinding wheel is trued and dressed. Refer to the angle and degree chart in Figure 55-10 for guidance while grinding.

2. Grind the side-relief and side-cutting edge angle at the same time. Hold the tool bit against the face of the wheel at the proper angle and gently move it back and forth. Cool it regularly in water because overheating will soften it. Cooling also makes the tool easier to hold with the fingers. Try to get a single smooth surface (Figure 55-11).

3. Grind the end-relief and end-cutting edge angles at the same time.

4. Grind the side-rake and back-rake angles at the same time.

Good tool-grinding requires patience and practice. Very often there are angle gauges and tool models available in the school shop to help you to learn tool-grinding skills. The ground tools are secured in **toolholders** for turning operations. There are straight, right-hand, or left-hand models, selected according to the requirements of each turning job (Figure 55-12).

ANGLE	DEGREES	ANGLE	DEGREES
END-RELIEF	6	END-CUTTING EDGE	20
SIDE-RELIEF	6	SIDE-CUTTING EDGE	15
BACK-RAKE	8	NOSE	65
SIDE-RAKE	14	NOSE RADIUS	1/8 INCH

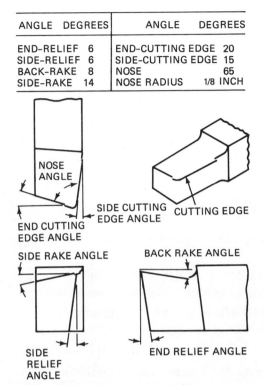

Figure 55-10 Typical angles of a roughing tool for turning mild steel.

Figure 55-11 The proper method for grinding lathe tools. (Courtesy of Clausing Industrial, Inc.)

Speed, Feed, and Depth of Cut

The movement of the carriage and the tool causes the cutting action to take place. There are three important terms which describe this action, as shown in Fig-

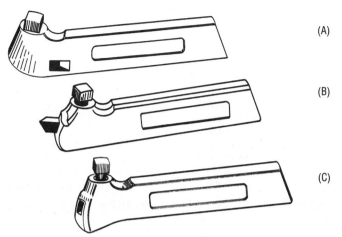

Figure 55-12 Toolholders: (A) left-hand, (B) straight, (C) right-hand.
(Courtesy of Clausing Industrial, Inc.)

Figure 55-13 The relationships between cutting speed, depth of cut, and feed.

ure 55-13. **Cutting speed** is the distance the workpiece moves past the cutting point in one minute, as measured around the surface. This is equivalent to the length of chip that would be removed in one minute. The table in Figure 55-14 shows the *spindle speed* in revolutions per minute (rpm) in order to get the correct cutting speed. Find out what spindle speeds are available on the lathe you are using. Adjust the belt or the speed control to get the correct speed.

Feed is the distance the tool moves along the bed with each revolution of the lathe. Generally, coarser feeds are used for rough turning and finer feeds for finish turning. These are set by changing the levers on the quick-change gearbox.

Depth of cut is the distance from the bottom of the cut to the uncut top surface measured at right angles to the machined surface. To reduce the diameter 1/4 inch

Material	Average cutting speed		Diameters and rpm			
	sfpm*	smpm**	1/2 in. (13 mm)	1 in. (25 mm)	1-1/2 in. (40 mm)	2 in. (50 mm)
low-carbon steel	100	30	800	400	266	200
tool steel	50	15	400	200	133	100
cast iron	75	23	600	300	200	150
brass	200	60	1600	800	533	400
aluminum	300	90	2400	1200	800	600

*surface feet per minute   **surface meters per minute

Figure 55-14 This table is used to determine the cutting speeds for various materials.

(6 mm), the depth of cut must be 1/8 inch (3 mm), as adjusted on the micrometer collar of the cross-feed. The beginner usually takes too small a depth of cut with too fine a feed. Practice this in order to learn how to do it correctly.

KNOWLEDGE REVIEW

1. What is meant by the *swing* of a lathe?
2. Name the five main parts of the lathe.
3. Explain the purposes and functions of live and dead centers.
4. Explain the relationships between cutting speed, feed, and depth of cut.
5. Make large wood or plastic models of some of the common cutting tools.
6. Prepare a bulletin-board display of the lathe parts.
7. Prepare a research report on the history of lathes.

Turning Between Centers

OBJECTIVES

After studying this unit, you should be able to:

- Set up a workpiece for turning between centers.
- Identify the tools for turning between centers.
- Locate and drill center holes.
- Face the ends of workpieces.
- Perform rough- and finish-turning operations.
- File and polish a workpiece.
- Calculate the tailstock setover.
- Calculate the rate of taper.
- Identify the tools for taper turning.
- Turn a taper using the tailstock offset method.
- Turn a taper using the tool post slide method.
- Turn a taper using the taper attachment.
- Identify the knurling tools.
- Knurl a workpiece.

KEY TERMS

center drilling	facing	lathe dog	rate of taper
center hole	filing	live center	rough turning
dead center	finish turning	plain taper attachment	setover
faceplate	knurling	polishing	

Metal stock must be rough-turned and finish-turned between centers as a first step in making many machine shop projects. This is a common lathe turning operation. The workpieces should be about 1/8 inch (3 mm) larger in diameter and about 3/4 inch (20 mm) longer than the completed work if the center holes are to be cut off. It needs to be only about 1/8 inch longer if the center holes are to be left in the finished work. Before turning, the work must be set up accurately between centers.

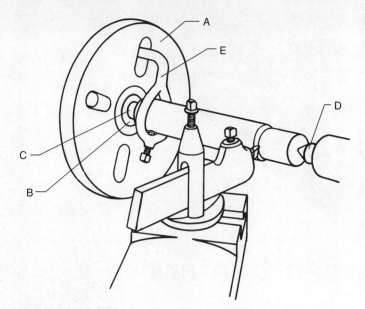

Figure 56-1 This is a typical turning between centers setup: (A) faceplate, (B) spindle sleeve, (C) live center, (D) dead center, (E) lathe dog.

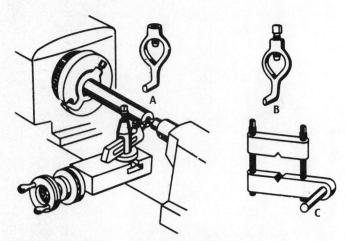

Figure 56-2 Three common kinds of lathe dogs: (A) bent tail safety screw, (B) standard bent tail, and (C) clamp.

Setup for Turning Between Centers

A workpiece held between centers is shown in Figure 56-1. This is a typical arrangement for such work. Note the following observations regarding this illustration: (1) a **faceplate** (or driving plate) is screwed onto the spindle of the headstock; (2) a *spindle sleeve* is inserted in the headstock spindle, and a **live center** is inserted into the sleeve. The live center is made of soft steel, and the point can be reshaped or dressed by turning; (3) a **dead center** is inserted in the tailstock spindle. The dead center is hardened steel, and the point can only be dressed by grinding. The workpiece is supported between the live and dead centers; and (4) a **lathe dog** is fastened to the workpiece to hold it between the centers. The tail of the

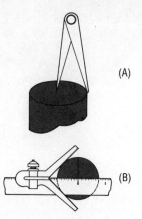

Figure 56-3 Locating centers by using the (A) hermaphrodite calipers and (B) the center head.

dog fits into a slot in the faceplate, so that the spindle power causes the workpiece to rotate. The *standard* and *safety screw* dogs are used for round workpieces (Figure 56-2). The safety dog has a headless screw and is less likely to catch in the operator's sleeve. The *clamp* dog is used to hold rectangular workpieces.

Locating and Drilling Center Holes

Center holes are the tapered openings which must be drilled into each end of the workpiece. The live and dead centers fit into these holes. Follow the directions below in order to properly locate and drill the holes.

1. File or grind the sharp burrs off both ends of the workpiece. This will help you to avoid cutting your fingers and hands while handling the stock.

2. Locate the centers by scribing two intersecting arcs with a *hermaphrodite caliper,* or by holding a *centering head* over the end and scribing two lines that cross (Figure 56-3).

3. Prick-punch and then center-punch a mark at each end of the workpiece.

4. Select a *combination center drill and countersink* (Figure 56-4). Fasten the workpiece in a drill press vise and carefully drill the center holes as shown in Figure 56-5. The holes also may be drilled on the lathe using a between-centers setup and a special steady rest mounted on the ways, or by mounting it in a chuck. This **center drilling** operation must be done carefully to avoid improperly placed holes which may cause the workpiece to wobble.

Facing the Ends of the Workpiece

Facing the workpiece is squaring the ends by making them true and flat. The proper setup for facing appears in Figure 56-6, and involves several operations.

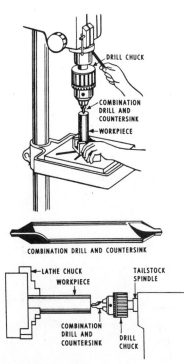

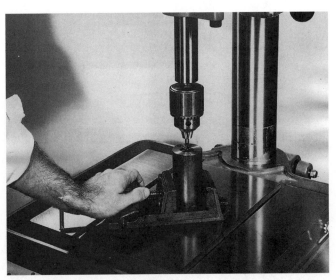

Figure 56-5 Drilling a workpiece center hole on a drill press. Be sure that the piece is locked in a vertical position. (Courtesy of Clausing Industrial, Inc.)

Figure 56-4 Combination drill and countersink for drilling center holes, and two drilling methods.

1. Move the tailstock until the dead center just touches the live center. Be sure the centers are aligned. If they are not, move the top casting of the tailstock toward or away from you.

2. Fasten the lathe dog to one end of the workpiece.

3. Move the tailstock assembly until the opening between centers is a little longer than the workpiece. Lock the tailstock in position.

4. Slip the tail of the lathe dog in an opening in the faceplate. Place the workpiece between centers. Apply a little lubricant to the dead center hole. Tighten the dead center until the workpiece is held snugly. It should not be so loose that the lathe dog clatters, or so tight that the dead center becomes scored or burned. Re-member that since metal expands when heated, you may have to readjust the dead center after two or three cuts.

5. Place a right-cut side-facing tool in a toolholder with the point well-extended.

6. Place the toolholder in the tool post, with the cutting edge at right angles to the centerline. Adjust this by eye.

7 Face the right end of the workpiece, placing your left hand on the carriage handwheel and your right hand on the cross slide handwheel.

8. For a rough cut, the tool is fed from the outside toward the center with the cross slide handwheel. For a finish cut, the tool is always fed away from the center hole toward the operator.

9. Move the carriage to the left until a light cut is taken. Then feed the tool. Reverse the workpiece in the lathe. Then face the other end.

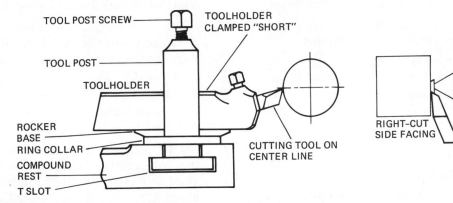

Figure 56-6 Positioning the toolholder for a facing cut.

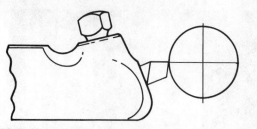

Figure 56-7 Note that the point of the cutting tool is set at the center of the workpiece.

Rough Turning

The purpose of **rough turning** is to quickly remove unwanted material. It is followed by smooth finish turning to the desired diameter. Follow the procedure below for rough turning.

1. Choose a right-cut roughing tool or roundnose tool and lock it in a straight toolholder.

2. Place the toolholder in the tool post, making sure that the toolholder does not extend out too far. The tool post should be at the left end of the T slot. The point of the tool must be on center (Figure 56-7). Turn the point a little away from the headstock to prevent accidental digging into the workpiece (Figure 56-8).

3. Adjust for correct feed and speed. Check to see how far the carriage can move before the lathe dog hits the tool post slide. Check to see that the carriage will move toward the headstock with power feed.

4. Adjust an outside caliper to 1/32 inch (1 mm) over finished size.

5. Turn on the power. Put your left hand on the carriage handwheel and your right hand on the cross slide handwheel. Move the carriage. Turn in the cross slide until a chip starts to form. Make a trial cut about 1/4 inch (6 mm) wide and deep enough to true up the workpiece.

6. Stop the lathe. Check the diameter with the outside caliper. You may have to make two or three roughing cuts.

7. Turn on the power. Engage the power feed lever.

8. Check the cutting action. Chips should come off in short pieces. There should be no chattering.

9. When half the cutting has been done, release the power feed. Then back out the cross slide. (Some machinists prefer not to back out the cross slide after releasing the carriage power feed. Instead, they stop the machine, remove the workpiece, return the carriage to the starting position, reverse the lathe dog, and machine the second end without changing the setting of the cross slide.)

10. Return the carriage to the starting position. Turn off the power and check the diameter. It may be

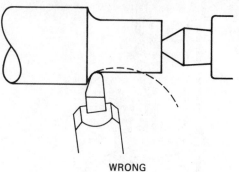

WRONG

(A)

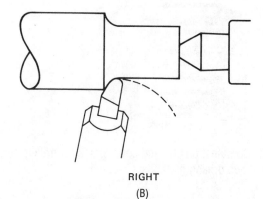

RIGHT

(B)

Figure 56-8 If the tool is set too far to the left (A) and becomes loose, it may dig into the workpiece. The proper setting is shown at (B), where it will swing safely out of the way.

necessary to make one or more additional cuts to turn to rough size.

11. Reverse the workpiece in the lathe. Rough-turn the second half to size.

Finish Turning

The finish cuts produce a smooth workpiece turned to the desired diameter. **Finish turning** is preferred as follows:

1. Use a right-cut finishing tool of the same shape as the roughing tool but with a smaller nose radius. The tool should also have a very keen cutting edge.

2. Place the workpiece between centers. Lubricate the dead center. Make a light trial cut about 1/2 inch (13 mm) long. Do not change the cross slide.

3. Check the machined surface with a micrometer to see how much stock is to be removed. Suppose the machined surface is still 0.006 inch (0.15 mm) oversize.

4. Set the micrometer collar to zero. Move the cutting tool to the right of the workpiece. Turn the cross slide in 0.003 inch (0.08 mm). Make a trial cut about 1/2 inch (13 mm) long. Then stop the machine and check the surface again with the mi-

Figure 56-9 Filing a workpiece on a lathe. Note the position of the hands and the rolled sleeve to avoid the lathe dog. (Courtesy of Clausing Industrial, Inc.)

crometer. If necessary, change the micrometer collar to get the correct diameter. Finish-turn the first half of the workpiece.

5. Place a soft copper or aluminum collar around the finished end of the workpiece to protect the smooth finish. Replace the lathe dog. Turn the second half to size. Sometimes, 0.002 to 0.003 inch (0.05 to 0.08 mm) of stock is left on the workpiece to allow for filing and polishing.

Filing and Polishing

A smooth, true finish, free of any tool marks, can be produced by **filing** and then **polishing** the workpiece, as described below.

1. Remove the tool post and adjust the lathe to a high speed. Use a mill file to take long, even strokes across the rotating workpiece. Always keep the file clean. Be careful to roll up your sleeves and keep your arm away from the revolving lathe dog (Figure 56-9).

2. To get a very smooth polish, apply a few drops of oil to a strip of fine abrasive cloth. Grasp it by the ends and hold it against the revolving workpiece, moving it slowly along the work. Do not wrap it around the work, as it can be ripped from your hands and cause a serious accident.

Taper Turning

The four common methods for turning tapers on the lathe are: (1) offsetting the tailstock, (2) using the tool post slide, (3) using the taper attachment, and (4) using a ground tool. The method used depends upon the length of the taper, the angle of the taper, and the number of pieces to be machined. These methods are described below.

Offset Tailstock Method

This simple way to cut a taper requires no special equipment and can only be done on a workpiece mounted between centers. It can be used, for example, to cut a taper on the handle of a steel machinist hammer. A disadvantage is that a steep taper may damage the live and dead center holes. As mentioned in unit 55, the upper member of the two-casting tailstock can be moved by means of adjusting screws. This will offset the dead center the distance necessary to produce the desired taper angle. Only external tapers can be cut by this method.

The correct amount of offset (or **setover**) is determined by a simple calculation. The hammer handle drawing in Figure 56-10 shows that the total hammer length is ten inches, the length of taper is four inches, the largest taper diameter is 3/4 inch, and the smallest taper diameter is 5/8 inch. These values must be converted to decimals in order to make the calculation. See the example below and follow this procedure:

1. Calculate the amount of setover. The setover (S) equals the total length of the workpiece (TL) times the large diameter (LD) minus the small diameter (SD) divided by the taper length (L) times 2.

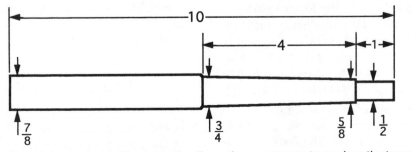

Figure 56-10 This drawing of a metal hammer handle shows the dimensions necessary to produce the taper.

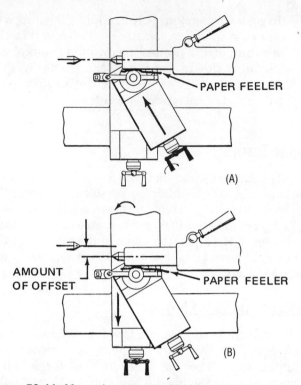

Figure 56-11 Measuring setover with the cross slide dial: (A) paper feeler against tailstock spindle; (B) cross slide backed off to the correct setover position.

Formula: $S = \dfrac{TL \times (LD - SD)}{L \times 2}$

Example: $S = \dfrac{10 \times (0.750 - 0.625)}{4 \times 2}$

$S = \dfrac{10 \times 0.125}{8}$

$S = \dfrac{1.250}{8}$

$S = 0.156$ inch

The result of this calculation means that the total setover is 0.156 inch, or 5/32 inch, and that the tailstock must be offset by that amount. You can now turn the taper on the hammer handle.

2. Begin the tailstock setover process by loosening the *bed clamp lever.* Adjust the *setover screws* (there is one on each side of the lower casting) to move the *upper casting* of the tailstock toward you. The result is that the small end of the taper will be at the tailstock end. Note that the *normal centerline* has been shifted to the taper *work center line.* The tool will now cut a taper on the workpiece.

A more precise technique for calculating setover is required when the setover cannot be accurately measured with a rule. To do this, fasten a *toolholder* in the *tool post* with the flat end toward the tailstock spindle (Figure 56-11). Place a strip of pa-

per over the end of the tool and crank the *tool post slide* until the paper touches the *tailstock spindle.* The slide is then cranked out the required setover distance. Now turn the *adjusting screws* until the spindle touches the paper strip.

3. Place the workpiece between centers as for straight turning. Keep the dead center well-lubricated or use a ball-bearing center. Select a thin roundnose tool bit and fasten it in a toolholder.

4. Start the cutting about 1/2 inch (13 mm) from the right end of the workpiece. If the workpiece has a very small diameter, be especially careful that it does not climb over the cutting edge and bend it.

5. Continue making several light cuts until the small end is the correct diameter. Measure this with an outside caliper or a micrometer. The taper may be filed and polished as necessary.

Some drawings will indicate the **rate of taper** for a project, such as 0.050 taper per inch. In this case, a different formula is used, where the total length of the workpiece (TL) times the taper per inch (TPI) is divided by 2.

Formula: $S = \dfrac{TL \times TPI}{2}$

Example: $S = \dfrac{10 \times 0.050}{2}$

$S = \dfrac{0.500}{2}$

$S = 0.250$ inch

The setover in this example is 0.250 inch, and the tailstock offset adjustment can be made as described above.

Taper in metric measurements is per centimeter, converted to millimeters. For a metric taper, calculate the setover as follows: Multiply the total length of the workpiece in centimeters times one-half the taper per centimeter in millimeters. The answer will be the setover in millimeters.

Setover = 1/2 taper per cm (mm) × TL

Tool Post Slide Method

For cutting short tapers and angles, set the tool post slide to the desired taper angle. Do the cutting by feeding the post slide by hand. The tool will follow the taper angle set on the tool post slide. Note the position of the tool post slide and the direction of the cutting in Figure 56-12.

Taper Attachment Method

This method is more convenient to use because the tailstock lathe center need not be offset. It is also good for turning many workpieces once the attachment has been set. A typical **plain taper attachment** is shown in Figure 56-13. There are many different models of this tool, as well

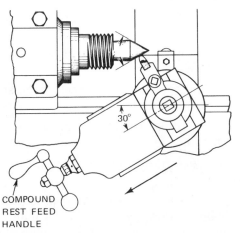

Figure 56-12 Note the 30-degree tool position to dress the point of a live center. The cutting is down in the direction of the arrow.

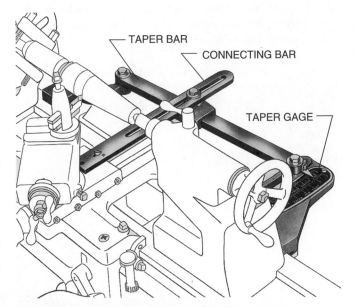

Figure 56-13 The parts of a plain taper attachment. (Courtesy of Clausing Industrial, Inc.)

as a line of special telescoping taper attachments. Each tool has its own setup procedures, and you should follow these directions carefully. To use this tool, the *taper bar* at the back of the lathe is set to the desired *taper gage* angle and locked. A *connecting bar* which slides along the taper bar is then clamped to the cross slide. As the carriage moves along the lathe ways, the cross slide (and tool) moves to duplicate the taper angle of the taper bar.

Ground Tool Method

This is the easiest method to use for cutting short tapers. It requires only that a tool is ground with a square nose, and that the tool be set at the correct angle to the workpiece (Figure 56-14). This setup is especially useful for small projects.

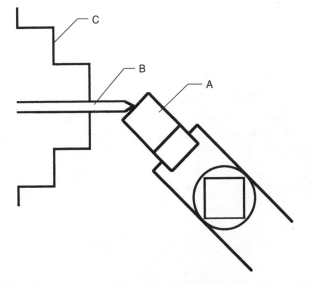

Figure 56-14 Here a square-nose tool (A) has been set at a 60-degree angle to cut a taper on a small workpiece (B). Note that the workpiece is held in a chuck (C).

Knurling

Knurling is press-cutting a straight or diamond-shaped pattern on the surface of the workpiece. The handles of some tools are knurled to prevent their slipping from your hand. Knurling can also improve the appearance of a project. A knurling tool consists of a holder with two hardened steel impression wheels that form the knurl. The process of knurling is done as follows:

1. Mark the beginning and end of the knurled area.

2. Adjust the lathe to a slow speed.

3. Place the knurling tool in the tool post, with the tool turned a little toward the headstock. The right side of the wheel should touch the workpiece first (Figure 56-15).

4. Move the carriage to the starting point near the tailstock.

5. Turn in the cross-feed until the wheel presses into the metal about 1/64 inch (0.5 mm). Then apply a little cutting fluid.

6. Turn on the lathe. Immediately use the automatic power feed.

7. Check to see if the tool is working correctly. A diamond-shaped knurling tool should cut a criss-cross pattern. If one wheel is not tracking right, release the cross-feed pressure and move the tool a little to the left.

8. If the tool is knurling correctly, increase the pressure and apply cutting fluid to the surface. Allow the tool to move across the work surface. Clean the wheels often with a wire brush.

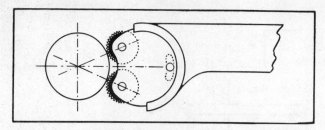

Figure 56-15 Using the knurling tool. The proper alignment of the tool and the workpiece is important, as shown.

9. As the tool reaches the end, turn off the power but do not release the automatic power feed.

10. Reverse the direction of the carriage. Apply a little more pressure to the wheels and run the tool back.

KNOWLEDGE REVIEW

1. Write a description of the process of turning between centers.

2. Explain the difference between a live center and a dead center.

3. What is the purpose of workpiece facing? Of center drilling?

4. Describe the differences among the three types of lathe dogs.

5. Describe the difference between rough and finish turning.

6. Describe the process of tapering and explain the four methods for doing this.

7. Write a description of the formula for tailstock setover.

8. Prepare a sketch and calculate the setover for this problem: A workpiece is 12 inches long and has a diameter of 1 1/4 inches. Three inches of one end is tapered to a diameter of 3/4 inch.

9. Calculate the rate of taper for the hammer handle in Figure 56-10.

10. Collect some objects in your school machine shop and try to determine which of the four tapering methods were used for each.

Machining Workpieces Held in a Chuck

OBJECTIVES

After studying this unit, you should be able to:

- Recognize and describe the uses of three- and four-jaw chucks.
- Install a chuck.
- Mount a workpiece in a chuck.
- Perform a facing operation.
- Recognize the three types of cutoff tools.
- Perform a cutoff operation.
- Understand the purposes of boring and reaming operations.
- Use taps and dies in lathe threading operations.
- Understand the principles of chasing or cutting threads on the lathe.
- Understand the uses of mandrels and collet chucks.

KEY TERMS

boring	draw-in collet chuck	independent chuck	six-jaw chuck
chasing	drilling	mandrel	universal chuck
cutting off	facing	reaming	

In addition to being held between centers, a workpiece can be held in a special clamping device called a *chuck,* which is screwed onto the headstock's threaded spindle nose. Many operations, such as facing, cutting off, drilling, boring, reaming, threading, and turning, can be done with this setup (Figure 57-1).

Kinds of Chucks

The three-jaw **universal chuck** (Figure 57-2) is the most common type. It is also the easiest to use because all three jaws move in and out together by turning a single screw with a chuck key. Because of this characteristic, it automatically centers a round or hexagonal workpiece. A **six-jaw chuck** is also self-centering, but has more holding power. The four-jaw **independent chuck** will hold any shape because each jaw is moved independently with the chuck key. There are two sets of jaws for each type of chuck, for outer and inner surface clamping, as shown in Figure 57-3.

Fig. 57-1 Machining metal on an engine lathe with the workpiece held securely in a chuck. (Courtesy of Clausing Industrial, Inc.)

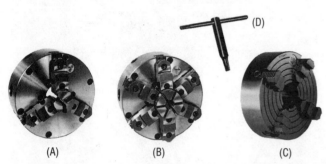

Fig. 57-2 Types of chucks: (A) three-jaw universal, (B) six-jaw, and (C) four-jaw independent. A chuck wrench is shown at D. (Courtesy of Clausing Industrial, Inc., and Pratt Burnerd America)

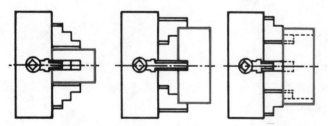

Fig. 57-3 The uses of the outer and inner types of chuck jaws.

Installing the Chuck

To install a chuck on a lathe, follow this procedure:

1. Remove the faceplate by turning it counterclockwise.

2. Force out the live center and sleeve with a knockout bar (Figure 57-4).

3. Wipe any dirt from the threads on the headstock spindle. Apply a few drops of oil.

4. Clean any dirt from the chuck threads.

5. Place a wooden cradle on the ways to protect them. Lift the chuck onto it (Figure 57-5).

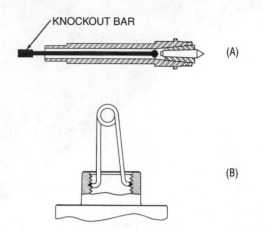

Fig. 57-4 Before installing the chuck, the live center must be removed with a knockout bar (A). It is also important to remove dirt from the chuck threads with a wire cleaner (B).

Fig. 57-5 Chuck installation is simplified by using a wooden cradle. (Courtesy of Sheldon Machine Company)

6. Lift the chuck by placing your hands or fingers in the center between the jaws. Never place your hands under the chuck. Turn it clockwise to start it on the spindle, and tighten it securely.

Mounting the Workpiece in a Chuck

On a three-jaw universal chuck, open the jaws until the workpiece slips in. Then tighten with the chuck key. *Always remember to remove the chuck key.*

If an independent-jaw chuck is used, open each of the four jaws an equal distance from the center using the circular guidelines on the chuck face. Insert the workpiece in the chuck, then tighten opposite jaws a little at a time until the workpiece is held firmly. Check to make

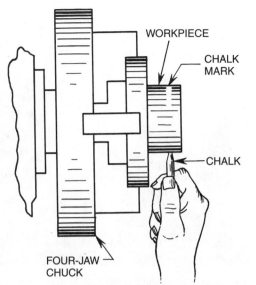

Fig. 57-6 Truing a workpiece in a four-jaw chuck. The chalk mark will indicate when the workpiece must be moved back slightly.

sure the workpiece is centered by holding a piece of chalk against it (Figure 57-6). Adjust the jaws accordingly.

Facing

The **facing** operation in chuck work is much the same as for turning between centers.

1. For rough cuts, choose a left-cut roughing tool. Cut from the outside of the workpiece toward the center. Remember to lock the carriage to the bed with the carriage-lock screw. Use the power cross-feed for large-diameter stock.

2. For finishing cuts, use a right-cut facing tool clamped in a left-hand holder. Adjust the holder until the cutting edge is at an angle of about 8 to 10 degrees to the workpiece face. If you set the toolholder at about 80 degrees, you will be setting the tool at the correct angle. Start at the center and feed the tool to the outside (toward the operator).

Cutting Off Stock

The **cutting off** operation is performed as follows:

1. Turn the workpiece to diameter.

2. Select the proper cutoff tool. Note that there are right, left, and straight cut styles, as shown in Figure 57-7.

3. Mark the location of the cut. Make sure that the blade is at right angles to the work and on center.

4. The speed should be about two-thirds that for turning.

5. Feed the cutting tool slowly and gently into the work. Apply plenty of cutting lubricant. Ease back on the tool as the work begins to separate.

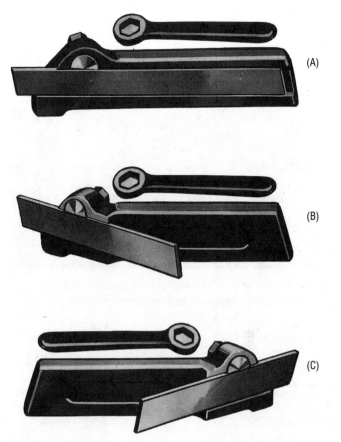

Fig. 57-7 Cutoff tools: (A) straight, (B) right-hand, and (C) left-hand. The tool wrench is also shown.

Drilling

The steps for **drilling** a workpiece are listed below.

1. Mount the workpiece in the chuck and face-cut it.

2. Remove the dead center and insert a drill chuck. Fasten a combination center drill and countersink in the drill chuck. Center-drill the location of the hole.

3. Select the size of drill needed. If it is large, first drill a pilot hole to remove excess metal and make the final drilling easier and more accurate. The pilot drill should be about one-half the size of the desired workpiece hole.

4. A drill chuck must be used for all smaller (under 1/2 inch) straight-shank drill bits. If it has a tapered shank, it can be inserted directly into the tapered hole in the tailstock spindle.

5. Adjust for the right speed as in drill press drilling. Larger drill bits require slower speeds. Feed the drill slowly into the workpiece, applying cutting fluid. Back the drill out several times to clean out the chips. Reduce the feed as the drill breaks through the back of the workpiece (Figure 57-8).

6. To make a hole of great accuracy, first drill the hole 1/32 inch (1 mm) undersize. Then bore to 1/64 inch (0.5 mm) undersize. Finally, machine ream it to

Fig. 57-8 Drilling a chuck-mounted workpiece with a taper-shank drill bit. (Courtesy of Clausing Industrial, Inc.)

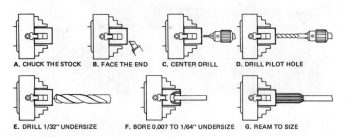

A. CHUCK THE STOCK   B. FACE THE END   C. CENTER DRILL   D. DRILL PILOT HOLE

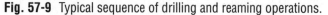

E. DRILL 1/32" UNDERSIZE   F. BORE 0.007 TO 1/64" UNDERSIZE   G. REAM TO SIZE

Fig. 57-9 Typical sequence of drilling and reaming operations.

the exact size. For most work, however, you can drill the hole 1/64 inch undersize and then hand- or machine-ream (Figure 57-9).

Boring

Boring is done to finish a hole that is not standard size, to trim out a hole in casting, or to finish a very accurate hole of any size. Various sizes and types of cutting tools are used for this. A *boring bar* with a *boring tool* is often used. The bar is mounted in a *holder* fixed to the tool post slide. Use the same speed as for turning, and slowly feed the tool with the carriage handwheel to avoid chattering.

Reaming

Reaming produces a hole that is accurate and smoothly finished. Hand reaming can be done on the lathe with the setup shown in Figure 57-10. The end of the reamer has a countersunk hole which fits on the dead center. The wrench keeps the reamer from spinning. Feed the reamer slowly by turning the tailstock handwheel. It is dangerous to use power feed for this hand-reaming operation—don't attempt to do it.

Machine reamers with tapered shanks can be inserted directly into the tailstock spindle. Adjust the lathe

Fig. 57-10 Hand reaming on the lathe. Note the use of the wrench to hold the reamer. (Courtesy of Clausing Industrial, Inc.)

for a very slow speed. Turn the tailstock handwheel slowly so that the reamer advances into the workpiece at a gentle, steady speed. Never turn the reamer backward.

Cutting Threads

The simplest way to cut *internal* threads is to use the hand tap setup shown in Figure 57-11. Remember that tap drills of the proper size must be used prior to tapping. *External* threads can be cut with a die held in a diestock, as shown in Figure 57-12. The usual precautions for using taps and dies are important here where the lathe is used as the source of power. Feeding too fast can result in broken tools and ruined workpieces. Use plenty of lubricating oil, and back off the tools frequently to clear the chips.

Chasing Threads

Thread cutting, or **chasing,** is an important engine lathe operation. It is the most accurate way of producing threads (Figure 57-13). It is done by establishing a ratio between the *headstock spindle rotation* and the *carriage movement.* For example, if the screw is to have a pitch of ten threads per inch, the spindle must rotate ten times while the carriage moves exactly one inch. The ratio between the spindle speed and the carriage speed is determined by a series of gears in the quick-change gearbox. These are set by control levers, knobs, and dials which can be shifted to select the desired gear combination. An *index plate* is located near the gearbox, and is marked with the number of threads per inch produced by each control position (Figure 57-14). The threading tool must be ground to an angle of 60 degrees, with about 5 degrees side clearance and 8 degrees front clearance. Both internal and external thread chasing

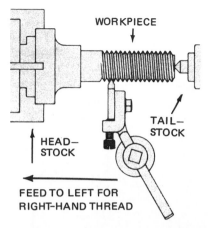

Fig. 57-13 The setup for chasing threads. Note that this workpiece is chuck-mounted, and the tailstock dead center is used for stability.

Fig. 57-11 A simple method of cutting internal threads using a tap secured in a tap wrench. (Courtesy of Clausing Industrial, Inc.)

Fig. 57-12 Using a die held in a diestock to cut external threads. (Courtesy of Clausing Industrial, Inc.)

can be done on workpieces mounted between centers or in chucks. Each type of lathe has its own directions for the thread-cutting gearbox setting. Carefully follow these directions and the instructions for carriage operation when chasing threads.

Turning a Workpiece on a Mandrel

Some jobs call for machining the outside diameter of a workpiece concentric with a hole which has been previously drilled in it. This operation requires the use of a **mandrel**, which is a slightly tapered shaft of hardened steel. It is important to choose a mandrel of the correct size. The size is stamped on the large end. Cover the mandrel with a

Fig. 57-14 The settings for thread-cutting are shown on the index plate. (Courtesy of Clausing Industrial, Inc.)

thin coat of oil. Press it into the workpiece with an arbor press (Figure 57-15). If an arbor press is not available, carefully drive the mandrel into the workpiece with a soft-face hammer. Turn the outside of the workpiece as shown in Figure 57-16.

Other Chuck-Turning Operations

Chuck-held workpieces can be turned to a desired outside diameter over their total length. This is done by turning half the length of the piece, removing it from the chuck, and then locking the turned end in the chuck to complete the turning. For long workpieces, use the tailstock center or a steady rest for added support. Shorter pieces can be cut safely without such support.

Draw-in collet chucks are the most accurate and convenient chucking devices, and are widely used for precision work. Their quick action allows for the easy and

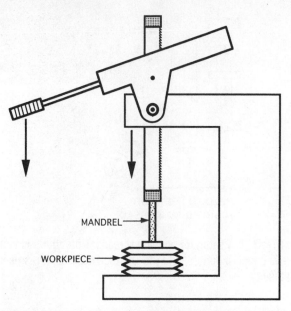

Fig. 57-15 Using an arbor press to insert a mandrel in a workpiece.

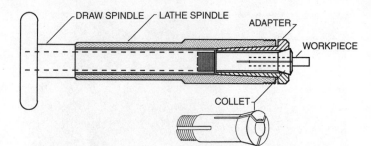

Fig. 57-17 A typical collet chuck setup showing the assembly of the parts.

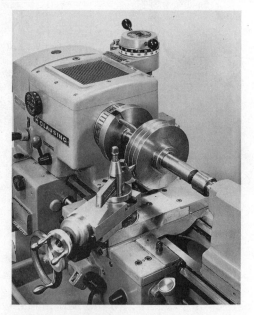

Fig. 57-16 Turning the outside of a workpiece mounted on a mandrel. (Courtesy of Clausing Industrial, Inc.)

Fig. 57-18 Turning a workpiece held in a collet chuck. (Courtesy of Clausing Industrial, Inc., and Pratt Burnerd America)

sure installation of workpieces. A simplified sketch of the collet chuck's parts and operation appears in Figure 57-17. A collet *adapter* fits into the lathe headstock spindle, and the *collet* fits into the adapter. The collet head is slotted to clinch the workpiece, and the tail end has fine external threads to lock onto the draw spindle. The hollow *draw spindle* has fine internal threads at one end, and it is run through the hollow headstock spindle to meet and screw onto the collet. The workpiece is inserted into the collet, and the draw spindle handwheel is turned to lock the work in the collet. This is a simple, fast, and accurate chucking operation (Figure 57-18). Collets come in a variety of sizes and shapes to accommodate round, square, and hexagonal stock. Be sure to select the correct collet for

the job. For example, if you wish to turn a piece of 1/2-inch round steel bar, select a round 1/2-inch collet.

KNOWLEDGE REVIEW

1. Explain the advantages of chucks in lathe operations.
2. Explain the differences in the jaw movements on a three-jaw universal chuck and a four-jaw independent chuck.
3. Describe the process of mounting a chuck on the lathe.
4. Why are taper-shank drills not used in drill chucks?
5. Write a research report on lathe speeds needed for drilling and reaming.
6. Describe the processes used to cut threads on a lathe.
7. What device should be used to press a workpiece onto a mandrel?
8. Prepare an oral report on the uses of collet chucks and how they work.
9. Visit a local machine shop and report on the various types of chucks used there.

Unit 58

The Shaper

OBJECTIVES

After studying this unit, you should be able to:

- Identify the important parts of the shaper.
- Level and true a workpiece in a vise.
- Perform a simple surface-cutting job on the shaper.
- Describe the process of cutting angles on a shaper.

KEY TERMS

parallels   shims   swivel vise   wedges

shaper   stroke

The **shaper** is a machine that cuts as a sharp tool is pushed through a workpiece. It is used for smoothing and shaping horizontal, vertical, angular, and curved surfaces. The cutting tool, similar to that used on the lathe, is locked in a toolholder which is clamped in a tool post attached to a heavy ram. The ram drives the cutting tool through the workpiece on the power stroke, and withdraws it on the return stroke. The workpiece is clamped in a vise which is fixed to the shaper table. The cutting tool and table can move vertically, and the table also can move horizontally under the cutting tool. The size of the shaper is determined by the maximum stroke in inches, which is about the same as the largest cube it can machine. Common sizes range from seven to sixteen inches. The main parts are shown in Figure 58-1.

The shaper is a relatively slow machine and has largely been replaced in industry by the faster and more versatile milling machine. However, shapers are still found in many maintenance shops, experimental laboratories, model shops, small industries, and school shops.

Holding Devices

For small jobs, the workpiece can be held in a **swivel vise** fastened to the shaper table. The top of the vise can be rotated so that the jaws are either at right angles or parallel to the ram movement. Several types of metal parts are used to adjust and level the workpiece in the vise (Figure 58-2).

Parallels raise the workpiece above the top of the vise jaws so that the tool will not cut into them. They are rectangular pieces of hardened steel, generally used in pairs. **Wedges,** or hold-downs, are similar pieces of

325

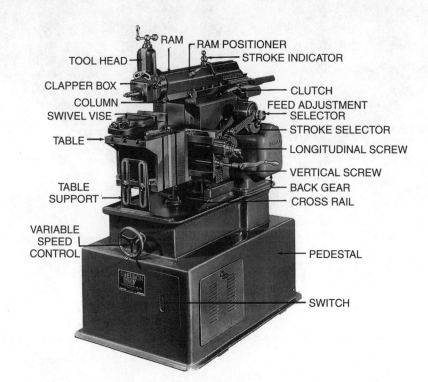

Fig. 58-1 The parts of a typical shaper. (Courtesy of Sheldon Machine Company)

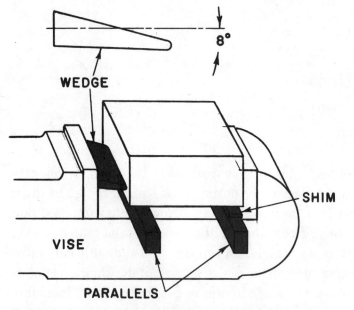

Fig. 58-2 Common accessories used for holding and leveling a workpiece.

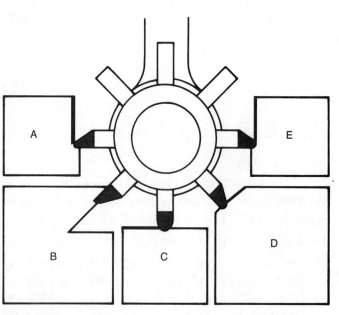

Fig. 58-3 Note how the swivel head can position the tool for various cuts: (A) vertical, (B) angular, (C) horizontal, (D) angular, and (E) vertical.

steel with the back edge beveled to an angle of about 3 degrees, and a rounded front edge. They are placed in the vise to hold the workpiece firmly against the parallels. **Shims** are bits of steel placed between the parallels and the workpiece for any required final leveling.

Cutting Tools and Toolholders

The toolholder can either be the type used on a lathe or the *swivel head* style shown in Figure 58-3.

The cutting tools are the same as those used on the lathe, except that the tool angles may have to be ground differently. Because the shaper does not feed laterally during the cut, the side-relief angle can be less than for a lathe tool—only about 4 degrees. There is no rocker under the toolholder, so the end-relief angle cannot be adjusted. It should also be about 4 degrees. If the cutting tool is held in the lathe tool-holder, the end relief must be ground at about 19 degrees to provide this necessary 4 degrees. A round-nose tool is used for most simple shaping.

Machining a Flat Surface

Producing a flat surface on a workpiece is a good beginning exercise, and is done as described below. Refer to Figure 58-1 to identify the machine controls. The descriptions of controls and their locations are for a typical shaper. Other machines may differ, but the general procedures are the same.

1. Clamp the workpiece in the vise with the longest side parallel to the jaws. If necessary, raise the workpiece with parallels. Sometimes a soft-metal rod is placed between the adjustable jaw and the workpiece to help hold it. Now tighten the vise securely.

2. Set the toolhead at zero, and center the *clapper box.*

3. Lock the cutting tool in the toolholder, and fasten the holder in the tool post fixed to the clapper. The toolholder should be turned a little to the right so that the tool will turn away rather than dig in if it slips (Figure 58-4).

4. Turn the tool slide (a moving part of the tool head) up as far as it will go to avoid too much overhang, which will affect the cutting action. The tool slide handle is located atop the tool head.

5. The table must be raised or lowered to get the workpiece at the proper height for machining. To do this, loosen the screws on the *table support* and move the table by turning the *vertical screw* crank. There should be at least 2 inches (50 mm) of clearance between the underside of the ram and the top of the workpiece.

6. Adjust for length of stroke. The **stroke,** or movement of the ram, must be longer than the work being machined. This length adjustment is controlled by the *ram positioner* and the *stroke selector.* The *stroke indicator* points to a scale on the top of the column to show the length of stroke. Turn the crank until the point of the tool clears the front of the workpiece by about ¼ inch (6 mm) and the back by about ½ inch (13 mm).

7. Turn the *variable speed control* wheel to set the machine speed. On belt-driven shapers, this is done by changing the belt positions. The speed should be faster for short strokes and soft material, and slower for long strokes and hard material.

8. Adjust for correct feed. The feed is the distance the table moves into the work after each cutting stroke. Generally, a finer feed is used with a heavy roughing cut and a heavier feed with a lighter or finishing cut. The feed is controlled by the *longitudinal screw* (or cross-feed) and must be set so that the table advances on the return stroke of the ram.

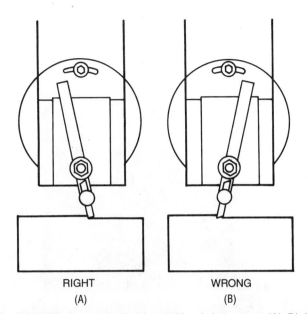

Fig. 58-4 The proper toolholder position is important. (A) *Right:* The tool swings away from the workpiece. (B) *Wrong:* The tool digs into the workpiece.

9. Stand at the front and a little to the right of the shaper. Turn the longitudinal screw to move the workpiece to the left and away from you. The cutting will then start at the right edge, or the edge nearest you.

10. Turn on the switch. Place your right hand on the down-feed tool slide handle, and your left hand on the cross-feed longitudinal screw crank. Turn the cutting tool (tool slide handle) down very slowly. Move the workpiece toward you until the cutting tool takes a chip about 1/8 inch (3 mm) deep (Figure 58-5).

11. Turn the cross-feed crank about an eighth to a quarter turn during each return stroke until three or four strokes have been made.

12. Stop the machine to check the surface and cutting action. If the tool is cutting properly, chips will curl away from a steel workpiece and crumble away from a cast iron workpiece.

13. Turn on the power. Engage the *automatic feed lever.*

14. When the first cut is complete, return the workpiece to the starting position. *Never cut on both forward and backward motions.*

15. Take additional cuts to within 1/64 (0.016) inch (0.5 mm) of the layout line on steel, or 0.005 inch (0.13 mm) on cast iron.

16. Increase the speed, use a finer feed, and take a lighter cut to complete the first surface.

Machining a Vertical Surface

The general procedure for producing a vertical surface on a workpiece is as follows:

Fig. 58-5 A cutting tool in the correct position for shaping. (Courtesy of Clausing Industrial, Inc.)

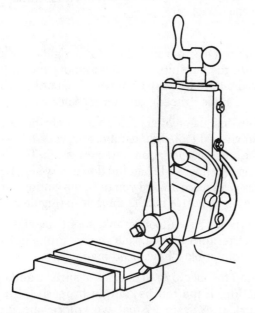

Fig. 58-6 Note how the clapper box is turned, or offset, for making a vertical cut.

1. Turn the vise with the solid jaw at right angles to the ram stroke.

2. Mark a layout line on the workpiece.

3. Clamp the workpiece so that the end to be machined clears the right side of the vise jaw. The workpiece should be positioned in the vise so that the down feed will have to move the shortest distance to complete the cut.

4. Turn the top of the clapper box away from the direction in which the cut is to be made (Figure 58-6). This is done to make the cutting tool clear the

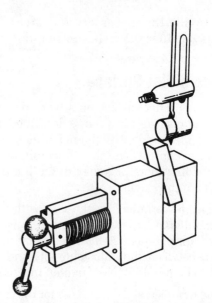

Fig. 58-7 Machine an angle by clamping the workpiece in the vise with the layout line parallel to the top of the vise jaws.

workpiece on the return stroke, and to keep it from digging in.

5. Use the left-cut, side-facing tool. Mount the cutting tool in the holder and the holder in the tool post.

6. Turn the tool slide up as far as it will go.

7. Move the table to the left until the cutting tool will clear the right end of the workpiece.

8. Turn the slide down to check it. The cutting tool must reach the bottom of the workpiece without too much overhang. Return the tool to the starting position.

9. Turn on the power. Put your left hand on the cross-feed crank and your right hand on the tool slide handle. Move the workpiece toward you and the cutting tool down until a chip forms.

10. Remove your hand from the cross-feed crank, then move the cutting tool down about one-fourth to one-half turn at the end of each cutting stroke.

11. Continue until the machining is complete.

Machining an Angle or a Bevel

The simplest way to machine angles or bevels is to hold the workpiece in a vise, with the layout line parallel to the top of the vise jaws (Figure 58-7). The angle or bevel is then machined in the same way as a horizontal cut.

Another method is to machine it like a vertical cut, except that the head must be set at the desired angle (Figure 58-8). Move the ram through one whole stroke, and check to see that the slide does not hit the column. Use a side-facing tool. Proceed with the shaping.

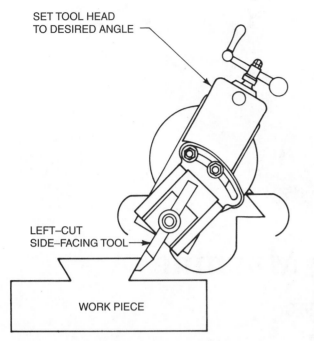

SET TOOL HEAD
TO DESIRED ANGLE

LEFT–CUT
SIDE–FACING TOOL

WORK PIECE

Fig. 58-8 Proper tool head setup for machining an angle.

KNOWLEDGE REVIEW

1. Explain how the shaper works.
2. Describe the uses of shims, parallels, and wedges in holding the workpiece.
3. In machining a flat surface, how long should the stroke be?
4. What kind of cutting tool is used for machining a vertical surface?
5. What does the term *stroke* mean in shaping operations?
6. What machine has replaced the shaper in modern industry?

Unit 59

The Milling Machine

OBJECTIVES

After studying this unit, you should be able to:

- Know the parts and functions of the horizontal mill.
- Know the parts and functions of the vertical mill.
- Understand the three movements of a milling machine table.
- Describe a column-and-knee machine.
- Understand the functions of the various milling machine controls.
- Recognize the various types of milling cutters.
- Understand the importance of cutting speeds and feeds.
- Perform a simple cutting operation on the vertical milling machine.

KEY TERMS

climb milling	conventional milling	knee	side milling
column	cutting feed	milling machine	straddle milling
column-and-knee machine	cutting speed		

The **milling machine** uses a rotating tool with two or more cutting edges to shape and smooth metal. Next to the lathe, it is the most useful and versatile machine in the shop. In this unit, only the basic uses of the milling machine are explained.

Kinds of Milling Machines and Their Parts

As discussed in unit 52, the two milling machines commonly found in school shops are the *horizontal mill* and the *vertical mill.* Both are called **column-and-knee machines.** This is because the *spindle,* or the part that supports the cutting tool and rotates, is fixed in the **column.** The *table,* part of the **knee,** can be moved longitudinally (back and forth), transversely (in and out or across), and vertically (up and down).

On the plain horizontal machine (Figure 59-1), the spindle is horizontal. The control mechanism is in the supporting column, and the table structure is fastened to the knee protruding from the column. The knee can also move up and down. The cranks are used to manually control all motions of the machine. Many industrial mills have automatic control features. This column-and-knee arrangement is typical of many milling machines. *Fixed bed mills* feature a solid, immovable knee and are common in industry to hold large, heavy workpieces. The horizontal mill also has special attachments for cutting gears.

On the vertical machine (Figure 59-2), the spindle is vertical. On smaller vertical machines, the head can

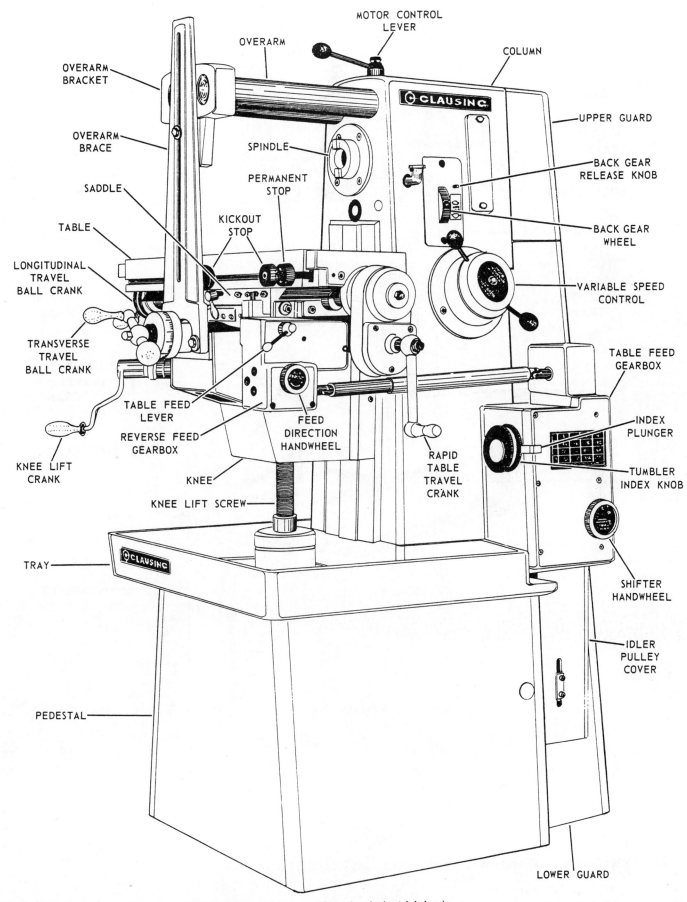

Fig. 59-1 Parts of a horizontal milling machine. (Courtesy of Clausing Industrial, Inc.)

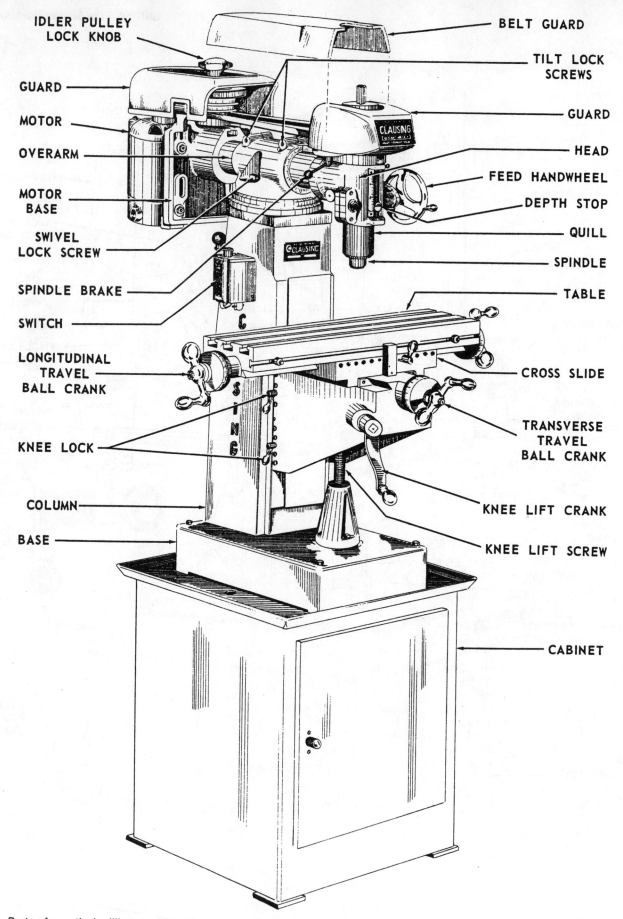

IDLER PULLEY
LOCK KNOB

GUARD

MOTOR

OVERARM

MOTOR
BASE

SWIVEL
LOCK SCREW

SPINDLE BRAKE

SWITCH

LONGITUDINAL
TRAVEL
BALL CRANK

KNEE LOCK

COLUMN

BASE

BELT GUARD

TILT LOCK
SCREWS

GUARD

HEAD

FEED HANDWHEEL

DEPTH STOP

QUILL

SPINDLE

TABLE

CROSS SLIDE

TRANSVERSE
TRAVEL
BALL CRANK

KNEE LIFT CRANK

KNEE LIFT SCREW

CABINET

Fig. 59-2 Parts of a vertical milling machine. (Courtesy of Clausing Industrial, Inc.)

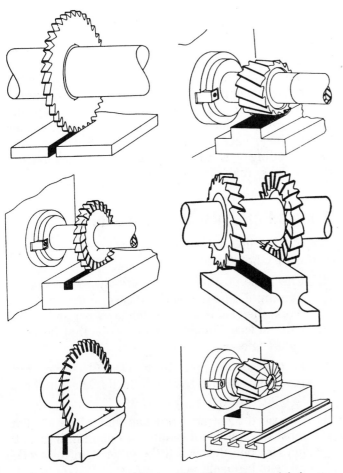

Fig. 59-3 Some common horizontal milling cutters and their uses.

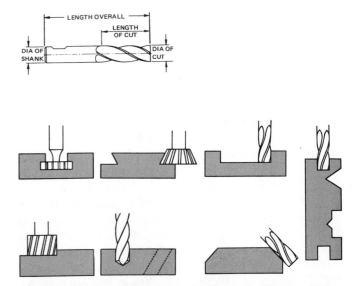

Fig. 59-4 A typical vertical milling cutter is shown here with its important dimensions. Also shown are common vertical milling cutter shapes and uses.

be turned 180 degrees in a horizontal plane. The head can also be adjusted in a vertical plane to any angle. The vertical machine requires less setup time than the horizontal. It can do a wider variety of operations with the workpiece clamped in the same position in the vise or on the table.

Milling Machine Controls and Operation

Machine, or spindle, speed is set by moving the belt, as on a drill press. The speed is expressed in revolutions per minute (rpm). On most small milling machines, there are eight speeds. The four-step pulleys provide four speeds. By changing the belt position on another set of two-step pulleys, four more speeds are obtained. Some machines have a variable speed-control lever. This permits the quick selection of speeds.

To adjust the knee, loosen the *knee lock lever* that holds the knee to the column. Turn the *knee lift crank* to lower and raise the knee. Always lock the knee in its new position.

To move the table toward the column, turn the *transverse ball crank* to the right. To move it away from the column, turn it to the left. This is called cross movement.

To move the table longitudinally in front of the column, turn the *longitudinal ball crank* at either end of the table. Some small machines also have a table power feed. On these, the *table feed lever* is moved to the right to make the table go to the right. Adjustable stops, or trip dogs, along the front edge of the table control the distance of table travel.

Care of the Milling Machine

Follow these machine care guidelines to ensure that your milling operations are safe and accurate.

- Keep the machine clean by brushing the chips away and wiping the table with a cloth.
- Always wipe the spindle nose before putting in an arbor, adapter, or cutter.
- Keep the machine oiled according to the chart supplied by the manufacturer.
- Never leave tools or equipment on the table. Never drop a tool on the table. Remember that this machine is a precision instrument.
- Handle the cutters and the arbor carefully, using a cloth to protect your hands.

Milling Cutters and Holders

There are many kinds, sizes, and shapes of milling cutters, most of which are made of high-speed steel. Carbon-steel cutters are also used on small machines. On a horizontal milling machine, an arbor or shaft is inserted into the spindle. The cutters are then slipped onto the arbor (Figure 59-3). On a vertical milling machine, the cutters are mounted directly in the spindle or in a collet, much like locking a drill bit in a drill press (Figure 59-4).

Cutting Speeds and Feeds

Cutting speed is the distance one tooth of the cutter moves as measured on the work in surface feet per minute (sfpm). In general, cutting speed is slower for harder materials and faster for softer ones. Cutting speed is not the same as machine speed. The spindle of a milling machine operates at a certain number of revolutions per minute (rpm). If you place a 2-inch (50-mm) cutter on the spindle and it makes one complete revolution, the tooth will travel 2 times 3.1416, or about 6 1/4 inches (160 mm). If a 1/2-inch (13-mm) diameter cutter is used, it will travel only 1/2 times 3.1416, or about 1 1/2 inches (40 mm).

Figure 59-5 shows the approximate cutting speeds for carbon-steel and high-speed-steel cutters. To find the spindle speed using customary measurements, use this formula:

$$\text{Formula: rpm} = \frac{4 \times \text{cutting speed in sfpm}}{\text{diameter of cutter in inches}}$$

Suppose you wish to machine a piece of mild steel with a high-speed-steel cutter three inches in diameter. The rpm would be 4 times 80 divided by 3, or about 107 rpm.

To find the spindle speed using metric measurements, where smpm means surface meters per minute, use this formula:

$$\text{Formula: rpm} = \frac{1,000 \times \text{cutting speed in smpm}}{3.1416 \times \text{diameter of cutter in mm}}$$

Suppose you wish to machine a piece of hard cast iron with a 75-mm high-speed-steel cutter. The rpm would be 1,000 times 18 divided by 75 times 3.1416, or about 76 rpm.

The machine speed, or rpm, is adjusted by moving the belt, as on a drill press. The **cutting feed,** the rate at which the workpiece moves under the cutter, depends on your own judgment. In general, feed is slower for heavy, rough cuts and faster for light, finishing cuts. When feeding by hand, one tends to move the table too slowly rather than too fast.

Using a Horizontal Milling Machine

Follow the procedure and guidelines below when cutting on a horizontal mill.

1. Check the position of the solid jaw of the vise by holding a square against the column. For most work, the vise is set with the jaws parallel to the column face. Some jobs require that they be set at right angles.

2. Fasten the workpiece in the vise. Place it on two parallels so that at least half the thickness is above the top of the jaws.

3. Choose the correct kind and size of cutter. For most work, the tool should be a plain milling cutter. Use one that is a little wider than the workpiece to be machined.

4. Choose an arbor with a hole of the same diameter as that of the cutter. Install the arbor in the spindle, using the draw-in bar to hold it firmly in place. Place

Fig. 59-5 A table of cutting speeds for various metals.

Materials	Milling cutter materials		
	Carbon-tool steel	**High-speed steel**	
	dsfpm	sfpm	smpm
alloy tool steel	—	28-40	8.5-12.2
tough alloy steel	20-26	40-52	12.2-15.8
medium alloy steel	26-31	52-65	15.8-19.8
cast iron—hard	26-31	52-65	15.8-19.8
SAE 1045 steel	31-38	65-79	19.8-24.1
malleable iron	31-38	65-79	19.8-24.1
SAE 1020 and C1018 (mild) steel	38-46	79-97	24.1-29.6
cast iron—medium	38-46	79-97	24.1-29.6
cast iron—soft	48-60	97-125	29.6-38.1
brass and bronze { medium	60-90	125-180	38.1-54.9
soft	90-135	180-280	54.9-85.3
aluminum and other light alloys	135-725	280-1500	85.3-57.2

sfpm = surface feet per minute

smpm = surface meters per minute

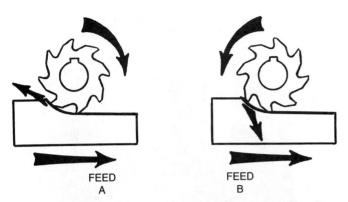

Fig. 59-6 Note the difference between (A) conventional and (B) climb milling.

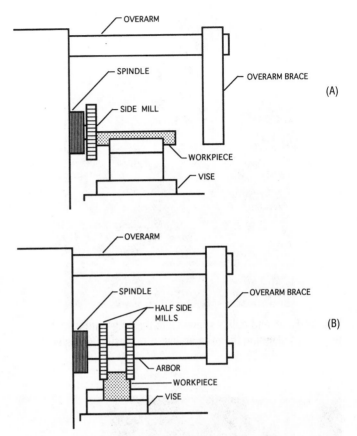

Fig. 59-7 Setups for (A) side and (B) straddle milling.

the cutter between the collars on the spindle so that it is about centered. Make sure the cutter is held firmly by tightening the arbor nut securely. Also, make sure that the *overarm* is clamped tightly.

5. Adjust for the right speed and feed. The feed should be opposite to the direction of cutter rotation. This is known as **conventional milling** (or up milling). Feeding in the same direction as cutter rotation is called **climb milling** (Figure 59-6). Never try to do climb milling on a small machine.

6. Move the table transversely until the workpiece is centered under the cutter.

7. Turn on the power, then move the knee until the rotating cutter just touches the workpiece surface. Now move the workpiece to the left of the cutter. Raise the knee for the correct depth of cut; about 0.015 inch (0.38 mm) for a roughing cut.

8. Feed the workpiece into the turning cutter.

9. After the first cut is complete, stop the machine and check the surface.

10. If more cuts are needed, lower the table one full turn. Then move the workpiece to the starting position. Raise the table one full turn plus the amount for the next cut.

Side-milling cutters are placed directly in the spindle, and require no arbor. They are used to machine vertical surfaces. In **straddle milling,** two side cutters are separated by spacers on the arbor to "straddle" the workpiece and thereby machine two vertical surfaces on it (Figure 59-7).

Using a Vertical Milling Machine

Follow the procedure and guidelines below when cutting on a vertical mill.

1. Choose an end mill that is a little larger in diameter than the width of the cut. However, if the cut-

ter is not wide enough, make two passes across the surface.

2. Mount the milling cutter in the spindle. Most small vertical millers have a collet for holding straight-shank end mills. Insert the cutter. Tighten it by turning the draw-in bar at the top of the spindle or by tightening the setscrews that hold the end mill in place (Figure 59-8).

3. Make sure the milling head is at right angles to the table, both vertically and horizontally.

4. Secure the workpiece in the vise.

5. Adjust for correct speed and feed.

6. Make sure the spindle is turning in the correct direction. A right-hand cutter should turn counterclockwise.

7. Move the workpiece until it is directly under the cutter.

8. Turn on the power, then raise the knee until the cutter just touches the workpiece.

9. Adjust the micrometer collar on the knee crank to zero.

10. Move the table to the right and raise the knee about 0.015 inch (0.38 mm).

11. Move the table until the cutting starts. Then feed it slowly, making a cut across the surface.

Fig. 59-8 An end mill is slipped into the collet and is held firmly by tightening the draw-in bar. (Courtesy of Clausing Industrial, Inc.)

Fig. 59-9 Using an end mill to machine a bevel on a workpiece. The head is set at 45 degrees. (Courtesy of Clausing Industrial, Inc.)

12. If much material must be removed, make several cuts.

Figure 59-9 shows how a bevel or chamfer can be cut with the workpiece in a vise and the head set at a 45-degree angle.

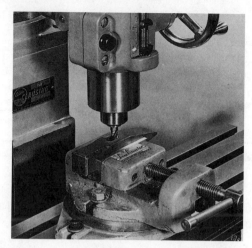

Fig. 59-10 Using an end mill to machine a handle opening in a hammer head. (Courtesy of Clausing Industrial, Inc.)

The vertical milling machine can also be used for accurate drilling, boring, and reaming (Figure 59-10). To do these operations, the spindle can be moved up and down with a feed handle, just as in the drill press.

KNOWLEDGE REVIEW

1. Name the two common kinds of milling machines.
2. Explain how the tables on these machines can be adjusted.
3. How do you change the speed on these machines?
4. Of what kinds of steel are milling cutters commonly made?
5. What is the difference between cutting speed and machine speed?
6. What is the difference between up milling and climb milling?
7. Name three operations that can be done with a vertical milling machine.
8. Write a research report on the use of milling machines in industry.
9. Visit a local industry and describe the kinds of parts which are made on milling machines.

Unit 60

Precision Grinding

OBJECTIVES

After studying this unit, you should be able to:

- Explain the processes of cylindrical, centerless, internal, and surface grinding.
- Identify illustrations of the above grinding processes.
- Identify the major parts of a surface grinder.
- Explain the terms associated with grinding wheels.
- Dress and true a grinding wheel.
- Install a grinding wheel.
- Perform a simple flat surface grinding operation.

KEY TERMS

aluminum oxide	dressing	magnetic chuck	silicon carbide
centerless grinding	grade	plain cylindrical grinding	structure
cubic boron nitride	grinding wheel	precision grinding	surface grinding
cylindrical grinding	grit size	resinoid bond	truing
diamond wheel	internal grinding	rubber bond	vitrified bond

In earlier units of this book, you were introduced to several kinds of abrasive machining, including grinding, polishing, and buffing. These are all material removal processes which employ abrasive grains as cutting tools. In this unit, you will learn about **precision grinding** and how it is used to finish metal workpieces to tolerances as close as 0.0002 inch (0.005 mm).

A basic function of this grinding is to produce smooth and accurate cylindrical external and internal surfaces, as well as flat surfaces. Grinding wheels and special machines are used in these operations. The particular surface to be ground determines the type of grinding operation and the kind of machine to be used. For example, if a flat surface is to be ground, the operation is surface grinding and the machine is a surface grinder.

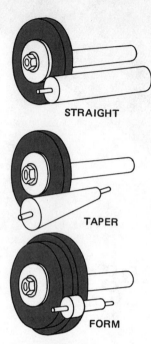

Fig. 60-1 Kinds of grinding that can be done on a plain center-type grinder.

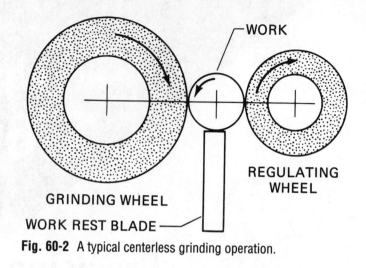

Fig. 60-2 A typical centerless grinding operation.

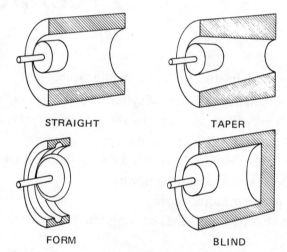

Fig. 60-3 Kinds of internal grinding.

Cylindrical Grinding

Cylindrical grinding deals with the grinding of exterior and interior surfaces of cylindrically shaped workpieces such as shafts and drums. It can be used to grind round bar stock or tapered workpieces.

Plain Cylindrical Grinding

In **plain cylindrical grinding** (or center type grinding), the workpiece is mounted between centers, as on a lathe, and rotates while in contact with the wheel. The wheel is fed into the work to begin the cut, and the workpiece then automatically moves along the wheel to complete the grinding. This is called *straight grinding*, and at the end of each pass the wheel is again fed into the work to begin another pass. On many machines, the headstock and tailstock are mounted on a swivel table to offset the workpiece for *taper grinding*. *Form grinding* (or plunge grinding) uses a shaped wheel to grind that shape on the workpiece (Figure 60-1).

Centerless Grinding

A second class of cylindrical finishing operations is **centerless grinding**. It too is used for grinding cylinders, except that the workpiece is driven by a *regulating wheel* and supported by a *work rest blade* (Figure 60-2). The advantages are that large hollow pieces can be ground, and there is no need to drill center holes in smaller pieces. In both cases, the automatic (robotic) handling of workpieces is thereby simplified.

On the centerless grinder, the pressure of the grinding action forces the workpiece down against the work rest blade and the regulating wheel. The regulator is usually made of a rubber-bonded abrasive. It serves as both a frictional driving and braking element, rotating the workpiece at a speed about equal to that of the wheel.

Internal Grinding

Internal grinding is a third type of cylindrical grinding, and is done to accurately finish internal surfaces or holes (Figure 60-3). It is also done to correct imperfections from previous operations, such as the distortion created by heat treating, or the rough surfaces produced by cutting tools. Straight or tapered holes, through holes or blind holes, and holes having more than one diameter can all be worked on internal grinders. Some machines have power feed and traverse feed, though the cycle is manually controlled. Others are fully automatic.

In a typical internal grinding setup, the hollow workpiece is mounted in a chuck, and the grinding

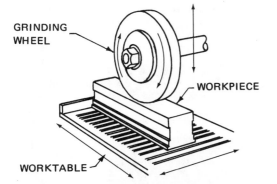

Fig. 60-4 The surface grinding operation.

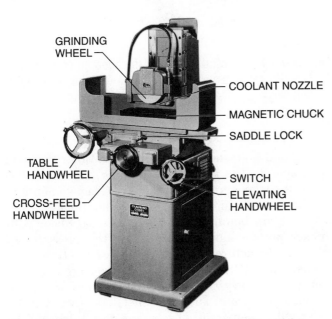

Fig. 60-5 The important parts of a surface grinder. (Courtesy of Clausing Industrial, Inc.)

wheel is mounted at the end of a shaft. The wheel and the workpiece rotate in opposite directions, and the wheel also moves in and out of the work to get cutting feed.

Surface Grinding

While all grinding is done on surfaces, the term **surface grinding** is usually used to describe the process of grinding a *flat* surface (Figure 60-4). Machines made for this purpose have either a horizontal or a vertical spindle, and a rotary or a reciprocating table. These arrangements make it possible to grind one or many parts at one time. The surface grinder is the type of grinder usually found in the school machine shop.

The most common kind of surface grinder has a horizontal spindle and a reciprocating table (Figure 60-5). The wheel is mounted so that it can be raised or lowered according to the size of the workpiece. It can be fed to the work either by hand or automatically.

Fig. 60-6 Many small parts can be held on a magnetic chuck for surface grinding. (Courtesy of Norton Company)

The table moves back and forth under the wheel using either hand or automatic controls. The table travel and speed are set before the job is begun. Such table speeds are high and may reach 60 surface feet (18.3 surface meters) per minute or more.

When possible, the workpiece is held on a **magnetic chuck** built into the table. For mass production, a number of identical pieces may be held on the chuck at one time (Figure 60-6). After grinding, the pieces should be run through a demagnetizer. Parts that would be injured by magnetization or that cannot be magnetized must be held by clamps, vises, or fixtures.

Straight-sided wheels with a flat cutting face are usually used in surface grinders. In grinding shoulders, a recessed wheel can be used. In grinding Vs and other shapes, wheels with formed faces are used.

Grinding Wheels

A **grinding wheel** is a multipoint cutting tool made of abrasive grains held together by a glue-like bonding agent. The *bond* acts as a toolholder while the *grain* acts as the cutting tool. The grain particles fracture away during the grinding process, to present new cutting edges for material removal.

Abrasives

Abrasive materials are crushed and sifted through screens to provide a variety of grit sizes for cutting applications. There are three common types of abrasives. **Aluminum oxide** (Al_2O_3) is a manufactured abrasive made by fusing aluminum bauxite ore in an electric furnace at a temperature of about 3,700°F. This fused mass is then crushed and screened. Aluminum oxide is a tough, sharp grain, generally used for grinding alloy, carbon, and high-speed steels; wrought iron; and hard bronzes.

Silicon carbide (SiC) is a manufactured abrasive made by converting pure silica sand and coke into SiC in an electric furnace at about 4,000 °F. This material is then crushed and screened to get the desired grit sizes. Silicon carbide is harder and sharper than aluminum oxide, and is used for grinding gray, chilled, and cast iron; nonferrous metals; and stone and ceramics.

Cubic boron nitride (CBN) is a manufactured abrasive which is nearly as hard as diamond. It resulted from research into the production of manmade diamonds by the General Electric Company. These abrasives are used to grind hardened carbon, tool and die, and stainless steels; and superalloys. They are tough, sharp, long-lasting, and expensive.

Bonds

Grinding wheel bond materials are mixed with abrasive grains, pressed into shape, and then solidified by heat and pressure. The result is a porous material in which the grains are held in a definite spacing arrangement. Ideally, the bond should be strong enough to hold the grains to the workpiece until they are dull, and then break down and release the dull grain. The result is that new, sharp grains are continually presented to the workpiece, in a kind of "self-dressing" operation. Different grinding operations require different types of bonded wheels.

There are four major bonding materials used to manufacture modern grinding wheels. Ceramic materials such as glass and clay are used in the **vitrified bonding** of abrasive grains. More than half of all grinding wheels made today have the vitrified bond. These wheels are rigid, strong, and porous; and rapid temperature changes have little or no effect on them. They are used for precision work at speeds up to 12,000 sfpm, and for general-purpose grinding.

The **resinoid bonding** material is a thermosetting resin, resulting in a tough, shock-resistant wheel which can be run at speeds up to 16,000 sfpm. They are especially useful for high-speed, rough grinding and cutoff work.

Rubber bonds are made of both natural and synthetic materials. Strong and elastic, they are used for thin cutoff wheels, and grinding and regulating wheels for centerless grinding. Common shellac also has limited use as a bonding material in applications where a high finish is desired.

Diamond wheels are made of metal-bonded, coated diamond grains, and are used on very hard materials and in the building and construction industries for ceramic and cement cutoff work.

Bond materials and the way in which grains are arranged affect grinding wheel performance in several ways. As shown in Figure 60-7, part of the bonding material *coats* the grains. The rest of the bond forms into

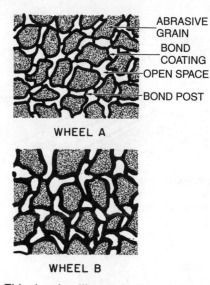

Fig. 60-7 This drawing illustrates the bonding of grains in a wheel. Wheel A is a softer grade than wheel B. Can you explain why?

posts that both lock the grains together to form the wheel, and hold them apart from each other to create open spaces. The amount of bonding material determines the wheel **grade,** or hardness. A large amount of bond makes a hard wheel; a small amount makes a soft wheel. More bonding material creates a wheel with thicker posts which can withstand the pressures of grinding. **Structure** refers to the spaces between the grains. Grains that are very closely spaced are said to be dense. When the grains are wider apart, they are said to be open. Generally speaking, metal removal will be faster with open-grain wheels; and the finish will be finer with dense wheels.

Grit Size

Abrasive grain size, or **grit size,** is determined by the number of meshes or openings per inch in the screen used to separate the grains. The sizing is as follows: coarse, 10 to 24; medium, 30 to 60; fine, 70 to 180; and very fine, 220 to 600. Fine-grain wheels are generally used for grinding hard materials, and coarse wheels for rapid metal removal on softer materials. Manufacturers of grinding wheels have coded markings to identify the different kinds of wheels (Figure 60-8). Some common styles and shapes of grinding wheels are shown in Figure 60-9. These are used for all types of grinding.

Installing the Grinding Wheel

Before installing the wheel on the spindle, look at it to see if it has any cracks or chips. Never use a defective wheel. You should also check for defects by giving

32A 46 - H 8 V BE

Abrasive	Grit Size		Grade	Structure	Bond Type	Norton Symbol
	Coarse	Medium	Soft	The structure	V = Vitrified	Letter or
Alundum = A	10	30	A	number of a	B = Resinoid	numeral or
16 Alundum = 16A	12	36	Thru	wheel refers	R = Rubber	both to
19 Alundum = 19A	14		H	to the relative	E = Shellac	designate a
23 Alundum = 23A	16	46		spacing of		variation or
32 Alundum = 32A	20	54	Medium	the grains		modification
38 Alundum = 38A	24	60	I	of abrasive;		of bond or
53 Alundum = 53A			Thru	the larger the		other char-
57 Alundum = 57A		Very	P	number, the		acteristic of
75 Alundum = 75A	Fine	Fine		wider the		the wheel.
NZ Alundum = NZ	70	220	Hard	grain spacing.		Typical
ZF Alundum = ZF	80	240	Q			symbols are
ZS Alundum = ZS	90	280	Thru			"P", "G",
37 Crystolon = 37C	100	320	Z			"BE".
39 Crystolon = 39C	120	400				
CBN = CB	150	500				
	180	600				

Fig. 60-8 A marking system for grinding wheels. (Courtesy of Norton Company)

the wheel a ring test. Insert your index finger in the wheel's center hole. Next, using only this finger, pick up the wheel. Tap the wheel lightly with a wooden hammer or the handle of a screwdriver. You should hear a clear metallic tone. A cracked wheel gives off a dull thud and should not be used.

When mounting the wheel on the shaft, be sure to use paper blotters on the inner side of the flanges. These will ensure an even pressure on the wheel when it is tightened in place. The blotters also help to dampen the vibration between the wheel and shaft while the machine is in use. Tighten the nut sufficiently to hold the wheel firmly. Do not overtighten, as this may damage the wheel.

Dressing and Truing the Wheel

Once the wheel is installed, it must be **dressed** to remove dirt and dull abrasive grains, and to smooth it by removing any grooves. **Truing** removes grains so that the surface of the wheel runs absolutely true to the shaft, to provide accurate grinding. The important process of dressing and truing is described below.

1. Insert the diamond nib shank as far as it will go into the diamond holder. This will eliminate vibration.

2. Tighten the square-head setscrew securely.

3. Raise the spindle so that the bottom of the wheel clears the top of the diamond.

4. Set the diamond to the left of the wheel centerline, as shown in Figure 60-10. Never set the diamond to the right of the wheel centerline. In that position it will grab and dig into the wheel, damaging both itself and the wheel. Also, never move the table to the right or left when dressing the wheel.

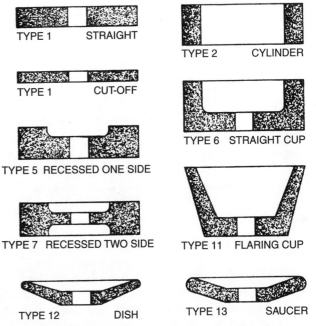

Fig. 60-9 Some types of grinding wheels.

5. Secure the diamond holder to the table or chuck.

6. Lower the spindle to a point where the wheel's high spot barely touches the diamond.

7. Using the cross-feed handwheel, pass the diamond back and forth across the wheel. Be sure to pass the diamond beyond the wheel face on each pass.

8. For a fine finish, use a slow pass and a down feed of about 0.001 inch (0.03 mm) per pass. For a rough finish, use a rapid pass and a slightly greater down-feed rate.

9. Continue down feeding until the sound of the diamond in contact with the wheel indicates that the diamond is cutting evenly across the full face of the wheel. Use very light cuts and a very slow pass in the finishing stages.

Using the Surface Grinder

The surface grinder can be used in machine shop operations such as finishing the surfaces of cast machine vises, C clamps, and tools. The following is a description of how the surface grinder is used:

1. Clean all grease and dirt from the workpiece. Dress and true the wheel if necessary.

2. Wipe the magnetic chuck with a clean cloth.

3. Center the workpiece on the chuck. Turn on the switch to hold it in place magnetically.

4. Adjust the table reverse dogs so that they clear the ends of the workpiece by about 2 inches (50 mm).

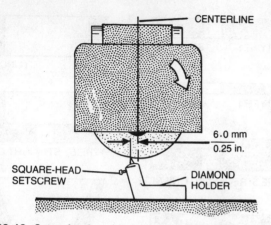

Fig. 60-10 Setup for dressing and truing a grinding wheel.

5. Turn on the coolant valve, if coolant is to be used. If you are dry grinding, start the dust collector. Set the speed control and start the table with the three-position switch.

6. Adjust the rate of table feed.

7. Turn on the power.

8. Hand-feed the table in until the workpiece is under the grinding wheel.

9. Turn on the power table feed.

10. Adjust the grinding wheel down until it is near the workpiece. Move the table cross-feed. Continue to feed the wheel down slowly until the grinder just touches the work. This should be the highest spot on the work surface. To make sure, feed the length of the workpiece under the wheel. Feed the grind-ing wheel down about 0.003 inch (0.08 mm) and start grinding.

11. Turn the cross-feed out about one-fourth the width of the grinding wheel just as the table changes direction.

12. Grind the entire surface. During grinding, do not let the wheel become soaked with coolant.

13. Dress the wheel. For a high finish, the last grinding should not remove more than 0.001 inch (0.03 mm).

14. After grinding, turn off the coolant and let the wheel run a few seconds to spin off any coolant that has collected.

15. Check the workpiece carefully to see if all areas are ground. Repeat the grinding if necessary.

KNOWLEDGE REVIEW

1. What tolerance can precision grinding achieve on metal workpieces?

2. Describe the different kinds of cylindrical grinding.

3. In centerless grinding, what guides the workpiece to the wheel?

4. Describe some applications of internal grinding.

5. What kind of chuck is often used in surface grinding?

6. What is a ring test?

7. Describe the processes of dressing and truing.

8. Write a report on precision grinding in industry.

Unit 61

Nontraditional Machining Methods

OBJECTIVES

After studying this unit, you should be able to:

- Recognize illustrations of, and describe the special techniques of, the nontraditional methods presented in this unit.
- Understand the meaning and importance of the term *noncontact* as it relates to metal cutting and forming.
- Describe the relationship between P/M and casting.
- Describe the relationship between CHM and photochemical milling.

KEY TERMS

chemical machining	explosive forming	magnetic forming	sintering
electrical discharge machining	friction drilling	maskant	spark forming
	gas forming	photochemical machining	ultrasonic machining
electrical discharge wire cutting	high temperature, short time extrusion cooking	powder metallurgy	waterjet cutting
electrochemical machining			

In addition to the basic methods of metal machining, many other *nontraditional* techniques have been developed to meet the special needs of modern industry. The newer methods use electricity, heat, chemicals, or some combination of these. One outstanding feature of such techniques is that they are *noncontact* methods where the tool never touches the workpiece. Many of these techniques are used for metal separation or removal, while others are used to form metal. These nontraditional methods add to, rather than replace, the basic procedures described earlier in this book. Some typical examples of these are presented in this unit.

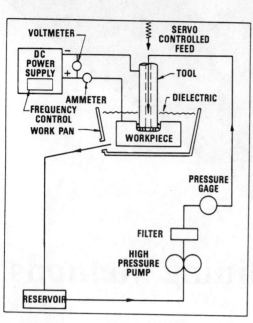

Fig. 61-1 Electrical discharge machining (EDM).

Electrochemical Machining

The process of removing metal from a workpiece by electrolytic action is called **electrochemical machining** (ECM). It is the opposite of electroplating. It works by passing a direct electric current between a tool (cathode) and a workpiece (anode) while both are immersed in a chemical solution (electrolyte). As the electric current travels from the workpiece to the tool, tiny particles of metal are carried away from the workpiece and into the electrolyte. The nose of the tool is shaped exactly like the shape to be produced in the workpiece. A bulge on the tool produces an exact depression in the workpiece. A depression on the tool produces an exact bulge on the workpiece.

ECM is used on metals that are difficult to machine. It can produce holes or cavities that have very complex shapes. Because it is a noncontact machining method, ECM can be used on thin, fragile metal workpieces.

Electrical Discharge Machining

Electrical discharge machining (EDM) removes metal by the erosive action of a controlled electric spark between a shaped tool (electrode) and a workpiece (anode) (Figure 61-1). A power supply provides a series of electric impulses at a certain rate and voltage. The tool and workpiece are submerged in a circulating dielectric fluid which acts as an electric insulator until the spark occurs. They are brought close together until an electrical path (arc) forms between them through the fluid. This permits a high-density discharge of current, thereby removing from the workpiece tiny particles which are washed away by the fluid. The process con-

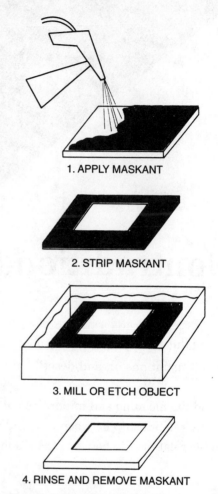

Fig. 61-2 Chemical machining (CHM).

tinues until a cut or cavity is formed in the workpiece. EDM is a noncontact machining process used primarily to produce complicated molds and dies for plastic injection molding and metal diecasting. An example of EDM can be found in the "In Focus" discussion on cereal making later in this unit.

An important variation of EDM is the **electrical discharge wire cutting** (EDWC) process, as shown in Figure 61-18. Here a traveling brass or tungsten wire (much like a band saw blade) uses sparks instead of teeth to cut intricate interiors for extrusion dies. Generally, EDWC is used to produce through-holes and solid shapes, while EDM is used to create cavities in a workpiece.

Chemical Machining

Chemical machining (CHM) is the removal of metal using acids in a deep etching process. This is similar to metal etching, which was described in unit 42. In CHM, the workpiece surface *not* to be machined is coated with an acid-resistant material called a **maskant** (Figure 61-2). The workpiece is then immersed in the acid which eats away the metal not covered with maskant. This method

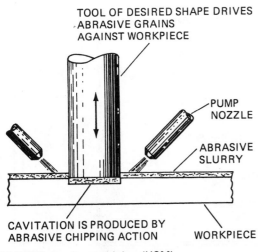

Fig. 61-3 Ultrasonic machining (USM).

is often used to remove metal from irregular surfaces of castings. The resulting surface will not be smooth because castings are porous.

Thin material can be completely eaten through by CHM. Complex shapes such as electronic circuit boards are made from thin metal sheets which are entirely masked except the outline of the circuit. This procedure is often called *chemical blanking.* When metal must be removed across a wide area to a very shallow depth, such as the skin on airplane wings, the process is called *chemical milling.*

A variation of CHM combines photography and chemical etching to produce precision parts. It is called **photochemical machining,** and the process begins with the preparation of a scale drawing two to one hundred times the size of the desired part. Next, the drawing is reduced photographically to the exact size of the final part. A thin metal workpiece is cleaned thoroughly and coated with a photochemical resist. An image of the reduced photographic negative is projected onto the sensitized metal workpiece, or plate, much as photographs are printed. The plate is then developed in a chemical solution that dissolves all the resist except that in the photo-exposed region. The plate is then acid-etched to dissolve all the unprotected metal, to produce the final part. The resist is now washed away, and the part is ready for use.

Ultrasonic Machining

Ultrasonic machining (USM) is material removal by the action of a thin slurry made of water and abrasive particles. In operation, a pump circulates the slurry between the face of the tool and the workpiece. A power unit drives the tool so that it moves up and down (oscillates) at about 20,000 cycles per second, driving the slurry against the workpiece with great force. The abrasive particles do the cavitation, or cutting, by chipping

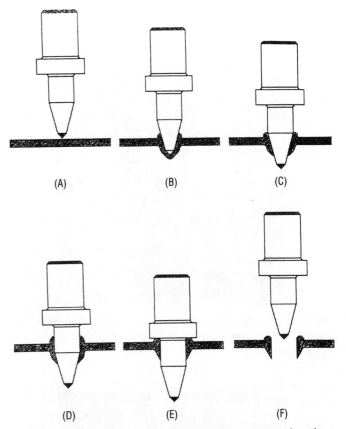

Fig. 61-4 Friction drilling. Refer to text for process explanation.

away small pieces of the material (Figure 61-3). Typical ultrasonic machining operations include: drilling, shaving, slicing, and cutting unusual punch and die shapes. Materials such as metal, ceramics, glass, and plastics can be shaped by ultrasonic machining.

Friction Drilling

In **friction drilling,** holes are produced in metals up to 1/4 inch thick by heat energy created as a rotating metal tool is forced against the surface of a workpiece. This is more properly a hole *forming* rather than a drilling process. The special tungsten tool is available in diameters from 1/8 inch to 1 inch and can be used in any drill press.

This chipless process is simple to perform (Figure 61-4). At *A*, in the figure, the rotating tool is pressed against the work, and the temperature at the tool point increases. At *B*, the contact area widens, and the rapid temperature rise melts the workpiece and allows the tool to advance. At *C*, the softened metal flows up the tool to form the upper bushing neck until the point breaks through. At *D*, the remainder of the displaced metal flows down below the workpiece surface to form the lower bushing neck. At *E*, the top of the bushing is flattened and trued by the collar of the tool. At *F*, the tool is withdrawn. The bushing created in friction

Fig. 61-5 Waterjet cutting can be used on many materials, including this candy bar. (Courtesy of Flow International Corporation)

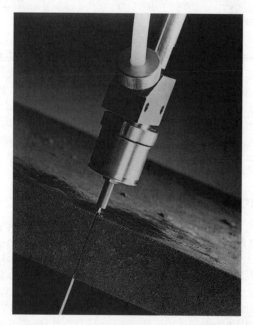

Fig. 61-6 This three-inch steel plate is being cut with an abrasive waterjet. (Courtesy of Flow International Corporation)

drilling can be tapped to receive threaded fasteners, or used for soldered tubing and other connections.

Waterjet Cutting

Waterjet cutting separates material with a powerful stream of water. These systems use a finely focused jet of water at pressures around 55,000 psi. The nonmetallic materials which can be jet-cut include: wood,

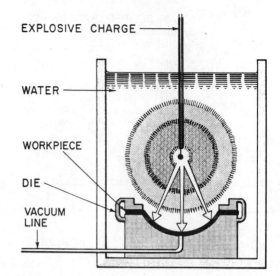

Fig. 61-7 Explosive forming uses the force of a controlled explosion to shape a workpiece.

plastic, cardboard, cloth, leather, fiberglass, and even food products (Figure 61-5). The noncrushing stream is dustless, clean, and generates no heat.

By adding abrasive powder to the waterjet, it is possible to separate dense plastic and glass, and cut steel plate up to 4 inches (100 mm) thick (Figure 61-6). This combination of water and abrasives in a pressurized stream is called *hydro-abrasive machining.*

Explosive Forming

Explosive forming is shaping metal by using dynamite or other explosives (Figure 61-7). The explosion bulges, or pushes out, the metal to an exact shape over a die. Equipment needed includes a tank of water, a die, a metal blank, and the explosive. The metal blank is clamped over the die cavity, and the air is then removed from the cavity by way of a vacuum line. Next, the explosive charge device is lowered into the liquid and ignited. The shock waves from the explosion force the metal into the die, shaping it in a split second. This method can be used to form very large workpieces, such as the domed top of a gasoline storage tank.

Spark Forming

Spark forming is making objects by forcing flat blanks into a die through the force provided by the vaporization of a small piece of special wire. An electrical charge is transmitted through heavy cables to the wire. When the electrical charge passes through the wire, its resistance is very great. This produces heat that causes the wire to vaporize. When a solid piece of wire is vaporized, its molecules expand rapidly. This rapid ex-

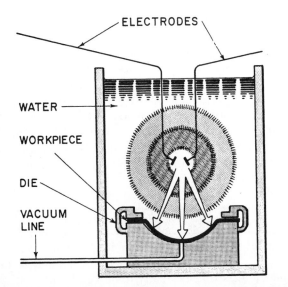

Fig. 61-8 Spark forming uses the force of a vaporizing wire to shape a workpiece.

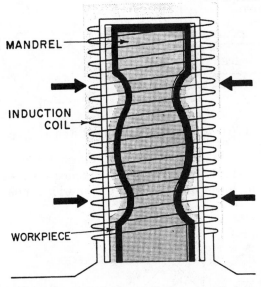

Fig. 61-9 Magnetic forming uses electromagnetic force to shape a workpiece.

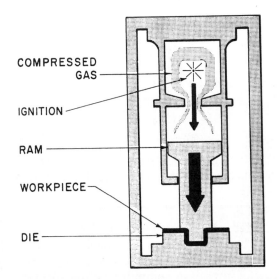

Fig. 61-10 Gas forming uses the force of gas combustion to form a workpiece.

pansion has the same effect as the explosion in explosive forming (Figure 61-8).

The time needed for forming by this method is from seven to twelve microseconds (millionths of a second). Tolerances are extremely fine. The spark method gives off less vibration than other methods. Products formed by the spark method include fuel tanks for rockets.

Magnetic Forming

Magnetic forming shapes metal workpieces through electromagnetic force. An induction coil is either wrapped around a workpiece or placed within it.

Electric current is sent through the coil, producing a strong magnetic force which presses the metal against a shaped mandrel (Figure 61-9).

Gas Forming

In **gas forming,** combustion drives a ram at high speed against a workpiece (Figure 61-10). The metal piece is first placed in the die. Gas behind the ram is then ignited, driving the ram down on the workpiece with great force. The ram and die form the metal to the exact shape needed.

Gas forming can be used to make products as different as metal dishes and parts for space vehicles. The metal to be formed can range from a sheet of aluminum to hot metal.

Powder Metallurgy

Powder metallurgy (P/M) is a process similar to casting, where product shapes are made from compressed metallic and nonmetallic powders (Figure 61-11). In a typical P/M process, a die is filled with mixed powders (A). The upper and lower punches move together to compact the mix (B). The compacted briquette or slug is ejected from the die (C). The slug is then **sintered,** or heated in a furnace, metallurgically bonding the particles below melting temperatures (D).

Powder metallurgy is used primarily for making both large and small parts using precision punches and dies. P/M can be used to make such special items as brake bands, where heat-resistant ceramic powders are mixed with metal powders to add toughness to the product. It also can combine two metals that do not alloy well together. For example, many electric switches

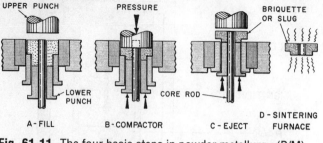

Fig. 61-11 The four basic steps in powder metallurgy (P/M). The core rod causes a hole to be produced in the workpiece.

contain nickel and silver. Nickel has good wearing qualities and silver is a good conductor. If these two metals were melted together to pour a casting, they would separate like oil and water. However, powder metallurgy makes it possible to combine them. P/M can be used to make oil-impregnated bearings which require no further lubrication. Most door lock mechanisms contain durable, precision P/M parts. In addition, the P/M process is economical because there is no scrap, and because further machining and finishing operations are generally not required.

IN FOCUS: *Making Dry Cereal*

The use of advanced machining technology processes is not restricted to making complicated parts for aircraft or computer equipment. For example, electrical discharge machining (EDM) is used to make the steel dies needed to produce the familiar dry, ready-to-eat breakfast cereals you enjoy each morning. These foods are typically made from mixtures of precooked cereal flour, bran, sugar, salt, vegetable oil, coloring, and flavoring. The process used to make such cereals is called **high temperature, short time (HTST) extrusion cooking.**

The HTST extrusion cooker is a drag screw which turns in a close-fitting extruder barrel (Figure 61-12). The cereal mixture enters the feeding zone of the barrel through a feed hopper. The screw drags the cereal into the kneading zone where it is mixed thoroughly. From there it moves into the cooking zone where pressure and the 266°F (130°C) temperature causes the cooking to take place. The cooked material is now a heavy, pasty mixture called the *melt*. Next, the screw forces the melt through a heated metal extrusion die which has an opening in the shape of the cereal product. At this point, much of the moisture in the cereal evaporates as it leaves the die.

Four of these extruder barrels are set into a steel die plate to increase the cereal production rate. As the melt emerges from the dies, a rotating knife slices

off cereal bits about 1/8 inch long (Figure 61-13). These bits are then further dried and cooled to become the crisp, tasty, ready-to-add-milk-and-eat cereal pieces to be enjoyed at the breakfast table.

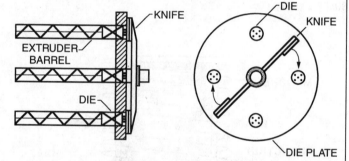
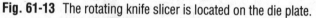

Fig. 61-13 The rotating knife slicer is located on the die plate.

The steel die parts and electrodes are shown in Figure 61-14. The EDM carbon electrodes are used to shape the lower die. One cuts the interior shape, as

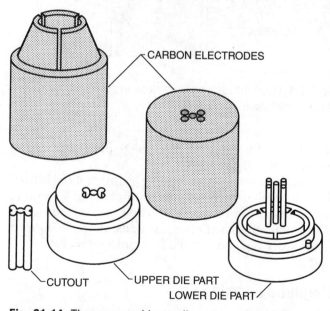

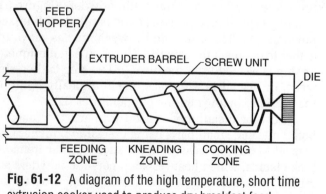

Fig. 61-12 A diagram of the high temperature, short time extrusion cooker used to produce dry breakfast food.

Fig. 61-14 The upper and lower die parts, and the EDM electrodes used to cut them.

IN FOCUS: *Making Dry Cereal* (continued)

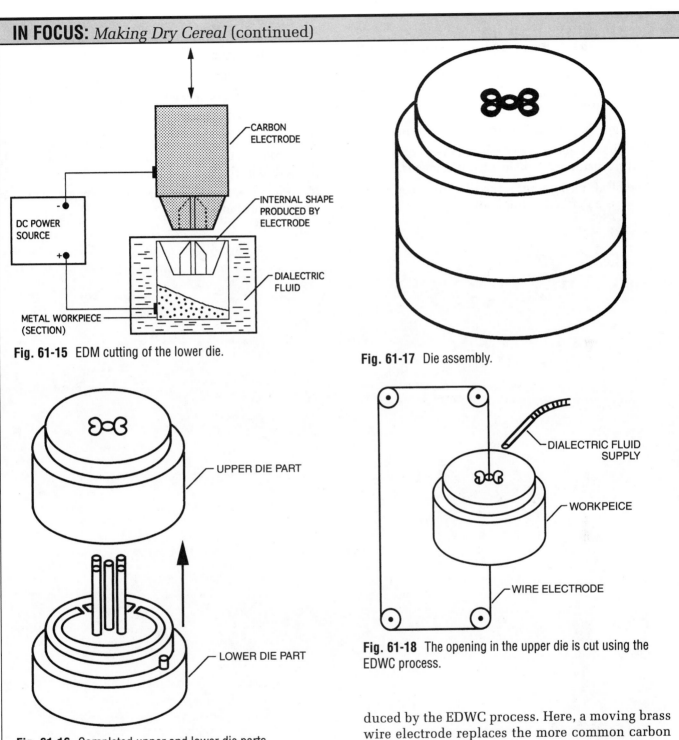

CARBON ELECTRODE

DC POWER SOURCE

INTERNAL SHAPE PRODUCED BY ELECTRODE

DIALECTRIC FLUID

METAL WORKPIECE (SECTION)

Fig. 61-15 EDM cutting of the lower die.

Fig. 61-17 Die assembly.

UPPER DIE PART

LOWER DIE PART

Fig. 61-16 Completed upper and lower die parts.

DIALECTRIC FLUID SUPPLY

WORKPEICE

WIRE ELECTRODE

Fig. 61-18 The opening in the upper die is cut using the EDWC process.

shown in Figure 61-15, and the other cuts the outer shape which has five rods. The upper and lower die parts (Figure 61-16) are then placed together to complete the die assembly shown in Figure 61-17.

The upper die is cut to shape by lathe turning, and the interior star-shaped cutout is pro-

duced by the EDWC process. Here, a moving brass wire electrode replaces the more common carbon electrode. The intricate cutout is separated from the die body by the computer-controlled wire (Figure 61-18).

As illustrated here, the technical processes of EDM, EDWC, and extrusion are used in the production of a common breakfast food. Dry pet foods and snack foods are among the many other products made using these processes.

KNOWLEDGE REVIEW

1. Draw a simple sketch and describe three of these five nontraditional machining processes: ECM, EDM, EDWC, CHM, and USM.

2. Why are they called *nontraditional* processes?

3. The above processes are referred to as *noncontact* cutting methods. Explain what this means.

4. Chemical machining is similar to what other process described in this book?

5. Draw a simple sketch and describe three of these four nontraditional forming processes: explosive, spark, magnetic, and gas.

6. Describe the steps of the P/M process.

7. Why is P/M a special kind of casting method?

8. Describe the waterjet cutting process.

9. Prepare a research report on some specific kinds of products made by one of the nontraditional methods described in this unit.

10. Prepare a research report on the chemistry used in chemical machining.

Unit 62

Heat Treating Metals

OBJECTIVES

After studying this unit, you should be able to:

- Describe heat treating.
- Explain how high temperatures affect the grain structure of metals.
- List the three basic steps in heat treating.
- Name the equipment needed for heat treating.
- Explain what hardness does to metal.
- Describe quenching.
- Give another name for tempering.
- Describe what tempering does to metal.
- Explain how the correct tempering heat can be determined.
- Identify what the tempering color in Fahrenheit and Celsius temperatures should be for twist drills.
- Explain the difference between the process of annealing and normalizing.
- Explain the purpose of case hardening.
- Name the materials that can be used for case hardening.
- Explain the procedure for case hardening.
- Name three methods of hardening metals used in industry.

KEY TERMS

annealing	drawing	hardening	normalizing
carburizing	ductility	heat treating	quenching
case hardening	flame hardening	induction hardening	tempering
critical temperature	grain structure	laser hardening	tensile strength
cyaniding	hardness	nitriding	

After you have machined a center punch, forged out a cold chisel, or made any other small tool, you will find that it is useless until it is heat-treated. **Heat treating** is bringing a metal workpiece to a high temperature to change its properties (Figure 62-1). The high temperature affects the metal's **grain structure,** a basic physical characteristic. Metal atoms are arranged in characteristic crystal structures. These, in turn, join together to form tiny grains arranged like closely fitting blocks. The effect of heat treating on these grains is usually to either harden or soften the metal. Careful heat treating can give the metal specific properties, such as hardness, toughness, and the like, needed to do a job. The basic steps in heat treating are: (1) *heating* to the right temperature; (2) *soaking,* or holding at this temperature for a certain length of time; and (3) *cooling* in a way that will give the desired results.

In the school shop, heat treating can be a very simple process that requires few tools. In industry, however, it is a highly scientific operation and takes special equipment. In this section, only the elementary information about heat treating steel is included. If you want specific information on a particular kind of steel, refer to library references.

Heat treatment can also be done on such nonferrous metals as aluminum, copper, and brass. The process of heat treating these metals is different, however, and will not be considered here. The basic heat-treating processes described in this unit are: hardening, tempering, annealing, and case hardening.

Fig. 62-1 Heat treating a part by inserting it in a furnace. Note the safety clothing. (Courtesy of Johnson Gas Appliance Company)

Fig. 62-2 A gas-fired heat-treating furnace that will operate at temperatures of 300 to 2,400°F. It can be used for hardening, tempering, and annealing small objects. (Courtesy of Johnson Gas Appliance Company)

Equipment for Heat Treating

All heat-treating processes use the same basic equipment:

- A *heat-treating furnace* is best, but a blowtorch, gas-welding torch, forge, or soldering furnace are all good sources of heat (Figure 62-2).
- The *quenching bath* can be a pail or other container of fresh water, tempering oil, or brine (salt water).
- *Forging tongs* are needed to hold the hot metal.

Hardening

Hardening is heating and then cooling steel to give it a fine-grained structure. This process reduces the steel's **ductility,** or ability to be deformed without breaking. It increases its **hardness,** or degree of firmness and strength. It also increases the steel's **tensile strength,** or the amount of stress that it can stand without breaking. Products are hardened to produce sharp-edged cutting tools, to make bearing surfaces wear better, to put the "spring" in a spring, and for many other reasons.

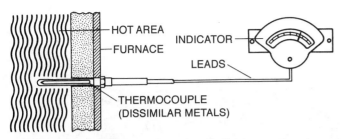

Fig. 62-3 The pyrometer accurately tells the temperature inside the furnace. The thermocouple is inserted in the back of the furnace with leads running to the indicator.

Hardening is done in a furnace fired by oil, gas, or electricity. The metal must first be heated and then rapidly cooled. As steel is heated, a physical and chemical reaction takes place between the iron and carbon in it. The **critical temperature,** or critical point, is the temperature at which the steel has the best characteristics. When steel reaches this temperature, which is somewhere between 1,400 and 1,600°F, it is in the ideal condition to make a hard, strong material if cooled quickly.

The critical temperature can be checked by testing with a magnet, by using a pyrometer, or sometimes by observing the color. When a piece of steel is below critical temperature, it is magnetic. When it reaches critical temperature, it is nonmagnetic. The *pyrometer,* an electric thermometer attached to the furnace, accurately registers the temperature in the furnace (Figure 62-3). Formerly, temperature was determined by observing the color of the hot metal. This is not a very accurate method, however. Even the estimate of an expert can be far off the true temperature.

After the metal reaches the critical temperature, it is **quenched,** or cooled by being plunged into oil, water, or brine. This is done so that the metal will retain the desirable characteristics. If the metal is allowed to cool slowly, it changes back to its original state. When hardened, the metal is very hard and strong and less ductile than it was before.

The Hardening Process

The procedure for hardening is as follows:

1. Light the furnace and let it heat up to hardening temperature.

2. Put the metal in the furnace and heat it to critical temperature. For example, heat high-carbon steel to about 1,475 to 1,500°F. (See unit 6 for an explanation of the AISI system of identifying steels.) For high-carbon steels, allow about twenty to thirty minutes per inch of thickness for the metal to come up to heat. Then allow about ten to fifteen minutes per inch of thickness for soaking at the hardening temperature.

3. Choose the right cooling solution. Some steels can be cooled in water, while others must be cooled in oil, brine, or air. *Fresh water* is used most often for quenching carbon steels because it is inexpensive and effective. *Brine* is made by adding about five to ten percent common salt to water. Brine helps produce a more uniform hardness. This is because it wets the parts all over more quickly. *Oil* is used for a somewhat slower quenching. It reduces the tendency of steel to warp or crack. Most oils used for quenching are mineral oils. *Air* at room temperature is used to cool steel by merely removing it from the furnace. The steel cools fast enough to harden.

4. Remove the hot metal with tongs and plunge it into the cooling solution. Agitate the metal by moving it about in a figure-eight pattern to cool it quickly and evenly. If the piece is thin, such as a knife blade, plunge the thin edge into the cooling solution as though cutting it. This will keep the project from warping. If one side cools faster than the other, it will surely warp. Never just drop the metal into the quenching bath. If you do this, the heat will create a coating of vapor around the metal that will keep it from cooling quickly.

5. Check for hardness. A correctly hardened piece will be hard and brittle and have high tensile strength. Test this by running a new file across a corner of the work. If it is hard, the file will not cut in or take hold.

Tempering

Tempering, or **drawing,** is reducing hardness and increasing toughness. It removes brittleness from a hardened piece and gives a more fine-grained structure.

Tempering is done by (1) reheating the metal, after it has been hardened, to a low or moderate temperature and (2) quenching it in air. The tempering heat can be determined by watching the pyrometer or by observing the *temper colors.* As the metal is heated for tempering, it changes color. You can tell by the color about when the correct heat is reached. Many project parts are completely tempered. Others, such as the cold chisel, are only partly tempered.

The Tempering Process

The procedures for tempering are as follows:

1. To temper the whole piece, put it in a furnace. Reheat it to the right temperature for producing the degrees of hardness and toughness needed (Table 62-1). Then remove the piece and cool it quickly. For example, heat water-hardening carbon steel to 425 to 590°F, and cool it in water.

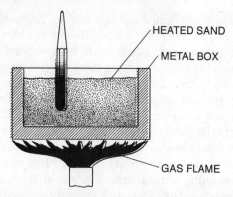

Fig. 62-4 Notice that the body of the tool has been inserted in the sand and that heat is being applied to the metal box.

2. To temper small cutting tools such as cold chisels, center punches, and prick punches:

   a. Harden the whole tool.

   b. Clean off the scale near the point or cutting edge with abrasive cloth.

   c. Heat a piece of scrap metal until it is red-hot. Place it on a welding or soldering table.

   d. Put the tool on the hot metal, with the point extending beyond it.

   e. Watch the temper colors. When the right color reaches the point of the tool, quench it in water.

3. Another method of tempering small tools is the following:

   a. Fill a metal box with sand.

   b. Heat the underside of the box with a gas torch (Figure 62-4).

   c. Clean the hardened tool with abrasive cloth. Place it in the sand, with the point sticking out.

   d. Watch the temper colors travel toward the point as the tool absorbs heat.

   e. When the right color reaches the point, remove the tool with tongs and quench it.

4. To temper a knife blade, clean one side of the blade with abrasive cloth. Pack the cutting edge with heat-resistant material. (Note that this has traditionally been done with wet asbestos. As this substance has been found to be hazardous to health, substitutes for it are now being used.) Heat the back of the blade with a Bunsen burner or welding torch until the right temper color runs toward the edge. Quench in water.

5. To harden and temper small tools at the same time, first heat the tool to critical temperature. Then put only the point of the tool in the quenching solution, and agitate the point (Fig-

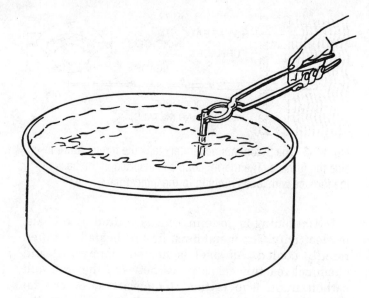

Fig. 62-5 Moving the point of the tool in the quenching solution.

ure 62-5). Remove the tool and quickly clean the point with abrasive cloth. Remember that the handle of the tool will be very hot. Now, watch the temper colors move toward the point. When the right color reaches the point, quench only the point of the tool.

Annealing and Normalizing

Annealing is softening metal to relieve internal strain and to make the metal easier to shape and cut. The metal is heated to the critical temperature and cooled *slowly*. The slower the cooling, the softer the metal becomes. Metals often develop stresses during manufacture. Such stresses can cause steel to warp and castings to warp or crack if not relieved. Metals must often be annealed when they come from the rolling mills or foundry. Annealing gives metals good grain structure and thus makes them easier to machine.

Normalizing is done to put steel in a normal condition again after forging or incorrect heat treating. The steel is heated above the hardening temperature, then cooled in air. This process is very similar to annealing.

The Annealing Process

You may want to anneal an old file or spring so that you can use the metal to make another project. To do this, heat the article to critical temperature (Table 62-1), then allow it to cool by one of the following methods: (1) place it in a pail of sand; (2) turn off the furnace and allow the article to cool in it; or (3) clamp the article between two pieces of hot metal and allow it to cool in air.

Table 62-1 Temper colors of common tools

Tools	Color	°F	°C
scriber, scrapers, and hammer faces	pale yellow	430–450	220–230
center punches, drills	full yellow	470	245
cold chisels, drifts	brown	490–510	255–265
screwdrivers	purple	530	275

Case Hardening

Case hardening is hardening the outer surface of ferrous-metal objects. This surface is the *case.* If you add a small amount of carbon to the case of low-carbon steel during heat treating, the case will become hard. However, the core will remain soft and ductile. Case hardening is done to produce parts such as screws for machines, hand tools, and ball and roller bearings.

Many methods of case hardening are used in industry. **Cyaniding,** a common one, involves placing steel in molten cyanide. This is very dangerous, however, and cannot be done in school shops. Another case-hardening process, **nitriding,** consists of soaking the part in an oven containing ammonia gas at about 950°F (510°C). The part is then allowed to cool slowly. In **carburizing,** a third industrial method, carbon is added to the steel from the surface inward. The carbon is obtained from one or a combination of the following: wood charcoal, animal charcoal, coke, beans or nuts, charred bone or leather, or Kasenit (trade name for a nonpoisonous coke compound).

The Case-Hardening Process

Carbon is added to steel by the pack, gas, or liquid-salt methods. The procedure for case hardening using the pack method of carburizing with Kasenit is as follows:

1. Put the project in a metal box or pot, with the Kasenit surrounding the project.

2. Place the receptacle in the furnace and heat it to about 1,650°F.

3. Leave the box in the furnace fifteen to sixty minutes. The mild steel will absorb carbon to a depth of as much as 0.015 inch.

4. Remove the box with its contents. Take out the project and quench it. Only the case will harden. The inner core will be relatively soft.

This procedure can be used on hammer heads, piston pins, and other items that must stand a good deal of shock and wear. Case hardening should never be done on products that must be sharpened.

Heat Treating in Industry

In industry, heat treating is a highly scientific process. It makes available steel with the best properties for each kind of product. The engineer, technician, and skilled metalworker must know a great deal about the principles of heat treating. The fundamental principles are the same as in the heat treating done in the school laboratory or shop. However, many special procedures and a great deal of specialized equipment are involved. Industry uses large continuous furnaces equipped to provide automatic control for temperature and time. The pieces to be heat-treated move on a conveyor to the furnace. They are held there for the right length of time, and are then plunged into quenching baths. Then the necessary tempering is also done.

Three interesting methods for hardening metal surfaces by heat treating are used in industry. They are: induction hardening, flame hardening, and laser hardening.

Induction Hardening

In **induction hardening,** the metal is put inside a coiled wire in which a current of low voltage and high amperage is flowing. A current is thereby induced on the surface of the metal, and it heats quickly. When the hardening temperature is reached, the current is turned off. The heated surface is rapidly quenched with a spray of water. The time required is only two seconds. Moreover, the area hardened can be kept very small. Only the surface of the metal is hardened. The core remains both tough and soft. The same equipment can be used for annealing, brazing, soldering, tempering, and other processes.

Flame Hardening

In **flame hardening,** the metal is heated with an oxyacetylene flame. Either the torch or the workpiece moves along slowly so that a thin surface layer of metal is heated. Cooling is done with a stream of water or by dropping the workpiece into a quenching tank, which hardens the surface quickly. With this process, only part of the workpiece is hardened. The remainder stays in an annealed condition. This process is particularly useful when only part of a large casting needs to be hardened. For example, on some metal lathes, the ways of the lathe bed are flame hardened.

Laser Hardening

In **laser hardening,** the metal is heated with a laser. This method is used, for example, on the area around the valve opening in a cylinder block.

The block is placed in a revolving table. A laser beam is focused on the valve opening as the table turns. After the hardening temperature is reached, the block is removed from the table and the surface is air-cooled.

KNOWLEDGE REVIEW

1. What is *heat treating?*
2. What are its three basic steps?
3. Name four good sources of heat for heat treating.
4. What liquids can be used to make a quenching bath?
5. What do you call the temperature at which steel has the best characteristics?
6. How can you tell when metal has reached the correct heat for tempering?
7. At what pace is metal cooled during annealing?
8. What is the condition of the core of a steel piece after case hardening?
9. List three industrial methods for hardening metal surfaces by heat treating.
10. Report on the methods used to heat-treat metals used in some familiar products.
11. Find out all you can about how case hardening is done in industry.

Unit 63

Hardness Testing

OBJECTIVES

After studying this unit, you should be able to:

- Describe hardness.
- Name four ways of describing hardness.
- Explain the Brinell method of hardness testing.
- Tell how the Rockwell test determines hardness.
- Name two Rockwell scales.
- Name the hardness test used mainly for research.
- Describe the principle on which the Shore-scleroscope test is based.

KEY TERMS

Brinell hardness test
hardness
indentation hardness

machinability
Rockwell hardness test
scratch hardness

Shore-scleroscope hardness test

Vickers hardness test
wear hardness

The term **hardness** suggests the solidity, firmness, and strength of a metal. It refers to strength, wearability, and resistance to erosion. Hard materials are difficult to cut or form into different shapes. Because hardness is not clearly defined, no single measure of hardness can be applied to all materials. Ways of describing hardness include:

- **Wear hardness**—resistance to abrasion, such as by sandblasting.
- **Scratch hardness**—measured in tests used by mineralogists. Heat treaters formerly pressed a new file across a corner of the metal. They could tell about how hard the metal was by the look of the scratch. You can estimate hardness in this way.
- **Machinability**—resistance to cutting or drilling, as determined by special tests.
- **Indentation hardness**—based on the fact that a hard object will dent a soft one. Indentation-hardness tests are the most widely used in testing most metals. They are nondestructive (although they do mar a finished surface), inexpensive, and easy to perform. The four tests most commonly used are the Brinell, the Rockwell, the Vickers, and the Shore-scleroscope. The test selected depends on the specific application and the condition and size of the metal.

Hardness tests are used to compare the hardness of similar metals under similar conditions. In all indentation-hardness tests, a specimen, usually flat, is placed on a rigid platform. An object called an *indenter* or *penetrator* is pressed into the specimen under load. Depending on the kind of test, a hardness reading is made from either the depth or the size of the dent made. The four basic hardness tests are described in this unit.

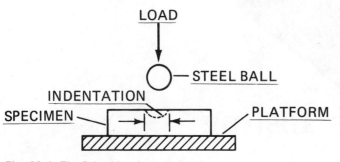

Fig. 63-1 The Brinell hardness test.

Brinell Hardness

The **Brinell hardness test** is a method used for testing specimens about 1/4 inch (6 mm) or more in thickness. The testing method is illustrated in Figure 63-1. The test consists of pressing a hardened steel ball into a metal specimen. It is customary to use a ball with a 10-millimeter (0.39-inch) diameter. The load depends on the material being tested. A load of 3,000 kilograms (6,614 pounds) is commonly used for medium-hard alloys. A load of 500 kilograms (1,102 pounds) is used for soft metals. The diameter of the impression made by the ball is measured with a microscope. It is then converted from a chart to a Brinell hardness number (BHN). The measured hardness is stated as 250 BHN or 100 BHN, for example. The narrower the dent, the higher the number and the harder the metal. This test is best for soft and medium-hard materials. The machine is shown in Figure 63-2.

Rockwell Hardness

In the **Rockwell hardness test,** the depth of penetration is used as a measure of hardness. It is fast and easy because the readings are shown directly on a machine. Smaller specimens can be tested better by this method than by the Brinell method.

Several kinds of indenters and loads can be used, depending on the specimen. The Rockwell B and C scales are most commonly used. The B scale uses a 1/16-inch (1.5-mm) steel-ball penetrator and a 100-kilogram (220-pound) load. The C scale uses a diamond-cone penetrator called a Brale, and a 150-kilogram (331-pound) load. The B scale is for testing materials of medium hardness. Its working range is from 0 to 100

Fig. 63-2 A Brinell hardness tester. (Courtesy of Tinius Olsen Testing Machines Co., Inc., Willow Grove, PA)

RB. If the ball penetrator is used to test material harder than about 100 RB, it might flatten. Also, the ball is not as sensitive as the cone to small differences in hardness (Figure 63-3).

The Rockwell superficial-hardness tester is a special-purpose machine developed for testing thin sheet metal. It operates on the same principle as the standard Rockwell machine. However, it uses lighter loads and a more sensitive measuring system.

Vickers Hardness

The **Vickers hardness test** is very precise. It is used mainly as a research tool. As shown in Figure 63-4, it is similar to the Brinell test. The penetrator is a square-based diamond pyramid. Loads may vary from 1 to 120 kilograms (2.2 to 265 pounds). The square impression made by the penetrator is measured with a microscope. The reading is taken from a reference table that gives the Vickers hardness number, such as 220 VHN or 220 DPH (diamond-pyramid hardness). Vickers test results are slightly higher than those from Brinell tests. Very thin sections of metal may be tested by using small loads.

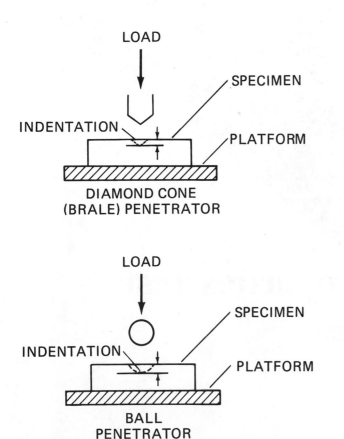

Fig. 63-3 The Rockwell B and C hardness tests.

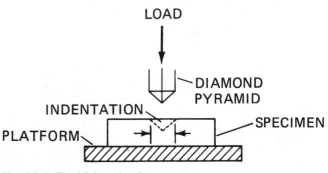

Fig. 63-4 The Vickers hardness test.

The Shore scleroscope is portable and can be used to test pieces too big for other testing machines. Another advantage of the Shore scleroscope is that it can be used without damaging finished surfaces. The main disadvantage of this machine is its relative inaccuracy. Samples without rigid backing and oddly shaped or hollow workpieces may give incorrect readings.

KNOWLEDGE REVIEW

1. What is the principle behind indentation-hardness tests?

2. In the Brinell test, what does it mean if the steel ball makes a wider dent in one specimen than in another?

3. In the Rockwell test, what is used as the measure of hardness?

4. What is the main use of the Vickers test?

5. In the Shore-scleroscope test, what is used as the measure of hardness?

6. Write a research report on industrial methods of testing hardness.

Shore-Scleroscope Hardness

The **Shore-scleroscope hardness test** is based on the following principle: If you were to place a mattress on the floor and drop two rubber balls from the same height, one on the mattress and one on the floor, the one dropped on the floor would bounce higher. This is because the floor is the harder of the two surfaces.

In the Shore-scleroscope hardness test, a diamond-pointed hammer is dropped onto the test piece, and the rebound, or bounce, is checked on a scale. The higher the rebound, the higher the number on the scale and the harder the metal.

Unit 64

Other Metal Properties Tests

OBJECTIVES

After studying this unit, you should be able to:

- Describe what is meant by *strained* metal.
- Name three ways metal can be changed in shape.
- Explain the movement of metal to form a new shape.
- Explain the tensile test for strength.
- Write the formula for stress in metric measurements.
- Explain ductility.
- Name two ductility tests.
- Describe compression strength.
- Explain shear strength.
- Write the formula for shear strength in customary measurements.
- Explain impact strength.
- Name two common tests for impact strength.
- Describe fatigue strength.

KEY TERMS

compression strength	elastic limit	plastic deformation	shear strength
deformed	elongation	plastic flow	strained
ductility	fatigue strength	plastic range	tensile strength
elastic deformation	impact strength		

In metalworking, when force is used to change the shape of a metal, the metal workpiece is said to have been **strained,** or **deformed.** Metals are usually very different after being strained. Strained metal, however, can be restored to an unstrained condition by heating to a high temperature.

Metal can change in shape in three ways when force is applied. First, the metal can change shape while the force is being applied but return to its original shape when the force is removed. This temporary change is called **elastic deformation.** It occurs, for example, when a coil spring is stretched. The spring returns to its original length as soon as the force is removed. Second, a metal can change shape when force is applied and remain in the new shape when the force is removed. This permanent change is called **plastic deformation.** It can occur, for example, when a coil spring is overloaded. Plastic deformation always takes more force than elastic deformation. The third way a metal can change shape is simply by *breaking.* More force is needed for breaking than for either elastic or plastic deformation.

Metal-forming operations change the shape of metal permanently, so they are classified as plastic deformation. The movement of the metal to form a permanent new shape is called **plastic flow.** The forces that cause plastic flow are greater than the forces that cause elastic deformation, but less than the force that causes breaking. The range between elastic deformation and the breaking point is called the **plastic range** of a material. All forming is done in the plastic range. In order to make sure that a metal can be formed, its properties must be tested. You have already learned about hardness in metal and how to test for it. Now you will learn about some other important properties and how to test for them.

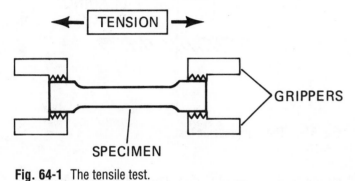

Fig. 64-1 The tensile test.

Tensile Strength

Tensile strength is the maximum *stress,* or force, that a material can withstand before it fractures. Testing for tensile strength is one of the best ways of evaluating metals. The test involves pulling a metal specimen; that is, placing it under tension (Figure 64-1). The specimen is mounted in a machine that applies a slowly increasing load (Figure 64-2). This load is measured in customary measurements as pounds of force and in metric measurements as newtons of force. It is recorded throughout the test. Stress is computed from the force and the original cross-sectional area of the specimen. Using customary measurements, stress in pounds per square inch is derived in the following way:

$$\text{Formula: stress (psi)} = \frac{\text{force (pounds)}}{\text{area (square inches)}}$$

For metric measurements, stress in pascals is derived using:

$$\text{Formula: stress (Pa)} = \frac{\text{force (newtons)}}{\text{area (square meters)}}$$

As the test progresses and the load increases, the specimen stretches until it breaks. This stretching is

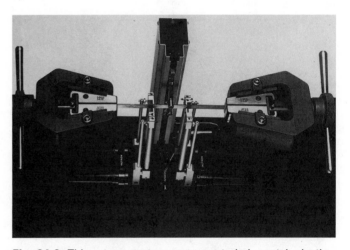

Fig. 64-2 This extensometer measures strain in metal, plastic, and rubber test samples. (Courtesy of Instron Corporation)

called *strain.* Strain is recorded continuously throughout the test. Strain expressed as a percentage is called **elongation.** Elongation is the percentage of increase in length of the test specimen.

If a metal part is stretched until it becomes permanently deformed, it can be just as useless as if it were broken. The greatest stress a metal can stand without permanent deformation is called the **elastic limit.**

Tensile testing has become fairly well standardized. The exact size and shape of the specimen depends on the product. Product examples may be bar, sheet, strip, plate, and tube stock. When tubes are tested, the actual tube is used if it is small enough. Specimens for testing sheet, strip, and plate also have standard dimensions.

The elongation that a metal specimen undergoes during a tensile test is a measure of its ductility. **Ductility** is the ability of a metal to be deformed without breaking. In other words, it is the ability of a metal to be formed by cold-working methods such as bending, deep drawing, spinning, or cold heading. A combination of strength and ductility in a metal is an excellent indication of its suitability for engineering uses (Figure 64-3).

Fig. 64-3 Because of its ductility, brass was used to manufacture this trumpet.

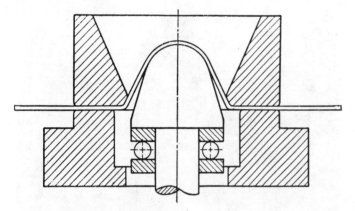

Fig. 64-4 The ductility test.

In the *Olsen ductility test,* a piece of sheet metal is deformed by a standard steel punch with a rounded end. The depth of the impression required to fracture the metal is measured in thousandths of an inch (0.001 inch). In the *Erichsen test,* the depth of impression at fracture is measured in thousandths of a millimeter (0.001 mm). The sheet is held between two ring-shaped clamping dies while a dome-shaped punch is forced against it until fracture occurs (Figure 64-4). The height of the dome is measured by a dial gage.

Compression Strength

Compression strength is the maximum compressive force a material can withstand before it fractures. It is measured in pounds per square inch (psi) in the customary system and in pascals (Pa) in the metric system.

A compressive test is run on the same machine and in the same general way as a tensile test. However, the force is applied in the opposite direction. In other words, the specimen is compressed instead of being pulled apart (Figure 64-5). However, the properties tested are similar. The major exception is that for ductile materials such as steels, a final compressive

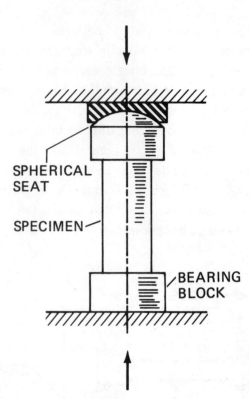

Fig. 64-5 The compression test.

strength cannot be obtained. This is because the sample keeps compressing until it is flat. Compression tests are usually run on materials such as cast iron or other similar metals.

Compression properties are used in the design of machine parts, structures, and aircraft parts. In many of these applications, selection of the material depends on both strength and stiffness.

Shear Strength

Shear strength is the maximum load a material can withstand when that load is applied vertically to the material's surface. It is measured in pounds per square inch in the customary system and in pascals in the metric system. An example of a material under shear stress is a sheet of paper being cut by scissors. The paper is fractured, or cut, because the blades of the scissors apply a stress that is equal to the shear strength of the paper. Bolts and rivets are often exposed to stresses that are similar to the cutting action of scissors. Material used for such parts must have good shear strength.

There are two kinds of shear testing: *double* and *single.* Double-shear strength is measured by mounting a bar specimen in a jig, as shown in Figure 64-6A. It is placed on the table of a tensile-testing machine. A rectangular cutter on the movable head of the machine contacts the specimen directly over the groove in the block. To reduce bending of the specimen, the clearance

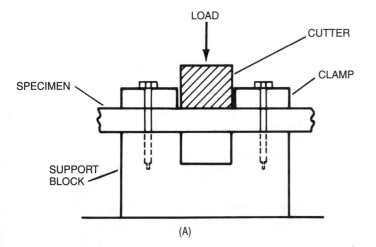

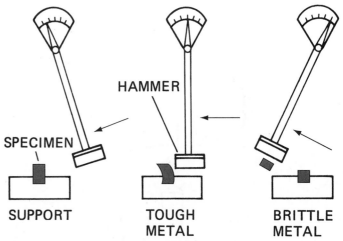

Fig. 64-7 An impact test.

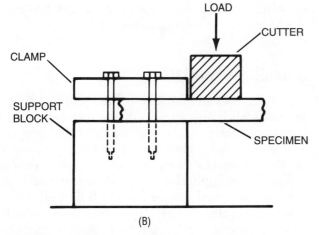

Fig. 64-6 (A) Double-shear and (B) single-shear tests.

between the cutter and groove is small, usually about 0.005 inch. For accurate results, the edges of both the cutter and the supporting block must be sharp. To find a specimen's double-shear strength, divide the load required to fracture the specimen by twice the specimen's cross-sectional area.

Testing for single-shear strength is the same as for double-shear strength except that only one end of the specimen is supported (Figure 64-6B).

Impact Strength

Impact strength is the ability of a piece of metal to withstand a hard blow or sudden shock without breaking. This strength is necessary for many applications, such as drilling equipment and automotive and aerospace engines, where shocks occur.

A tensile test is a good measure of metal toughness. However, in this test, the load is increased slowly, not suddenly as in a blow. Thus, it cannot measure impact strength. Special impact tests have been developed

to measure the toughness of materials under shock loads. Two of the most common are the *Izod test* and the *Charpy test.* Both are based on the same principle: A pendulum strikes a sharp blow on a specimen, bending or breaking it (Figure 64-7).

Results of the tests are reported as impact strength, measured in foot-pounds (ft-lb) in the customary system and in newton-meters (N m) in the metric system. This strength is computed from the weight of the pendulum and the difference between the rise of the free-swinging pendulum and the rise of the pendulum that has struck a specimen.

Both Izod and Charpy specimens are notched. The notch is included because a plain flat bar of a ductile material will not generally break with a single blow in these tests. The notches are always carefully machined to standard specifications.

The specimen, method of support, and energy of the pendulum are all quite different for the Izod and Charpy tests. Thus, the test used must always be identified, and results from the two tests cannot be compared. In the Izod test, the specimen is clamped in a vertical position as a cantilever beam. The Charpy specimen is a simple beam supported at the ends (Figure 64-8). These tests are most useful for comparing materials of similar strengths and for evaluating the effects of heat treatment, processing variables, or temperature on the same type of material.

Fatigue Strength

Fatigue strength is the ability of metal to withstand changing loads without breaking. Such loads occur, for example, in the buffeting of airplane wings and in the pushing and pulling of the connecting rods in piston engines. These push-pull applications of force and on-off applications of pressure are called *cyclic* loads.

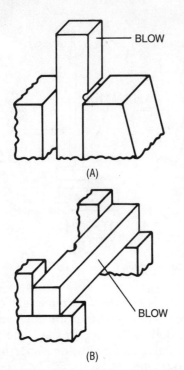

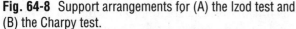

Fig. 64-8 Support arrangements for (A) the Izod test and (B) the Charpy test.

During the useful lives of machines that move rapidly and parts that are subjected to severe vibrations, load changes may occur billions of times. Metal fatigue is thought by many to be the most common cause of failure. In fatigue tests, failure is always a brittle fracture. *Fatigue life* is the number of cycles a material can withstand at a specified stress without failure.

There are a variety of fatigue tests, depending on the kind of load applied and the shape of the specimen. One of the most common is the *cantilever test*. This utilizes the Krouse testing machine to test sheet, strip, and thin plate. One end of the specimen is held rigidly while the other is free to move. A rotating cam moves the free end of the specimen up and down until fatigue occurs and the specimen breaks (Figure 64-9). Stress and number of cycles are recorded automatically.

Other Material Testing Methods

The tests described above are directed at measuring material properties. A number of other nondestructive tests are also used to determine the quality of metal parts. *Ultrasonic* inspection uses high frequency sound waves to detect internal flaws in a workpiece. The sound waves, which reflect back from any defects, are converted to an electronic signal which appears on a screen. *Radiographic* inspection involves the use of X-rays to find defects in metal parts. In this process, the workpiece is exposed to the radiation, and the resulting image appears on a screen or film. *Fluorescent pene-*

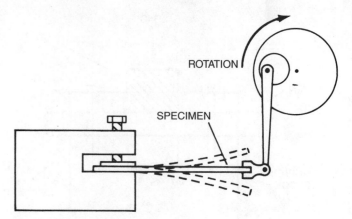

Fig. 64-9 The cantilever fatigue test.

trating dyes can be brushed on a metal part. The part is then dried and dusted with a powder that draws the dye out of any cracks. When exposed to ultraviolet light, the cracks appear as bright fluorescent lines.

Selecting Materials

Industrial and engineering designers select product materials from information found on computer data bases or other published lists of material properties. The information found in this and the previous unit provides general descriptions of material characteristics and their applications. In industry, the basic product material specifications must be followed by discussions with materials specialists and suppliers, to learn about properties, characteristics, process technology, availability, and costs. Materials selection is very important, whether one is making a metal project in the school, or designing a product in industry.

KNOWLEDGE REVIEW

1. What is the difference between elastic deformation and plastic deformation?

2. In tensile testing, what is applied to the metal specimen?

3. In a compressive test, in what direction is force applied to the metal specimen?

4. What is shear strength?

5. Name the two tests for impact strength.

6. What kind of loads can a metal with fatigue strength withstand?

7. Write a research report on accidents or disasters caused by metal fatigue or other metal failure.

Unit 65

Introduction to Welding

OBJECTIVES

After studying this unit, you should be able to:

- Define *welding*.
- Name two kinds of fusion welding.
- Identify another name for oxygen cutting.
- Describe arc welding.
- Describe the use of electrodes.
- Describe forge welding.
- Explain the basic steps of oxyfuel gas cutting.
- List the safety rules to follow when welding.
- Tell how many people are employed as welders.
- List six kinds of welding occupations.

KEY TERMS

arc welding	gas welding	resistance welding	welding analyst
electrode	inspector	welder	welding engineer
flame cutting	metallurgist	welder-fitter	welding operator
forge welding	oxyfuel gas cutting	welding	welding technician
fusion welding	oxygen cutting		

Welding is fastening metals together using intense heat and sometimes pressure (Figure 65-1). It is a very economical process and is used in both manufacturing and repair work. When the operation involves heat alone, it is called **fusion welding.** Two kinds of fusion welding are gas welding and arc welding. Using heat and pressure is called **resistance welding.** Closely related to welding is **oxygen cutting,** also called **flame cutting.** In this process, operators use torches to cut metal to shape and size. There are also other kinds of welding. Gas-tungsten arc, submerged-arc, and many others were designed to meet special technological needs. These methods are described in unit 68.

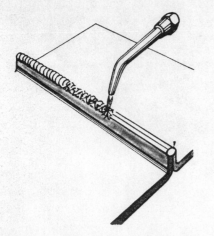

Fig. 65-1 Welding is joining metal parts with heat and sometimes pressure. This illustration shows gas welding.

Fig. 65-3 This table structure was assembled with gas-welded joints.

Fig. 65-2 Practicing the gas welding process. Note the torch and the filler rod. (Courtesy of The Linde Air Products Company)

Fig. 65-4 Welding boat sections together to provide a tough, durable joint. (Courtesy of Aluminum Company of America [ALCOA])

Fusion Welding

Gas welding, one kind of fusion welding, uses a hot flame with a temperature of 6,300°F (3,482°C) that is produced by burning oxygen and acetylene in a welding torch (Figure 65-2). The welder heats the metal part by holding the flame near the metal until a molten puddle forms. Then the welder applies a welding rod to build up the weld. Gas welders must know how to choose the right torch tip and welding rod. They must also know how to adjust regulators to get the right flame and how to use the torch properly to get a strong, clean weld as shown in Figure 65-3.

Arc welding, another kind of fusion welding, uses electric power from either an electric generator or a transformer. An **electrode,** which is similar to the welding rod used in gas welding, provides the metal filler for the weld. One cable from the power source is connected to the metal welding table or to the object itself. The other is connected to the holder that holds the electrode. The welder first strikes an arc by touching the metal part to be welded with the electrode. Then the welder withdraws the electrode a short distance from the metal surface, creating an electric arc. The arc produces a heat of about 9,000°F (4,982°C). This heat melts the metals and also the electrode. The welder moves the electrode along and feeds it into the joint. When the electrode is used up, a new one is slipped into the holder and welding continues. An arc welder must know how to adjust the electric power, how to choose the right electrode, and how to do the welding (Figure 65-4).

Resistance Welding

In resistance welding, which uses both heat and pressure, the heat is generated by resistance to the flow of electric current. In manufacturing, resistance welding is done with a machine that brings metal parts together under heat and pressure. The operator merely sets the controls to the correct electric current and pressure for the job and then feeds the work into the machine. Some of the most common kinds of resistance-welding machines are *spot welders, seam welders,* and *portable spot-welding guns.*

Another pressure-welding process is **forge welding.** Here, a blacksmith's hammer is used to apply enough force to the heated metal parts to join them together.

Oxygen Cutting

In oxygen cutting, a hot flame is used not to fuse metal, but to cut. One kind of oxygen cutting is **oxyfuel gas cutting.** In this process, a cutter directs a flame of oxygen and fuel gas on the workpiece until the metal reaches a red heat. Then the cutter increases the amount of oxygen in order to burn or cut the metal. The cutter guides the torch by hand to cut along a marked line. Cutting torches can also be mounted in a machine that follows the layout line.

Safety

It is important to wear proper equipment when welding. Every welder should have eye protection and some safety clothing. Never weld without eye protection. This includes special dark glasses. Follow the safety rules given in each unit on welding.

Occupations in Welding

There are well over 480,000 welders and oxygen cutters. This number is expected to increase by over thirty percent in the next ten years. Welders are employed in automobile and aircraft plants, construction industries, metalworking shops, and all kinds of maintenance and repair work. Courses in welding are offered in many public schools. Welding can also be learned on the job. If a weld must meet a certain standard, the welder must first pass a qualifying examination.

There are many different kinds of welding jobs. A **welding operator** operates special welding machines. A **welder** does arc, gas, and special welding. A **welder-fitter** sets up welding work for others to do. Blueprint reading is important in this job. **Welding technicians** work with engineers. They do experimental welding work, run laboratory tests, and build special equipment. **Welding analysts** and **inspectors** examine welded pieces to see if they are correct. If there are flaws, they must find out why. **Welding engineers** and **metallurgists** do experimental and design work on new welding methods and equipment.

KNOWLEDGE REVIEW

1. What is *welding?*
2. What kind of welding uses heat alone? What kind uses heat and pressure?
3. What are two kinds of fusion welding?
4. What is the name of the process in which a hot flame is used to cut metal?
5. Name some industries that employ welders.
6. Make a report on one of the welding processes not discussed in this book.
7. Write a report on occupations in welding.

Unit 66

Gas Welding, Cutting, and Brazing

OBJECTIVES

After studying this unit, you should be able to:

- Identify another name for gas welding.
- Explain what the letters OAW mean.
- Describe the equipment needed for gas welding.
- List the safety rules to follow when gas welding.
- Describe the kinds of welding rods and fluxes used when gas welding.
- Describe how to light and adjust the flame for gas welding.
- Describe how to make an edge weld without using a rod.
- Describe how to run a bead using filler rod.
- Describe how to make a butt weld without using a rod.
- Describe how to do oxyfuel gas cutting.
- Explain the purpose of brazing.
- List the major steps in brazing.

KEY TERMS

brazing	gas welding	oxyfuel gas cutting	torch soldering
bronze welding	oxyacetylene welding		

Gas welding, or **oxyacetylene welding** (OAW), uses an extremely hot flame produced by burning oxygen and acetylene to melt metal. Brazing is a closely related process. Gas welding is the easiest welding operation to learn (Figure 66-1). This process is used in almost every school metalworking shop. You will be able to do a fairly good job of welding or brazing low-carbon steels if you master the basic procedure for gas welding.

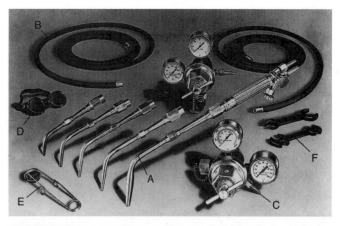

Fig. 66-3 An oxyacetylene welding outfit: (A) a torch with five tips, (B) a hose, (C) two regulators, (D) goggles, (E) a sparklighter, and (F) wrenches. (Courtesy of The Linde Air Products Company)

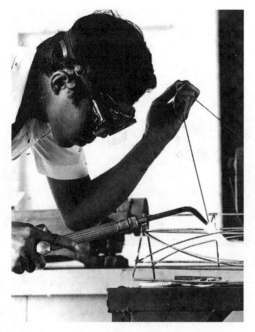

Fig. 66-1 Gas welding is a common joining process learned in the metal shop.

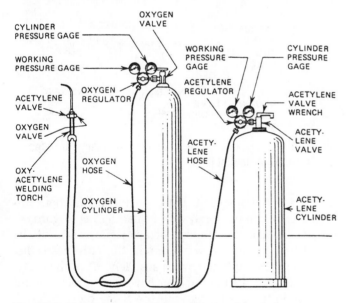

Fig. 66-2 The welding outfit assembled and ready for use.

Equipment

Gas-welding equipment consists of a cylinder of *acetylene,* a cylinder of *oxygen,* two *regulators,* two lengths of *hose* with *fittings,* and a *welding torch* (Figure 66-2). The acetylene hose is red. The oxygen hose is green or black. The welding torch is made of copper to give off heat rapidly. The tips come in many sizes. Other items needed are a *welding table* with a top covered with firebrick; a *sparklighter* to light the torch; an *apparatus wrench* to fit the various connections on the

regulators, cylinders, and torches; and suitable *goggles* to protect your eyes (Figure 66-3).

Safety Rules

Always obey the following safety rules when welding.

- Never use any oil or grease on or around a welding outfit. Grease and oil can cause a fire when they come in contact with oxygen.
- Always use a sparklighter to light the torch. Never use matches.
- Never try to light the torch with both valves open.
- Always wear goggles when welding.
- Never let anyone watch while you are welding.
- Always hang the torches in the right place. Never lay them on the welding table.
- Always turn off the torch when you have finished a weld.
- Test for leaks by smearing soapsuds around the fittings and hose connections. Bubbles indicate leaks.

In welding, never guess, experiment, or assume anything. Make sure you know how to weld safely before you start.

Welding Rods and Fluxes

Two kinds of welding rods are used in beginning welding: mild-steel and brazing. A *mild-steel welding rod* usually has a copper coating. This coating prevents rusting. In general, use a smaller rod for light welding and a larger rod for heavy welding (Table 66-1). For example, using customary measurements, on 1/16-inch sheet steel you would use a 1/16-inch steel welding rod.

Table 66-1 Recommended rod sizes for oxyacetylene welding

Thickness of metal	Rod diameter in inches
18 gage and lighter	1/16 1/16
18 to 16 gage	1/16 to 3/32
16 to 10 gage	1/32 to 1/8
10 gage to 3/16 in.	1/8 to 5/32
1/4 in. and heavier	3/16 to 1/4

For 1/8-inch sheet steel, use a 3/32-inch rod. If the steel is 1/4 inch thick or more, a rod about 1/8 to 3/36 inch in diameter is used.

A *brazing,* or *bronze, rod* is used for brazing. It has a yellow-gold color. Brazing requires a *flux,* a chemical compound that keeps scale, or oxide, from forming. Borax is good, but a commercial brazing flux is better.

Adjusting the Welding Outfit and Lighting the Torch

If your welding outfit is not already assembled, follow the assembling instructions supplied with it. To prepare for welding, do the following:

1. Choose the correct size of tip for the welding to be done. Generally, a small tip is used for thin metal, a medium tip for medium thickness, and a larger tip for heavy welding. For example, tips with 1/32-inch openings are best for the thinnest metal. The booklet that comes with the welding outfit tells you what size tip to use. Use the wrench that comes with the welding outfit to remove the new tip to check its threads. Also make sure that the hole at its end is clean and round. Then fasten the tip in place.

2. Put on a pair of goggles, leaving them over your forehead.

3. Make sure that the crossbars on both pressure regulators are screwed out until they are loose.

4. Carefully turn the cylinder valve on the top of the oxygen tank until it is wide open. Always open the valve gradually. With the tank wrench, open the valve on the acetylene about one-half to one and one-half turns—never any more. The cylinder pressure gages on the regulators show the amount of pressure in both tanks. Never stand directly in front of the regulators.

5. Turn in the regulator crossbar on both the oxygen and acetylene until about five pounds of pressure show on the working pressure gages of both. The exact pressure needed can be found in the booklet supplied with the welding outfit.

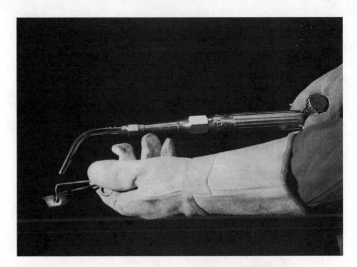

Fig. 66-4 Using a sparklighter to light a torch. Never light it with a match or another student's lighted torch. Always wear your goggles when gas welding. (Courtesy of The Linde Air Products Company)

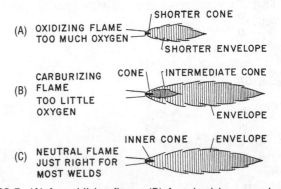

Fig. 66-5 (A) An oxidizing flame. (B) A carburizing, or reducing, flame. (C) A neutral flame.

6. Open the acetylene valve on the torch about 1/4 inch. Light the torch with a sparklighter (Figure 66-4). Adjust the amount of acetylene gas until the flame just jumps away from the tip. This means that there is slightly too much acetylene. Turn the acetylene back a little bit. You have the correct amount when the gas burns with the flame blowing away from the tip. Then turn on the oxygen a little at a time until you get a *neutral flame,* which is the best flame for most welding. This flame has equal amounts of oxygen and acetylene. It has a full envelope, or outer flame, and a sharp inner cone. If there is too much acetylene, the flame has an intermediate cone between the envelope and inner cone. This is called a *carburizing,* or a *reducing,* flame. If there is too much oxygen, the envelope and the inner cone are greatly shortened, and the torch gives off a hissing sound. This is called an *oxidizing* flame (Figure 66-5).

7. When turning off the torch, always close the acetylene valve first and then the oxygen valve. This

Fig. 66-6 A practice edge weld made without a rod. Note the tent form in which the weld is made.

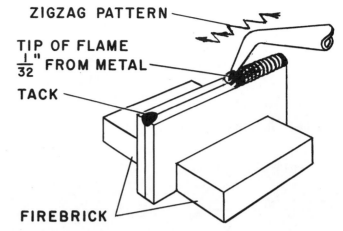

Fig. 66-7 Edge welding without a rod.

prevents a slight backfire, or pop. It also keeps the tip of the torch free of soot. When you have finished welding for the day, first turn off the torch, then close the valves on the acetylene and oxygen cylinders. Next, open the acetylene torch valve to drain the hose. Then, release the crossbar of the acetylene regulator. Last, close the acetylene torch valve. Follow the same steps in the same order for the oxygen.

Making Practice Welds

Before you weld a project, you should make some practice welds to learn to operate the equipment.

To make an *edge weld without rod:*

1. Use two pieces of 1/16-inch or 16-gage scrap sheet steel about one inch wide and three to four inches long.

2. Place the two pieces of scrap steel together on edge or in the form of a little tent (Figure 66-6). Put goggles over your eyes.

3. Hold the torch in your right hand, with the inner cone about 1/32 inch from the metal. Zigzag it back and forth to tack the pieces of metal together at one end.

4. Start welding at the other end, working along the edge with a zigzag torch movement (Figure 66-7). If you burn through the metal, change to a smaller tip or hold the torch at a slightly lower angle. The greatest amount of heat is applied to the surface when the flame is at right angles to it.

5. Move the flame from one piece of metal to the other, forming a puddle. Then move the torch along from one side to the other as you move it for-

ward. You should practice this a few times until you can join the two pieces together with a weld that has smooth, uniform ripples.

To run a *bead without rod:*

1. Choose a piece of metal about 1/8 × 2 × 4 inches. Place this on the firebrick.

2. Use a tip of about No. 2 size. Adjust for a neutral flame. Remember to place the goggles over your eyes.

3. Hold the tip at an angle of about 45 degrees to the metal, with the tip of the inner cone about 1/32 to 1/16 inch away from the surface.

4. Form a puddle. Then move the torch slowly forward in a weaving or semicircular motion.

5. Make sure that the forward motion and the weaving motion are even to make the bead smooth and regular. If you hold the tip of the torch too long in one place, it will burn a hole through the metal. If you hold the torch too close to the metal, the flame may go out.

To make a *butt weld with rod:*

1. Cut two pieces of scrap steel about 1/8 × 3/4 × 5 inches.

2. Turn a 1/16-inch flange on one edge of each piece.

3. Place the pieces with the flange edges together and facing up, about 1/32 inch apart on the welding table.

4. Hold the torch with the point of the inner flame about 1/8 inch above the flange at one end.

5. Move the torch back and forth in a slight arc until the metal melts and tacks together.

6. Start at the other end to melt the flange to form a bead. Weave the torch back and forth a little as you work from one end to the other. Keep the molten puddle of metal running to form a smooth, even bead.

To run a *bead with added filler:*

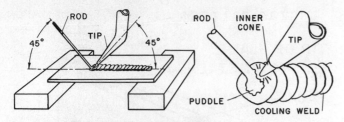

Fig. 66-8 Running a bead with added filler.

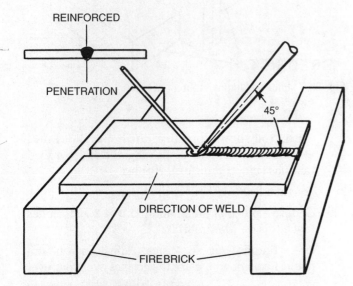

Fig. 66-9 Welding a flat steel plate.

1. Choose a piece of scrap stock about 1/8 × 2 × 4 inches.

2. Select a 3/32-inch mild-steel welding rod.

3. Start forming a puddle, and continue heating the metal in one place without melting a hole. Use a semicircular or weaving motion. If a hole melts through the metal, you have held the flame in one place too long.

4. After the puddle starts to form, add the welding rod to the middle of it (Figure 66-8). The rod should add about 25 percent more material. Remember to keep the rod in the puddle, not above it, and to direct the flame on the metal, not on the end of the rod. If the tip of the flame is too close to the metal, small blowholes form and the torch may backfire.

5. If you apply the heat mostly to the rod instead of to the metal, the weld will not have good penetration. In this case, the rod merely melts and drops onto the metal, rather than fusing with it.

To make a *butt weld with rod:*

1. Cut two pieces of 1/8-inch scrap stock and place them on the table 1/16 inch apart on one end and 1/8 inch apart on the other end.

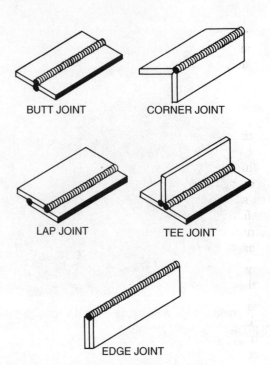

Fig. 66-10 Common kinds of joints or welds.

2. On metal thicker than 1/8 inch, one edge must be ground to form a V. Tack the two ends together.

3. Select a piece of 1/8-inch mild-steel welding rod.

4. Form a puddle on one end. Then begin melting the rod into the puddle to build up the weld about 25 percent.

5. Weave the torch back and forth, moving the rod with just the opposite movement to form the bead (Figure 66-9).

Five basic kinds of welded joints are shown in Figure 66-10. Practice making them with and without rods. Most metal projects you make will require only a small weld to join the parts.

Oxyfuel Gas Cutting

Oxyfuel gas cutting of ferrous materials, especially mild steel, is done with oxygen and the fuel gas acetylene. First, a flame of oxygen and acetylene is used to heat the steel to a bright red color. Then, a stream of oxygen is directed onto the heated area. Where the oxygen touches the heated steel, rapid oxidation causes burning, which cuts the steel apart.

The cutting torch has valves to control the flow of oxygen and acetylene during the heating stage. It has another lever to control the flow of oxygen during the burning stage. Different-sized tips are available for cutting different thicknesses of metal.

Much oxyfuel gas cutting is done with a device that holds the cutting torch and guides it along the work at a steady speed. Large cutting machines have multiple torches that work in unison, cutting several pieces of the same shape at once.

Brazing

Brazing is used to repair metal and to make joints almost as strong as welded joints, without actually melting the metal. It is sometimes also called **bronze welding** or **torch soldering.** The process is somewhat like hard soldering (see unit 43). Rods and fluxes are available for doing low-temperature welding or brazing on aluminum, copper, and other metals. Following is a description of the brazing process:

1. Know the metal you are brazing. Choose the right brazing rod for the purpose.

2. Cover the work area with *firebricks.* Ordinary bricks contain moisture and can explode under high heat. Protect the surface under the firebricks by covering it with heat-resistant material.

3. Thoroughly clean the metals to be joined. Depending on what must be removed, use a wire brush, sandpaper, and/or *noncombustible* solvent.

4. Before starting work, put on gloves and dark safety glasses. Have handy a fire extinguisher in case of fire and a pail of water to cool tools. Also have on hand tongs or long-nose pliers for lifting or turning the work.

5. Heat the metals to be joined until they start to get red. Since heavy metal takes longer to get hot, apply heat first to heavier pieces. The torch tip should be about ½ inch from the metal surface.

6. If an uncoated brazing rod is being used, dip it in flux.

7. Put the end of the brazing rod in the torch flame. When the joint and rod are hot enough, metal from the rod will flow easily onto the joint. Move the torch flame back and forth over the joint. Repeat these steps as often as necessary. Remember never to melt the rod directly with the torch. Also remember that the metal will weaken if you overheat it. The finished weld should be bright and clean.

8. After the work has lost much of its heat, it can be cooled in a pail of water. However, it is better to let the work cool naturally.

KNOWLEDGE REVIEW

1. Name the basic equipment needed for gas welding.

2. What should you always wear when welding?

3. What size welding torch tip is used for welding thin metal? For heavy welding?

4. Which is the best flame for most welding?

5. What should you do if you burn through the metal when welding?

6. What happens if you apply heat mostly to the welding rod instead of to the metal?

7. List five basic kinds of welded joints.

8. What are the two stages in oxyfuel gas cutting?

9. What does welding do to the metal that brazing does not?

10. Make a sample display of the common kinds of welds.

Unit 67

Arc Welding

OBJECTIVES

After studying this unit, you should be able to:

- Identify another name for arc welding.
- Describe the equipment and accessories needed for arc welding.
- Name two kinds of welding machines.
- Describe electrodes.
- Explain two kinds of polarity.
- Name the organization that sets standards for electrodes.
- Describe five types of welded joints.
- Explain the steps in practicing arc welding.

KEY TERMS

arc welding	electrode	negative polarity	shielded-metal arc welding
butt weld	fillet weld	positive polarity	tee weld
corner weld	lap weld	rectifier	transformer welder
edge weld	motor-generator welder		

Arc welding, also called **shielded-metal arc welding** (SMAW), is joining metal by means of heat from an electric arc. The pieces to be welded are placed in position, and the intense heat of the electric arc applied to the joint melts the metal. At the same time, more metal is added to the joint and mixed with the melted base metal. When all this metal cools, it becomes one solid piece.

Equipment and Accessories

There are several basic kinds of arc welders. **Transformer welders** provide alternating current (AC) for welding. The most popular size for school shops and farms has a 225-ampere (A) capacity (Figure 67-1). Manufacturing plants use transformers with larger outputs, from 200 to 500 A. A transformer reduces the voltage of

ordinary current from the power line and increases its amperage so that high heats can be created. The transformer welder is the most economical type of welder. It is also easy to maintain. **Motor-generator welders** provide direct current (DC) for welding. Transformers can also provide direct current when equipped with a **rectifier.** Direct-current welders can do certain jobs alternating-current welders cannot do. All welding ma-

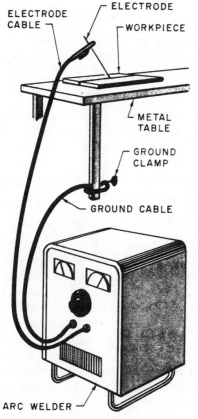

Fig. 67-1 A student learning to arc weld. (Courtesy of The Lincoln Electric Company)

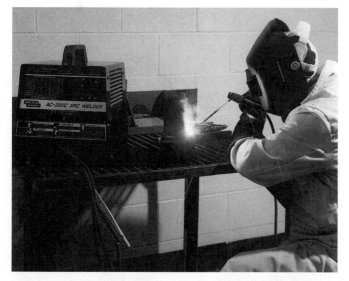

Fig. 67-2 A typical arc welding setup.

chines have devices for changing the amount of heat to weld different thicknesses of metal.

Two *cables,* or *leads,* carry the current from the welder to the work and back to the welder (Figure 67-2). One, the *electrode cable,* is connected to the *electrode holder.* The electrode holder grips the electrode. The

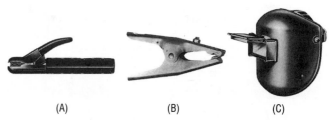

Fig. 67-3 Arc welding equipment: electrode holder (A), ground clamp (B), and helmet or shield (C). (Courtesy of The Lincoln Electric Company)

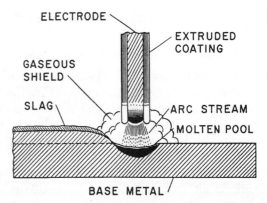

Fig. 67-4 Base metal and the metal from the electrode are fused together by the intense heat of the electric arc. The flux floats impurities to the top to form a slag.

other cable, the *ground cable,* is connected to a *ground clamp.* This is attached to the welding table or to the workpiece. An *all-metal welding table* is needed for practice welding and for small jobs that can be placed directly on the table. A *shield* protects the face and eyes from the rays of the electric arc and from heat and molten metal (Figure 67-3). *Tongs* or *pliers* are used for handling hot metals. A *chipping hammer* is used to remove slag from the weld bead. A *wire scratch-brush* is needed to clean the bead after chipping.

Electrodes

Electrodes are devices used to conduct and emit an electric charge. Those used in arc welding are made of metal wire or rod covered with a hard chemical-flux coating. Their size is determined by the diameter of the core wire. The chemical-flux coating serves several purposes:

1. Part of the flux burns up with the intense heat. This forms a blanket of gas that shields the molten metal from the surrounding air. This keeps the oxygen and nitrogen in the air from forming impurities in the weld.

2. Part of the flux melts, mixes with the weld metal, and floats impurities to the top to form a slag. This slag covers the bead as a crust and protects it from the air (Figure 67-4).

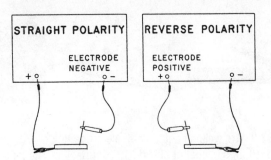

Fig. 67-5 Straight polarity and reverse polarity in welding.

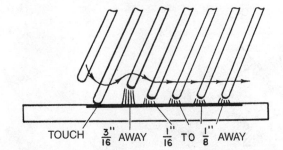

Fig. 67-6 Strike the arc as you would a match.

3. It slows the rate at which the metal hardens in cooling.

4. It affects the way the arc acts during welding.

For DC welding, you must attach the electrode cable to the correct output terminal, or pole. When the electrode cable is connected to the negative terminal of the DC welder, there is *straight,* or **negative, polarity.** When the electrode cable is connected to the positive terminal, there is *reverse,* or **positive, polarity** (Figure 67-5).

The American Welding Society has established classifications for standardizing and identifying electrodes. There is also a uniform color code. For example, an electrode that is satisfactory for general welding of mild steel is AWS No. E6013. The *E* means that it is used for the electric arc. The *60* shows the minimum tensile strength (60,000 pounds per square inch). The *13* indicates that the electrode can be used in all positions for either AC or DC welders. It is also identified by a brown secondary spot. This electrode will produce a soft arc and rather light penetration. The penetration is deep enough, however, for most work.

In purchasing electrodes, you must consider the following: (1) size of electrode needed—diameter of core wire; (2) kind of welder on which it is to be used—AC or DC; and (3) the base metal to be welded—steel, cast iron, nonferrous, and so forth.

Safety Rules

Always obey the following safety rules when arc welding.

- Dress correctly. Wear gloves and work clothing that protect your skin from heat, spatter, and arc rays.
- Wear a shield with the right filter glass. Keep it in the up position when you are not welding.
- Make sure that the people around you do not watch the arc.
- Keep away all flammable materials, such as gas and oily rags.
- See that the floor around you is clean.

- Wear safety goggles when you do not have your shield in place, especially for chipping, grinding, and similar operations.
- Always hold your hand slightly above a piece of metal before touching it to see if it is hot.
- Make sure that the welding area is well ventilated.

Procedure for Practice Welding

Before welding a project, make some practice welds to learn how to operate the equipment.

1. Choose a piece of scrap steel about 3/16 inch thick. Make sure it is clean. Place it on the welding table.

2. Adjust the welder to the right amperage. Set the voltage if a DC welder is used. Use about 100 to 105 A for a 1/8-inch electrode and about 140 to 145 A for a 5/32-inch electrode. Generally, the larger the electrode, the higher the amperage.

3. Turn on the machine.

4. Drop the shield over your face before you start the arc.

5. To start the arc, grasp the electrode holder firmly. Lower it until the tip is about one inch from the base metal. Now, lean the electrode at an angle of about 25 degrees in the direction of travel. Start the arc by scratching the tip of the electrode on the surface of the metal, much as you would strike a large match (Figure 67-6). There will be a sudden burst of light and spark. Withdraw the electrode a little. The current will jump across the gap, creating the arc. If you hold the electrode on the metal too long, it will fuse with the base metal and stick tight. *If this happens, without raising the shield, break the electrode off quickly by twisting it or releasing it from the holder.*

6. After the arc is started, move it slowly and evenly along the workpiece. Try to maintain an arc length of 1/16 to 1/8 inch. You can steady the electrode holder by holding your free hand under the wrist of the hand holding the electrode. You can also steady your elbow against your body or the table as you feed the

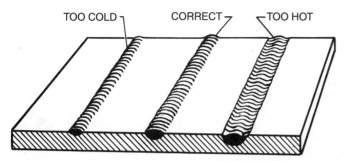

Fig. 67-7 Beads made with an arc that is too cold, with a correct or normal arc, and with an arc that is too hot.

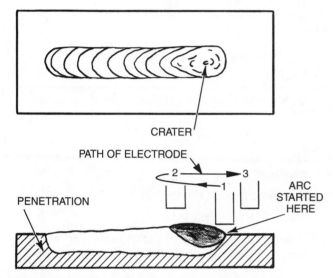

Fig. 67-8 Picking up the weld after inserting a new electrode in the holder. Note that the arc is (1) started slightly ahead of the completed weld, (2) brought back to the crater, and then (3) moved forward.

electrode. Check the arc length by watching the electrode and listening to the sound. When the right setting is used and the correct arc length is maintained, there is a crackling sound like eggs frying.

7. Move the electrode along smoothly and evenly. If you move the electrode too fast, there will be a thin bead and no strength or penetration (Figure 67-7). If you move it too slowly, there will be too large a bead with a lot of waste metal. Continue to maintain the correct speed until the electrode is melted down to a length of 1/2 inch. Then pull it away to break the arc.

8. To restart the bead, first chip the slag from the crater at the end of the bead and brush off the excess metal. Start the arc a little ahead of the bead, bringing it back to pick up the weld. Then continue forward (Figure 67-8).

9. Study the bead after you have finished it to make sure you have done it correctly. After you have run a bead, try it again with a weaving motion. This is

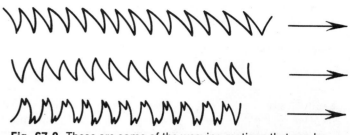

Fig. 67-9 These are some of the weaving motions that can be done in arc welding.

necessary when you must cover a wider area to weld a joint. The proper motions are shown in Figure 67-9.

Types of Welded Joints

Five basic kinds of welds are used to join metals. These are the same as those used in gas welding (see Figure 66-10 on page 372).

1. A **butt weld** can be made in metal up to 1/8 inch thick without any previous preparation. If the metal is 1/8 to 1/4 inch thick, bevel one side to form a 60-degree V. For metal from 1/4 to 3/8 inch thick, grind a V-bevel from one side, leaving a 1/16-inch shoulder. Pieces thicker than 3/8 inch should be ground from both sides to form two Vs. To make the weld, first tack one end of the two pieces about 1/8 inch apart and the other end 1/16 inch apart. A butt weld on thinner material can be made with one pass. On thicker material, it may be necessary to make two or more passes.

2. The **tee,** or **fillet, weld** is used to join two pieces of metal at right angles.

3. The **lap weld** is made in somewhat the same way as a tee weld. Two pieces of metal are lapped and tacked together. Then the weld is made along the corner.

4. The **corner weld** is used to join two pieces of metal to form a corner.

5. The **edge weld** is used largely on light metal to join two pieces that are flush.

Table 67-1 suggests the right electrode size and welding current for various kinds of welds.

Ventilation for Arc Welding

Because arc welding produces injurious fumes, it must always be done in well-ventilated areas. In some factories and school shops, arc welding is done in special welding booths containing exhaust ducts, work tables, and welding connections.

Table 67-1 Electrode size and welding current for various types of welds

plate thickness*	1/32	3/64	1/16	3/32	1/8	3/16	1/4	5/16	3/8	1/2+
For tee and lap welds in flat positions										
electrode size*	—	5/64	3/32	1/8	5/32	5/32	5/32	5/32	5/32	5/32
machine setting**	—	45	65	100	140-150	140-150	140-150	140-150	160-170	170-180
For butt welds in flat positions										
electrode size*	1/16	5/64	3/32	1/8	5/32	5/32	5/32	5/32	5/32	—
machine setting**	30	45	65	100	140-150	140-160	140-160	160	160	—
gap*	none	none	none	1/32	1/32	1/16	1/16	1/8	1/8	—
For corner and edge welds										
electrode size*	1/16	5/64	3/32	3/32	1/8	5/32	5/32	5/32	5/32	—
machine setting **	30	40	65	70	100	140-160	140-160	140-160	160-180	—

*In inches. **In amperes. The machine settings given are approximate. It may be necessary to set machine higher or lower, depending on welder's skill and welding conditions.

KNOWLEDGE REVIEW

1. What is *arc welding?*

2. What kind of arc welder is cheapest and easiest to maintain?

3. What are arc-welding electrodes made of?

4. What protective gear should you wear when arc welding?

5. How long should the arc be?

6. Name five kinds of welds.

7. Make a display board of the common arc-welding electrodes. Show what each kind is used for.

Unit 68

Industrial Welding Methods

OBJECTIVES

After studying this unit, you should be able to:

- Explain the harmful effects of air on hot metal.
- Explain gas-metal arc welding.
- Describe the GTAW process.
- Explain the submerged-arc welding procedure.
- Name two things needed to do plasma-arc welding.
- Explain electron-beam welding.
- Describe pulsed-laser welding.
- Name three industrial resistance-welding processes.

KEY TERMS

electron-beam welding	gas-metal arc welding	plasma-arc welding	seam welding
explosive bonding	gas-tungsten arc welding	projection welding	submerged-arc welding
flash welding	inert gas	pulsed-laser welding	

You have learned about gas welding and arc welding in the school shop. These processes are also used in industry. However, industry also uses many other special welding processes in manufacturing metal products. These methods are used to weld parts that cannot be welded easily with the gas or arc methods. Industrial processes are often automatic, including robotic welders (Figure 68-1).

One of the problems in welding is the harmful effect air can have on the hot metal. Chemical reaction of the metal with gases in the air can weaken the finished joint. Oxide films, for example, can form on the metal and prevent a good bond. Several methods are used to protect the weld metal from contamination by air until the metal has cooled. The most widely used methods consist of covering the weld metal with a flux or with an **inert gas,** one that will not react with the metal. The method used to apply the flux or inert gas varies with the process. Some of these industrial welding processes are described below.

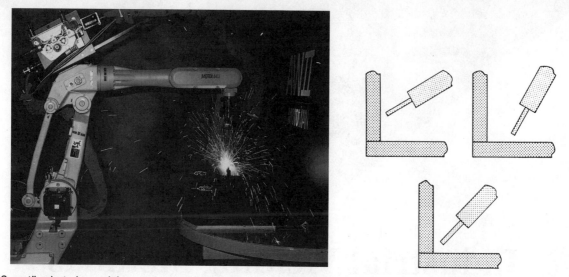

Fig. 68-1 "Smart" robots have vision systems to "read" the part positions and locate the weld area, as shown in the diagram. (Courtesy of Motoman, Inc.)

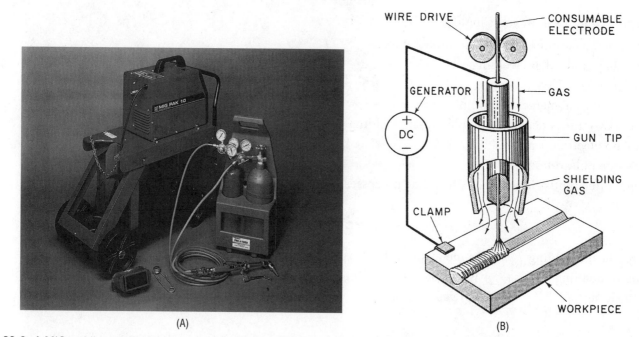

(A) (B)

Fig. 68-2 A MIG welding outfit (A), and a diagram of the GMAW process (B). (A is courtesy of The Lincoln Electric Company)

Gas-Metal Arc Welding

Gas-metal arc welding (GMAW) uses the heat of an electric arc between the workpiece and a consumable electrode. The process is a kind of gas-shielded arc welding. No flux or coating is used on the filler metal. Welding takes place in a "shield" of gas that protects the metal from the air. The process has also been called *MIG*, for *metal inert gas* welding. GMAW is used on aluminum and magnesium and for much production welding.

High speed can be achieved by gas-metal arc welding. The filler metal is a continuous coil of small-diameter wire fed automatically into the joint. Since no flux is used, there

is no slag deposit to be removed from the weld bead. The process can be used with automatic welding equipment.

In the GMAW process, power-driven rolls feed the wire through a flexible tube to the welding gun. A hose carries shielding gas, usually argon or a mixture of argon and helium, from a pressurized tank to the gun nozzle. Electrical cables connect the workpiece and filler metal to the power source. During operation, the filler metal slides against an electrical contact inside the feed tube to maintain the circuit. A trigger on the handle of the gun controls both the wire feed and the gas flow. Water is sometimes circulated through the gun to cool it (Figure 68-2).

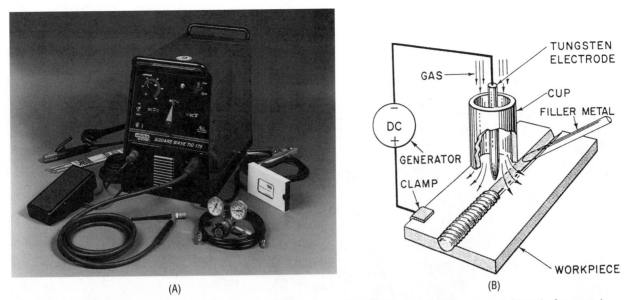

Fig. 68-3 A TIG welding outfit (A), and a diagram of the GTAW process (B). (A is courtesy of The Lincoln Electric Company)

Gas-Tungsten Arc Welding

Gas-tungsten arc welding (GTAW) is done with the heat from an electric arc discharged between the workpiece and a nonconsumable electrode made of the metal tungsten. The GTAW process, like GMAW, is gas-shielded arc welding, using an inert gas and no flux. The difference between the two processes is in the electrodes used. The GTAW process has also been called *TIG,* for *tungsten inert gas* welding. Tungsten is used for the electrode because it resists high temperatures and has good electrical characteristics. Welding can be precisely controlled and done in all positions with the GTAW process. Because the electrode does not melt, the arc is extremely steady. GTAW can be done with or without filler metal. The process is used to join thin pieces without filler metal.

The equipment used for GTAW is similar to that used for GMAW. The electrode holder, like the GMAW gun, is connected to a supply of shielding gas. The process is similar to oxyacetylene welding. An arc supplies the heat and the filler rod is worked by hand (Figure 68-3).

Submerged-Arc Welding

Submerged-arc welding (SAW) uses the heat from an electric arc; a shield of loose, grainy flux; and filler metal supplied by an electrode wire to weld heavy workpieces. This process is much like GMAW in that it uses an automatically fed electrode wire of filler metal. The difference is the flux blanket.

In the SAW process, automatic equipment feeds the filler metal into the joint. The power source is connected to the base metal and filler metal. The filler metal slides against an electrical contact inside the tubular electrode holder. The flux is normally stored in a hopper and poured into the joint through a tube mounted in front of the electrode holder (Figure 68-4).

Plasma-Arc Welding

Plasma-arc welding uses an inert gas and a transferred, constricted arc. This process replaces GTAW in a number of industrial applications. It offers greater speed at a lower current, better weld quality, greater arc stability, and less sensitivity to process variables. A suitable gas—argon or argon mixed with hydrogen or with helium—is heated and ionized, or charged, into a *plasma.* It flows through the nozzle, protecting the electrode and providing the desired *plasma jet.* However, since the flow rate of the plasma gas is kept relatively low to avoid turbulence and too much displacement of the molten metal, a supplementary shielding gas is required (Figure 68-5).

In plasma-arc welding, the plasma jet produces a hole, called a *keyhole,* at the leading edge of the weld puddle. The arc passes completely through the workpiece. As the weld progresses, surface tension causes the molten metal to flow in behind the keyhole to form the weld bead.

A plasma-arc welding torch is operated with reverse polarity and a water-cooled copper electrode, or with straight connections and a tungsten electrode at currents up to 450 amperes. Power cables at the top of the torch supply power and cooling water to the electrode. Filler metals are used with the plasma arc where more than one weld pass is needed.

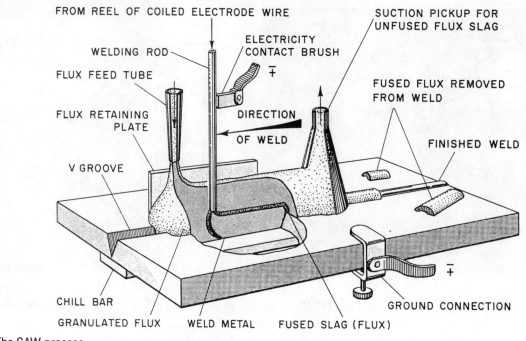

Fig. 68-4 The SAW process.

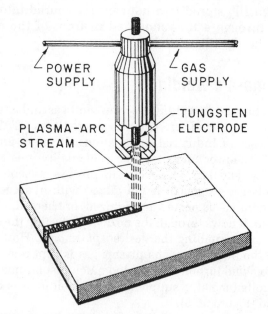

Fig. 68-5 Plasma-arc welding diagram.

Electron-Beam Welding

In **electron-beam welding** (EBW), the heat needed to melt the joint edges is supplied by an electron gun. This gun can produce temperatures of more than 2,000,000°F. The gun focuses a beam of electrons onto the joint line parallel to the existing interface (Figure 68-6).

It can concentrate a large amount of energy in a spot with a diameter of about 0.01 inch or less. Electron-beam welds usually do not require any filler wire. The process is economical for repairing close-tolerance parts.

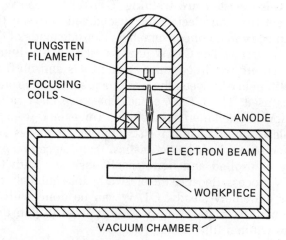

Fig. 68-6 The EBW process.

EBW can be used on materials with thicknesses ranging from about 0.0015 inch to about 2 inches. It is usually done in a high-vacuum chamber, which minimizes distortion. Lightweight fixtures are needed to hold the parts in place. The flat welding position is commonly used. EBW is either high-voltage (75,000 to 150,000 volts) or low-voltage (15,000 to 30,000 volts).

The weld area must be very clean before welding. This is because the process allows very little time for the escape of any gaseous impurities during the welding. Two passes may be needed with very thin materials to avoid undercutting. The underside of an electron-beam weld sometimes has an undesirable contour that must be removed.

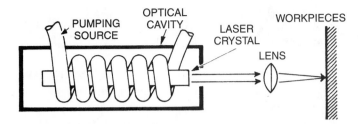

Fig. 68-7 Pulsed-laser welding.

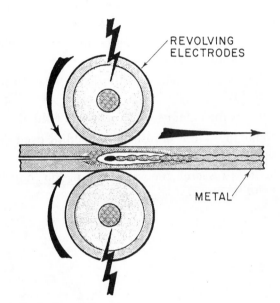

Fig. 68-8 Seam welding.

Pulsed-Laser Welding

In **pulsed-laser welding,** the energy of a laser light beam is focused at the area to be welded (Figure 68-7). With each pulse, the workpiece is moved to refocus the light one-half spot away. The pulse is normally one to ten milliseconds long. The metal solidifies before the next pulse, so the next area to be welded is not affected by heat from the previous weld. Because very little heat is generated, there is little shrinkage and distortion. Laser welds generally have about one-tenth the distortion of electron-beam welds.

Since there is no contact with the work, difficult joints can be welded. Also, welds can be made through glass or plastic. The process is effective on small electronic materials, since laser spot welds can be less than 0.001 inch in diameter.

Explosive Bonding

Explosive bonding uses the force of an explosion to join surfaces. Although some melting occurs, joining does not depend on fusion. The joint has an extremely thin bond area.

The detonation velocity of an explosive must be such as to produce the needed impact velocity between the two metals. The explosives are generally detonated with a standard commercial blasting cap, often along with a line-wave generator for larger charges. When an explosive with a high detonation velocity is being used, the explosive and metal must be separated by a rubber or acrylic buffer.

Explosive bonds are usually wavy or rippled. With metals that form brittle compounds, shrinkage can lead to cracks. Though surface preparation does not have to be as careful as with other processes, surfaces should be smooth to within 0.001 inch in the area of the bond.

The widest use of explosive bonding has been to join flat sheet to plate. Tubular transition joints, seam and lap bonds, rib-reinforced structures, and spot bonds also are made. Explosive spot bonding is particularly useful because the small explosive charges can be packaged and used with a hand-held tool.

Industrial Resistance-Welding Processes

Spot welding, described in unit 36, and many other kinds of resistance welding are used in industry. Industrial spot welding is used to assemble sheet metal articles when a sealed joint is not needed. A series of welds, similar to riveting, is used along the lap joint. Several spot welds are sometimes made at the same time by multiple pairs of electrodes. The space between the welds depends on the strength needed. This is a speedy, economical production method.

Seam Welding

Seam welding is similar to spot welding except that rollers are used instead of electrodes. These allow much more rapid welding (Figure 68-8).

Flash Welding

Flash welding is used to butt-weld bars, tubes, and sheets. The two workpieces are clamped in dies that carry the electricity. The workpieces themselves act as electrodes. As the dies move together, the workpieces touch, and the welding flash takes place. The molten ends are joined by heat and pressure (Figure 68-9).

Projection Welding

Projection welding is a method of producing spot welds from projections on the joint surfaces (Figure 68-10). The parts, held under pressure between electrodes, contact each other only at the projections. The projections melt to form spot welds. In projection welding, two or more spot welds can be made at the same time with one pair of electrodes.

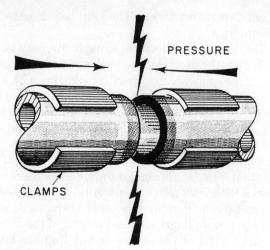

Fig. 68-9 Flash welding.

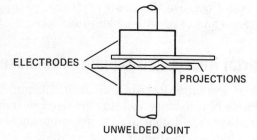

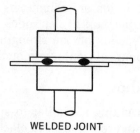

Fig. 68-10 Projection welding.

KNOWLEDGE REVIEW

1. How does the electrode used in the GMAW process differ from that used in the GTAW process?

2. What are other names for GMAW and GTAW?

3. How does submerged-arc welding differ from regular arc welding?

4. What is the hole produced in plasma-arc welding called?

5. What temperatures can be produced in electron-beam welding?

6. What is the duration of the light pulse in laser welding?

7. What is the chief use of explosive bonding?

8. Name four kinds of resistance welding used in industry.

9. Write a research report on one of the industrial welding processes.

Unit 69

Industrial Assembly Methods

OBJECTIVES

After studying this unit, you should be able to:

- Tell another name for the assembly process.
- Describe adhesion.
- Describe cohesion.
- Explain how pressure is used in assembling products.
- Describe interlocking devices.
- Identify when soldering is used in industry.
- Describe the injected-metal method of assembly.
- Name two major industries that use adhesives for assembling products.

KEY TERMS

adhesion	cohesion	injection assembly	mechanical fastening
assembly	fastening	interlocking device	

Assembly, or **fastening,** is an important step in the manufacture of products. A bicycle, for example, is made up of many parts that must be assembled into a final usable machine. Industry is constantly searching for improved, lower-cost methods of joining parts. Not only the standard fastening methods—riveting, soldering, and welding—are used in industry, but also hundreds of very unusual methods. Most metal products can be assembled in more than one way.

All assembly methods can be grouped under three basic kinds of fastening: adhesion, cohesion, and mechanical. **Adhesion** is assembly with a material such as an adhesive or solder that holds parts together. **Cohesion** is creating an actual chemical or metallurgical bond between the parts, as in welding. **Mechanical fastening** is using a device such as a screw or rivet to fasten parts together. Some of the more interesting assembly operations are described in this unit.

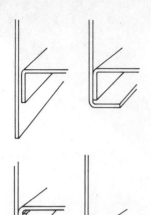

Fig. 69-1 Standing end-lock seam; a typical way of holding two parts of sheet metal together.

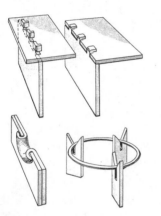

Fig. 69-2 Some methods of attaching parts without special fasteners.

Interlocking Devices

The sheet metal seam is a typical example of an **interlocking device** for holding parts together (Figure 69-1). Industry uses a wide variety of techniques for fastening sheet metal parts. Twisted tabs or wire, snap-on rings, and crimping methods are but a few ways to assemble such parts without adding any other materials (Figure 69-2).

Industrial Innovations

Product manufacturers continue to create newer, better, and more efficient fastening systems, many of which lend themselves to automatic production. Several of these are discussed below.

TOX Technique

The *TOX technique* joins metal parts by means of an upsetting-pressing process. By studying the illustra-

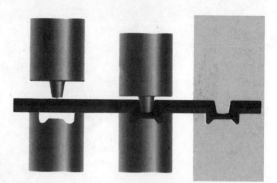

Fig. 69-3 The TOX joint process. As shown, the punch presses the metal sheet into the die to complete the joint. (Courtesy of TOX-Pressotechnik Inc.)

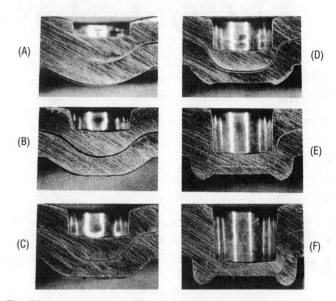

Fig. 69-4 Forming the TOX round joint: (A) penetration, (B) upsetting, (C) emergence of upper contour, (D) filling out of ring groove, (E) lateral spreading of punch side material, and (F) finishing TOX round joint. (Courtesy of TOX-Pressotechnik Inc.)

tion in Figure 69-3, you can see how a shaped punch presses the metal pieces into a shaped die to create the joint. The metal surfaces are permanently joined without damage. A microphotograph of the process is shown in Figure 69-4. The movement of the metal to form the joint can be clearly seen. A typical industrial part example appears in Figure 69-5.

Removable Rivets

The unique design of a *removable rivet* allows the dismantling of access panels in blind applications where access to the back of the panel is inconvenient or impossible. No special tools are necessary (Figure 69-6). Note the insert position of the rivet, and that it can be locked with simple finger pressure. The tip of a ball-point pen can be used to depress the center of the rivet. This releases the locking mechanism to allow rivet removal. These removable rivets are ideally suited for many applications, such as computer and electronics cabinets, and recreational vehicle (RV) utility panels.

Fig. 69-5 An industrial part showing TOX round joints. (Courtesy of TOX-Pressotechnik Inc.)

Versa Stud Inserts

The *Versa stud insert* consists of a threaded stud surrounded by a collapsible sleeve. During installation, the stud is inserted into a blind hole (Figure 69-7). As it is drawn down with the tool (much like a pop rivet), the sleeve collapses on the blind side of the hole, clinching the fastener permanently to the part. The stud fits standard nuts, for part assembly and disassembly. This insert can be used without damage to plastic and wood, as well as metal parts.

Injection Assembly

A special application of die-casting technology in metal fastening is called **injection assembly.** One such example is the *Fishertech process,* which involves injecting molten metal, under pressure, into a fixture holding two metal parts. In Figure 69-8 you will see a metal gear and a shaft. The parts are placed in a special assembly fixture which contains a mold cavity. Molten

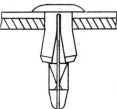

INSERT POSITION
(LEGS PARALLEL)

LOCKED POSITION
(LEGS FLANGED)

REMOVED POSITION
(LEGS PARALLEL -
ENABLING REMOVAL)

Fig. 69-6 The removable rivet. (Courtesy of ITW Fastex)

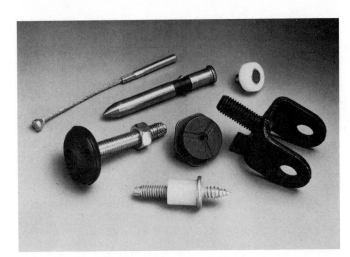

Fig. 69-7 A stud insert installation tool is shown at left. Typical inserts are shown on the right. (Courtesy of CAMCAR Textron)

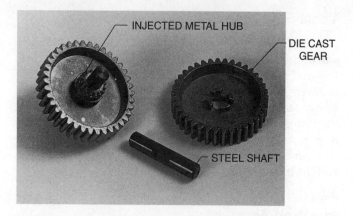

Fig. 69-8 A metal gear and shaft with an injection-molded hub.

zinc is injected into the cavity, and the zinc quickly hardens as a hub to join the gear and shaft. Note in this example that the hub also becomes a second gear component for this assembly. Other examples include electric motor armatures and braided cable end nubs.

KNOWLEDGE REVIEW

1. Describe the three basic kinds of fastening.

2. What are some ways of assembling parts without adding any fastening materials?

3. Describe the TOX, removable rivet, and stud insert assembly methods.

4. Describe the injection assembly process.

5. Examine some of the machines and equipment in your school shop and identify some of the methods used in their assembly.

Unit 70

Introduction to Metal Finishing

OBJECTIVES

After studying this unit, you should be able to:

- Identify four purposes for metal finishing.
- Name three groups of finishes.
- Describe corrosion.
- Describe tarnish.
- Tell how rust compares with corrosion.
- Describe several kinds of corrosion other than tarnish and rust.
- Name three metals that are highly resistant to corrosion.
- Describe how steel can be protected from corrosion.
- Tell how to design a product to avoid corrosion.
- Name three kinds of stainless steel.

KEY TERMS

anode	finishing	oxidation	stainless steel
cathode	galvanic action	rust	tarnish
corrosion			

Metal **finishing** is the final process in the manufacture of a metal product. A finish is applied to metals for a variety of reasons and in many different ways. Finishes are used for one or more of the following purposes:

- *To improve the appearance of the product.*

An example is the attractive chromeplating on appliances, tools, and other products.

- *To prevent corrosion.*

An example is the finishes that are applied to an automobile body to keep it from rusting.

- *To cover a less expensive metal with a thin coating of a more-expensive one.*

An example is table silverware that is made of nickel silver or brass, and plated with silver.

- *To improve the wearing quality of surfaces.*

An example is the superfine grinding done to the moving parts of an engine.

Fig. 70-1 Parts of this racing car are chromeplated or enameled. (Courtesy of Raychem Corporation)

Fig. 70-2 This steel roofing has been eaten away by rust, a kind of corrosion.

All methods of finishing fall into one of three groups: (1) adding *coating* material to the surface, such as in painting and electroplating; (2) *coloring* a metal with heat or chemicals, such as in anodizing; (3) giving some *mechanical treatment* to the surface, such as by shot peening, sand blasting, or polishing.

Look at the metal products you use every day. You see the industry uses many kinds of finishes, often two or more on the same product (Figure 70-1). Almost every type of finish is applied to one or more parts of an automobile.

What is Corrosion?

One of the reasons for applying a finish to metal is to stop corrosion. **Corrosion** is the deterioration, the slow eating or wearing away, of metal. You may have seen a tarnished copper dish. **Tarnish** is the staining and discoloring that takes place when raw metal is exposed to air. It is a kind of corrosion. You have also seen a rusty steel bolt or piece of sheet steel. **Rust** is another kind of corrosion (Figure 70-2).

Water in some form is necessary for corrosion to take place. The rate of corrosion is affected by use and other conditions. Corrosion results in expensive damage and represents a great technological challenge.

Causes of Corrosion

There are many kinds of corrosion besides tarnish and rust. Corrosion takes place, for example, when un-

like metals are joined. It can also be caused by stress and fatigue of metal. Electric currents can cause metals to corrode. Many metals corrode more rapidly when used around acids or salts. Corroding also starts in cracks and crevices or where parts have been assembled. Air pollution can contribute to corrosion. The two most common causes of corrosion are oxidation and galvanic action.

Oxidation

Oxidation is the common deterioration of metals exposed to air, water, or acid (Figure 70-3). Oxidation results when the oxygen in the air reacts with the surface of the metal. The oxidation process and the resulting corrosion are similar in the common metals. However, the color that results differs. The color will usually be blue-green for copper, black for lead, white or gray for aluminum and zinc, and reddish brown for most kinds of steel.

Most people think only of the rusting of steel as a corrosion problem. Actually, all metals will corrode under certain conditions. For instance, aluminum reacts with air to form an oxide coating. This coating completely covers the metallic aluminum and prevents further oxidation. However, if the oxide is prevented from forming or if it is continually being removed, the aluminum will corrode very quickly. Even stainless steel corrodes when used around many acids. It is important to remember that metals vary in their ability to resist corrosion.

All water is not equally corrosive. Distilled or very pure water is the least corrosive. Fresh water is more corrosive. Seawater is the most corrosive.

Fig. 70-3 This cutaway illustration of an engine shows how rust and scale build up around the cylinder walls. The water in the cooling system causes it to fill up with a corrosive scale.

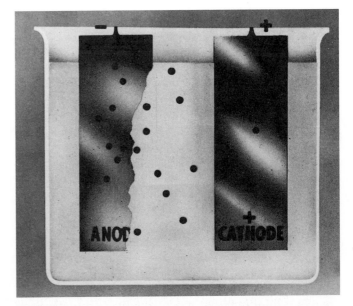

Fig. 70-5 The flow of electricity between the anode and the cathode in metal corrosion. Note how the anode is wearing away.

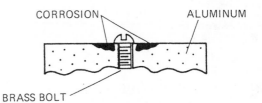

Fig. 70-6 Galvanic action taking place between two different metals causes corrosion.

Fig. 70-4 Staining or eroding metal by placing it in acid. This is what happens in chemical coloring or etching.

Sulfuric acid is a corrosive substance often encountered in industry. It can be present in the air and will corrode metals quickly. Other acids—hydrochloric, nitric, and hydrofluoric—will also attack metals. This occurs most rapidly if the metal is dipped in the acid (Figure 70-4).

Galvanic Action

Galvanic action causes corrosion when two different metals touch one another or are placed in the same vat of water. The metals behave as though they are part of a galvanic cell, or wet battery. The relationship between the rusting of a steel bolt and the action of a battery may not seem obvious. However, the same chemical action is present.

It has been proved that corrosion usually involves a small, localized electric current. In a battery, an electric current is produced by placing two materials, one of which is usually a metal, in a chemical solution. When the circuit is completed, one material, called the **anode,** dissolves. An electric current flows through the solution to the other material, the **cathode** (Figure 70-5).

A piece of metal in contact with moisture behaves like a small battery. The presence of particles, impurities in the air, or contact with some other metal (Figure 70-6) produces a slight difference of voltage, causing a very small electric current to flow. Galvanic corrosion is common whenever two different kinds of metal touch. The corroded metal is always the anode.

Protection from Corrosion

Some metals, such as platinum, titanium, and stainless steel, are highly resistant to corrosion. Unfortunately, these are all quite expensive. Therefore, carbon steel, even though it corrodes rapidly when exposed to the elements, is still the most widely used metal in the world. Carbon steels are usually protected by enamel, paint, zinc, or electroplating. Some types of modern structural steel produce a hard, dense coating of rust that prevents further rusting of the steel. It also eliminates the need for painting or rustproofing structures made of this steel, such as buildings and bridges. Aluminum is usually anodized to help protect the surface.

Zinc, aluminum, tin, and lead can be applied to steel by the hot-dip method. Zinc is the most widely used coating. Steels coated by this method are referred to as *galvanized* or *hot-dip galvanized.* Hot dipping produces a continuous coating that is the thickest available from any conventional process. This thickness gives maximum protection against corrosion. Zinc coatings tend to be somewhat rough, especially as thickness increases. Aluminum coatings are relatively smooth. Both zinc and aluminum coatings quickly develop an oxide that dulls the surface.

Zinc and cadmium coatings are commonly applied to the surface of steel by the electroplating process. A thin, tight, smooth, uniform coating of high purity and good ductility results.

Designing to Avoid Corrosion

Good design includes selecting the right material. Not only the physical and mechanical properties of a material, but its resistance to corrosion as well, must be taken into consideration. A designer has to know the fundamental principles of corrosion as applied to a particular material.

When the probable cause and type of corrosion has been determined, a variety of methods can be chosen to prevent or minimize corrosion. These include:

1. Painting, coating, or insulating dissimilar metals—the best and least-expensive method for most purposes.
2. Using metal combinations close together in the galvanic series.
3. Avoiding combinations in which the area of one metal is relatively small.
4. Avoiding irregular stresses in design.
5. Using materials made of a metal or an alloy most likely to resist the corrosive environment.
6. Adding suitable chemical inhibitors to a corrosive solution.
7. Keeping dissimilar materials as far apart as possible and avoiding fastening them together by uninsulated threaded connections.
8. Revising the design and insulation to protect against currents.

Stainless Steels for Corrosion Prevention

The **stainless steels** are alloys of iron with chromium, with chromium and nickel, or with chromium, nickel, and manganese. They have one chief characteristic in common—resistance to corrosion. It is the chromium that makes these alloys corrosion resistant.

KNOWLEDGE REVIEW

1. List four reasons for finishing metals.
2. List the three kinds of metal-finishing methods.
3. What is *corrosion?*
4. What are the two main causes of corrosion in metals?
5. Under what conditions does galvanic corrosion take place?
6. What are zinc-coated steels called?
7. What alloying metal allows stainless steels to resist corrosion?
8. Write a report on how industry prevents corrosion on bridges and steel buildings.
9. Prepare a display on the types of metal finishing.
10. Prepare a display showing examples of corrosion in various metals.

Paint and Lacquer Coat Finishes

OBJECTIVES

After studying this unit, you should be able to:

- Describe a simple temporary finish.
- Name two kinds of enamels or lacquers.
- Tell how to apply enamel and lacquer with a brush.
- Describe how to apply a wrinkle finish.
- Explain how to do antiquing.
- Tell how to finish metal with bronze powder.
- Describe how to spray enamels and lacquers.

KEY TERMS

antiquing	filler	primer	wax
bronze powder finish	lacquer	spray finishing	wrinkle finish
enamel	opaque	transparent	

Finishing metal by coating it will preserve it and make it more attractive (Figure 71-1). In this unit, you will learn of several ways of applying **opaque** (or light-blocking) and **transparent** (or light-admitting) coatings.

Wax Finishes

The simplest temporary finish for projects used indoors is **wax.** Warm the metal slightly and apply a coat of paste wax. Let it dry. Then rub it briskly with a clean cloth. This finish will wear off with time and handling and must be reapplied as needed.

Enamels and Lacquers

A colored or transparent finish can be applied with **enamels** or **lacquers.** Before applying the finish, make sure the surface is smooth and free of dirt and grease. Lacquer thinner is a good cleaner to use. Clean galvanized steel with vinegar or special cleaners made for this purpose.

Before applying opaque enamels or lacquers to metal, you must use a **primer,** or first coat. It will bind and adhere to the metal, giving a good base. (Transparent finishes need no primer.) A red iron oxide primer is used for most opaque finishes. A zinc chromate primer is good for exterior finishing. Aluminum surfaces that will be exposed to salt water are covered with

Fig. 71-1 Metal projects are coated to preserve them and make them more attractive. This letter holder has a satin black lacquer finish.

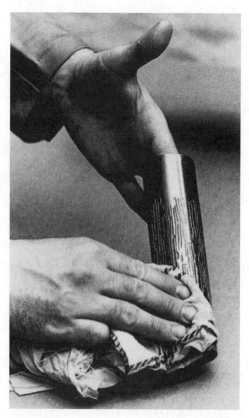

Fig. 71-2 Clean the project carefully before finishing it.

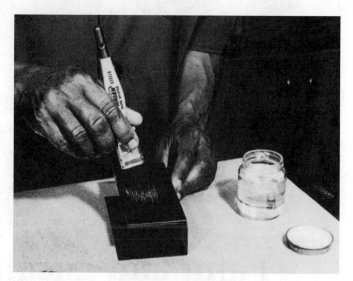

Fig. 71-3 Apply lacquer with light, even strokes.

zinc chromate. However, ordinary lacquers cannot be applied over zinc chromate.

On some metal projects, especially castings, a **filler** may be needed. This is a finish used to fill in the holes in a porous surface before an enamel or lacquer is applied.

Applying Enamels and Lacquers with a Brush

The key to good metal finishing is to have a clean project, a clean finish, and a clean brush. You should work away from drafts and dust.

To apply clear lacquer:

1. Make sure the project is clean and spots are removed (Figure 71-2).

2. Make sure the surface is satisfactorily completed.

3. Use clean paper or cloth to handle the project.

4. Warm the metal in an oven if one is convenient. Heating makes the lacquer flow smoothly. Too much heat, however, makes lacquer boil.

5. Use a good brush and apply lacquer a little at a time (Figure 71-3).

6. Do not pass over an area a second time.

7. Allow the project to dry in a clean, warm room for an hour or two.

8. Apply a second coat as needed.

9. Clean the brush in lacquer thinner, which is the solvent for lacquer.

To apply enamel:

1. Clean the project carefully.

2. Apply primer or filler as needed. Allow it to dry.

3. Brush on enamel with even strokes.

Fig. 71-4 Clean brushes after each use. Use the right solvent for each finish.

4. Allow it to dry for several hours. (Follow the instructions on the can.)

5. Apply second and third coats as needed.

6. Clean the brushes in paint thinner or lacquer thinner (Figure 71-4). Either of these is a good solvent for enamels.

Wrinkle Finish

Many metal goods, such as furniture, heaters, pumps, and novelties, have a **wrinkle finish** that can easily be duplicated in the shop. This finish has the advantages of covering minor defects and of drying quickly. Choose wrinkle-enamel finish in the color you want, and follow the procedure below.

1. Apply a heavy coat to the surface of the metal with a brush, by spraying, or by dipping.

2. Let this coat dry in the air for about twenty minutes, then place the object in an ordinary baking oven.

3. Heat to 180°F.

4. Bake the object from thirty to forty-five minutes.

5. Turn the oven up to 300°F for a short time to harden the finish.

Black Antique Finish

For a project that has been peened, a very popular finish is **antiquing.** Apply a coat of black lacquer. Let it dry, then rub the surface with abrasive cloth. This will

Fig. 71-5 Spraying is done by holding the can about twelve inches from the project.

highlight the metal, leaving black recesses. Apply a coat of clear lacquer. If you use black paint instead of black lacquer, you must use clear varnish as a finishing coat.

Bronze Powder Finish

Many decorative articles that look like brass, aluminum, or silver are really mild steel covered with a thin coat of **bronze powder.** This powder can be bought in liquid form or mixed with clear lacquer, then applied like paint. Another effect can be obtained by applying a clear lacquer or varnish and then blowing the powder on while the surface is still wet. Still another method is to apply colored lacquer and then blow on the powder. This gives a two-tone appearance.

Spraying Lacquers and Enamels

Most industrial finishing is done with spray equipment. **Spray finishing** can also be done in the school shop. The simplest method is to use the spray cans available at paint shops. Follow the procedures for preparing the project (cleaning and so forth) listed above. Hold the spray can about twelve inches from the project (Figure 71-5). Move the can from side to side, and spray a light coat with each pass. Do not spray long in any one area. This will cause runs to appear. After spraying, let the project dry. Apply more coats as needed, or touch up spots you have missed. When you are through, hold the can upside down and press the button for a few seconds. This will clear all spray from the nozzle cap.

Spray guns are also used in the school shop. This type of spraying takes much practice. Your instructor

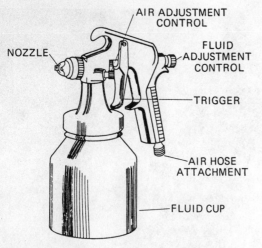

AIR ADJUSTMENT
CONTROL

NOZZLE

FLUID
ADJUSTMENT
CONTROL

TRIGGER

AIR HOSE
ATTACHMENT

FLUID CUP

Fig. 71-6 The spray gun, showing the controls and parts.

must show you these procedures. There are different kinds of spray equipment. The most common type of spray gun is shown in Figure 71-6. Spraying must be done in a properly ventilated spray booth. Other items needed include an air compressor and air-and-fluid, or paint, hoses. During spraying, the gun should be held perpendicular to the workpiece and six to eight inches away. Arm motion should be free, similar to the stroke used in spray-can finishing. It is important to clean the equipment properly after use.

KNOWLEDGE REVIEW

1. What is a primer? What is a filler?

2. Why should metal be warmed before being lacquered?

3. What kind of project looks good with a black antique finish?

4. What can you use to make mild steel look like brass, aluminum, or silver?

5. What happens if you spray lacquer too long in one spot?

6. Write a report on industrial finishing, describing the methods used to clean the products before finishing.

Unit 72

Electroplating and Aluminum Anodizing

OBJECTIVES

After studying this unit, you should be able to:

Describe electroplating.

Explain the principle of electroplating.

List the equipment necessary for electroplating.

Tell what electroplating solution to use.

Name two metals commonly used for electroplating.

Describe the aluminum anodizing process.

KEY TERMS

aluminum anodizing   electroplating   pickle   plating bath

Electroplating is using electricity and chemical solutions to cover a metal object with a thin coat of some other metal. For example, the brass water faucet in Figure 72-1 is *chromeplated* to protect it from corrosion. One reason for plating, then, is for protection. Another metal plating is *tin plate,* which is mild steel sheet with a coating of pure tin. It is applied either by dipping in molten tin, or by electroplating. *Galvanized* coatings are similar to the tin plating process, but zinc is used instead of tin. *Terneplate* is tin-lead coated sheet steel, done by immersing in a molten bath. These plated mild steel sheets are commonly found in school shops.

The principle of electroplating is simple. An electric cell is created using the object to be plated as the *cathode,* or negative pole. The *anode,* or positive pole, is usually the metal to be deposited. Both are placed in a **plating bath,** which is a solution containing salts of the metal to be deposited. As an electric current is passed through the cell, the metal particles separate from the plating bath and are deposited on the article to be plated. These particles are replaced by metal that either comes from the plating bath or is dissolved from the anode (Figure 72-2).

Electroplating adds beauty to the product and protects its surface. Also, plating an article can be less expensive than making it entirely of the metal used for plating. Much silverware is actually made of a fairly inexpensive metal, such as brass or nickel silver, that has been plated.

In commercial electroplating, many different metals can be used in metal-to-metal coating. These include copper, chromium, nickel, gold, silver, and zinc. For decorative plating used for tableware and jewelry, chromium, silver, and gold are common choices. For industrial electroplating, especially of automobile parts, the most common plating is a coat of nickel followed by a coat of chromium. The simplest kind of plating that can be done in the shop is with copper, cadmium, and nickel.

Fig. 72-1 Chromeplated water faucets are common in the home and the shop. (Courtesy of The Chicago Faucet Company, Des Plaines, IL)

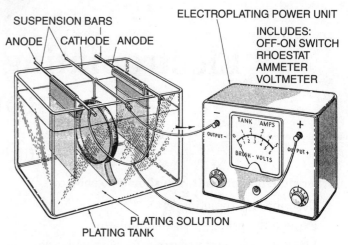

Fig. 72-2 The electroplating process.

Electroplating Equipment and Materials

A reliable and easily controlled source of direct current is needed. For small plating setups, a *storage battery* or *dry cell* can be used. However, it is better to use an *electroplating power unit* that permits control of the current and voltage. Other equipment needed includes:

- *Plastic or ceramic electroplating tanks.*

Tanks are needed for each different kind of plating solution. Another tank is needed for cleaning the metal. All tanks should be made of plastic or ceramics.

- *Small-tank suspension-bar sets.*
- *Anodes.*

Anodes must be the same kind of material as the plating solution. Common ones are cadmium, copper, and nickel.

- *Plating solutions.*

You can mix your own plating solutions, but it is much better to purchase the bath concentrates. These usually contain no dangerous chemicals and are safely packaged in plastic bottles or paper containers. The concentrate is mixed with water. The resulting solution can be stored either in the plating tanks or in plastic or glass bottles.

Preparing Metal for Electroplating

Metal to be electroplated must be free of all grease, dirt, oxides, and other foreign matter. *This is essential for good plating.* The first step is to polish and buff the metal to remove all blemishes and to improve the surface luster. Next, dip the article in an acid pickling solution to remove any grease or dirt. A good acid **pickle** can be made by adding one part muriatic, or hydrochloric, acid to one part water. *Always pour the acid into the water, never the reverse.* Acid should be kept in a glass, rubber, or ceramic jar. After pickling, rinse the metal in cold water. It must be so clean that water on the surface will not bead, or separate into small drops. Never touch the metal with your bare hands after it has been cleaned. Handle it with clean rubber gloves or a piece of clean paper.

Metals for Electroplating

Cadmium, one of the soft metals, is used as a rust-resistant coating. When polished, cadmium looks much like silver and is just as hard. To do cadmium plating, first mix the plating solution in a plating tank, and then suspend the cadmium anode from one of the suspension bars. Suspend the article to be plated from the other bar, the cathode. The voltage needed is from 1 to 3 volts, depending on the solution. Plating will start immediately. Voltage and current control are important. It is best to start with a low voltage of about 1 to 1 1/2 volts and not more than 2 to 2 1/2 volts.

Copper is used both for coating and as a base metal. It is harder than cadmium but softer than nickel. *Nickel* plating is also important. Nickel is hard metal and is used for a rust-resistant coating. It is also used as an undercoating prior to *chromium* plating, as on automobile parts.

Aluminum Anodizing

A thin coating of oxide on aluminum helps protect it from corrosion. However, a heavier coating can be produced by a very simple electrochemical process called **aluminum anodizing.** This oxide film, about 0.01 inch in thickness, is very hard and protects the surface from abrasive and corrosive action. Many varieties of oxide finishes can be produced through the anodizing process. They may be made either clear or colored by dyes. The color and hardness of the film can vary with the type of aluminum that is anodized.

The Aluminum Anodizing Process

The usual commercial anodizing process is as follows (Figure 72-3):

1. The surface is thoroughly cleaned, rinsed, and dried.

2. A finish is applied. The surface of the metal can be given a texture by applying a scratch-brush finish with polishing and buffing wheels or a chemical finish that causes etching.

3. The aluminum is cleaned and again rinsed with water.

4. The actual anodizing is done by a process similar to electroplating. The aluminum object is suspended in an electrolyte, and current is passed through it. This causes a dense aluminum oxide coating to be built up on the surface. This is a conversion coating formed from the metal itself and bonded to the surface. The properties of the film depend on the aluminum alloy used and on the electrolyte composition and concentration, temperature, time, current, and voltage. The thickness

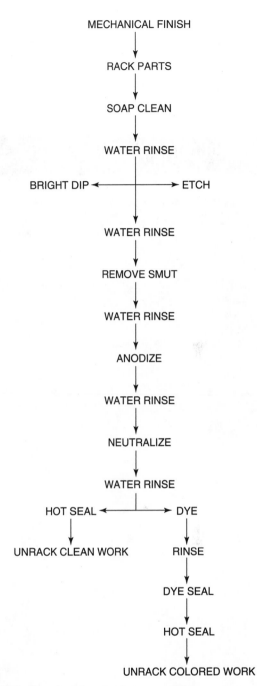

Fig. 72-3 Flowchart for aluminum anodizing.

of the film is limited by the fact that it grows from the inside out.

5. The article is neutralized to stop the process. It is then ready for use or for dyeing.

A colored coating can be produced on anodized aluminum by introducing *inorganic* pigments within the pores of the newly formed oxide coatings. These pigments are especially suited to exterior use, since they weather well and are reasonably colorfast to sunlight. The pigments give an evenly colored surface and can produce a variety of useful decorative effects. However, the range of colors is limited.

Organic dyestuffs produce more brilliant colors than inorganic. They are used to give aluminum coatings a metallic luster. The colors have a clarity and depth that are striking and unusual. Since these dyes are not colorfast to sunlight, they are not suitable for exterior use.

KNOWLEDGE REVIEW

1. What is electroplating?
2. What is a plating bath?
3. Which metals are usually used for electroplating in the shop?
4. Describe the anodizing process.
5. Write a report on electroplating in the automobile industry.

Unit 73

Metal Enameling

OBJECTIVES

After studying this unit, you should be able to:

- Explain enameling.
- Identify what enameling is used for in industry.
- Describe the tools and materials needed to do enameling.
- Describe how to enamel on copper.
- Describe jewelry findings.

KEY TERMS

enameling   jewelry findings   porcelain enamels

Enameling is applying a permanent, glassy finish to the surface of metal. This finish is made of grains of glass, ceramic chemicals, and other colorants. It is fused to the metal by high heat.

Enameling is used in industry to give many metals the good qualities of a glass surface. For example, enameled metal resists acid, heat, erosion, and stains. Kitchen utensils, stoves, refrigerators, bathtubs, and many other objects are covered with enamel. Jewelry, trays, and bowls are typical projects (Figure 73-1).

Tools and Materials

You will need the following materials and equipment for your enameling projects:

- *Copper;* 18 gage (1 mm) or thicker is best because it is easy to shape and will not warp in enameling.
- An *electric enameling kiln* or oven that will heat to about 1,500°F is needed (Figure 73-2). A welding torch or gas burner can also be used.
- A commercial *pickling solution* can be used for copper. A solution can also be made by adding one part nitric acid to two parts water.

 SAFETY RULE: Always put the water in a glass or earthenware container and add the acid to it. Never add water to acid.

- *Metal enamels* come in many colors and are either transparent (light-admitting) or opaque (light-blocking). Wash enamels in water before using them. Transparent colors can be mixed together or fired one over the other. If opaque enamels are mixed together, they give a salt-and-pepper effect. They cannot be applied one over the other. Threads and lumps can also be used for interesting effects.

Fig. 73-1 Two striking pieces of enameled art. The wall piece measures 24" × 48" and is made of woven strips of copper. The interesting bowl is nine inches high. (Courtesy of Vivian B. Kline)

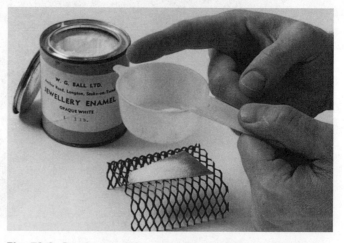

Fig. 73-3 Dusting on the enamel. Dust on a fine, uniform coat of dry enamel. Begin at the edge and work around the piece toward the center. Tap the shaker gently.

Fig. 73-2 A metal-enameling kiln.

- An *adhesive* is needed to hold the enamel in place before it is fired. Examples of adhesives include commercial gum solution and light machine oil.
- Other materials needed include: files, 3/0 steel wool for polishing, salt shakers or 80-mesh sifters, tongs or enameling forks, enameling racks, tweezers, and a small glass tray.

Enameling on Copper

To give copper an enamel coat, use the following procedure.

1. Make sure the kiln is ready for use. Turn it on about thirty minutes before the firing.

2. Cut and form the copper to shape. Smaller pieces should be domed slightly. Enamel shows up much better and is much less likely to crack or chip on a domed surface than on a flat surface.

3. Clean off all impurities on the copper. To remove grease, place the metal in the preheated kiln for a few seconds. Then clean it in the pickling solution. The metal can also be cleaned with a cloth and a paste made of salt and water.

4. Rinse the copper under cold water, then place it on paper towels to dry. Never touch the clean metal with your fingers.

5. Spray or brush on a thin coat of adhesive. No adhesive is needed on a flat piece of copper. Place the workpiece over clean paper on a piece of cardboard or metal that is smaller than the workpiece.

6. Select a base color of enamel. Dust it on as you would apply salt from a shaker (Figure 73-3). Start at the edge and work toward the center. Be sure you cover the surface completely. Always pour the excess enamel into a storage jar. If a gum solution is being used as an adhesive, spray the piece with the solution again. Let the enamel dry before firing.

7. Check to see if the kiln is at the right temperature, about 1,450 to 1,500°F. If it is, it will show a yellow-orange color. With a wide spatula, lift the piece directly into the kiln. If you use an enameling rack, lift the piece onto the rack first and then place the rack in the kiln (Figure 73-4). Enameling can also be done with a torch (Figure 73-5).

8. Carefully watch the piece, checking it every fifteen seconds. It takes two to three minutes for the

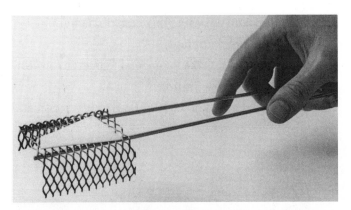

Fig. 73-4 Use a wide-blade spatula to lift the dry enameled piece into the kiln or onto an enameling rack or firing holder.

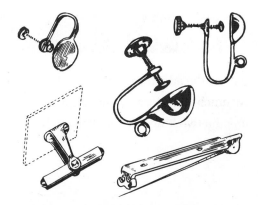

Fig. 73-6 Common findings used in making jewelry.

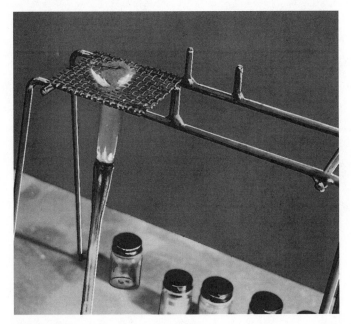

Fig. 73-5 Enameling with a torch.

Fig. 73-7 These aluminum pots and pans have porcelain enamel finishes. (Courtesy of Aluminum Company of America [ALCOA])

enamel to melt. As soon as this happens, move the piece to the cooler part of the kiln or take it out. The piece will be cool enough to touch after about ten minutes.

9. Clean the enameled surface with a commercial cleaner, or remove the oxidation with emery, crocus cloth, or steel wool. Then dip it in a pickling solution. If you wish, you can add more color or design by applying the adhesive again and then using a different color enamel. You can also add a design by placing bits of silver, copper, glass, or thread on the surface before applying more enamel. Fire the piece again.

10. File the edge smooth to prevent chipping. Clean the enameled surface with tripoli, rouge, or another buffing compound (see unit 44). Then apply a coat of clear lacquer.

Attaching Jewelry Findings

Jewelry findings, such as cufflink backs, ear screws, or bar pins, can be attached with either a heatless solder or epoxy cement (Figure 73-6). However, a better job can be done by using soft solder. Soft soldering should be done before the lacquering.

Industrial Enameling

Porcelain enamels are used to coat products such as metal bathtubs, sinks, and showers; cabinets and doors; and pots, pans, and other kitchen utensils. This finish is hard, durable, abrasion-resistant, waterproof, easy to clean, and attractive. This type of finish is made from glassy, inorganic powders which are fused to metals at temperatures above 800°F. Application methods include dipping, electrodeposition, and spraying. The attractive aluminum kitchen pans and pots shown in Figure 73-7 are good examples of porcelain enameled products.

KNOWLEDGE REVIEW

1. What are the advantages of enameled metal?
2. How much heat does enameling need?
3. Name the two kinds of metal enamels.
4. Why must an adhesive be used in enameling?
5. How is enamel applied?
6. How long does it take for enamel to melt in the kiln?
7. What are jewelry findings?
8. Make a report on commercial uses for metal enameling.

Unit 74

Industrial Finishing Methods

OBJECTIVES

After studying this unit, you should be able to:

- Describe the roll-coating and curtain coating processes.
- Explain the automatic and airless spraying systems.
- Describe three methods of plastic coating.

KEY TERMS

airless spraying	curtain coating	fluidized bed coating	roll coating
automatic spraying	dip coating	knife coating	

Industry uses many of the same finishing processes as you do in your project work. Liquid paints and lacquers, and dry powders, are used to produce tough, durable, and attractive finishes on manufactured products. Industry takes special care in cleaning workpieces to ensure quality finishes. The methods include: solvent and acid cleaning, vapor degreasing, and abrasive blasting. The purpose of this unit is to present some of the more common industrial finishing methods.

Spraying Systems

Industrial spraying is a common method of applying finishes, as well as cleaners and adhesives, in a range of applications. Some hand-spraying operations were described in unit 71, and these are also widely used in small job shops. However, **automatic spraying** systems are common in mass production situations where large numbers of similar or identical parts are being processed. Flat metal panels are conveyor-fed to a finishing compartment. Here they are sprayed with a nozzle head that moves transversely (back and forth) to coat the panels. Normally, the conveyor then would carry the panels to an oven or hot-air drier.

Airless spraying is a method of forcing a coating liquid through a spray nozzle by means of hydraulic rather than air pressure. This eliminates the wasteful coating "fog" created by the air pressure of a standard spray gun. In electrostatic air systems, the workpiece carries a negative charge to attract the positively charged coating material (Figure 74-1).

Roll Coating

Roll coating is a method of applying a liquid coating material to one or both sides of flat workpieces (Figure 74-2). Rollers feed the finish to the panels, which is followed by drying or heat curing.

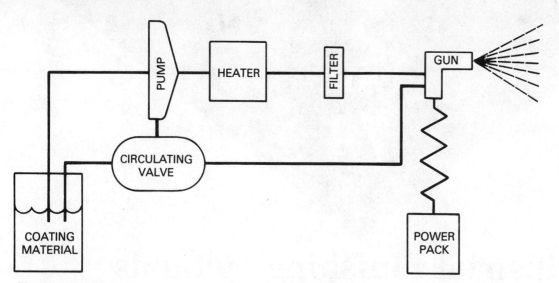

Fig. 74-1 A diagram of an airless spray system. The part being finished is electrically charged to attract the finishing material.

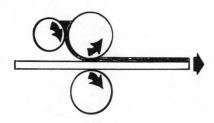

Fig. 74-2 The roll-coating process.

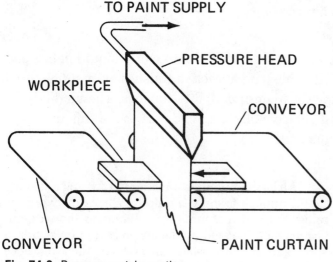

Fig. 74-3 Pressure curtain-coating.

Curtain Coating

The **curtain-coating** process is used on both flat and slightly curved surfaces. A coating liquid, usually under pressure, moves from an upper hopper as a curtain through which workpieces are conveyor-fed (Figure 74-3). The pieces are often preheated to speed the

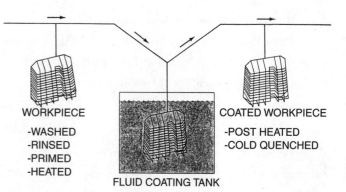

Fig. 74-4 Plastic dip-coating.

drying time. Coating material which does not strike the workpiece drops into a lower hopper and is fed back to the upper hopper for reuse. This is an efficient and economical finishing technique.

Plastic Coating

Plastic powders are often fed to workpieces moving on a carrier belt, and then raked with a coating knife to control the thickness of the powder coat. This is called **knife coating,** and is followed by transfer to a curing oven.

Heated metal furniture parts also can be finished by immersing them in a **fluidized bed** of plastic powder held in suspension by air. This provides a "boiling," constantly moving mass of powder which totally coats the parts. This is also followed by oven curing.

In the **dip-coating** process, workpieces are dipped in liquid plastic and then heat cured (Figure 74-4). Metal workpieces of any shape can be finished with this machine.

KNOWLEDGE REVIEW

1. Describe three methods of plastic coat finishing.

2. What is the purpose of the knife in knife coating?

3. Write brief descriptions of the roll-coating and the curtain-coating finishing methods.

4. Select a number of interesting metal articles found in the home and present a class report on the methods used to apply their finishes.

5. Write a research report on industrial workpiece finishing techniques.

SECTION 4

MANUFACTURING METAL PRODUCTS

The technical skills and knowledge learned by machine operators are directed to the goal of making things. However, making is but one part of the manufacturing process. The purpose of this section of the book is to describe those engineering and planning activities also needed in product manufacture. Beginning with some manufacturing basics, the discussion moves to information on product development, or product engineering (Figure S4-1). Next, the matter of production planning and control is explained and illustrated, followed by an examination of quality control. The very important methods of human resources management also are presented, in order to better understand that industry cannot operate without people. The picture of manufacturing organization is completed by studying how the marketing of metal products is planned and carried out. Finally, there is an opportunity to learn about manufacturing in the school shop, in order to practice what you have learned.

Fig. S4-1 Product development includes experimenting with new processes. This technician is working on new methods of anodizing aluminum. (Courtesy of Aluminum Company of America [ALCOA])

Unit 75

Introduction to Manufacturing

OBJECTIVES

After studying this unit, you should be able to:

- Describe the evolution of the lathe as an important machine tool.
- Describe the keys to modern manufacturing.
- Describe the organization of a typical industry.
- Define the term *industry*.
- Explain the differences between the essentials and the functions of industry.

KEY TERMS

automatic operation	human resources	manual operation	production planning
computer integration	industry	marketing	quality control
corporation	interchangeability	powered operation	semiautomatic operation
fixture	jig	product engineering	services
goods			

From their earliest history, people have made things. Tools, weapons, cooking utensils, clothing, and shelter were "manufactured" in one way or another. Early humans fashioned tools out of stone, bone, and wood, and later out of metals. These first goods were made by hand with the simplest of tools (Figure 75-1). Gradually, machines were developed to make products for personal use, for trade, or for sale. For example, the illustrations in Figure 75-2 chart the evolution of lathe-turning operations from the primitive to the modern.

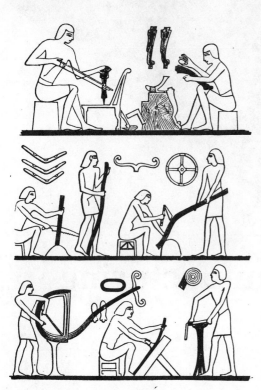

Fig. 75-1 Egyptians of the 13th century B.C. created practical tools to make furniture and implements.

The Evolution of Machines

The history of the lathe can be used as an example for discussing the evolutionary stages of machine tools. The first lathes were manually operated, with improvements over time leading to powered operation, semiautomatic and automatic operation, and finally to computer integration.

Manual Operation

Manual operation is a term used to describe the workings of very early lathes, where human muscles provided the energy to rotate a wooden workpiece. (The lathe is the oldest machine tool.) Simple machines such as the foot-powered treadle were used in the 14th century. On these machines, the work was positioned between centers mounted on a wooden bed structure. The motion was supplied by the foot treadle and a rope wrapped around the workpiece, and the tool was fed by hand to shape the piece. Such manual operations were tiring to the operator, and workpiece accuracy was low.

Powered Operation

Powered operation became possible with the aid of steam engines in the late 1700s. The power was transmitted to the workpiece by means of crude gears, wheels, pulleys, axles, and belts. These improved

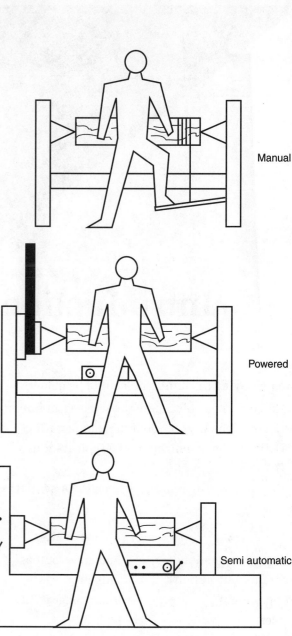

Manual

Powered

Semi automatic

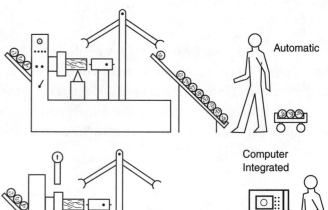

Automatic

Computer Integrated

Fig. 75-2 Typical phases of the evolution of the lathe.

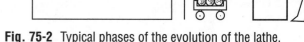

Fig. 75-3 This semiautomatic lathe was an efficient and accurate machine for its day. (Courtesy of South Bend Lathe)

Fig. 75-4 A modern lathe with computer controls. (Courtesy of Clausing Industrial, Inc.)

power methods freed the lathe operators from the manual energy task. They could now concentrate on more careful hand tool manipulation, which led to more accurate workpieces. Whereas early lathes were restricted to wooden workpieces, these steam-powered machines also could turn metal stock. In the late 1800s, the development of the movable carriage, lead screw, and toolholder made possible a new level of accuracy and productivity.

Semiautomatic Operation

The next level of development came with the electric-powered **semiautomatic** lathe, where workpieces were installed manually, but the speed, feed, and tool movement were directed by machine controls. These improvements came about in the early 1900s, and continued well into the middle of the century (Figure 75-3).

Automatic Operation

Automatic machines came about with progress in control technology. This began with cam and lever devices, and stop and limit switches. Then hydraulic and electric systems were developed; with the invention of the transistor (1947) and the silicon chip (1960) providing the basis for electronic controls. The word *automation* began to be used to describe these developments. Numerical controls were available in the 1960s, and computer-assisted mechanisms began in the early 1970s (Figure 75-4). This led to the automatic loading, positioning, and unloading of workpieces. Machinists had to learn an entirely new range of skills in order to work with these systems.

Computer Integration

Metal-cutting technology is now in the stage of **computer integration,** where functions such as product measurement and inspection, material handling, factory management, and movement of workpieces to other machines are computer controlled. More details of these modern systems are presented in unit 85.

Keys to Modern Manufacturing

The development and improvement of machine tools was one important event leading to manufacturing as we know it today. There also were several other key achievements which made it possible.

Measuring Tools and Standards

The earliest measuring instruments were developed in Africa, the Middle East, and southern Europe. Later, accurate and reliable industrial scales, micrometers, plug and ring gages, verniers, and gage blocks were perfected by technicians in America and Europe. Standard sizes of measurement units, materials, and parts were vital to the advancement of manufacturing. Only with the availability of these standards and tools was the theory of interchangeable parts possible.

Interchangeable Parts

Interchangeability means that each of the several parts of a product must be made within certain definite dimensions so that it fits in any one of the final products. It is impossible to make parts exactly perfect. The limits of sizes to which parts can be made must be carefully specified. For some kinds of products, individual parts may vary as much as 1/2 inch (13 mm) in size and still be satisfactory. For other products, such as certain preci-

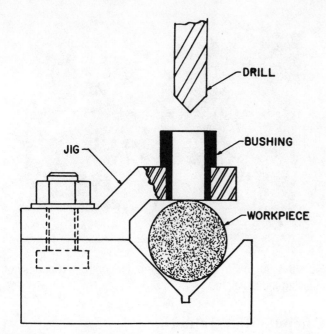

Fig. 75-5 A drilling jig. Note how the bushing guides the drill bit.

Fig. 75-6 A drilling fixture. Note how the simple device holds the workpiece for accurate drilling.

sion parts of a jet engine, the parts must be accurate to within seven-millionths of an inch (0.000178 mm). Part interchangeability could not occur until accurate measuring tools were available, and special material holding devices, called jigs and fixtures, were developed.

Jigs and Fixtures

A **jig** is a device for holding a workpiece and guiding a tool. A simple drilling jig is shown in Figure 75-5. The jig can be moved with the work. A **fixture,** on the other hand, is fastened (fixed) to a machine tool and holds the workpiece in the proper position for machining (Figure 75-6). Note that it does not guide the tool, it holds the work. Eli Whitney, the inventor of the cotton gin, improved the use of special jigs and fixtures to produce 10,000 muskets for the U.S. Army in the early 1800s. This event proved two things: that semiskilled gunsmiths could produce a reliable product, and that mass production of almost any product was possible with jigs and fixtures.

Industry

The machines and devices describe above led to the development of the factory system of production, or the manufacturing *industry*. Here, materials and tools and people were organized and arranged so that quality goods could be produced at a reasonable cost. This grew into a new method of identifying and managing manufacturing operations.

An **industry** may be defined as all the work needed to produce certain types of goods or services and to make those goods and services available to

those who need them. **Goods** are material things such as bicycles, cars, and computers. **Services** are nonmaterial, helpful, needed activities. Examples include the radio and television services of the entertainment industry; the savings, checking, and investment services of the banking industry; and the telephone, mail, and computer network services of the communications industry. This book is mostly concerned with the goods-producing manufacturing industries, as shown in Figure 75-7.

Essentials of Industry

A modern manufacturing (or production) industry needs three basic resources in order to mass produce things (Figure 75-8). *Material* resources are the physical substances such as wood, metal, plastic, ceramics, water, earth, air, and chemicals that are transformed into products. There is not an endless supply of these, and people must learn to use them wisely. *Human* resources are the people who design, make, sell, and maintain the products, and manage the factories that create them. People use machines in their work, but the machines need to be tended by humans. Many different skills are needed in production, and each is very important. *Capital* resources are the tools, machines, factories, energy, and transport systems used to process materials into usable goods. All these resources must be carefully organized to produce goods.

These same essentials apply to the service industries. Can you explain how they apply, and why this is so?

Organization of Industry

The factories where products are made are operated by a team of people made up of skilled workers and managers. The managers are responsible for organizing

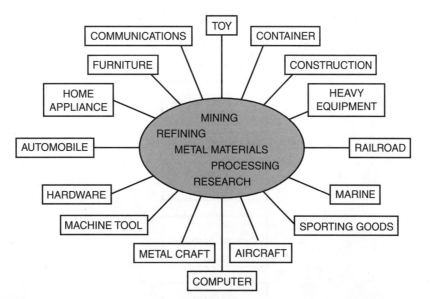

Fig. 75-7 These industries begin from a base of metal materials, and all are involved with making metal products.

PRODUCTION
RESOURCES

MATERIAL

HUMAN

CAPITAL

Fig. 75-8 The essentials of industry are the resources needed to make it work.

the resources of industry. They plan the products to be made, purchase raw materials, train workers to do jobs, organize the production line, and sell the goods. The skilled workers are responsible for operating the production machines accurately, efficiently, and safely.

The mass production of goods is a big business, and it must be organized and operated efficiently. The type of organization used in most large industries is the **corporation.** In a corporation, a number of people pool or invest their money and talents and start a business. The investors are the owners of the business. They are called *stockholders* because they own the stock, or shares, of the company. There are often hundreds or thousands of stockholders in a corporation. The stockholders elect a board of directors to advise the corporation. The president of the corporation is charged with running it efficiently and at a profit. The president is assisted by one or more vice presidents, but a general manager has direct charge of the factory operations. The general manager supervises several managers, each in charge of a special department. A manager is assisted by a supervisor, who directs the work of several operators

or skilled workers. A typical organization chart of a corporation is shown in Figure 75-9.

There are other kinds of business organizations, such as proprietorships and partnerships. Jobs with the same duties may have different names in different businesses. The department or divisions in a production organization are often called the *functions* or *essentials* of that industry.

The Functions of Industry

The functions of industry are, in fact, the steps a product goes through from the mind of the inventor to the hands of the consumer (Figure 75-10). These are: product engineering, production planning, quality control, human resource management, and marketing.

In the first phase, **product engineering,** a product is invented, designed, or developed by design teams. This activity is often called research and development, or R & D. In another important activity, special tools, such as jigs, fixtures, and dies, are designed and made and tested. In **production planning and control,** the product is studied to find out what operations are needed to make it. This is much like a plan of procedure. These operations are then fixed into a production schedule, as part of setting up the production line. **Quality control** is setting and maintaining acceptance standards for the products. In the next phase, the **human resources** are selected, trained, and supervised. **Marketing,** the final function of industry, is getting the quality product to the customer, and seeing that the customer is satisfied with it. Each of these functions is described in greater detail in the following units. The systems chart in Figure 75-11 shows the sequence of input-process-output results in the manufacture of quality metal products.

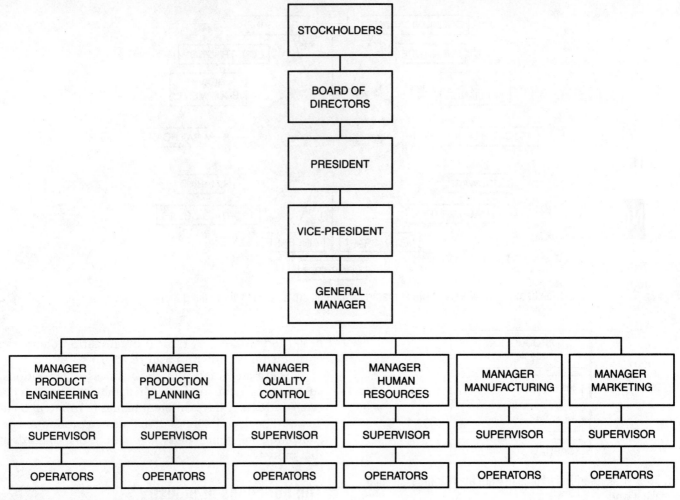

Fig. 75-9 A typical industrial organization chart.

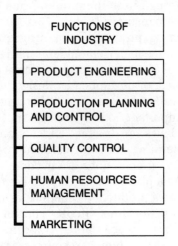

Fig. 75-10 The functions of industry.

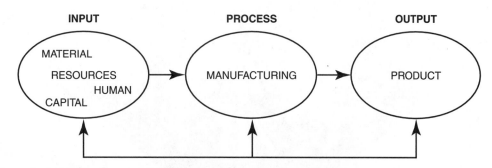

Fig. 75-11 A systems chart showing the manufacturing input, process, and output stages.

KNOWLEDGE REVIEW

1. Write an essay on the five stages of the development of the lathe.

2. Explain the three keys to modern manufacturing.

3. Write an essay on the organization of industry.

4. Prepare a corporate organization chart for a company you might set up in your metalworking class.

5. Define the term *industry*.

6. Describe the difference between the goods and services industries.

7. Describe the three essentials of industry.

8. Describe the five functions of industry.

9. Write a report on how a typical product is mass produced. Examples are bicycles, toys, sports equipment, tools, or machines.

Unit 76

Product Engineering

OBJECTIVES

After studying this unit, you should be able to:

- Explain the purposes of product engineering.
- Describe the purpose of product design.
- Describe the purpose of production tooling.
- Explain the differences among jigs, fixtures, and dies as a part of production tooling.

KEY TERMS

market research   product design   production tooling

Product design is the most important part of the manufacturing process. All the other planning, control, and marketing functions relate to it. Product engineers and designers are responsible for creating products and planning how to make them.

Product Design

The manufacturing process begins with the need to make a product which the consumer requires or wants. **Market research** is important in helping a company to decide whether to design a new product or improve an old one. This research involves *surveying* consumer needs; *evaluating* user suggestions and complaints; and *studying* new product trends. This information will give direction to the work of the product design team.

The product redesign example in unit 12 dealt with the need to improve an *existing* product, the exercise machine. Another example is the Buell motorcycle shown in Figure 76-1. This was a special type of redesign, where the designers wished to improve a standard cycle by making it more visually appealing through restyling, and making it safer and more user-

friendly. The key to the design of this product was the *innovative* effort to build a sportbike that was an extension of the rider, with the focus on *simplicity* by using fewer parts that do more than one job.

The designers began by concentrating the greatest amount of mass as close to the center of the bike as possible. This resulted in reduced frame stress with a lower center of gravity for more responsive handling. They also determined that any flex or movement in the frame reduced rider control, especially under off-road conditions. This problem was resolved by creating a very rigid frame. They also reduced the weight of all components not supported by springs. This lower unsprung weight allows tires to maintain road contact and traction on irregular surfaces. The end result of this redesign program is a safer, better-handling, and very elegant driving machine.

Fig. 76-1 This striking S2T Thunderbolt motorcycle is safe, user-friendly, and visually appealing. (Courtesy of Buell Motorcycle Company)

Fig. 76-2 The inline roller skate is an example of a totally new and very successful product.

On the other hand, inline roller skates (Figure 76-2) were a totally *new* product, invented by someone who thought the idea was good; one that sports enthusiasts would embrace. It was an excellent new product, and a very successful one. Today, these skates are used for street hockey; as a means of transport by students; and as a form of exercise by people young and old.

Product design, therefore, is an important part of the product engineering function of industry. Companies often call this design function *research and development,* because it involves a great deal of research effort, creative design activity, and gradual development through analysis and improvement. As stated in unit 12, the collaborative design team is the modern and ef-

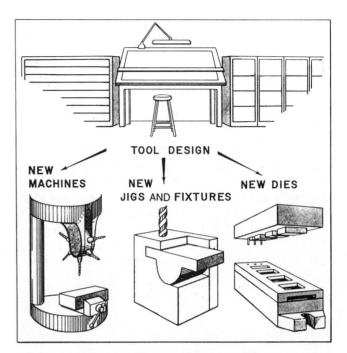

Fig. 76-3 Production tooling involves the design of new machines, jigs and fixtures, and dies, as shown here.

fective way to determine what products a company needs, and how to develop them.

Production Tooling

After the final design has been selected, the factory must plan how to make the parts and assemble them into the product. This involves first inventorying the existing factory machines and equipment, and then designing and building, or purchasing, machines they need but do not have. These production engineers and technicians also must design the jigs, fixtures, dies, and other devices needed to produce and assemble product parts (Figure 76-3). This important function is called **production tooling,** or tooling up for production by collecting all the tools needed to make the product.

Jigs and fixtures were discussed in unit 75. Other important tools which must be developed are *dies,* an example of which is shown in Figure 76-4. Whereas jigs and fixtures are tool-guiding and workpiece-holding

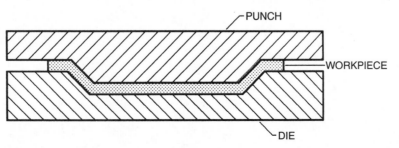

Fig. 76-4 A typical punch and die (called a die-set) used to form a shallow pan.

devices, dies are forming tools used to give shape to products. Note in the example that a punch and die are used in combination to form a shallow dish or pan. Other dies are described and illustrated in unit 37.

Another valuable tool is the template, described earlier in unit 14. This simple tool permits one to trace the outline of a part for its speedy reproduction.

In summary, tooling up consists of three basic activities: (1) deciding what existing machines and tools can be used; (2) ordering or designing/making other needed machines and tools; and (3) supervising the setting up of tools and machines for production. One of the primary aims of the production tooling department is to determine how the skilled work force can be integrated with the tools, machines, and equipment to produce high quality products at a reasonable cost.

KNOWLEDGE REVIEW

1. Describe the two main responsibilities of a production engineering department.
2. What is *production tooling*?
3. List the three basic tasks in tooling up.
4. Give some examples of jigs, fixtures, and dies found in your school shop.
5. Prepare a report on the production tooling needed for a new product, such as a bicycle or a boat.

Unit 77

Production Planning and Control

OBJECTIVES

After studying this unit, you should be able to:

- Describe the basic production planning activities.
- Describe the basic production control activities.
- Explain the meaning of *routing*.
- Prepare a route sheet.
- Prepare a process flowchart.
- Prepare a project flowchart.
- Explain the meaning of *scheduling*.
- Prepare a Gantt chart.
- Explain the meaning of the term *just-in-time*.
- Define the terms *dispatching, monitoring, corrective action,* and *replanning*.
- Describe the basic plant layout procedure.

KEY TERMS

bill of materials	just-in-time	production control	replanning
corrective action	monitoring	production planning	route sheet
dispatching	plant layout	project flowchart	routing
Gantt chart	process flowchart	purchasing	scheduling

Once the product has been designed and tooled up, it must be readied for production. This step is called **production planning and control,** and its purpose is to ensure that people and materials and machines are properly organized so that a product can be manufactured. The *planning* part of this function of industry involves advanced purchasing, routing, and scheduling to utilize the plant and its personnel effectively. The *control* aspect involves dispatching, monitoring, and related work. It is a means of keeping track of the manufacturing activity and, when necessary, taking corrective action (Figure 77-1).

In a factory, materials and parts move together on assembly lines. There are many kinds of such lines—a typical one is shown in Figure 77-2. Note that the various parts join at different stages of manufacturing until the automobile is complete. Simple products such as a metal table, with few parts, are much easier to produce than an automobile with many parts. However, the kind of planning that is necessary is the same in both examples. Good planning means that the right material in the right amount arrives at the right place at the right time. It ensures uninterrupted manufacturing; efficiently and at a minimal cost.

Thousands of cars, such as the model shown in Figure 77-3, are produced on these assembly lines to meet the demand of the thousands of consumers who want them. A racing car, on the other hand, is a one-of-a-kind product made to the specifications of the design team and driver. The production planning and control programs are obviously different for the two cars.

There may be several weeks or even months of planning before a product can be made. To begin, market forecasts and product plans must be studied. This information is needed to decide how many products to make, and when periods of heavy demand can be expected. For example, most people buy motorboats in the spring and summer. It would be unwise to schedule watercraft production in late summer, for the boats must be on the market well before then. Working with this information, the several important tasks of production planning and control can begin, as described in this unit. Review unit 11 to see the similarities between planning a project to be made in class, and how industry does it.

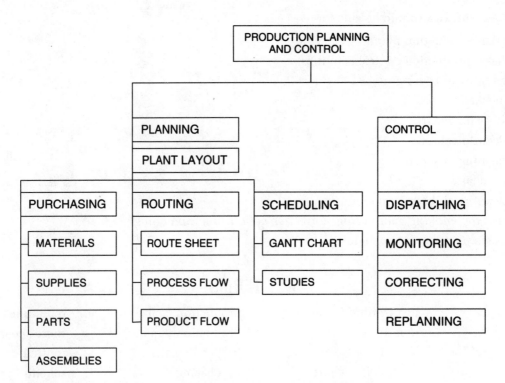

Fig. 77-1 The important elements of production planning and control are shown on this chart.

Production Planning

Purchasing, routing, and scheduling are the important elements of production planning.

Purchasing

Determining and ordering the materials and supplies needed to manufacture a product is the responsibility of the purchasing department. A typical activity is the writing of **bills of materials,** which specify quantities and kinds needed. The bills indicate the needed *supplies,* such as abrasives and finishing materials. Bills also provide information for ordering materials and supplies from warehouses, and stocking them to meet production demands. They also are necessary when placing orders for *parts* and *assemblies* made by a different company at a different location. For example, tires (parts) and automobile radios (assemblies)

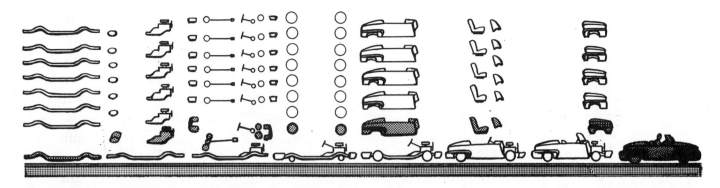

Fig. 77-2 A typical automobile assembly line.

Fig. 77-3 Automobiles such as these are mass produced. (Courtesy of Gilmore Classic Car Club Museums)

generally are made by outside (or outsource) suppliers. All of this information is given to the company purchasing department for action. A typical bill is shown in Figure 77-4.

Many companies now use a newer purchasing system called **just-in-time.** Its purpose is to eliminate much of the warehousing of materials, supplies, parts, and subassemblies. This is done by requesting that orders be delivered on the specific date they are needed in the production program. In other words, they arrive at the factory just in time to be used.

Routing

The next important step is to determine the operations and equipment needed to perform a job. Industry refers to this as *methods engineering,* and the **process flowchart** is a valuable aid to their work. This chart is a

BILL OF MATERIALS	PRODUCT NAME:				WRITTEN BY:			
					DATE:			
PART NAME	DIMENSIONS			MATERIAL	NO. PCS.	UNIT MEAS.	UNIT COST	TOTAL COST
	T	W	L					

Fig. 77-4 A typical bill of materials.

PROCESS FLOWCHART		WRITTEN BY:		
		DATE:		
PRODUCT NAME:	FLOW BEGINS:		FLOW ENDS:	
PART NAME:	PART NO.		PAGE____ OF____	
TYPE OF PROCESS: PART MANUFACTURE		SUBASSEMBLY	FINAL ASSEMBLY	

PROCESS SYMBOLS ◯ Operation (changing or working a material)　▢ Inspection (examining a part for quality)　▽ Storage (keeping a part safely until needed)

⇨ Transportation (moving a part from one place to another)　◻ Delay (holding a part until next activity)

OPER. NO.	STA. NO.	PROCESS SYMBOL	DESCRIPTION OF PROCESS	MACHINE OR TOOL	TOOLING NO.
1		◯⇨▢◻▽			
2		◯⇨▢◻▽			
3		◯⇨▢◻▽			
4		◯⇨▢◻▽			
5		◯⇨▢◻▽			
6		◯⇨▢◻▽			
7		◯⇨▢◻▽			
8		◯⇨▢◻▽			
9		◯⇨▢◻▽			
10		◯⇨▢◻▽			
11		◯⇨▢◻▽			
12		◯⇨▢◻▽			
13		◯⇨▢◻▽			
14		◯⇨▢◻▽			
15		◯⇨▢◻▽			
16		◯⇨▢◻▽			
17		◯⇨▢◻▽			
18		◯⇨▢◻▽			
19		◯⇨▢◻▽			
20		◯⇨▢◻▽			

Fig. 77-5 A process flowchart.

graphic representation of the sequence of all operations, transportations, inspections, delays, and storages which occur during a process. The sample chart shown in Figure 77-5 makes it easy to follow the "flow" of the actions which take place during the making of a part. A **product** or **project flowchart** is valuable as a way of seeing the total picture of a product's manufacture. They generally are used only for simple items, because they can become very complicated for large products with hundreds of separate parts. A sample project flowchart for a punch set is found in unit 81.

Using information from the process flowchart, a **route sheet** is prepared for every part of a product. These are similar to the plans of procedure described in

unit 11. This sheet is a list of each operation, and the equipment, used to make a product part. In short, it is an indication of the route or path of a shop order. Machine operators use this information as a guide to their work. The sheets should be accurate so that no process errors occur during the work. A sample route sheet is shown in Figure 77-6. In some small commercial machine shops these are often called *job sheets*.

Scheduling

The next planning step is scheduling, which is setting up a timetable for when an operation is to be performed, or when work is to be completed. Routing is the

PRODUCT NAME:	MATERIAL:	DRAWING NO:	ROUTE SHEET	
PART NAME:	WRITTEN BY:	UNITS REQ'D #	DATE:	PAGE OF

OPER. NO.	OPERATION DESCRIPTION	TOOL—MACHINE—EQUIPMENT DESCRIPTION	NOTES

Fig. 77-6 A route sheet.

"where" of production, and scheduling is the "when." Schedules help shop supervisors or foremen determine when work will be started and ended, to make the most efficient use of people and equipment. This is especially important when many parts must come together for final assembly. In industry, work study technicians determine how much time it takes to perform each operation. For example, if a steel hammer is to have a knurled handle, a work study person can learn by observation how much time it will take to install the workpiece, set and adjust the knurling tool, knurl the handle, and remove the work from the lathe.

A common timetable used in scheduling is the **Gantt chart** (Figure 77-7). This simplified form shows the job, its operations, and days and hours worked. This is prepared from the information listed on the route sheet. Industry uses variations of these charts to make time and motion *studies* of people and operations, for work simplification and to improve the manufacturing process.

Production Control

The important elements in controlling the production process are: dispatching, monitoring, taking corrective action, and replanning.

Dispatching is issuing orders for production work and is the first step in production control. This is done by studying schedules, meeting with operators and supervisors, and then issuing route sheets. Dispatching is authorizing the start of an operation on the shop floor.

Supervising and tracking the actual operator performance is called **monitoring.** This is necessary to discover machine breakdowns, material delays, operator errors, or other problems which can affect production schedules.

Any problems discovered in monitoring are solved by taking some **corrective action.** A key to good production control is the ability of supervisors to immediately correct things when they go wrong. This may range from getting additional equipment to retraining operators to changing route sheets. It can mean hiring extra people or authorizing overtime. Such corrections or changes are important if the production schedules are to be met.

Changes in market conditions, production methods, or the availability of better equipment can all result in a new manufacturing plan for a product. This is known as **replanning,** and is not to be considered a corrective action. It can be as simple as changing the colors of bicycles to rearranging machines around a computer control center.

Plant Layout

A **plant layout** is a plan of the most effective arrangement of the physical facilities, equipment, and personnel for the manufacture of a product. Such a plan

PRODUCT NAME:		WRITTEN BY:		GANTT CHART		
PART NAME:		DATE:		NOTES:		
OPERATION		MON	TUE	WED	THU	FRI

Planned Work: ▬▬▬
Actual Work: ▬▬▬

SYMBOLS USED

A–operator absent   M–material holdup
G–green (inexperienced) operator   R–machine repair
I–poor instuctions   T–tools lacking
L–slow operator   V–holiday

Fig. 77-7 A Gantt chart.

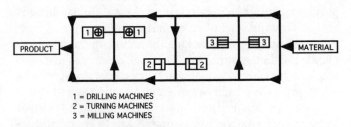

1 = DRILLING MACHINES
2 = TURNING MACHINES
3 = MILLING MACHINES

Fig. 77-8 A typical plant layout, where parts can be moved efficiently to and from machining areas.

goes beyond the placement of machines. It must also take into account where raw materials and supplies enter the factory and where the finished product leaves it. Space must be provided for part storage and machine maintenance areas, and for quality control stations. A good plant layout minimizes the travel distances of subassemblies and maximizes the use of floor space. There are many types of layout schemes—a typical one is shown in Figure 77-8. Modern factories have automatic, computer-controlled material and part handling systems. These will be described in unit 85.

KNOWLEDGE REVIEW

1. Describe the major tasks in production planning.

2. Describe the major tasks in production control.

3. Describe the uses of these production planning documents: bills of materials, flowcharts, route sheets, and Gantt charts.

4. List the important points to consider when preparing a plant or shop layout.

5. Write a research report on Henry L. Gantt, who developed a common work progress chart. Find out what other similar charts are used in scheduling.

6. Visit a local industry and see what kinds of production planning sheets and charts are used there.

Unit 78

Quality Control

OBJECTIVES

After studying this unit, you should be able to:

- Explain the meaning of the term *quality control.*
- Describe the importance and purpose of quality control in manufacturing.
- Explain the three major elements of quality control.
- Describe the various sampling methods.

KEY TERMS

acceptance	final inspection	in-process inspection	quality control
complete inspection	gaging	inspection	random sampling
destructive testing	in-coming inspection	noise	specifications

When you and your friends go to a restaurant for a hamburger and french fries, you expect the food to be fresh, tasty, and well-prepared. The hamburger must be hot and cooked properly, and served on a warm bun. You also want the fries hot and crisp. If the food is not to your liking, you probably won't go back there again. The manager realizes this and so trains the employees to cook the food the way most people like it. This is an example of quality control in food preparation. The manager or the company sets certain standards (rules) of quality in order to satisfy the customers.

Things made in factories must also meet quality standards or they will not be useful and consumers will not buy them. **Quality control** (or QC) can be defined as those activities which prevent defective articles from being produced, and if they are produced, prevent them from reaching the market. *Quality assurance* is another term for quality control.

Quality control begins with the design of the product and continues through each step of planning and production. A good collaborative design team builds quality into the product at the design stage. They make sure it is safe, and easy to make, use, and maintain. Planners and producers build quality checkpoints into their production program. Quality is the responsibility of every employee. A good QC program benefits a company in many ways: it reduces the amount of waste; it saves the expense of replacing or repairing defective products; it protects consumers by preventing unsafe products from reaching the market; and it makes products that satisfy the customer's needs. A quality company understands that quality goes into a product as it is being made. It cannot be tacked on later. This is why managers should not rely on inspection alone to ensure quality.

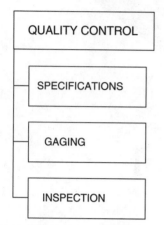

Fig. 78-1 Shown here is a chart which lists the main functions of a quality control program.

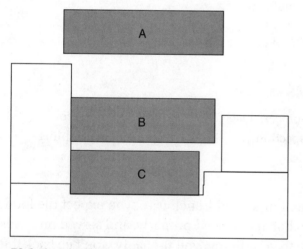
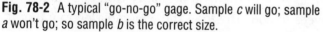

Fig. 78-2 A typical "go-no-go" gage. Sample *c* will go; sample *a* won't go; so sample *b* is the correct size.

Three Phases of Quality Control

The three steps, or phases, of a QC program are: specifications, gaging, and inspection (Figure 78-1), as described below.

Specifications

The detailed descriptions and requirements of the standards for materials, supplies, parts, and products are called **specifications.** Typical examples might include requirements about size, kind of material, function, and shape. For example, you might specify that a screw shaft for a drill press vise must be 1020 carbon

Fig. 78-3 This engineering technician is testing metal samples to ensure that their properties meet the material specifications. (Courtesy of Aluminum Company of America [ALCOA])

steel, with a diameter of 0.50 inch and a 6.00-inch length, and have 1/2-13 threads. You might further state you will accept shafts that are 0.0625 inch shorter or longer in length. This specifies the amount of error you will accept (or tolerate) in each piece to ensure interchangeability of parts. This *tolerance* would be written as ±0.0625″ (plus or minus 0.0625 inch). Tolerance is necessary because tool wear and operator error can make it impossible for all the shafts to be exactly the same size. Good judgment must be used when establishing tolerances. Stricter tolerances than necessary should be avoided because they may increase production costs.

Gaging

Gaging in QC refers to the special devices or tools required to measure the accuracy of parts. A "go-no-go" gage is an example of such a tool (Figure 78-2). By slipping a part into the gage, you can tell at a glance whether or not it is the right size. There are many types of these gages used for industrial QC applications, such as plug and ring gages. Accurate QC measuring machines also are used in industry. The technician in Figure 78-3 is using a compression tester to check the properties of metal test samples. It is the responsibility of the QC staff to design and build, or to purchase, such devices.

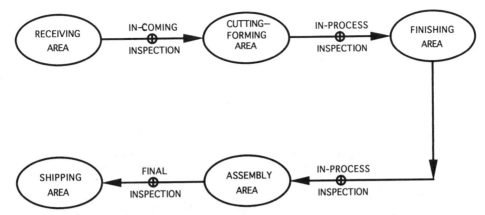

Fig. 78-4 This chart shows how inspection stations are set in the production line to ensure quality.

Inspection

A continual **inspection,** or examination or monitoring, is necessary to ensure that acceptance standards (specifications) are being met. Such acceptance activities include:

1. Inspecting in-coming raw materials and parts to be used in production.

2. Inspecting goods in-progress.

3. Inspecting finished goods.

Inspection is therefore an **acceptance** activity. The chart in Figure 78-4 illustrates how inspection stations are located in the production line. The **in-coming** (or receiving) inspection of purchased items is very important. Parts and materials may be damaged in shipment, or the wrong material or size may be sent. Sometimes items are received which do not meet specifications. This inspection can be done by sampling or visual spot-checking, depending on the nature of the materials or parts.

In-process inspection takes place during part production. It may be an *informal* inspection in which the worker checks the items simply by examining and gaging them. In a *formal* inspection, a specially trained and qualified person from the QC department does the checking at critical points in the production and assembly process. Formal inspection is expensive. Care must be taken not to do more inspecting than necessary.

Final inspection ensures that the products meet specifications and work the way they are supposed to work. Any malfunctions or unacceptable items must be repaired or replaced. This is why in-process inspection is so important. Below-standard parts can be detected before assembly. This final inspection is made by qualified QC people. It is the goal of the QC program to have products in acceptable form at this final stage. With good in-coming and in-process inspections, there should be few rejects at this point. The familiar inspection labels are attached to products at the final inspection stage of QC (Figure 78-5).

Inspection Methods

The techniques used in product inspection vary according to what is being made. In a good QC program, every automobile or bicycle made is inspected thoroughly. However, it is not practical to examine every vitamin pill or steel washer. The kind of product being made determines the method by which it is inspected. The three main inspection methods are: complete, random sampling, and destructive.

Complete inspection (or 100% inspection) is expensive and means that every part or product is carefully examined, often by several people. Complex products such as machinery, vehicles, and appliances must be one hundred percent inspected. Even though this method is very reliable, it is not foolproof. There may be oversights due to human error, hidden material flaws, or problems with the inspection equipment.

In **random sampling,** a small number of items are taken from a larger batch and tested. The items are taken at random; that is, without any particular pattern or plan. For example, from a batch of ten thousand golf balls, any one hundred may be selected for testing. If the specifications state that not more than ten percent of a sample may be substandard, then ninety or more of the golf balls must meet the specifications in order for the whole batch of ten thousand to be accepted. If fewer than ninety meet "specs," the entire batch is rejected as scrap. Most product inspection is done by sampling methods. This method is reliable and far less expensive than complete inspection.

Destructive testing is an inspection method in which product samples are destroyed in order to assure their quality. Electric fuse samples must be destroyed to see if they do, in fact, work. Chairs are also inspected this way. Samples are "torture-tested" for strength and durability, and they are destroyed in the process. The test results for the samples indicate the strength and durability of the entire batch.

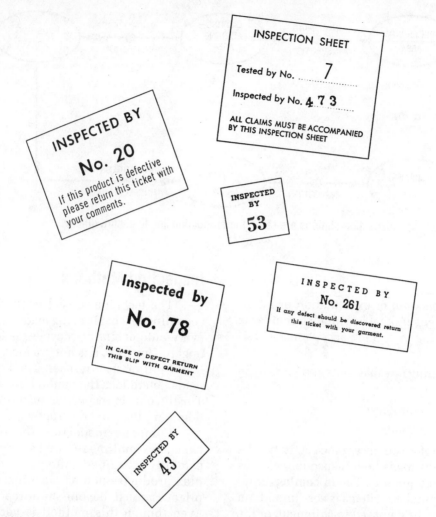

Fig. 78-5 Quality control labels such as these often are placed with the product after final inspection.

Planning for Quality Control

In unit 77, you learned about production planning and control. It is during this planning phase that the QC inspection stations are set up as part of the production plan. These stations are shown on the flowcharts with an inspection symbol. Remember that production is monitored to ensure that the work is being done correctly, at the right time, and to specifications. Recall also that one of the functions of production control is to take corrective action if needed. In the event quality standards are not being met, corrective action must be taken promptly to avoid wasting time and materials.

Every care must be taken to *prevent* quality errors, but also to correct them as needed. Some typical prevention activities include:

- studying defect or rejection records;
- examining customer complaints;
- improving inspector training; and
- conducting employee morale meetings.

While product quality is the responsibility of the entire company, the key to any successful QC plan is the individual worker on the production and assembly line. Assuming that the equipment and processes for doing work are adequate to meet the quality standards, it is the worker's attitude that makes the difference. Workers who are not well-prepared or who do not care about the quality of their work can cost a company dearly. They may turn out work that is not acceptable or that is just barely acceptable rather than of high quality.

To encourage a positive attitude about doing quality work, some companies use incentive plans. These plans can include things like bonuses, time off, and special awards for excellent work. In some companies, the workers help make decisions about how a job should be done through *quality circles.* When people know that their opinions and knowledge are valued, they care more about helping their company achieve its product quality goals.

IN FOCUS: *Industrial Quality Control Techniques*

The information and examples presented in this unit are simple, basic quality control methods. Modern industry uses these same methods but in a much different way. Because of the complexities of producing things such as airplanes or computers, industry must use more complicated QC techniques. Perhaps the earliest and best example of these newer methods is the statistical process control (SPC) system developed by Dr. W. E. Deming. SPC is a tool for building quality into a product as it is being made. It emphasizes setting rigid specifications and controls, and then continually improving them.

Another interesting, effective, and useful technique was created by Dr. Genichi Taguchi. According to the Taguchi Method, quality is the *loss* passed by the product to the user when the product is delivered. Customers suffer loss whenever they buy a faulty product because they are deprived of its use. If a product causes no loss, it is of high quality. If it causes high loss, it is of low quality. In QC, the focus should be on those losses that result from the deviation of the product's functional characteristic (what is expected of the product) from its target value (the ideal state of product performance from the user viewpoint).

For example, assume that the functional (or use) characteristic being studied is the force required to turn the faucet handle on a kitchen sink. The target value of the force is five pounds. Any losses due to the deviation (or movement away) of the force from the five-pound target are losses due to functional variation. A person who buys and installs this faucet expects it to work. If it is hard to turn, or won't turn at all, the person suffers a loss because he or she can't use the kitchen sink. The undesirable features that cause such variations from target values are called **noise** factors, and there are three types:

1. *Outer* noise refers to environmental conditions, such as changes in temperature and humidity, or the presence of dirt. A copying machine may jam because of moist paper or because of dirt in the system.

2. *Inner* noise is a malfunction between parts caused by deterioration. The jamming in the copy machine may be caused by a worn paper-feed mechanism.

3. *Between product* noise is caused by variations among product parts, supposedly made according to rigid manufacturing specifications. The jam may be a result of an improper paper-feed mechanism assembly or adjustment.

An automotive seat belt that fails to retract fully and gets caught in the car door as it is being closed is an example of noise. So are ceramic tiles that crack during firing, leaky faucets, and excessively noisy home furnace fans and refrigerators. The aim of QC is to design and manufacture products that are robust with respect to the three noise factors. The word *robust* implies that product functional characteristics are not affected by such noise or nuisance variations. Noise sources can be *off-line* manufacturing activities (occurring outside the actual production line), such as product design or process selection; or they can be *on-line* activities (occurring during or after product making), such as part-making or customer service. Obviously, if quality can be assured at the off-line stage, the need for on-line reworking will be prevented or reduced. The Taguchi Method, complex as it is, combines quality engineering and statistical techniques to reduce the causes of functional variations. One of its greatest advantages is to readily identify the source of a product fault and correct it immediately.

KNOWLEDGE REVIEW

1. Describe the term *quality control.*

2. Who in a company is responsible for quality control?

3. What are *specifications* and why are they written?

4. Describe a simple gage and explain its use.

5. Describe the various methods used for quality control inspections.

6. Write a research report on one of the industrial quality control techniques.

7. Visit a local industry and report on the QC methods observed there.

Unit 79

Human Resources Management

OBJECTIVES

After studying this unit, you should be able to:

- Describe the functions of a company human resources department.
- Understand how industry recruits and hires workers.
- Know the purposes and procedures of labor relations.
- Describe the importance of company public relations programs.

KEY TERMS

benefits and services	employee relations	health and safety	stockholder relations
collective bargaining	employment	labor relations	supervision
community relations	government relations	public relations	training
day-to-day relations	grievances		

Human beings design things, plan how they will be produced, and make them. They also sell and service them. (And they also consume the products.) People are assisted in these activities by tools, machines, and computers. But no product can reach the user without *people,* who are the most important part of the manufacturing enterprise (Figure 79-1). The purpose of this unit is to describe how this vital human resource fits into the process of making things.

Every company has a human resources, personnel, or industrial relations department whose function is to select, protect, and manage its people. One aim of this department is to attract and keep qualified employees. Another goal is to create and maintain good will between the company and its workers, and with the public. The things a company does to fulfill these goals can be classified into three programs: employee relations, labor relations, and public relations (Figure 79-2).

Fig. 79-1 People are the most important resource in product manufacturing. (Courtesy of Aluminum Company of America [ALCOA])

Fig. 79-2 Chart of human relations programs.

Employee Relations

Whether a company employs ten or a hundred or a thousand workers, these people must be trained and managed effectively to keep things running smoothly. This can only be done by tending to the employee needs, which range from hiring and training to safety and services. The chart in Figure 79-3 outlines these functions.

Employment

Before people can be hired, a company must decide what kind of workers it needs. This decision is based upon the kind and numbers of products to be made by that company. Once this decision has been made, *job descriptions* are prepared. These identify the type of work to be done, the necessary education and training, and the skills and knowledge required. The process of finding workers who meet the requirements of the job is called *recruiting.* Recruiting can be done in several ways. People (called recruiters) from the company may go to schools to talk with interested students.

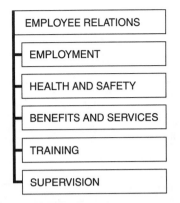

Fig. 79-3 Chart of employee relations programs.

Or the company may ask an employment agency to find qualified workers. Jobs may also be advertised in local papers or in trade or professional magazines. Recruitment is also done within the company, to allow its people to advance to better jobs.

People interested in the job complete job application forms. These forms ask for information about the applicant's education, work experience, and so forth. A typical application form is shown in Figure 79-4. Applicants with the right backgrounds may be asked to take some tests to find out who is best qualified for the job. They may also be requested to come for a job interview. During the interview, applicants are asked about their backgrounds, career goals, and experience. Salaries or wages are discussed, as well as promotion opportunities. The company's benefit package, including sick leave, retirement, insurance, and vacations, is explained. Applicants have the opportunity to ask questions about the job and opportunities for promotion and growth. The interview, along with the other information, helps the employer decide whether an applicant is qualified for the job, and how well he or she will get along with the other workers.

Training

New employees are introduced to the job by their supervisor. They also receive a general orientation about company policies and procedures. The new workers are often given a handbook which contains helpful information about the company and the way it operates. If training is required, they are enrolled in an appropriate program.

One kind of training is called *on-the-job training.* For example, a new machinist needs to become familiar with the equipment used and the schedule of jobs. He or she learns these things from a supervisor or co-worker while working on the job.

People learning highly skilled work may be placed in an *apprenticeship* program. This program includes classes as well as actual work on the job. Apprenticeship

TRAIL BIKE COMPANY

KIND OF WORK APPLIED FOR

FIRST CHOICE _____ YEARS OF EXPERIENCE _____

SECOND CHOICE _____ YEARS OF EXPERIENCE _____

DATE AVAILABLE TO START WORK _____ WILLING TO WORK ANY SHIFT? YES ___ NO ___

NAME (PRINT) _____ SOC. SEC. NUMBER _____

      (LAST)   (FIRST)   (MIDDLE)

ADDRESS _____ TELEPHONE NO. _____

    (STREET AND NO.)   (CITY)   (STATE)   (ZIP CODE)

ARE YOU A CITIZEN OF THE UNITED STATES? _____   IF NOT, DO YOU HAVE THE LEGAL RIGHT TO REMAIN PERMANENTLY IN THE UNITED STATES? _____   IF YOU ARE EMPLOYED MAY WE CONTACT YOUR EMPLOYER?   YES ☐   NO ☐   EMPLOYER TELEPHONE NO. _____

HAVE YOU EVER BEEN CONVICTED OF A CRIME? (OTHER THAN TRAFFIC VIOLATIONS)   YES ☐   NO ☐   IF SO GIVE DETAILS _____

UNITED STATES MILITARY SERVICE   DATES OF SERVICE   FROM _____ TO _____   BRANCH OF SERVICE _____

RATING AND TYPE OF WORK _____

LIST SERVICE SCHOOLS ATTENDE _____

INSTITUTION	NAME AND LOCATION OF SCHOOL		MAJOR FIELD OF STUDY
HIGH SCHOOL		GRADE COMPLETED 9 ☐ 10 ☐ 11 ☐ 12 ☐ DID YOU GRADUATE YES ☐ NO ☐	
COLLEGE		YEARS COMPLETED 1 ☐ 2 ☐ 3 ☐ 4 ☐ DID YOU GRADUATE YES ☐ NO ☐	
OTHER TRAINING		HOW LONG ____ ____ YRS. MONTHS	

(1) PRESENT OR LAST PLACE EMPLOYED _____

ADDRESS _____

CITY _____ ZIP CODE _____

KIND OF WORK _____

DATES OF SERVICE FROM _____ TO _____

REASON FOR LEAVING _____

(2) SECOND LAST PLACE EMPLOYED _____

ADDRESS _____

CITY _____ ZIP CODE _____

KIND OF WORK _____

DATES OF SERVICE FROM _____ TO _____

REASON FOR LEAVING _____

(3) THIRD LAST PLACE EMPLOYED _____

ADDRESS _____

CITY _____ ZIP CODE _____

KIND OF WORK _____

DATES OF SERVICE FROM _____ TO _____

REASON FOR LEAVING _____

ARE THERE ANY ADDITIONAL COMMENTS YOU WOULD CARE TO MAKE REGARDING YOUR ABILITIES, EXPERIENCE, OR SPECIAL SKILLS?

_____

_____

_____

SIGNATURE OF APPLICANT _____ DATE _____

Fig. 79-4 This is a typical application form used by industry when it recruits people.

Fig. 79-5 Skilled technicians are an important part of modern manufacturing. (Courtesy of Harley-Davidson Motor Company)

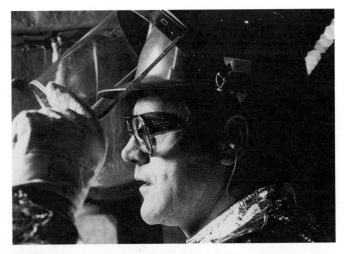

Fig. 79-6 How many pieces of worker safety equipment can you identify in this photograph? (Courtesy of Aluminum Company of America [ALCOA])

programs often last from three to six years. During this time, the worker is paid and the wage rate increases as he or she progresses in the program. Another type of training is the *cooperative* program. Some of you may participate in this kind of program through your school. In this case, a student attends school for part of the day and works the other part at a job approved by the school.

Modern manufacturing requires skilled people to design, plan, make, assemble, test, and market its many products. These teams of managers, supervisors, and skilled operators are an important part of the production process. The technicians shown in Figure 79-5 are assembling the spoked wheels of Harley-Davidson motorcycles. A company cannot be successful without this human resource.

Supervision

After people have been trained to do a job, their supervisor will observe them to see that the job is being done properly. If mistakes are being made, they must be corrected. Some workers need special attention and extra training. The supervisor also identifies those workers who are doing an especially good job and considers them for promotion.

Health and Safety

All workers must learn safety in their jobs to avoid injury, and prevent the waste of materials and parts. Accidents are costly, and all companies have safety programs to help prevent them. Workers are required to attend in-plant safety classes, and generally will receive first-aid training. People who run these programs are called safety engineers or safety supervisors. They sponsor safety campaigns, observe workers, and try to spot dangerous work habits or operations. Safety is an important part of a company's training program. Also,

the use of the proper clothing and gear is required and should be used (Figure 79-6).

The health and well-being of workers are also important. Workers who are not feeling well or who are working in unpleasant conditions may not be very productive. Managers work with union representatives to provide a safe work environment. Many companies have a medical office where people can get emergency help. Large companies have a full-time medical staff to care for those who are injured or become ill while at work.

Benefits and Services

Benefits include such things as paid holidays and vacations, retirement and sick leave plans, insurance, and service awards. These are called fringe benefits.

Services are usually of a social, cultural, or recreational nature. They can include company sports programs, parties for special occasions, annual picnics, discount tickets to special events, a regular newsletter, the use of the cafeteria for special events, and many other "extras." Such services help to develop a close relationship between workers and managers. Benefits and services help boost morale and increase loyalty because they show the workers that management cares about them.

Labor Relations

The second program in the human resources department is that of labor relations. In companies where the workers belong to a union, it is important to have good working links among managers, workers, and union representatives. Normal programs with the union usually include: negotiating labor contracts, handling grievances, and day-to-day operations with workers and union officials (Figure 79-7).

Fig. 79-7 Chart of labor relations programs.

Negotiating Labor Contracts

Union labor works under a contract, which is a binding agreement between workers and management. Representatives of the union and the company bargain to reach an agreement about working conditions, benefits, and wages. This process of offering proposals and counterproposals is known as **collective bargaining**. When the union and management representatives agree on the terms, the contract is put to a vote by the union members. If the majority ratify (approve) the contract, it goes into effect. If the workers reject the contract, more bargaining is called for.

Sometimes when agreement on a contract cannot be reached, the union calls for a *strike* (work stoppage). At times, a *mediator* (a go-between) is called in to help both sides work out their differences. When these negotiations are not successful, it may be necessary to accept *binding arbitration*. In this case an arbitrator is assigned to listen to the concerns expressed by both the union and the company. After hearing both sides, the arbitrator makes a decision to settle the dispute. This decision is final and must be accepted by both sides.

Handling Grievances

When disagreements arise over details in the contract, a grievance (complaint) may be filed (Figure 79-8). The contract spells out the procedure to follow in registering and processing a grievance. Usually, the worker registers the complaint with the shop steward. The shop steward is a fellow worker who serves as the official union representative for the worker. The complaint is presented to the company by the shop steward, who follows it until a solution is reached. In most cases, grievances can be settled without going to binding arbitration.

Day-to-Day Relations

The shop steward is in daily contact with members of the management team. Together they attempt to resolve minor issues of concern before they grow into big problems. Each side should be reasonable in its requests and demands, and be willing to compromise to resolve issues. A mutual feeling of trust and respect should be developed between both parties and their representatives. A harmonious, productive working environment depends on trust, respect, and willingness to work out problems. Any time a strike occurs, the two teams have failed in their responsibilities.

EMPLOYEE GRIEVANCE
TRAIL BIKE COMPANY

Dept. _____ Date _____ Time _____ A.M.
P.M.
Nature of Grievance _____

Signed _____ Clock No. _____
Committee Person _____
Reported to _____ Supervisor

Disposition by Supervisor _____

Date _____

Grievance Satisfactorily Settled	Referred to

Disposition by _____

Date _____

Grievance Satisfactorily Settled	Referred to Management-Shop Committee Meeting

Disposition by Management _____

Date _____ Signed _____

Grievance Satisfactorily Settled	Appealed

Fig. 79-8 A typical grievance form used in labor relations programs.

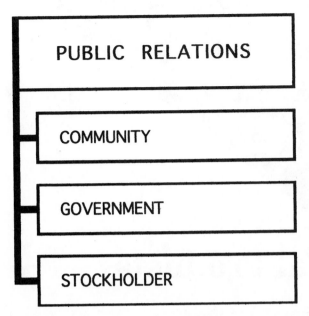

Fig. 79-9 Chart of public relations programs.

Public Relations

A third human resources program deals with public relations. The function of this program is to establish and maintain strong ties between the company and the local community, the government, the owners (shareholders), and the general public. Whereas the programs linked to employees and labor unions are considered *internal*, public relations programs are *external* (Figure 79-9).

Community

It is important for a company to establish and improve its image in both the local and national communities. To do this, it contributes to local charities, sponsors sports teams, or provides funds to build a new hospital or recreational center. The company regularly gets publicity for the special accomplishment of an employee. Such actions promote a good image for both the company and its workers.

On the national level, the company makes educational materials, product literature, and videotapes available to the public through its marketing communications office. They help publishers and authors by supplying pictures and technical booklets for use in their books. They run ads which tell the story of how their products support our improved lifestyle. Companies often help sponsor national or state events which appeal to the general public. Activities like these help the company gain public support and serve society at the same time. Everyone gains from their public-spirited actions.

Government

The company may work closely with government leaders at all levels to support a cause which is in its best interest. Lobbying can be very worthwhile and should not be considered bad or improper when conducted for the benefit of a large number of people. The public relations people also work closely with federal and state safety organizations.

Stockholders

Naturally, the company wants to have good relations with its stockholders. One way to do this is for the company to make a reasonable profit. However, success in developing a good public relations image is also important to the stockholders. The good program includes convincing the stockholders that the company's most important mission is to make high quality, safe, and usable products, and not just money for dividends.

KNOWLEDGE REVIEW

1. Briefly explain how companies hire employees.
2. List some typical fringe benefits which companies provide their employees.
3. What is the purpose of a labor union? How does it operate in a company to protect its members?
4. Write a research report on the history of the labor movement in America.
5. Explain the duties of the public relations department.
6. Study the employment opportunities section of your Sunday newspaper. Make a list of the kinds of metalworking-related jobs you find there.
7. Write a letter of application for one of the jobs found in the Sunday paper.

Unit 80

Marketing Metal Products

OBJECTIVES

After studying this unit, you should be able to:

- Describe the term *marketing*.
- Understand the purpose of market research.
- Differentiate between consumer and industrial goods.
- List the various aspects of packaging and advertising.
- Describe the purpose of a distribution system.
- Explain the importance of servicing in a marketing program.

KEY TERMS

advertising	marketing	packaging	servicing
distribution	market research		

Very simply, **marketing** is moving goods from the product makers to the product users. It involves getting the right things to the right people, in the right condition, at the right time, in the right amount, and at the right price. Meeting these conditions will result in satisfied customers, which is the goal of any marketing program. Marketing begins long before a product is ready for sale. In fact, market research, a vital part of marketing, contributes information that is used in the design of the product, and also in the advertising and distribution programs. Companies often will develop a complex marketing program, starting with the customer, to ensure that no detail is overlooked (Figure 80-1).

In manufacturing, both industrial goods and consumer goods are marketed. *Industrial* goods are sold to other manufacturers so that they can make consumer goods. Materials such as steel and plastic, product parts, and factory machines, are all considered industrial goods. Things which are sold to the public for daily use are *consumer* goods, including clothing, hardware items, and sports equipment.

Effective marketing involves many people doing a variety of jobs. Their work can be clustered into five major tasks (Figure 80-2): market research, packaging, advertising, distribution, and servicing. Some companies do all of these things themselves. Others hire marketing or advertising agencies which specialize in one or more of these functions.

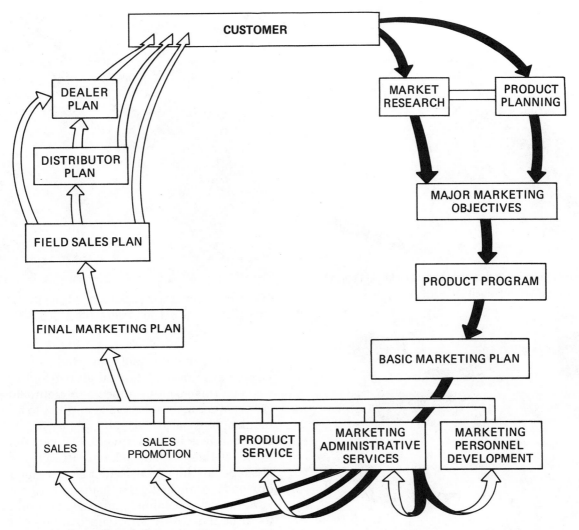

Fig. 80-1 An example of a total marketing program.

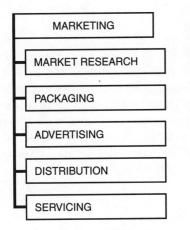

Fig. 80-2 A chart of marketing activities.

Market Research

Collecting data about a product and who will use it is called **market research.** This information can be obtained by conducting surveys, by studying records and trends, and by test marketing. By interviewing people, and mailing surveys and questionnaires, companies can collect information, which will assist product planners and advertisers. In test marketing, a new product is introduced in a limited area. Usually, this area is typical of the market the company wants to reach. Through such regional marketing, the company can learn how well its product may be accepted without having to distribute a large number of products.

Packaging

The design and development of a product's package, or its **packaging,** are often as important as the product itself. For example, a plastic bottle for liquid detergent, or an aerosol can for spray paint, are special packages which must work properly. Otherwise, the product is of no value because it is not accessible. Packages have five major purposes: they identify the product; they protect it; they hold it; they display it; and

Fig. 80-3 This machine is designed to form and glue corrugated boxes. (Courtesy of Moen Industries)

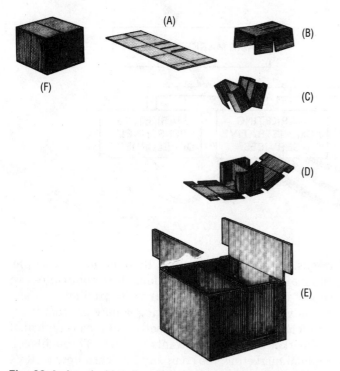

Fig. 80-4 A typical box forming sequence performed by the machine in Figure 80-3. The letters refer to the descriptions in the text. (Courtesy of Moen Industries)

Fig. 80-5 A convenient package for a familiar transparent tape. (Courtesy of 3M Commercial Office Supply Division)

machine in Figure 80-3 can produce up to twenty boxes per minute. A typical gluing/folding sequence done on this machine appears in Figure 80-4. Note at *A* that a hot melt adhesive is applied to a flat die-cut blank of boxboard. Next, at *B,* the legs of the blank are folded down into an inverted U position. The "U piece" is then folded back onto itself at *C* to form the H-shaped divider. The divider is now glued at *D* to the box body wrap shell, also made of boxboard. Finally, at *E,* the wrap is folded and glued around the divider to form the sturdy shipping container. The box is shown at *F* with the lid folded into position. Similar packaging machines have been developed to form plastic toothpaste tubes, aluminum soft drink cans, and boxes for breakfast cereal.

Packaging is closely linked with advertising because it also serves an advertising function. The package must be designed to aid in the sale of the product. It should be made of suitable materials and be attractive. It must contain information that clearly identifies the product, its contents, and its manufacturer. The sample package in Figure 80-5 not only holds and advertises the tape, but also provides a handy dispenser for it.

Advertising

Through **advertising,** the public is made aware of products and services which are available. Advertising can also be done for ideas or causes, such as "Pitch In" or "Drive Safely" or "Don't Do Drugs." Product advertising can do many things. It can inform the public about the product; point out its special features; state the cost of the product; or tell where to buy it. The many

they provide a means for shipping and storing. A package goes through the same research and development as does the product, to ensure that it will meet these needs and function as it should.

The package manufacturing industries design innovative and functional packages, as well as the machines to make them. For example, durable containers are created from corrugated boxboard materials and special forming machines. The speedy and reliable

Fig. 80-6 This sign advertises the availability of restaurants at an interstate highway exit.

ways to advertise include: newspapers, magazines, television, radio, and billboards (Figure 80-6).

Two main features in an advertisement are the product name and its trademark. These help identify the company and distinguish its product from anything similar offered by competitors. A trademark is a word, name, or symbol which a company uses to identify it and its products. Usually it is registered with the U.S. Patent and Trademark Office. It is illegal for a manufacturer to use someone else's trademark. One well-known trademark is shown in Figure 80-7. Note that it is followed by a circled letter R, meaning that the trademark is registered. A TM after the trademark means that it has not been registered, but it is still a trademark. Trademarks can appear on everything from products and advertising copy to airplanes and sporting goods.

Fig. 80-7 This registered trademark is applied to the products manufactured by the company. (Courtesy of Worth, Inc.)

Distribution

The process of getting goods from the manufacturer to the consumer is called **distribution.** There are many activities associated with distribution services. A company ships its products to a central storage warehouse (Figure 80-8), where they in turn are loaded onto trucks, trains, boats, and airplanes for distribution to its buyers or users. Both human-operated and computer-guided lift vehicles are used to retrieve the stored goods for shipment (Figure 80-9).

Technical representatives (tech-reps) regularly visit the users, such as hardware stores and factories, to inform them of new products or to restock their shelves.

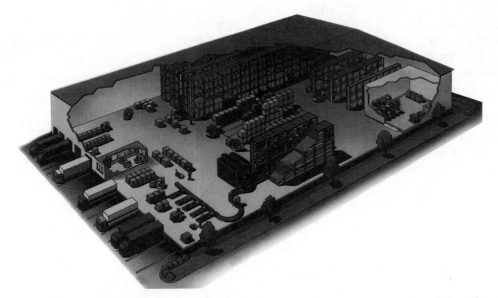

Fig. 80-8 A modern warehousing system. Goods are received, stored, and shipped with speed and precision. (Courtesy of Catalyst International)

Fig. 80-9 Safe, convenient, and efficient reach trucks store and retrieve warehoused goods. (Courtesy of Material Handling Associates, Inc.)

These tech-reps must know the products well and be able to answer any technical questions. For example, a drug company representative keeps doctors informed about the company's newest medicines. A person who sells tools visits repair shops and small factories to explain a new line of power tools.

These people must present the product in a convincing manner. They need to be able to explain financing arrangements and make the sale. Once the sale is made, the appropriate paperwork, such as receipts or order forms, must be completed. Orders are sent to the company for shipment from warehouses. These details are all part of the training which salespeople receive. Some of these representatives receive a salary, while others work for a commission which is usually a percentage of the sales price. The commission is often an incentive to create more sales. The more products sold, the greater the salesperson's income. Some companies pay a combination of salary and commissions.

Selling is the process of exchanging the product for money. It is done in shops by the salespeople who help you select a new mountain bike or a pair of blue jeans. This is the final stage of distribution—getting the product into the hands of the customer.

Servicing

As stated in unit 12, serviceability, or maintenance, begins at the design stage. It should be built into the product. Things sometimes go wrong with products due to customer use or misuse, and they must be repaired or replaced. A company must provide a convenient and reliable network of service agencies for its customers to take care of these needs. Remember that faulty products cause a loss to customers, because they lose the use of them. The company has a responsibility to service products quickly to compensate a user for his or her loss.

Most companies protect their customers from this loss by providing a product *warranty*. This assures the buyer that the product will perform as advertised by the maker. For example, you might purchase an electric hand drill that is guaranteed to operate satisfactorily for one year, against any defects due to faulty construction or materials. If the tool fails, due to no fault of the user, the company will repair or replace it free of charge. Customer satisfaction is important to a manufacturer, because an unhappy customer will probably never again purchase any of that company's products.

KNOWLEDGE REVIEW

1. What is *marketing*?
2. Explain the importance of marketing in product manufacture.
3. Name and explain the main marketing functions.
4. Write a report on market research methods.
5. Prepare a marketing plan for the project your class will mass produce.

Unit
81

Mass Production in the School Shop

OBJECTIVES

After studying this unit, you should be able to:

- Form a classroom corporation.
- Plan a project for mass production.
- Prepare bills of materials, flowcharts, route sheets, and Gantt charts.
- Learn and perform jobs as part of the class mass production activity.
- Work as a team member to ensure a smooth-running class mass production activity.

KEY TERMS

dispatching   monitoring   pricing

A school shop activity that is a popular and rewarding learning experience is the mass production of projects by students. This provides an opportunity to put into practice all the skills, technical knowledge, and manufacturing management information learned in the metalworking course. Some of these programs are organized under such titles as Junior Engineers or Junior Achievement. Others are operated by an individual teacher and the class (Figure 81-1).

Getting Started

To begin the activity, the first order of business is to form a corporation and elect or appoint the officers. Study the chart in Figure 75-9 for guidance. To do this correctly, you will have to work closely with your instructor, who will often act as president or chairperson of the board of directors. It is the job of this person to offer advice on the best way of operating your company. Work closely with your teacher on each phase of the work. Do not appoint the production line supervisor and operators at this time. They should be chosen later, when you know what product is to be made and what special skills will be needed.

Product Engineering

The manager of product engineering must work with the supervisor and operators. They should all work as a team to select three or four suitable project designs and make prototypes of them for presentation to the class. A number of project ideas can be found in section VIII of this book, or the class may elect to design something totally different. After class discussion, the final project is selected. For purposes of illustration, assume that you will be making the punch set shown in Figure 81-2. Study project drawings and prototypes carefully to become familiar with the work to be done

Fig. 81-1 Planning is an important part of a class mass production experience.

Fig. 81-2 The punch set is a good machine shop project.

TOOL BODY

Ø 0.75 MILD STEEL
ALL CHAMFERS 30°

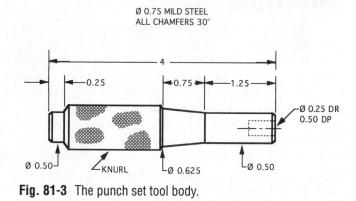

Fig. 81-3 The punch set tool body.

TOOL POINTS
Ø 0.25 X 1.5 DRILL ROD
HARDEN AND TEMPER

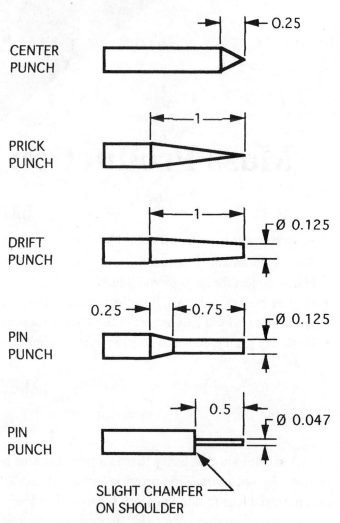

CENTER PUNCH

PRICK PUNCH

DRIFT PUNCH

PIN PUNCH

PIN PUNCH

SLIGHT CHAMFER ON SHOULDER

Fig. 81-4 The punch set tool points.

(Figures 81-3, 81-4, and 81-5). The tool points are suggested shapes. You should feel free to design some others to add to or replace these. As shown in Figure 81-5, the toolholder is made of wood. You may wish to change the design to include a metal hanger (Figure 81-6) so that the set is movable rather than fixed to a tool panel. Or, a metal holder can be designed to replace the wooden model. This is the kind of design decision-making that should be done in the product engineering part of the activity. This punch set is essentially a lathe project, but the wooden holder will require woodworking equipment such as a band saw, drill press, and sander.

The product engineers must work with the production planners to determine how many units will be

TOOL BODY AND POINT HOLDER
1 X 2 X 6 BIRCH

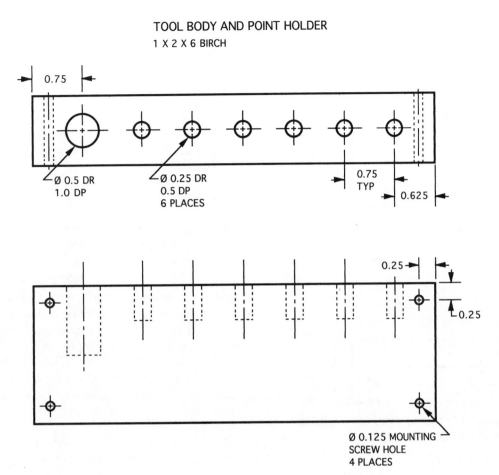

Fig. 81-5 The tool body and point holder.

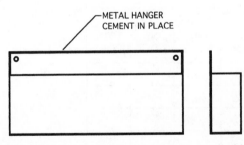

Fig. 81-6 A metal hanger can be added to the toolholder as a design change.

Toolmaking

The toolmaking group is a very important part of the production engineering team. Each part to be made must be analyzed to determine what kinds of special jigs, fixtures, and gages will be needed to make the work easier and more accurate. A good way to begin is to have the toolmakers and other operators make the punch parts. From this activity they can determine the kinds of tools which will be needed. Remember that

produced. If there are twenty students in the class, you must make at least that many punch sets.

you will be making a large number of punch sets, and that the tooling is necessary for efficient production. Very often the tools will be modified or improved as the part-machining takes place. Machine operators will frequently suggest changes, as they are the people using the tools.

One helpful tool is a sheet metal gage used to measure the length of the machined tool point prior to the cutoff operation (Figure 81-7). Another, a drilling jig tool, is used to locate the holes to be drilled in the wooden holder for the tool body and tool points (Figure 81-8). These are but two examples of tools needed for the punch set. You should study the other punch parts to develop other required tooling. Remember that lathe setups can be a part of tooling. For example, some punch points have tapers, and the tool post slide can be set at the proper angles to cut these tapers.

Production Planning

The manager of production planning and the other members of the team must prepare the *bill of materials* (Figure 81-9). Then a *process flowchart* must be pre-

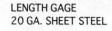

LENGTH GAGE
20 GA. SHEET STEEL

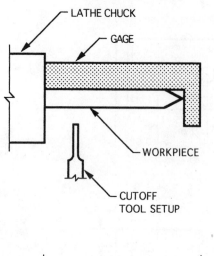

LATHE CHUCK

GAGE

WORKPIECE

CUTOFF
TOOL SETUP

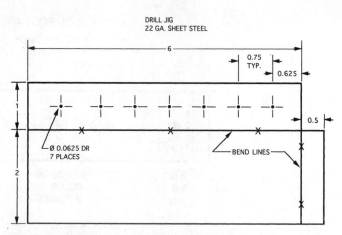

DRILL JIG
22 GA. SHEET STEEL

6

0.75
TYP.

0.625

1

0.5

2

Ø 0.0625 DR
7 PLACES

BEND LINES

Fig. 81-8 This simple drill jig is used to locate the holes to be drilled in the wooden toolholder.

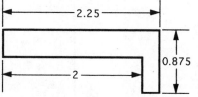

2.25

2

0.875

Fig. 81-7 This gage is made of sheet metal and is used to measure the lengths of the tool points.

pared for each part, such as the sample shown in Figure 81-10. This will help to identify the required operations and equipment. Next, a *route sheet,* which is a plan of the procedures for making the parts, is written for each part. Follow the information listed on the process flowchart. A sample route sheet for the center punch tool point is shown in Figure 81-11. You will have to prepare additional route sheets for each of the other parts. It is wise to have one or two students make trial runs of the part machining so that any errors or problems can be found and eliminated. For example, the operators

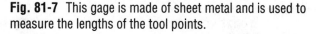

BILL OF MATERIALS	PRODUCT NAME: Punch set			WRITTEN BY: *J. Lind*			
				DATE: *March 18, 1998*			
PART NAME	DIMENSIONS			MATERIAL	NO. PCS.	UNIT MEAS.	UNIT COST	TOTAL COST
	T	W	L				
Tool body	0.75		4	*Mild steel*	20	*in.*		
Tool point	0.25		1.5	*Drill rod*	120	*in.*		
Holder	1	2	6	*Birch*	20	*Bd. Ft.*		

Fig. 81-9 A bill of materials for the punch set project.

PROCESS FLOWCHART		WRITTEN BY:	*J. Perez*		
		DATE:	*March 12, 1998*		

PRODUCT NAME: *Punch set*	FLOW BEGINS:	*Storage*	FLOW ENDS:	*Storage*

PART NAME: *Center punch point*	PART NO.:	PAGE *1* OF *1*

TYPE OF PROCESS ☐ PART MANUFACTURE ☐ SUBASSEMBLY ☐ FINAL ASSEMBLY

PROCESS SYMBOLS

◯ Operation (changing or working a material)   ☐ Inspection (examining a part for quality)   ▽ Storage (keeping a part safely until needed)

⇨ Transportation (moving a part from one place to another)   D Delay (holding a part until next activity)

OPER. NO.	STA. NO	PROCESS SYMBOL	DESCRIPTION OF PROCESS	MACHINE OR TOOL	TOOLING NO.
1		◯⇨☐D▼	*Obtain drill rod*		
2		◯➡☐D▽	*Move to lathe*		
3		●⇨☐D▽	*Mount rod in lathe*	*Lathe*	
4		●⇨☐D▽	*Measure point length*	*Gage*	
5		●⇨☐D▽	*Set tool post angle*	*Lathe*	
6		●⇨☐D▽	*Cut taper*	*Lathe*	
7		●⇨☐D▽	*Remove from lathe*		
8		◯⇨■D▽	*Inspect*	*Visual*	
9		◯➡☐D▽	*Move to cutoff lathe*	*Lathe*	
10		●⇨☐D▽	*Cut to length*		
11		●⇨☐D▽	*Remove from lathe*		
12		◯⇨■D▽	*Inspect*	*Visual*	
13		◯➡☐D▽	*Move to storage*		
14		◯⇨☐D▼	*Store for assembly*		
15		◯⇨☐D▽			
16		◯⇨☐D▽			
17		◯⇨☐D▽			
18		◯⇨☐D▽			
19		◯⇨☐D▽			
20		◯⇨☐D▽			

Fig. 81-10 A sample process flowchart.

will find that it is better to use a long piece of drill rod, rather than short pieces, when making the tool points.

After the routing sheets are completed, a *schedule* or Gantt chart is made. By studying the materials and routing sheets, you can set up a timetable for the work. You can estimate the time needed to make each part, or time studies can be made to get more accurate figures. Your instructor can be very helpful here. Look at the sample schedule in Figure 81-12. The days are marked in five-minute segments. You may choose any interval that is convenient for the length of the class period. You should refer to this chart as production begins. Keep it up to date, so you can tell how you are doing.

A *project flowchart* is then made (Figure 81-13). Note that every operation is listed, and each inspection and assembly point identified. From this, you can plan the flow of materials through the shop. While a plant layout plan is necessary in industry, it generally is not practical or possible to move machines around the school shop.

PRODUCT NAME: Punch set	MATERIAL: Drill rod	DRAWING NO: 81-4	ROUTE SHEET	
PART NAME: Center punch point	WRITTEN BY: D. Gregg	UNITS REQ'D. 20	DATE: 3/19/98	PAGE 1 OF 1
OPER. NO.	OPERATION DESCRIPTION	TOOL—MACHINE—EQUIPMENT DESCRIPTION	NOTES	
1	Mount 12 " drill rod	3—jaw chuck; lathe		
2	Measure point length	Length gage		
3	Set tool post 45° angle	Lathe; tool		
4	Cut taper	Lathe; tool		
5	Remove rod from lathe			
6	Transport to cut off lathe		New location	
7	Cut off to length	Lathe; tool		

Fig. 81-11 A sample route sheet for the center punch tool point.

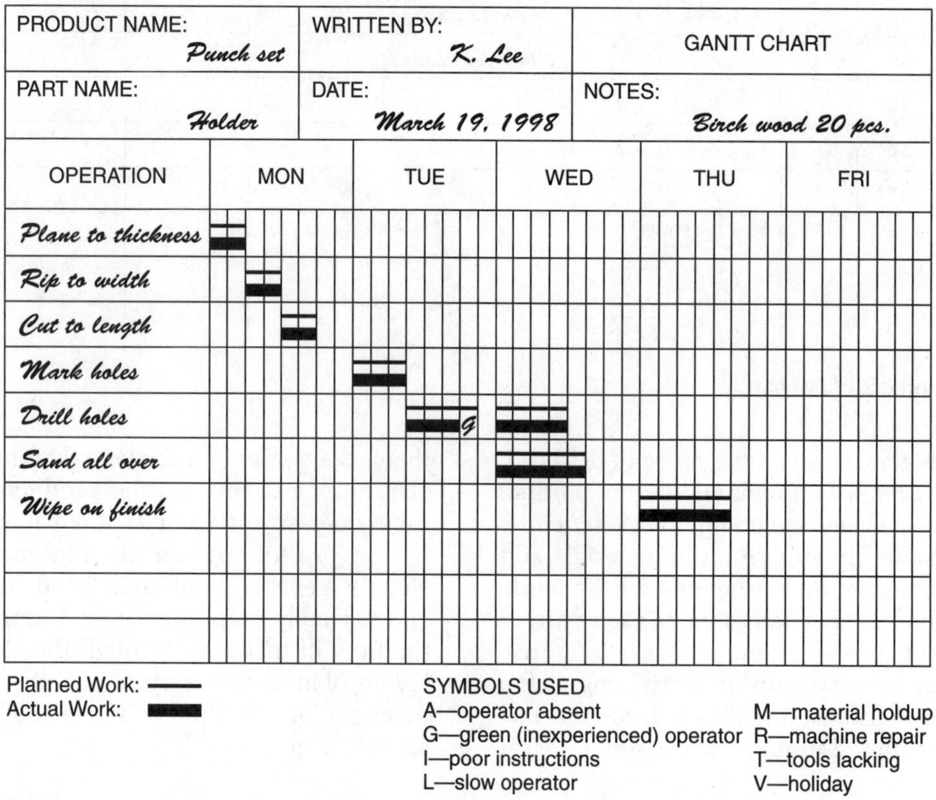

PRODUCT NAME: Punch set	WRITTEN BY: K. Lee	GANTT CHART			
PART NAME: Holder	DATE: March 19, 1998	NOTES: Birch wood 20 pcs.			
OPERATION	MON	TUE	WED	THU	FRI
Plane to thickness					
Rip to width					
Cut to length					
Mark holes					
Drill holes		G			
Sand all over					
Wipe on finish					

Planned Work: ——
Actual Work: ▬▬

SYMBOLS USED
A—operator absent
G—green (inexperienced) operator
I—poor instructions
L—slow operator
M—material holdup
R—machine repair
T—tools lacking
V—holiday

Fig. 81-12 A sample Gantt sheet, used in scheduling the production work.

CENTER
PUNCH POINT

▽ DRILL ROD
⬂ TO LATHE
◯ MOUNT ROD
◯ MEASURE POINT
◯ SET POST ANGLE
◯ CUT TAPER
▢ INSPECT
◯ REMOVE
⬂ MOVE TO CUTOFF LATHE
◯ CUT TO LENGTH
◯ REMOVE
▢ INSPECT
⬂ MOVE
▽ STORAGE

Fig. 81-13 This project flowchart presents a total picture of the flow of materials and operations for the punch set project.

Quality Control

Quality control procedures should be discussed with the instructor. Because this is a simple project with no moving parts, quality control is important but not difficult. You should write out the standards of acceptance you desire. These will help the inspectors to be aware of what flaws or errors to watch for. Here is a list of errors that might occur and may need correcting:

- rough machining cuts
- burrs not removed or sharp edges present
- holder holes crooked or out of line
- poor tapers
- incorrect point sizes
- poor finish or scratches
- improper heat-treating

Human Resources Management

With the production planning completed, the next step is to select machine operators and other personnel. You will want all of the class members to take part. Study the routing sheets, schedules, and flowcharts. Determine how many and what kinds of workers you will need for each job. You may wish to formally interview and hire people. Invite the class members to apply for jobs. Then work with your instructor to train them in the right ways to do the various jobs. Remember to include safety instruction.

Production Control

When the planning and training functions are finished, the production-line work may begin. Supervisors are very important here. They will have to make sure that safety procedures are followed and that the work is being done correctly and on schedule. Your teacher can be helpful to you.

Dispatching is issuing work orders to start production. In school shop mass production activities, this only requires a meeting of the managers and supervisors to determine that production can indeed begin. Supervisors will now distribute route sheets and have the operators begin the part-machining, as well as **monitor** the work to make certain that it moves smoothly. Any problems should be identified and solved, such as machine breakdowns or slow operators. *Corrective actions* may include requirements for a different machine, operator retraining, or a better flow of raw materials. Again, your instructor can help with these control activities.

Marketing

The marketing function is simple if school projects are to be given only to class members. In this case, the cost of the projects can be computed to determine the price to be charged to each student. However, if they are to be sold to other people, you will also have to solve the problems of packaging, advertising, distribution, and selling.

The **price** is computed by collecting data on material costs, overhead, and labor costs. *Material* costs are very easy to collect. Study one of the supply catalogs or the shop materials price lists to get the costs of the materials you used in manufacturing. *Overhead* costs should include a reasonable estimate of the heat, light, water, and power needed for manufacturing. Include a figure for a fair rental cost of the shop and the equipment and tools used. Finally, add in the costs for *labor*. Your instructor and school business office can help with this.

Several of the team members can develop packages for the product if these are needed. Remember that the package should protect the product and make shipping and handling easier. You may also wish to prepare advertising posters. Set up a sales booth or sell door to door in your neighborhood. Remember to include marketing costs in your final cost figures. With the final accounting, you can determine if you made a profit.

A Final Word

This is your production activity, and it should be planned to meet your special needs. You may decide that the suggested planning sheets are too complicated, or are unnecessary for the project you elect to produce. You might want to contract out the part-machining to select teams of operators on a job-shop basis, or to organize small production groups to do the work. Be as creative in your shop organization as you were in your project design and selection. The main goal is to have a meaningful and interesting mass production experience.

KNOWLEDGE REVIEW

1. Describe the process of selecting a project to mass produce.
2. Describe the functions of the product engineering, production planning and control, quality control, and marketing groups.
3. List the typical paperwork produced by the production planning team.
4. List some possible steps when selecting class members for specific jobs.
5. Pricing is based on what cost figures?
6. Write a research report on how to start a corporation.
7. Visit a local industry and compare its manufacturing operations with those which took place in your class.

SECTION 5

COMPUTERS IN MANUFACTURING

All modern industries, from the smallest to the largest, use computers as aids to their manufacturing systems. Product design and planning; factory management; materials movement, processing, and storage; and many other functions all depend heavily on computers for efficient operations. Modern computer systems can link designing with making, by working from a common data or information base (Figure S5-1). Humans continue to be the key element in automated factories by providing the ideas and skills necessary to operate these computers and plan for their wise use. In the next several units we will explore this amazing technology and learn how it is used in manufacturing.

Figure S5-1 Computers join the designing and making activities in modern manufacturing. (Courtesy of Tricon Industries, Inc.)

451

Unit 82

Computer Basics

OBJECTIVES

After studying this unit, you should be able to:

- Understand the operation of a simple computer.
- Describe the three types of computers.
- Explain the basic parts of a computer.
- Describe computer input devices.
- Describe computer output devices.
- Know the difference between computer hardware and software.
- Know the difference between application programs and operating systems programs.

KEY TERMS

analog	digitizing tablet	machine language	output devices
application program	disk	memory	plotter
bit	hardware	microchip	printer
byte	hybrid	microprocessor	RAM
central processing unit	input devices	monitor	ROM
CD-ROM	joystick	mouse	software
computer	keyboard	operating systems program	storage
digital			

A **computer** is an electronic machine that can handle information and make calculations at an amazing speed. The current machines are powerful, convenient, and easy to use. For example, the model shown in Figure 82-1 features 8MB of RAM; a 1.2GB hard drive; a large, comfortable keyboard; and a 14-inch color monitor. From your personal experience with these machines, you know that they can be used to play exciting and challenging video games. Office workers use them to type letters and keep financial records. They are a help to farmers in keeping records of their livestock and crops. People in industry design and plan the manufacture of products with the aid of computers. You and every other young person must become computer literate by learning how to use these remarkable machines.

Fig. 82-1 This modern computer is a powerful, convenient, and user-friendly machine. (Courtesy of Apple Computer, Inc.; photograph by John Greenleigh)

Types of Computers

There are three kinds of computers in use today: analog, digital, and hybrid.

Analog

The **analog** computer processes continuously variable information, such as speed or temperature or time. It expresses one physical quantity in terms of another. For example, an analog wristwatch indicates time by the movement of the hands around the dial. Here one quantity (time) is measured in terms of another (the position of the dial hands) (Figure 82-2). Because the hands move constantly, there is never really any exact time. Analogs do not measure in precise, separate units; they measure work continuously.

These analog instruments can process data where several quantities vary continuously over a period of time. For example, they are used in airplane cockpit simulators because they can present the relationships of speed and altitude, and thus imitate the behavior of real aircraft. Most of these are special-purpose machines, usually designed for engineering and scientific tasks.

Digital

Digital computers are general-purpose machines that work with whole numbers (digits) to solve problems. Referring to the wristwatch example, a digital watch expresses time directly in terms of the numbers displayed on the watch face. The digital computer is the most common type of computer; the kind found in factories, schools, businesses, and in the home. The information in this unit is therefore given over to a discussion of digital computers.

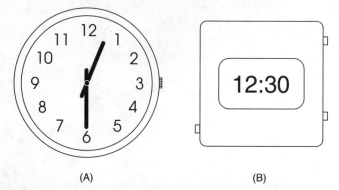

Fig. 82-2 The analog watch (A) and the digital watch (B).

Unlike analog computers, digital computers require that data be in precise amounts. It responds only to exact signals, and not to varying signals. The signals for digital computers are electric pulses called **bits.** The bits can be in only one of two forms: on or off; they either exist or they don't. The number *1* represents an on bit, and *0* (zero) means an off bit. Because the computer tracks only these two numbers, it is called a *binary* system. The word *bit* is a shortened form of the term *binary digit.* Patterns of bits are used to form characters called **bytes.** (One byte can contain eight bits.) Computers recognize a value for each byte. For example, in one computer language, the letter *A* is represented by 11000001. However, a byte does not always indicate a letter; it may also mean a number or a punctuation mark.

The patterns of ones and zeros feed data to the computer and form a special **machine language.** It is easily understood by the computer, but not by humans. Therefore, other languages have been developed that are more like the human language. One of the most popular early programming languages was BASIC (Beginners All-purpose Symbolic Instruction Code). Created in the 1960s, it was based on English words, punctuation marks, and algebraic notations. BASIC has generally been replaced by two other programs, "C" and Pascal. People write programs (or computer instructions) in one of these languages, and the computer then translates these into machine language.

Computers range in size from large supercomputers to the small, portable *microcomputer.* Most of the common school, office, and home machines are microcomputers. One popular type of portable computer is the *laptop,* which is small and powerful, and can be folded like a small brief case (Figure 82-3). Note the convenient built-in trackpad, which in this machine replaces the familiar mouse.

Hybrid

A **hybrid** computer combines the analog and digital operations, such as in numerical control machining systems (Figure 82-4).

Fig. 82-3 An efficient and portable laptop computer. (Courtesy of Apple Computer, Inc.; photograph by John Greenleigh)

Fig. 82-4 A hybrid computer is commonly used in automatic milling operations. (Courtesy of Cincinnati Milacron, Inc.)

Computer Hardware

There are five major physical parts of a computer system, known as the **hardware.** These components are: the central processing unit, memory, storage, input devices, and output devices (Figure 82-5). The nonphysical computer element is the instructional program, or the *software.*

Central Processing Unit

The **central processing unit,** called the CPU, is the brain of the computer. Here all the control functions

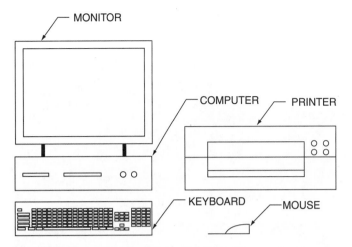

Fig. 82-5 The parts of a common computer.

Fig. 82-6 A computer chip (A) and the package that contains and shields the chip (B) are shown here. The quarter is about 24 mm in diameter, and the chip is 4 mm square.

take place, and all the data are processed and held. In a microcomputer, the CPU is contained on a single **microchip,** which is a quarter-inch-square (or circular) wafer of silicon holding over a million tiny transistors (Figure 82-6). A CPU that is contained on a single chip is called a **microprocessor.**

Memory

Computer memory is also located on chips in the CPU, but these chips are different from the microprocessors. **Memory** chips hold information for quick access by the CPU. There are two types of memory. **RAM** (random-access memory) stores information temporarily, while the computer is working. Data can be loaded (written) or retrieved (read) from RAM. However, once the computer is shut down, any data in RAM are erased. For example, when you load a program into a computer, you transfer it from the disk onto the RAM chip. When you shut down the com-

Fig. 82-7 A typical floppy disk.

Fig. 82-8 This operator is using a digitizing tablet and a stylus. (Courtesy of Grob Werke, GmbH)

puter, the program is automatically erased and must be reloaded to be used again.

The power of a computer is expressed in the number of bytes of RAM it contains. Microcomputers typically have from 8MB to 16MB of RAM. One MB (megabyte) equals approximately one million bytes.

ROM (read-only memory) chips provide for the permanent storage of information, but only the CPU can read it. ROM is put in at the factory and usually cannot be erased. Information can be read from ROM, but no additional information can be added to it.

Storage

It is not practical for large amounts of data to be kept inside the computer. For one thing, if someone loads a program into RAM, it will be lost when the computer is shut down. Another form of memory, the magnetic **disk,** is therefore used for long-term mass storage. One common type is a very thin (floppy) plastic disk, 3.5 inches in diameter and coated with tiny metal particles called *oxides* (Figure 82-7). When a special magnetic head passes over the disk, it arranges the particles into a precise digital pattern. When you insert a floppy disk into the computer disk drive, and give the computer the proper instructions, the information in the RAM is transferred to the disk for storage. This disk may then be ejected from the computer for safe, "backup" storage.

Input Devices

The special tools needed to feed information (or data) into the computer are called **input devices.** The most common of these is the **keyboard,** which closely resembles the ordinary typewriter keyboard. Most computers also have a separate numeric keypad alongside the keyboard for the quick and convenient entry of numbers. Several other special keys are found on the computer board: a *delete* key to erase characters which

have been typed; *arrow* keys to move the cursor around the screen; and *control* keys to be used with other keys to call up special symbols.

A **digitizing tablet** is an electronic drafting board. A drafter uses an *electronic pen* or stylus, which is wired into the tablet, to send signals to the computer to move the cursor around the screen. This allows the drafter to create a drawing on the computer screen via the tablet (Figure 82-8). A *puck* with a plastic lens and crosshairs also can be used to transfer an existing paper drawing to the screen.

A **joystick** is familiar to anyone who has played computer games. It can "steer" a car or direct the firing of "bullets" at an enemy from outer space. Joysticks also are used to direct the screen cursor to desired positions to create a computer image.

Another input device is the **mouse.** This hand-controlled tool is rolled on a surface to move the screen cursor to a desired position. It is fast, precise, and easy to use.

One of the newer input devices is the **CD-ROM,** which looks like the compact disk (CD) you use to play music. The CD-ROM contains much more information than a floppy disk. For example, a complete encyclopedia and dictionary can be placed on one disk. Computers must have a special CD-ROM optical scanner to read the disk. They can only be used to input data; a computer cannot write information onto a CD-ROM disk.

Output Devices

The drawings and letters which result from computer processing must be made available to the user. These data can be moved out of the computer or "outputted" in several ways. The most common output device is the **computer monitor,** or display screen. Every time you enter data, they come up on the monitor as visible "soft copy." If you wish a hard copy of your

Fig. 82-9   A small, convenient, and reliable inkjet printer. (Courtesy of Apple Computer, Inc.; photograph by John Greenleigh)

Fig. 82-10   Plotters draw ink lines on paper rolls to reproduce large maps and drawings. (Courtesy of CalComp)

work, it must be printed. Hard copies of data are made with a *dot matrix, inkjet,* or *laser* **printer.** The reliable, efficient, and compact inkjet printer shown in Figure 82-9 produces laser-quality prints. Large drawings and maps usually require a **plotter,** which actually draws lines on paper with a pen (Figure 82-10). Recording data on disks for storage is also an output method.

A special kind of output is the instructions a computer can send directly to a computer-controlled milling or turning machine. This technology is further described in unit 85.

A summary illustration of computer hardware appears in Figure 82-11.

Software

A computer can do nothing without a set of instructions called a *program.* The program, called **software,**

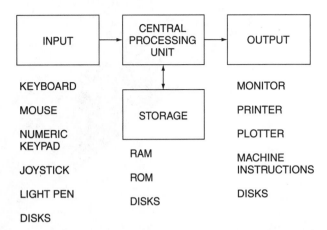

Fig. 82-11   The computer hardware elements are shown here as part of an input-process-output diagram.

tells the computer what to do with the data it receives. There are many of these programs, each written to perform different design, drafting, and processing tasks.

There are two classes of programs. The **application program** is a software package that allows the user to play games, type letters, prepare mechanical drawings, or solve engineering design problems. Application programs always reside in the computer's RAM, and not in ROM. They are transferred to RAM from a disk. For example, when you want the computer to act as a word processor, you select the disk with the word processing application program on it, and transfer that program to RAM.

The **operating systems program** manages the computer and its input, output, and storage functions. For example, it is used to transfer an application program from a disk into the computer's memory; or to transfer application program data from the computer's memory to a magnetic disk for storage.

KNOWLEDGE REVIEW

1. Sketch the main parts of a computer and write a short paper on how it operates.

2. Take a tour of your school and report on the kinds of computers you find and what they are used for.

3. Explain the difference between a bit and a byte.

4. Explain the difference between analog and digital computers.

5. What is an application program? What is an operating systems program?

6. Explain the meanings of the terms *RAM* and *ROM.*

7. What is the difference between a chip and a disk?

8. List some typical input, output, and storage devices.

9. Prepare a research report on supercomputers.

Unit 83

Computer-Aided Design

OBJECTIVES

After studying this unit, you should be able to:

- Explain the meaning of the term *computer-aided design.*
- Describe the four main CAD functions.
- Explain the purposes of rapid prototyping.
- Describe two examples of rapid prototyping.
- Explain the process of sheet metal layout and nesting.

KEY TERMS

automated drafting	engineering analysis	kinematics	rapid prototyping
computer-aided design	geometric modeling	LOM process	stereolithography

In addition to the board and instrument drawing methods described earlier in this book, most modern industries also use computers in their design work. As you learned earlier, a computer is an electronic machine that makes calculations and processes information at very high speeds. Its use in drafting and design work is called **computer-aided design,** or CAD. These CAD systems are interactive, meaning that the user directs (interacts with) the machine to create drawings or solve engineering design problems. The person supplies the input, the computer processes the data, and a drawing is the output (Figure 83-1).

In the CAD process, the operator feeds data to the computer through a keyboard. The computer reads and follows these instructions because it has been programmed to do so. The results are displayed on the computer screen or monitor. This graphics display is made up of the same geometric elements (points, lines, circles, and planes) used to make a mechanical drawing with instruments. By using the keyboard or other special controls, the user also can command the computer to change a drawing in some way. For example, the drawing can be made larger or smaller, moved to another part of the screen, crosshatched or sectioned, or changed from a three-view to a pictorial drawing. Metric or customary dimensions can be added, and the size scale can be revised.

It takes a drafter about the same amount of time to produce a drawing using instruments as it does with a computer. The advantage in using a computer is that changes can be made quickly without erasing and redrawing. CAD systems obviously give designers tremendous advantages in their work. If you are planning a drafting career, you should be competent in both board and computer systems.

Fig. 83-1 Inkjet plotters can produce bright, detailed images such as this racing car. (Courtesy of CalComp)

CAD Functions

A CAD system operates by using the application programs described above. These programs are used to perform four main CAD functions: geometric modeling, engineering analysis, kinematics, and automated drafting.

Geometric Modeling

A **geometric modeling** program permits the user to create the shape of an object as either a wireframe model (Figure 83-2) or a solid model (Figure 83-3). The computer changes these models into digital (numerical) information and stores it for later use. The modeling function is important because it serves as a basis for engineering analysis and automated drafting. It can also be used to create numerical control instructions for computer-aided manufacturing (CAM), which is discussed in unit 85.

Engineering Analysis

The geometric model information is stored in the computer. The user can command the computer to figure the surface area, weight, volume, or other characteristics of the part model. This is **engineering analysis,** and it also includes performing a *finite element analy-*

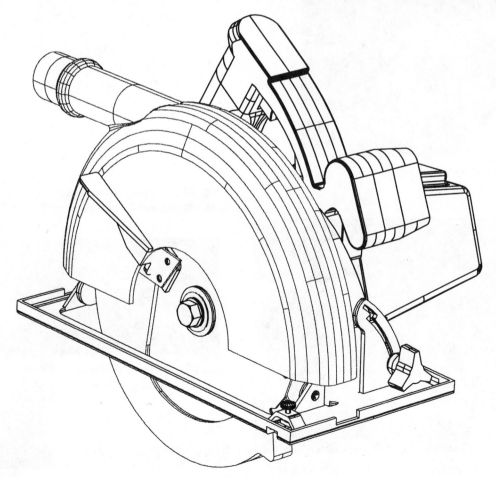

Fig. 83-2 A wireframe drawing of a circular saw. (Courtesy of SDRC)

sis (FEM) of the part. In this method, the part is broken down graphically into a "mesh" of sections, or elements. The computer then uses these elements as a base to figure stresses and deflections (bending) in the part. This analysis would take many hours of work by ordinary "hand" methods. The computer can do it in minutes, or seconds, and more accurately. A FEM of a cordless screwdriver body is shown in Figure 83-4.

Kinematics

Some CAD programs can show the movements and clearances of such moving parts as automobile door hinges or hood linkages, the bucket on an endloader, or a farm plow blade. The study of such motion is called **kinematics.** Before computers, the actions of such mechanisms often were determined with pin-and-cardboard models. These trial-and-error methods were often inaccurate and could take many hours of

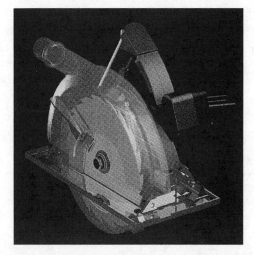

Fig. 83-3 A solid model drawing of the circular saw. (Courtesy of SDRC)

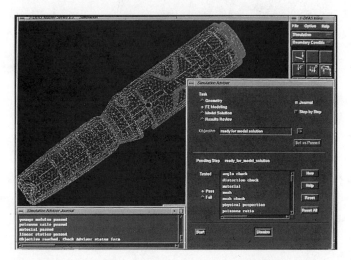

Fig. 83-4 This FEM drawing of a cordless screwdriver is used to analyze the strengths and weaknesses of the plastic body structure. (Courtesy of SDRC)

work. A typical kinematic analysis of a front-end loader mechanism appears in Figure 83-5.

Automated Drafting

Automated drafting programs are used to generate (draw) up to six views on the display. For example, it might draw the front, top, and right side of an object. A design change in one view is automatically made in every other view. Detail drawings, assembly drawings, and pictorials can be made in this way, complete with dimensions. These drawings also may appear as a part of a set of analysis images (Figure 83-6).

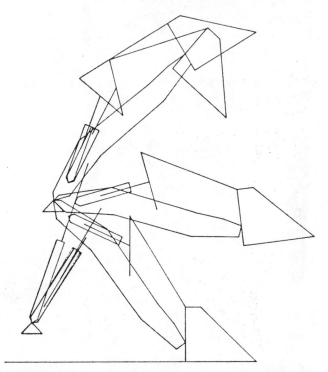

Fig. 83-5 Kinematic drawings are used to check the motions of objects, such as this front-end loader bucket. (Courtesy of SDRC)

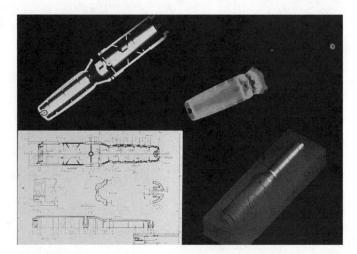

Fig. 83-6 An engineering drawing of the cordless screwdriver, with additional structural views. (Courtesy of SDRC)

A hard copy can be made of the drawing on a plotter, as mentioned earlier.

Rapid Prototyping

Parts of units 11 and 12 dealt with the subject of preparing experimental product models and mockups to test preliminary design results. These artifacts are called *prototypes,* and often are prepared by model makers employing such methods as numerical control (NC) machining, or hand modeling from clay, wood, or plastic foam. To reduce costs and speed up model-building operations, a number of newer prototyping techniques are used, whereby CAD plans are quickly transformed into solid objects and not merely 3-D drawings. The term used to define this concept is **rapid prototyping,** and it is an important CAD application with a number of advantages. Most important, it gives the designer a 3-D solid object to hold and evaluate. Other ad-

vantages include speed, accuracy, and the elimination of design errors. Two of these prototyping systems are described below.

LOM Process

The **Laminated Object Manufacturing** (LOM) process is a development of Helisys, Inc. It produces solid objects by bonding and laser cutting layers of sheet material such as papers, plastics, or composites. The system uses CAD data inputted into the LOM machine computer (Figure 83-7). A cross-sectional slice is produced, and the laser cuts the outline of the slice. A new layer of material then is bonded to the top of the previously cut layer. Another cross section is prepared and cut, and the process continues until all the required layers are laminated to create the finished piece (Figure 83-8). The solid object is now available for further analysis and possible modification (Figure 83-9). Objects can be machined and finished for use as molding and casting patterns, or for test models for the automotive, aerospace, medical, computer, and electronics industries. The golf club heads shown in Figure 83-10 are an example. The LOM-created mold was filled with wax to create a pattern for investment casting. (Refer to unit 51 for information on investment casting.) Fifty of these heads were produced for golf course testing purposes.

Stereolithography

Stereolithography rapid prototyping was developed by 3D Systems, Inc. The process combines the technologies of computers, lasers, optical scanning, and photochemistry by using plastics which change from liquid to solid when exposed to ultraviolet light. The

Fig. 83-7 A LOM rapid prototyping machine. (Courtesy of Helisys, Inc.)

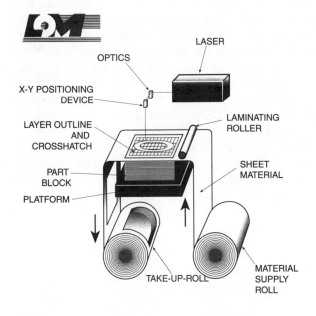

Fig. 83-8 A diagram of the LOM process. (Courtesy of Helisys, Inc.)

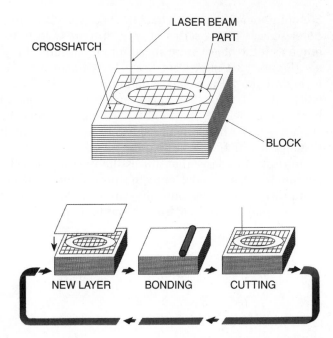

Fig. 83-9 Analyzing a LOM solid workpiece. (Courtesy of Helisys, Inc.)

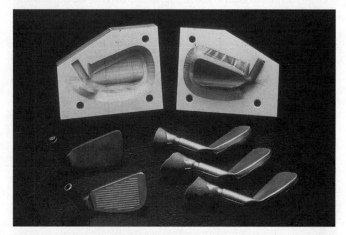

Fig. 83-10 This golf club head mold was produced by the LOM process. (Courtesy of PML)

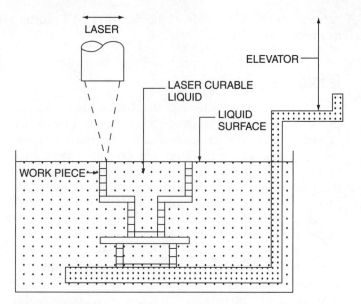

Fig. 83-11 A diagram of the stereolithography rapid prototyping process.

StereoLithography Apparatus (SLA) converts 3-D computer images into a series of very thin cross sections, much as if the object were sliced into thousands of layers (Figure 83-11). A laser beam then traces a single layer onto the surface of a vat of liquid polymer. The intense spot of ultraviolet light causes the polymer to harden precisely at the point where the light hits the surface. The hardened material is then lowered a small distance below the remaining liquid, recoating it with more resin. The laser then traces the next layer of the computer image model onto the now liquid surface. The process continues until the complete 3-D model is built, one layer at a time, from the bottom up. The completed part is removed from the vat after the last layer is formed, and undergoes a final curing with an ultraviolet light. SLA product applications include medical bone implants, electrical discharge machining electrodes, nozzles for metal spraying, cellular phone cases, and automotive wheel rims.

Pattern Development

CAD systems also create flat patterns for products that are made by folding sheet materials. For example, the layout for a sheet metal box is designed as a flat pattern, cut out, and then folded to form the box. Some of these software programs are capable of developing a pattern from a 3-D CAD model; specifying the type of seam or joint; determining the bend allowances according to the thickness of the sheet material; and then "nesting" the patterns on a sheet for economic cutting. An example of such a layout appears in Figure 83-12. This same design procedure is used to lay out cardboard toothpaste cartons, clothing patterns, television cases, and sheet metal ductwork.

There are many uses for CAD in product design and analysis in industry. Young people with mechanical interests and abilities should explore a career in this exciting and challenging field (Figure 83-13).

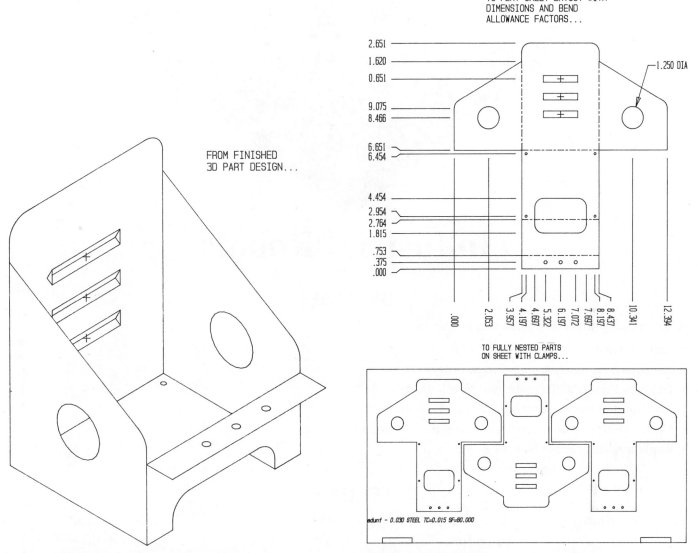

TO FLAT SHEET LAYOUT WITH
DIMENSIONS AND BEND
ALLOWANCE FACTORS...

FROM FINISHED
3D PART DESIGN...

1.250 DIA

TO FULLY NESTED PARTS
ON SHEET WITH CLAMPS...

adunf - 0.030 STEEL TC=0.015 SF=60.000

Fig. 83-12 This sheet metal layout and nesting program was developed from the wireframe object, as shown here.

Fig. 83-13 CAD careers in industry can be interesting and rewarding. (Courtesy of SDRC)

KNOWLEDGE REVIEW

1. Define *CAD*.

2. What is a plotter used for?

3. Explain the four main functions of CAD and give examples of each.

4. What is the purpose of rapid prototyping?

5. Explain the LOM process and the SLA process.

6. Describe the purpose of pattern layout and nesting in sheet metal product development.

7. Prepare a research report on current developments in CAD.

Unit 84

Industrial Robots

OBJECTIVES

After studying this unit, you should be able to:

- Explain what a robot is.
- Describe how robots are used in industry.
- Describe the parts of a robot, and how it works.
- List and describe the four basic types of robots.
- Understand the difference between a servo and nonservo robot.
- Explain what vision robots do.

KEY TERMS

articulated robot	manipulator	robot	spherical robot
control unit	power unit	servomechanism	work envelope
cylindrical robot	rectangular robot		

A robot is a mechanical device that can be programmed to weld metal chair parts, spray paint bicycle fenders, move parts and materials, and perform many other industrial operations. Humans also can do these things. However, there are several good reasons for using robots in industry.

- They can repeat the same dull, routine job for long periods of time with great precision and without getting tired.
- They can handle very heavy, very hot, or very toxic workpieces without being damaged.
- They can work in hot, noxious, hostile environments without being damaged.

Obviously, robots can and should be used to replace humans in the performance of tasks that people cannot or should not do. Robots are widely used in factories to do these and other jobs (Figure 84-1).

The term **robot** comes from the Czech word *robota,* which means "serf" or "forced laborer." The word was first used in the play *R.U.R.* (Rossum's Universal Robots) written in 1921 by Karel Capek. The first commercial robot used in the United States was one developed for the Atomic Energy Commission in 1958 to handle radioactive materials. The first robots used on an automobile assembly line were installed in 1962. Since that time, there has been an explosion in robot use in the metalworking industry.

Some robots are simple "pick and place" devices. They pick up a part, whether a tiny light bulb or a large

Fig. 84-1 This jointed arm robot is used to move workpieces from one milling machine to another. (Courtesy of Cincinnati Milacron, Inc.)

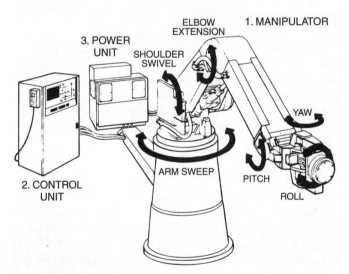

Fig. 84-2 The basic parts and movements of a typical robot. (Courtesy of Cincinnati Milacron, Inc.)

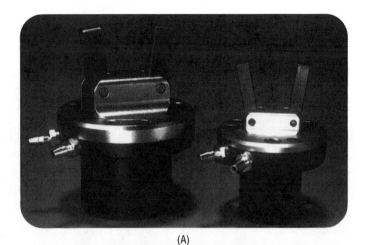

Fig. 84-3 Two-finger grippers are shown at A, and a suction cup panel holder at B. (Courtesy of PFA, Inc.)

truck axle, and place it somewhere else. These are called *nonservo* robots, and are relatively easy to program and operate. (A **servomechanism** is an automatic control system that uses feedback to direct the action of a robot.) Nonservo robots move until their limits of travel are reached. Very often, simple stop and go switches are used to control these travel limits. They can perform easy, repeatable tasks at high speeds, and are less expensive than servo robots.

Servo-controlled robots can move anywhere within the limits of their work area (or work envelope), and are therefore ideal for jobs such as spray painting or arc welding. They perform with smooth motions and

accurate speed control, exactly duplicating the movements of the human arm. Such robots are more complex, more difficult to program, and more expensive.

Parts of a Robot

The three main parts of a modern industrial robot are: a manipulator, a control unit, and a power unit (Figure 84-2). The **manipulator** is a human-arm-like structure, with a shoulder, elbow, arm, wrist, hand, and sometimes fingers. This complex arrangement can duplicate almost any human arm movement, resulting in a machine that can weld or paint or assemble parts. The robot's fingers are called *grippers* (Figure 84-3), and can be designed for pick-up or holding operations. Other operations are performed by suction cups, used to pick up and install automobile glass; or magnetic plates, used to move metal panels to a spray painting rack. A special end-of-arm device is the *end-effector*, which can be a welding or drilling head, or a spray or glue nozzle. These and other industrial robot manipulators can be programmed to perform many different tasks (Figure 84-4).

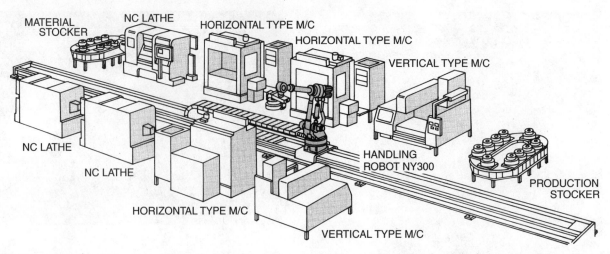

Fig. 84-4 This robot picks a workpiece from the material stocker, moves it to and from the several lathes and milling centers, and deposits the completed piece on the production stocker. (Courtesy of Motoman, Inc.)

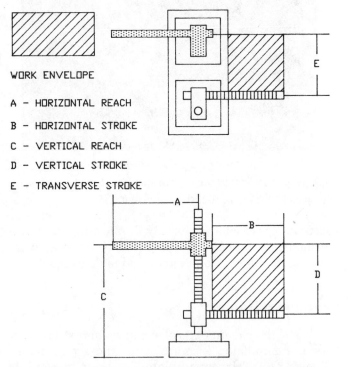

WORK ENVELOPE

A – HORIZONTAL REACH

B – HORIZONTAL STROKE

C – VERTICAL REACH

D – VERTICAL STROKE

E – TRANSVERSE STROKE

Fig. 84-5 A rectangular robot with work envelope shown.

All but the simplest of mechanical robots are directed by a **control unit.** The typical robot has a small computer to guide its movements, and a special computer program is prepared for each job to be done. The **power unit** supplies the hydraulic, electric, or pneumatic energy (or combinations of the three) to move the manipulator, and is guided by computer information sent from the control unit. Hydraulic units are generally large, powerful, slow-moving, and messy, performing heavy tasks such as lifting cast iron engine blocks. Electric units are fast, quiet, clean, and very accurate systems often used for precision assembly jobs. Pneumatic units are driven by compressed air, and are generally weak and not too accurate, but quick and inexpensive. These are ideal for the speedy, pick-and-place movement of parts.

Types of Robots

All robots do not look alike, nor do many of them resemble the comical C3PO or R2D2 models from the popular *Star Wars* films. These machines have different performance tasks, and are therefore designed with different work areas. The reach limits of a robot manipulator is called its **work envelope,** which is different for each type of robot.

Rectangular Robots

The **rectangular robot** can move in vertical, horizontal, and transverse directions with simple straight-line motions. As shown in Figure 84-5, this forms a rectangular work envelope which determines the limits of the robot reach. The arm can move up and down, in and out, and back and forth (transversely). The range of these robots can be extended by placing them on a movable track.

Cylindrical Robots

The vertical and horizontal motions of the **cylindrical robot** (Figure 84-6) are similar to those of the rectangular, but the base can rotate (sweep) to form a work envelope in the shape of a cylinder. For example, the robot shown in Figure 84-7 picks up a completed aluminum automotive wheel from the turning chuck, revolves, and then loads the chuck with a new workpiece.

Spherical Robots

A **spherical robot's** reach and sweep are the same as those of the cylindrical robot. However, its vertical movement results from the pivoting action of the arm

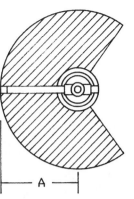

WORK ENVELOPE

A — HORIZONTAL REACH

B — VERTICAL REACH

C — VERTICAL STROKE

D — HORIZONTAL STROKE

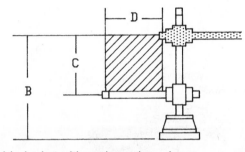

Fig. 84-6 A cylindrical robot with work envelope shown.

Fig. 84-7 A robot arm is shown here unloading and loading automotive wheels in a turning center. (Courtesy of Cincinnati Milacron, Inc.)

(Figure 84-8). As can be seen, this forms a sphere-shaped work envelope or pattern.

Articulated Robots

The fourth type of robot is the **articulated** or jointed-arm (Figure 84-9). This has a horizontal swing pattern like that of the cylindrical and the spherical models. The unique movement in the vertical and hor-

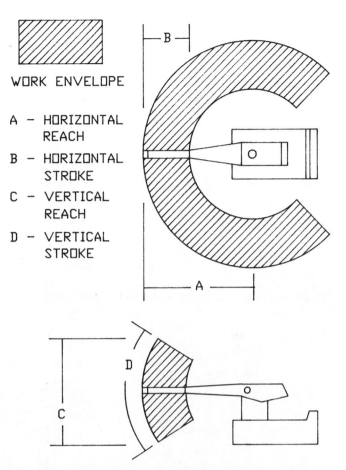

WORK ENVELOPE

A — HORIZONTAL REACH

B — HORIZONTAL STROKE

C — VERTICAL REACH

D — VERTICAL STROKE

Fig. 84-8 A spherical robot with work envelope shown.

izontal reach planes is made possible by the motions of the jointed arm, which creates the unusual work envelope. This is the most sophisticated and expensive type of robot, and is widely used to spray paint automobiles.

Industrial Applications

A *machine vision system* is a robotic application that consists of a video camera linked to a microcomputer. The camera is placed, for example, above a pair of robot welding heads (Figure 84-10). The video images are analyzed by the computer to locate the welding area. A typical search diagram is shown in Figure 84-11. The robots are then directed to begin the welding at a set point, and end it at another. These sophisticated systems also can detect dirty, crushed, or mislabeled packages passing a camera, which are then pushed off the production line. Other industrial applications include checking the correctness of parts and assemblies.

Humans and Robots

Robots will not replace people in many industrial jobs because they cannot easily react to unforeseen events and changing conditions. Robots are usually employed to complement people on the assembly line. The robots

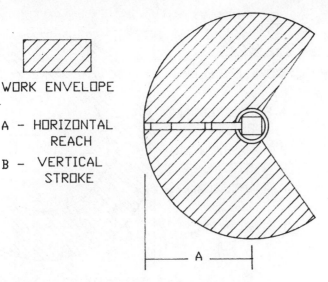
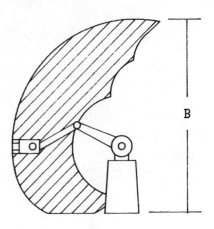

Fig. 84-9 An articulated robot with work envelope shown.

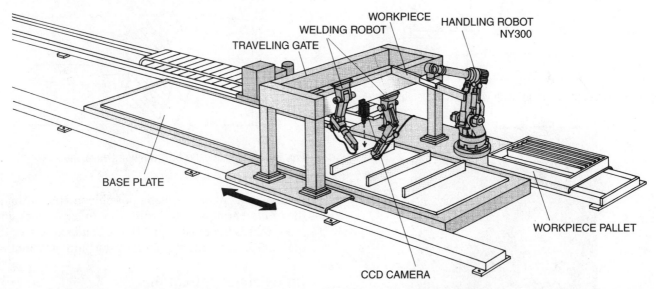

Fig. 84-10 An application of a machine vision system. Note that a handling robot places the workpieces on the base plate. The camera locates the weld area and the other two robots do the welding. (Courtesy of Motoman, Inc.)

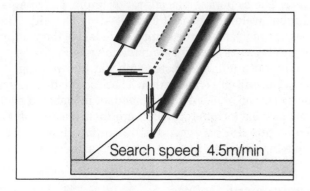

Fig. 84-11 In this search function diagram, the scanner locates a point on the horizontal workpiece and a second point on the vertical workpiece. It then locates the weld area between the two points. (Courtesy of Motoman, Inc.)

should do the heavy, unsafe, repetitive jobs such as welding, grinding, spray painting, stacking, feeding machines, and loading and unloading parts and supplies. People do the tasks that require human motor and perceptual skills, eye-hand coordination, planning, judgments, and evaluations. People and robots must be used wisely for the efficient operation of the manufacturing industries.

KNOWLEDGE REVIEW

1. What is a *robot,* and where does the word come from?

2. List some good reasons for using robots in industry.

3. List the parts of a robot and describe their functions.

4. List the four main types of robots and explain their differences.

5. What is a *work envelope?*

6. Describe the three kinds of power drives used on robots.

7. Which type of power drive must be used if the robot is to pick up heavy loads?

8. Describe a typical machine vision robotic application.

9. Why will people always be needed in a manufacturing plant?

10. Prepare a research report on the future of robots in industry.

11. Divide the class into small work groups, and explore the possible uses of robots in the home.

Unit
85

Computer-Aided Manufacturing Systems

OBJECTIVES

After studying this unit, you should be able to:

- Define the terms *CAM, CNC, FMS,* and *CIM.*
- Describe the operations of machining centers and cells.
- Describe a typical sheet metal FMS operation.
- Understand what lasers are and how industry uses them.
- Explain the uses of AS/RS and AGV.
- List some important factory management systems.

KEY TERMS

automated storage and retrieval systems

computer-aided manufacturing

computer-integrated manufacturing

computer numerical control

distributed numerical control

factory management systems

flexible fabrication system

flexible manufacturing system

laser

machining center

manufacturing cell

numerical control

process planning

The use of computers to plan and monitor the processing of a designed part, and to support its related plant operations, is called **computer-aided manufacturing** (CAM). Some of its major functions include: the numerical control of machine tools, process planning, and factory management (Figure 85-1). The aim is to provide an efficient manufacturing environment, and the computer is the central part of this.

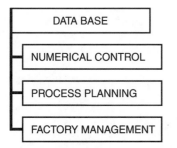

```
┌─────────────────────────┐
│        DATA BASE        │
└─────────────────────────┘
  ├──┌──────────────────────────┐
  │  │    NUMERICAL CONTROL     │
  │  └──────────────────────────┘
  ├──┌──────────────────────────┐
  │  │     PROCESS PLANNING     │
  │  └──────────────────────────┘
  └──┌──────────────────────────┐
     │    FACTORY MANAGEMENT    │
     └──────────────────────────┘
```

Fig. 85-1 The elements of a computer-aided manufacturing system.

Numerical Control

Directing the operation of machine tools by means of coded commands is called **numerical control,** or NC. In the past, these instructions were stored on either punched paper tape or on magnetic tape, and fed to the machine. The machining of a part could then be done automatically. When the job changed, the program of instructions was changed. This change capability gave NC a great advantage, for it is much easier to write changes into the control program than to change the production equipment. The more advanced and current systems employ **computer numerical control** (CNC), where the machine tool contains a microcomputer which stores the information and controls the machine. The most sophisticated system is **distributed numerical control** (DNC), where several CNC machines are linked to a central computer.

A typical NC turret lathe machining example appears in Figure 85-2. The workpiece is mounted in a chuck. The turret swings from one position to the next, all according to NC instructions. The piece is drilled, bored, reamed, recessed, counterbored, and tapped, at these positions. Some workpieces may require only the drilling, reaming, and tapping operations, and the program can be rewritten to do this. The NC control of machine operations is logical, efficient, and automatic. Most modern equipment has CNC controls, but the program theory is the same.

Process Planning

Determining how a part will be made (or processed) in a factory is called **process planning.** Whereas NC directs the operation of a single machine, process planning involves the detailed breakdown of part production steps, and then the selection of the best way to do each step. For example, there are several ways to produce (or generate) a round hole in a piece of sheet metal: drilling, punching, laser cutting, abrasive jet cutting, gas torch cutting, and acid etching. The planner must consider all the possible hole-producing processes and select the best according to the equip-

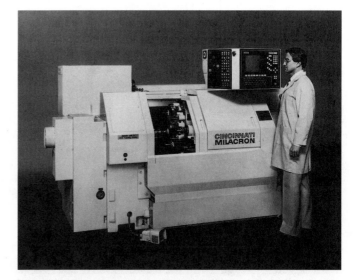

Fig. 85-2 The turret on this lathe rotates to present a tool and then moves to the workpiece to do the cutting according to NC instructions. (Courtesy of Cincinnati Milacron, Inc.)

ment available in the plant, the type of product, and time and cost. This is like writing a plan of procedure, but on a larger scale.

Process planning is therefore involved with changing raw materials into usable products. Modern manufacturing uses several newer systems to do this, as described below.

Machining Centers

A basic unit in any computerized material processing system is the **machining center.** Such a center is in fact a highly sophisticated version of a common machine tool such as a mill, lathe, or drill press. This special equipment differs from the ordinary in that machine operations can be performed automatically, with little or no human participation. All process functions are computer controlled, from the actual cutting to tool changing and the movement of workpieces to and from the machine. These automatic machines include horizontal or vertical milling centers, turning centers, and drilling centers, among others.

One style of *horizontal milling center* is shown in Figure 85-3. A key to the operation of such a center is the *automatic tool changer,* or ATC. The ATC shown has three main parts. The *chain-type tool magazine* is a movable unit with pockets to hold the machining tools. The tools can be drills, taps, reamers, or a variety of milling cutters. A *tool-changer arm* is mounted on a tool-changer column, and the arm pivots to select a needed tool from one of the chain pockets. The column then swings around to the machine head *spindle* to install the tool.

In a typical tool change sequence, the computer reads ahead in the workpiece machining program as the

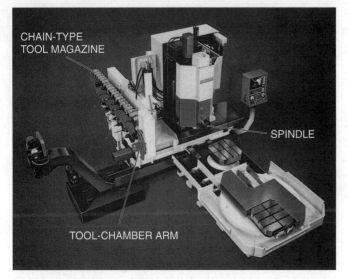

Fig. 85-3 A horizontal milling center with a chain-type tool magazine. (Courtesy of Cincinnati Milacron, Inc.)

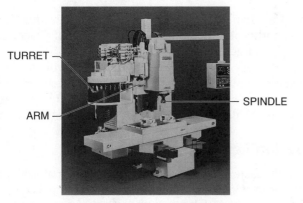

Fig. 85-4 A vertical milling center with a turret-type tool magazine. (Courtesy of Cincinnati Milacron, Inc.)

cutting is taking place. It commands the tool magazine to move the next needed tool to the pick-up position, where it is grasped by the changer arm. Upon completion of the present cutting operation, the computer issues a command to retract the workpiece from the machine spindle area. Now the tool change can take place. The changer column swings to the machine head position. The arm pivots to retrieve the just-used tool from the spindle, and again pivots to install the next needed tool. The column then swings to the magazine positions, and the arm pivots to return the just-used tool to the proper pocket. The workpiece moves back into position and the cutting continues. This process is repeated as the computer issues a command for another needed tool and the necessary chain movement, until the part machining has been completed. Any number of industrial parts can be machined on these centers, such as automotive transmission cases and engine blocks. Some ATCs have *turret-type tool magazines,* as shown in Figure 85-4.

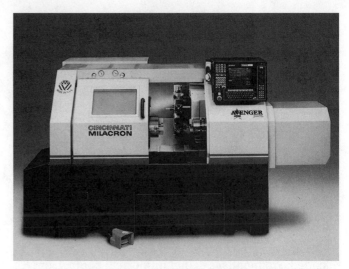

Fig. 85-5 A CNC turning center. Note the attached computer unit. (Courtesy of Cincinnati Milacron, Inc.)

A *turning center* differs from an ordinary lathe in that it has one or two turrets, which automatically present needed tools for turning operations (Figure 85-5). The tool turret can be mounted at the end, in front, or behind the workpiece. The workpiece generally is placed in the headstock chuck by a robot, or it can be installed manually. Parts can be turned to shape by a series of facing, sizing, bevelling, grooving, drilling, and shouldering operations. The turrets revolve to present the tool needed for the desired operation.

Manufacturing Cells

A number of machining centers, usually serviced by a robot and dedicated to the production of a family of related parts, is called a **manufacturing cell.** In such a system, an automated guided vehicle (AGV) is sometimes used to deliver workpieces to and from a material station. A robot loads a workpiece into a turning center where the part is machined. The robot then moves the workpiece to the milling and drilling centers for the necessary machining. Completed parts are then moved to the measuring center where they are checked for accuracy. In operation, the various centers of the cell are all filled for continuous machining, and the robot can move workpieces from center to center as required. An example of a cell serviced by a central robot is shown in Figure 85-6.

Most automated manufacturing involves the use of machining cells and centers, generally fed by robots. Guided vehicles may or may not move materials and parts; this is often done by human-guided transport devices.

Flexible Manufacturing Systems

Other computer-controlled systems have been designed and tested, but are not in general use. The **flexible manufacturing system** (FMS) is an example. This is

Fig. 85-6 An illustration of a manufacturing cell. Note the central handling robot. (Courtesy of Cincinnati Milacron, Inc.)

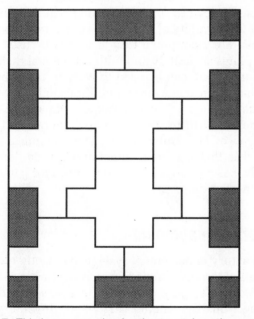

Fig. 85-7 This is an example of a sheet metal nesting program. The shaded areas are scrap cuttings. Note how the parts "nest" for efficient material use.

an arrangement of machines, inspection stations, material transport devices, and a common computer to produce parts randomly from a select family. It is a special type of machining cell whose main advantage is high productivity and product variety.

An FMS is complex and expensive, and most manufacturing operations cannot justify its use. One exception is the **flexible fabrication system** (FFS), used to make sheet metal products. The theory of sheet metal FFS is that several types of case products, such as tool boxes, fishing tackle boxes, and electric drill cases, can be nested on a metal sheet (Figure 85-7). The cases are known as a part *family,* for they all are sheet metal boxes which must be cut and then bent to shape. The system is designed to unload a sheet, feed it to the punching center, cut the parts, and separate and transport the cut parts, all under computer control.

A typical FFS line is shown in Figure 85-8. In operation, a suction gripper selects one sheet from the stack and carries it to the punching center. The sheet is positioned and held securely with finger clamps. Rotary punch and die sets are mounted in the tool head, with the punch above the table and the die below. The head holds die sets of many shapes, such as round, square, rectangular, or diamond. The computer directs the selection of the dies and the punching of the sheet according to the nesting program. The punch moves down through the metal sheet and into the die to do the cutting. The finger grips move the sheet in any horizontal direction under the punch. When the parts have been cut to shape and separated by shearing, they drop off the table and onto a material conveyor which sorts and moves them to the bending center. The flexibility of the machine lies in its ability to cut any shape, randomly, within a part family. One type of punching center is shown in Figure 85-9. Laser, plasma-arc, and abrasive-jet cutting also can be done on these machines.

Additionally, an FFS is used to cut the cloth pieces for blue jeans and baseball jackets. Cloth sheets are stacked on a table, covered with a plastic film, and compressed by a vacuum hold-down device. A sharp knife then moves up and down (reciprocates) at high speed to cut the fabric, following the nesting patterns prepared by the computer.

The most advanced and complicated flexible manufacturing system, and one not yet widely used, is called **computer-integrated manufacturing** (CIM). In

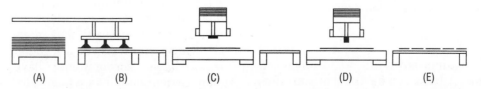

Fig. 85-8 A diagram of a sheet metal FFS: (A) metal sheets, (B) robot sheet mover, (C) punching center, (D) slitter, and (E) parts removal and sorting.

Fig. 85-9 A typical automatic CNC punching center. Laser or plasma-arc cutting can also be done on these machines. (Courtesy of TRUMPF Inc.)

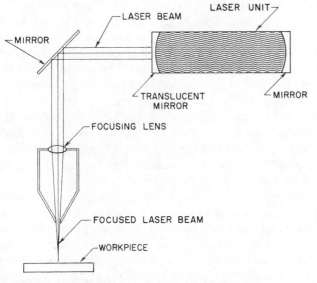

Fig. 85-10 A diagram of a typical laser welding system.

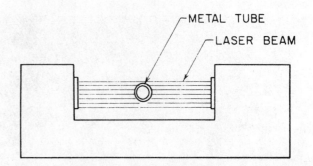

Fig. 85-11 Laser micrometers are used for precise measurement of industrial parts.

this system, all phases of manufacturing, from the design stage through final inspection, are controlled by a computer. Because CIM is very expensive, only the largest manufacturing companies can afford the system. CIM includes all high-technology advancements, such as: computers, robots, lasers, automated storage and retrieval systems, computerized material handling using driverless transport vehicles, numerically controlled machining centers, and automatic inspection devices.

Laser Technology

Lasers are an important addition to advanced machining systems. A **laser** is a device that amplifies or strengthens light to create a narrow beam that can burn holes in metal workpieces, or weld them to-

gether. A typical unit is shown in Figure 85-10. The word *laser* is derived from its technical description: light amplification by stimulated emission of radiation. Laser theory is based upon the ability of light energy particles to stimulate each other to cause additional particles to be formed, each having the same wavelength and direction of travel. The resultant beam has many applications. While there are several types of lasers, the gas and the crystal are most often used in industrial material processing.

Laser devices are important because they have such a wide variety of uses in advanced manufacturing systems. For example, a CO_2 laser can be used to cut metal plates accurately and with little metal distortion. This same beam can be used to weld or to scribe lines and shapes on metal products. Projected laser beams provide for the precise measurement of parts, to ensure that they fall within specifications (Figure 85-11). This is a type of laser micrometer. The advantage of this method is that it is noncontact and can be done very quickly. Lasers also are commonly used for parts alignment in precision assembly.

Factory Management

Factory management systems tie together (coordinate) all of the operations of a manufacturing organization. *Materials and procurement systems* include purchasing, quality engineering, and inventory and warehouse management. *Work management systems* involve production planning and scheduling, engineering change control, project management, and document access control. *Financial management systems* keep records of corporate budgets, accounts, and assets. *Safety and compliance systems* track personnel and qualifications data, material safety sheets, maintenance, employee health and safety, and air and emissions regulations. As you can gather from the above listings, large corporations have many complex management activities which must be linked by computers to ensure efficient plant operations. Two examples are presented below.

Fig. 85-12 This pallet load of boxed products is being moved with a fork-lift vehicle. (Courtesy of GATX Logistics, Inc.)

Automated Storage and Retrieval Systems

Materials movement, storage, and distribution can be a matter of simply moving things, as needed, by human beings (Figure 85-12). This is the way it is done in most factories. In more complex plants, this task is done by using **automated storage and retrieval systems** (AS/RS). This system is a combination of equipment and computer controls that automatically handle materials and products quickly and accurately. AS/RS can deliver raw material to a machining center, move parts to an assembly area, and carry finished, packaged products to warehouses for storage and distribution. Automated guided vehicles (AGVs), for example, may carry workpieces to machining areas (Figure 85-13). These battery-powered carts are guided by wires imbedded in the floor, or by lines of reflective paint. Sensors on the AGV follow the wires, and a computer directs the route of the vehicle. In a modern factory, a number of AGVs are operating continually to carry stock, parts, and prod-

Fig. 85-13 AGVs carrying workpieces to various work centers. Note that the workpieces are mounted on special holding fixtures. (Courtesy of Cincinnati Milacron, Inc.)

ucts. The example shown in Figure 85-14 is a theoretical design of such a system.

Management Information Processing

Working from the computer data bank, machine maintenance can be scheduled, and documents can be delivered to technicians for action. Such documents are based on records of machine use, inventories of replacement equipment, and operator reports of machine breakdown. Process and assembly information is similarly handled (Figure 85-15). The control of material and supply inventories is based on product bills of materials and warehouse records. The use of the computer in these activities makes it possible to keep track of the many management documents, and to be certain that they are delivered to the people who need the information.

1 Four Milacron T-30 CNC Machining Centers

2 Four tool interchange stations, one per machine, for tool storage chain delivery via computer-controlled cart

3 Three computer-controlled carts, with wire-guided path

4 Cart maintenance station

5 Parts wash station, automatic handling

6 Automatic Workchanger (10 pallets) for online pallet queue

7 One inspection module — horizontal type coordinate measuring machine

8 Three queue stations for tool delivery chains

9 Tool delivery chain load/unload station

10 Four part load/unload stations

11 Pallet/fixture build station

12 Control center, computer room (elevated)

13 Centralized chip/coolant collection/recovery system (----- flume path)

↰ Cart turnaround station (up to 360° around its own axis)

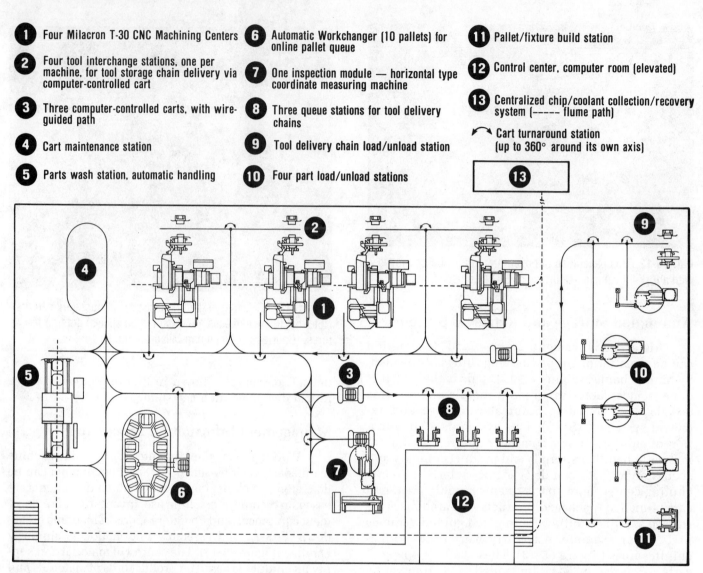

Fig. 85-14 A diagram of a flexible manufacturing system showing the AGV paths. (Courtesy of Cincinnati Milacron, Inc.)

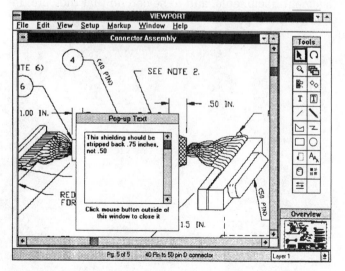

Fig. 85-15 A computer document image of a connector assembly. (Courtesy of The Indus Group, Inc.)

IN FOCUS: *Making Ball-Point Pen Tips*

The familiar ball-point pen is a useful, reliable, and inexpensive writing tool (Figure 85-16). It works when you need it, and it seldom fails. Accuracy in making the pen tip is the key to both reliability and writing quality. As a rule, the seat that holds the roller ball must have an accuracy of plus or minus 0.0000787 inch (0.002 mm). The drilled holes range in size from 0.0196 inch (0.5 mm) to 0.0471 inch (1.2 mm). The tip also must be inexpensively made because you throw the pen away when it runs out of ink. A thousand of these tips cost about six dollars.

The brass pen tip is made on an automatic rotary transfer machine (Figure 85-17), which is a combination drilling machine and lathe. The machine

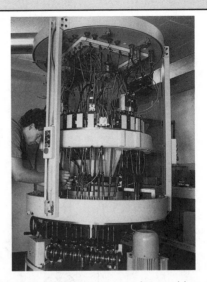

Fig. 85-17 An automatic rotary transfer machine. (Courtesy of Mikron Corp. Monroe)

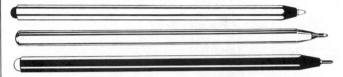

Fig. 85-16 Examples of throw-away ball-point pens. (Courtesy of Mikron Corp. Monroe)

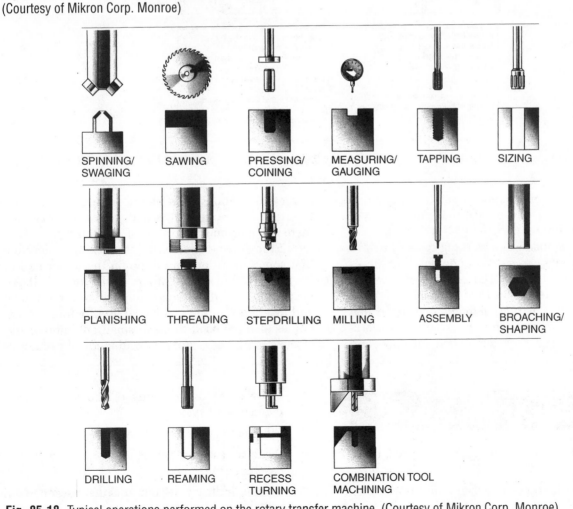

SPINNING/ SWAGING   SAWING   PRESSING/ COINING   MEASURING/ GAUGING   TAPPING   SIZING

PLANISHING   THREADING   STEPDRILLING   MILLING   ASSEMBLY   BROACHING/ SHAPING

DRILLING   REAMING   RECESS TURNING   COMBINATION TOOL MACHINING

Fig. 85-18 Typical operations performed on the rotary transfer machine. (Courtesy of Mikron Corp. Monroe)

Making Ball-Point Pen Tips (continued)

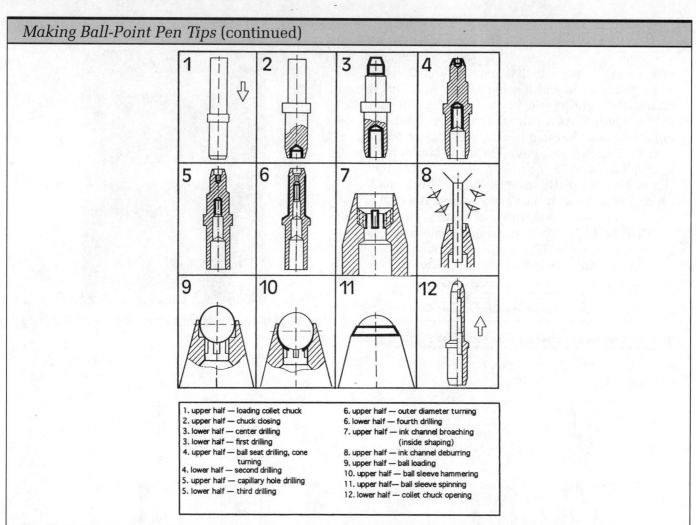

1. upper half — loading collet chuck
2. upper half — chuck closing
3. lower half — center drilling
3. lower half — first drilling
4. upper half — ball seat drilling, cone turning
4. lower half — second drilling
5. upper half — capillary hole drilling
5. lower half — third drilling
6. upper half — outer diameter turning
6. lower half — fourth drilling
7. upper half — ink channel broaching (inside shaping)
8. upper half — ink channel deburring
9. upper half — ball loading
10. upper half — ball sleeve hammering
11. upper half— ball sleeve spinning
12. lower half — collet chuck opening

Fig. 85-19 These twelve operations are necessary to produce the ball-point pen tip. (Courtesy of Mikron Corp. Monroe)

also is capable of many other operations, such as spinning, broaching, and planishing (end turning or end facing), as shown in Figure 85-18.

A pen tip is made in a working cycle of twelve steps (or operations), at a rate of one hundred pieces per minute (Figure 85-19). The action begins with a precut brass workpiece measuring 0.45 inch (11.4 mm) long by 0.10 inch (2.54 mm) in diameter being loaded in a special collet chuck. This chuck holds the workpiece so that it can be worked from both the upper and lower tool positions. This feature makes it unnecessary to turn the piece over repeatedly, resulting in faster, more efficient, and more accurate machining.

Study the other steps of the working cycle. Note that some operations are done on the upper half of the brass workpiece and others on the lower half. After completion, the tips are attached to the ink tube and inserted into the pen barrel. Rotary transfer machines also are used to make small automotive and electronic parts, gas nozzles, and medical needles.

KNOWLEDGE REVIEW

1. Write brief descriptions of: CAM, NC, CNC, FMS, and CIM.

2. What is the difference between a machining center and a manufacturing, or machining, cell?

3. Write a short description of the laser, and list some ways it is used in manufacturing.

4. Write short descriptions of the various factory management systems.

5. What is the difference between AS/RS and AGV?

6. Write a research report on the manufacture of baseball uniforms.

7. Visit an industry in your area that uses machining centers or cells. Prepare an oral report of your visit.

8. Prepare a bulletin board display of machining centers and cells.

SECTION 6

CONSTRUCTION SYSTEMS

What is the construction industry? Stop to think about how many construction projects you see every day. A new house is taking shape in the vacant lot around the corner. The highway that takes you from your home to the nearest shopping center is being widened to accommodate increasing traffic. A huge new high rise is changing the landscape across from your school as it rises higher and higher. Chances are that the utility section of your city is putting in new sewer lines. Construction contributes to our economic advancement through both public and private projects. The construction industry is the largest single industry in the United States. It employs about fifteen percent of the labor force (about one in every six jobs) and accounts for about fifteen percent of our gross national product (GNP).

The construction industry is divided into two major areas: public and private projects. Public projects include roads, highways, bridges, tunnels, airports, power plants, dams, railroads, pipelines, water treatment plants, schools, hospitals, and many other structures. Private projects generally involve shelters (homes, duplexes, low- and high-rise apartments, condominiums, mobile homes, boats, cabins, and temporary shelters). Private construction also includes office buildings, banks, warehouses, repair shops, and other commercial buildings.

Careers in Construction

The construction industry employs people in over seventy different careers; from architects and engineers to common laborers. There are many different kinds of skilled workers in construction who use metal tools and materials: plumbers; electricians; heating, ventilating, and air-conditioning workers; plus those who apply metal siding and use metal framing materials. For more information on specific occupations, check out the *Occupational Outlook Handbook* from your library.

Tools and Materials in Construction

The process of constructing a structure includes the use of the materials, the tools, and the machines needed to complete it. The materials are: concrete, wood, metal, plastics, glass, ceramics, brick, stone, and many others. The most common metals used are steel, stainless steel, copper, and aluminum. The tools needed to build a structure include various hand tools, portable power tools, and fixed machines such as the table saw. The heavy machinery needed includes: trucks, bulldozers, cranes, scrapers, graders, backhoes, and trenchers. Welding machines are necessary for assembling high-rise buildings of structural steel.

Unit 86

Shelter

OBJECTIVES

After studying this unit, you should be able to:

- Name eight different kinds of shelters.
- Describe how to plan a shelter.
- Name two methods of building basement walls.
- List two materials that can be used for framing.
- Explain the difference between a floor truss and a ceiling truss.
- Name three places where metal is used in roof construction.
- Describe the major metal used in wiring.
- Identify the kind of solder needed to seal copper pipes.
- Explain why stainless steel sinks do not rust.

KEY TERMS

planning   shelter

Shelter (residential construction) includes all kinds of structures in which people live, such as: single family homes, duplexes, apartments, condominiums, cooperatives, mobile homes, motels, hotels, houseboats, and recreational vehicles. Shelter units are built in two ways: (1) on the job, using the "stick" method of adding one piece of construction material at a time, or (2) manufactured in factories. Manufactured housing includes mobile homes, panelized units, and sectional housing. Mobile homes are built on a heavy metal frame so that the completed unit can be towed to the mobile home park by truck. In the United States, nine of ten living units are constructed primarily of wood. However, even a wood home has about six tons of metal found in such items as the furnace, air-conditioning unit, water heater, plumbing and electrical systems, and appliances. Homes with active solar units require a great deal more metal, such as copper and steel. A diagram of the solar heating process is shown in Figure 86-1.

Before any shelter units can be constructed, various kinds of materials must be available. These building materials—including wood, metal, plastic, and ceramics—are used to create attractive modern homes, such as the one shown in Figure 86-2.

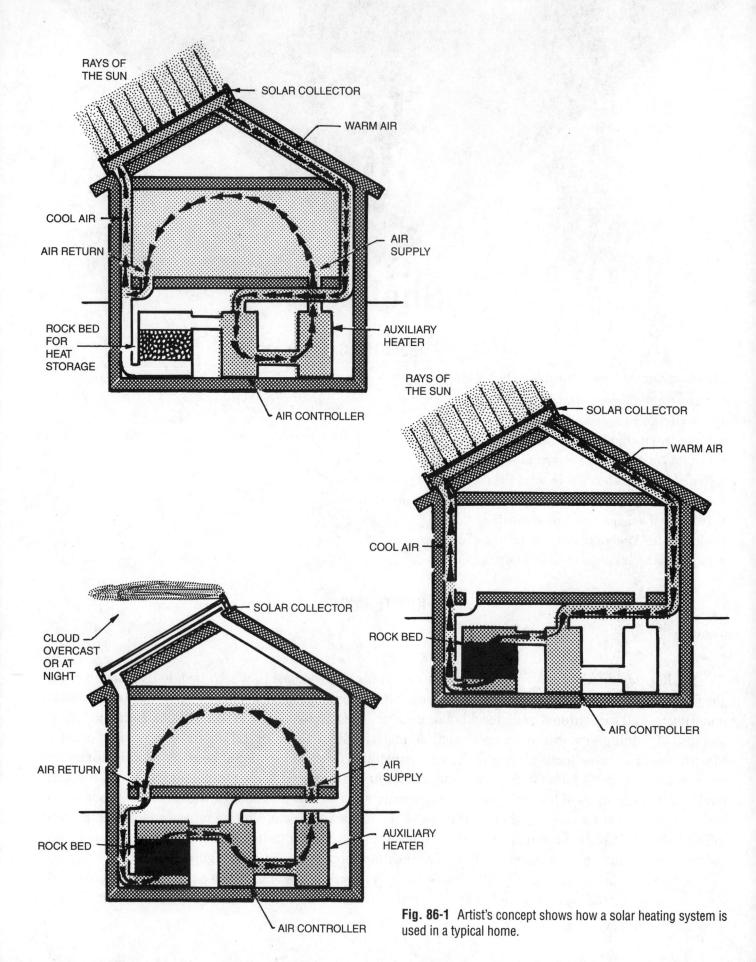

Fig. 86-1 Artist's concept shows how a solar heating system is used in a typical home.

Fig. 86-2 Modern homes are built of a variety of materials. (Courtesy of Norman F. Carver, Jr., architect and photographer)

Fig. 86-4 A backhoe is one of many pieces of heavy equipment used in home construction.

rials used. **Planning** starts with the idea and the needs of the people who will live in the shelter. The architect puts these on paper in the form of plans or prints. Computers can be used to develop the plans, make changes as necessary, and produce the final prints. These prints will show how the shelter will look, arrangements of the rooms, and materials needed. Figure 86-3 shows a finished house and floor plans. The plans for the shelter must have a *list of specifications* that details the information needed to build it. The builder or contractor uses this information to complete the project. The builder is responsible for hiring workers, including subcontractors such as plumbers, electricians, and roofers. The builder also orders the materials and supervises every step in the construction.

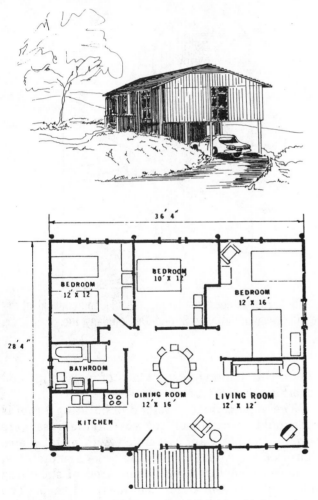

Fig. 86-3 A functional, pleasant rural home, with floor plan.

Planning a Shelter

The kind of shelter to be built depends on many factors, including: location, the number of people who will live in the shelter, climate, finances, and the mate-

Clearing the Land and Building the Basement

The land must be cleared using heavy machinery such as bulldozers. The topsoil is moved into a big pile that can be used later for landscaping. Shelters are built with or without basements. Some shelters have only a concrete slab foundation which rests directly on the ground. For shelters with basements, wood stakes are pounded into the earth to show where the corners of the structure will be. These are joined by string to show the outline. Then the opening is dug by a backhoe or similar equipment (Figure 86-4). Cement masons build double walls (forms) of wood that run around the inside of the hole. These forms hold the concrete for the footings. The footings are the bottom parts of the foundation walls. The concrete (a mixture of cement, gravel, sand, and water) arrives in huge concrete trucks ready to pour the foundation. The concrete is allowed to dry, and the wood forms are removed. The basement walls are built

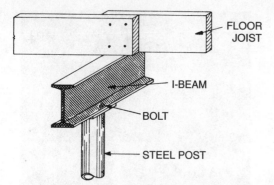

Fig. 86-5 A steel post is used to support a steel I-beam that in turn supports the floor joists or trusses.

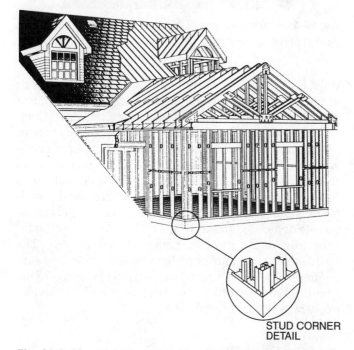

Fig. 86-6 Metal framing is used in building many houses. Note the openings in the studs for the passage of electric wiring. (Courtesy of Dale/Incor, Inc.)

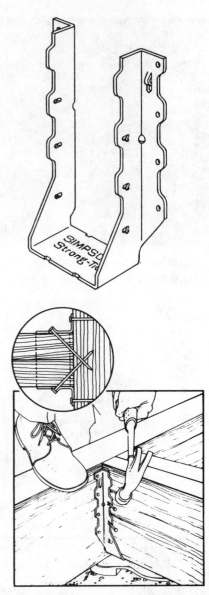

Fig. 86-7 Joist hangers are strong, reliable, and convenient to use. (Courtesy of Simpson Strong-Tie Company, Inc.)

of either steel-reinforced concrete or concrete block. After the basement is completed, one or more steel I-beams supported by steel posts support the floor joists or trusses (Figure 86-5).

Framing Methods

After the floor joists or trusses are installed, they are covered with plywood or some other manufactured boards. The walls are built using either wood or metal studs. Steel or aluminum framing materials are often preferred where termites are a problem (Figure 86-6). If the framing materials are of wood, the parts are assembled with nails using a pneumatic nailer or stapler. To

reinforce the framing, many kinds of metal clips and hangers are used (Figure 86-7).

The most common method of building a roof is to use roof trusses of metal or wood. If the trusses are of wood, the parts are assembled with metal plates (Figure 86-8). After the framing work is complete, the house is enclosed with some kind of sheathing such as plywood or other manufactured sheet. The roof is completed by applying shingles, with metal flashing at the chimney.

Completing the Exterior

After the rough enclosure is complete, the windows and doors are installed. Many outside doors are

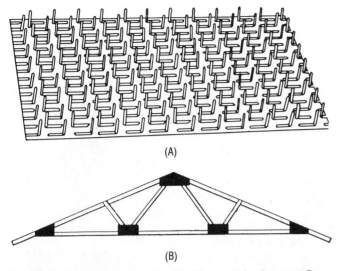

Fig. 86-8 A typical frame plate (A). The truss is shown at B, with the plates appearing in black.

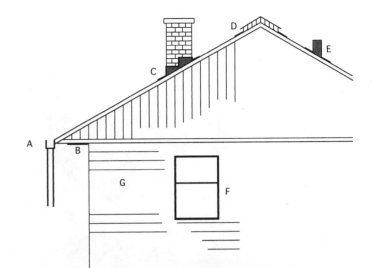

Fig. 86-9 The use of metal components on the exterior of a house: aluminum gutters and downspouts (A), aluminum eave vents (B), copper chimney flashing (C), galvanized steel ridge vent (D), lead sheet vent cover (E), aluminum storm windows (F), and aluminum siding (G).

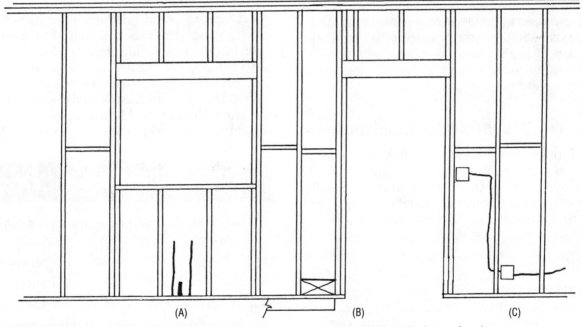

Fig. 86-10 Plumbing (A), heating (B), and electric (C) components are roughed into the house framing.

made of steel for greater security. The windows often have aluminum frames. The exterior of the house is covered with siding made of wood, metal (aluminum or steel), shingles, brick, stone, or other materials. Many other metal components are also used on the exterior of the house (Figure 86-9).

Electrical System Installation

A temporary electrical line is installed in the beginning of the construction process so that workers have electricity to operate the necessary power tools such as table saws, portable saws, drills, and joiners. When the framing is complete, the electrician installs and connects the main switch box to the outside source of electricity. The electrician drills holes in the framing materials to install coated copper wire from the main switch box to the metal or plastic outlet boxes. After the interior is finished, the electrician returns to install switches, fixtures, appliances, and power to other devices such as the furnace and air-conditioning unit (Figure 86-10).

Fig. 86-11 Stainless steel was invented in England by accident. The metallurgist added 14 percent chromium to the steel, which prevents rusting. Typically, stainless steel sinks contain steel with 18 percent chromium and 9 percent nickel. (Photos courtesy of Elkay Manufacturing Company)

Furnace and Air Conditioning Installation

The furnace is usually installed after the basement is completed. The air conditioner is usually located at the side of the structure. After the framing is completed, ducts (usually of sheet metal) are installed to carry the heat and cool air to the various rooms in the house.

Plumbing

The plumber installs all the pipes that bring water, gas, and sewer lines into the structure. The pipes may be made of copper, galvanized iron, or plastics. Copper is usually preferred for water pipes. Lead-free solder must be used to connect the pipes to fittings. The rough plumbing is completed while the framing is still exposed. After the interior is complete, the plumber returns to install the fixtures (Figure 86-11).

Completing the Interior

Insulation must be installed in the walls and ceilings. There are two major ways of completing the interior walls. One method is to nail metal lath to the studs and then apply plaster (Figure 86-12). The more com-

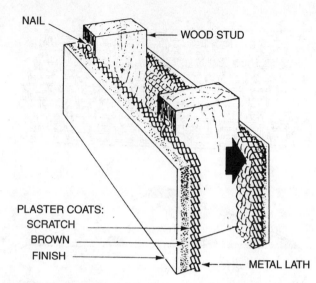

Fig. 86-12 A cross section showing the use of metal lath plus plaster.

mon method is to apply sheets of drywall. The sheets are fastened with metal nails or screws. The joints are smoothed over with tape and a joint compound.

Carpenters complete the interior by installing the interior doors, trim, and cabinets. The cabinets have metal hinges, drawer slides, and handles, as well as other metal components. Painters add color and finish to the walls and ceilings. Finished flooring or carpets are installed. The structure must be landscaped and the driveway and walks poured. Metal fences and gates are often added to provide safety and decoration to the house.

KNOWLEDGE REVIEW

1. What factors must be considered before building a shelter?

2. Why are a set of prints needed to build a structure?

3. Name two methods of building basement walls.

4. What two kinds of materials can be used for framing?

5. Name three metal items used to complete the roof.

6. Why should metal studs be used in certain areas of the country?

7. Name seven items of metal used in building a structure.

8. Design and build a model home.

Unit 87

High-Rise Buildings

OBJECTIVES

After studying this unit, you should be able to:

- Describe the invention that made high-rise buildings possible.
- Identify two methods of construction for high-rise buildings.
- Explain how much hurricane winds can sway the top of a very tall building.
- Name three methods of joining structural steel.
- List the three major components used in framing a high-rise building.
- Identify where aluminum is used in high-rise buildings.
- Describe space frame systems.

KEY TERMS

construction management   general contractor   space frame system   subcontractor

The invention of the elevator in the mid–nineteenth century (Elisha G. Otis developed the first passenger elevator in 1857) made it possible to build high-rise structures for commercial, residential (condominiums, apartments, and hotels), and multipurpose needs (Figure 87-1).

Many high-rise buildings have shops on the first few floors, a hotel on the next ten or more floors, and apartments or condominiums on the top floors. High-rise buildings must be strong enough to resist winds of hurricane force. Hurricane winds must not sway the top of a tall building more than eight or nine inches. Everyday gusts of winds of thirty miles per hour will cause a movement of less than one inch. Architects know how to design a building to meet the highest standards of safety. Often before a building is constructed, a model is tested in a special wind tunnel. The tunnel simulates the two main features of wind current; namely, greater force as altitude increases and wind which is gusty rather than steady. When problems are found in the model, changes must be made in the design to strengthen the frame, install thicker glass, and apply thicker exterior curtain walls.

Low-rise commercial buildings (not more than two or three stories) are built for offices, malls, motels, schools, and similar structures. The building may have a steel frame and supporting walls of steel, glass, wood, poured concrete, or concrete block (Figure 87-2).

Fig. 87-1 High-rise buildings dominate a city skyline. Note the structural welder at work. (Courtesy of American Iron & Steel Institute)

Fig. 87-2 A steel frame building under construction. Note that an articulating and telescoping boom vehicle carries workers to the structure. (Courtesy of Snorkel)

Preconstruction Planning

The exact purpose and needs for a particular building must be determined before plans can be made. For example, a banking concern might decide to build a large office building. Architects are hired to make sketches and perhaps a model of the building, which can be used to raise the money needed to complete the structure.

Site Selection

Before construction can begin, a building site is chosen. In many large cities, old structures are torn down to provide the site. Federal, state, and local governments may already own land on which to build public buildings. Utilities must be available for sewer, water, gas, and electricity.

Construction Financing

After basic plans are complete, the owner must find the money needed for the structure. An industry may use its own funds to build the building. A state or local unit may issue municipal bonds to cover the costs. Developers often go to banks or insurance companies to obtain the money.

Architectural Design and Estimating

After the site has been selected and financing obtained, an architectural firm is hired to design the building. Architects work closely with the people who will own or use the facilities.

Once the final plans are ready, they are sent out for bids to construction companies. These companies then prepare detailed drawings of the building. They need these drawings to get estimates from subcontractors for heating, ventilating, air-conditioning, electrical, and plumbing systems. When all these bids are in, the construction company will bid on the project. At this point, the company will hire a **general contractor.** The general contractor appoints a management team and hires the subcontractors and office personnel needed to complete the project.

Subcontractors must be selected to work on a specific part of the building. They may do this at one time or over a period of time. For example, the craft people who install utilities must work when the building is in its construction stages. They also must come back at various times until the final installation of the utilities is completed.

Construction Management

Before a building can be started, utilities such as water, sewer, and electrical power must be available at the site. Also, roads must be constructed, particularly if the building is on raw land. **Construction management** teams oversee the selection of subcontractors. They also develop the scheduling of the project. Once the construction is started, they coordinate the entire operation.

Completing the Permanent Building

Once the planning stage is complete, actual construction of the building begins. This includes: site analysis; excavation; building a foundation; and construction of the basement, shell, and interior of the building.

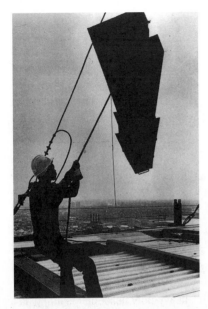

Fig. 87-3 Moving a piece of structural steel in place using a crane. (Courtesy of U.S. Department of Labor)

Site Selection and Layout

The site of the building must be carefully checked, especially when constructing large buildings. Large buildings require soil that can support the structure. Site selection is also based on the use of the building.

Excavation

The type and depth of the excavation is shown on the building prints. Large earth-moving equipment removes the earth for foundations and basements. Bulldozers and power shovels may be used.

Completing the Foundation

The foundation of the building must support the substructure as well as the building itself. For light buildings, the foundation may consist of footings and piers or foundation walls. For heavy buildings, piles, caissons, and other deep-wall foundations are needed. A *pile* is a long post of steel and/or concrete that is driven into the ground by pile drivers. The crane in Figure 87-3 is lifting a foundation section. These piles support the footings. A *caisson* is a shell or box filled with concrete. It is somewhat like a concrete pile but much larger. After these supports are in place, *forms* are built for footings and piers. Metal reinforced concrete is poured into these forms, making them strong enough to support the rest of the building's structure.

Completing the Basement Walls

Once the foundation is hardened, forms are built for the poured concrete that will make the basement walls. These same kinds of forms are used to build the shell of a building that is to be constructed of rein-

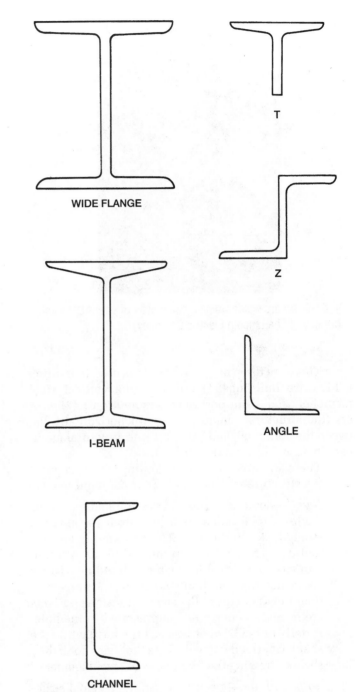

Fig. 87-4 Rolled steel structural shapes used to construct steel frame buildings.

forced concrete. Once the forms are in place, ironworkers called *rod people* place reinforced steel rods inside the forms before the concrete is poured.

Completing the Shell of the Building

The shell (also known as the frame or skeleton) consists of steel members that are riveted, bolted, or welded together. Rolled structural shapes are the most important metal products in high-rise buildings. These products include I-beams, wide-flange beams, channels, angles, T's, and Z's (Figure 87-4).

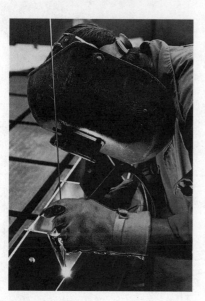

Fig. 87-5 An arc welder joining two pieces of structural steel. (Courtesy of The Lincoln Electric Company)

Fig. 87-6 This heavy-duty helicopter is lifting an air handling unit onto a building. (Courtesy of C. L. Mahoney Company)

There are three major components used in framing a high-rise building: (1) *columns* (the vertical steel members used to support the floors and roof), (2) *girders* (the structural members that run horizontally between the columns), and (3) *fillers* (the structural members that run between the girders).

There are three methods of joining structural steel so that a structure is strong and can resist wind forces.

1. *Rivets*—devices of steel fabrication made in three grades of steel: soft carbon, high strength, and high strength structural steel. The first kind is for normal use. The second and third kinds are used to join structural steel for high-rise buildings. Holes are drilled that are about 1/16 inch (1.5 mm) larger than the diameter of the shank of the rivet. Rivets are heated to cherry red and inserted in the hole. A dolly or buck is held against the head while the shank is formed using dies. As the rivets cool, they shrink, drawing the two pieces of steel together.

2. *Bolts and nuts*—used to assemble the steel structures. The bolts and nuts are made of high tensile strength steel. Holes must be drilled in the structural steel and then bolts of the correct size installed. The nut is tightened to the proper tension.

3. *Arc welding*—another method of joining steel members together. Welders must work in high places to complete the welds (Figure 87-5).

Aluminum is used for many parts of a high-rise building, including structural shapes and sheets. Plain aluminum sheets are used for flashing, roofing, drains, weather stripping, air ducts, and ventilators. Sheet aluminum is treated to improve appearance and increase its resistance to weathering. Aluminum sheet can also be an-

odized; an electrical and chemical process that increases the thickness of natural oxides. The process makes it possible to dye the surface in bright colors. Treated aluminum sheets are often used for *curtain walls,* which cover the outer surfaces of a structural steel building.

The shell of the building can be constructed of cast-in-place concrete. This kind of skeleton is assembled by building forms, installing reinforcing rod, and pouring the concrete one floor at a time. A movable crane is placed at the center of the building. Forms are built, rod people install steel rods for support, and liquid concrete is brought in by truck. It is hoisted from ground level by the crane and poured into the forms. After each floor level is completed, the crane is raised and the next levels are poured. Skyscrapers may be built forty floors or more with this method. Very often, large industrial helicopters are used to lift air cleaning systems onto the roofs of buildings (Figure 87-6).

Completing the Interior

The interior of the building is sprayed with insulating material before anything else is done. The interior itself may be natural, such as poured concrete walls, or it may be finished with stone or plaster. The interior is divided into rooms by installing metal studs and drywall. All high-rise building plans provide openings for elevators. These may be installed or roughed in as the building is being built or after the skeleton or frame is completed.

Space Frame Systems

On your next visit to a modern shopping mall, airport, or convention center, look up at the open metal structure that supports the building. You cannot help

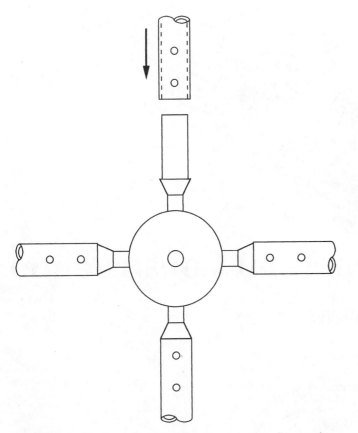

Fig. 87-7 In this type of node connector, the tubes fit over the node arms.

Fig. 87-8 This dramatic football stadium is built of a node and tube structure. (Courtesy of Busch Industries, Inc.)

but be amazed at the intricate latticework that looks very much like a piece of steel sculpture. These are called **space frame systems.** They consist of special metal connectors which hold long bars and tubes to create the form. The example shown in Figure 87-7 is a forged steel sphere (called a *node*) containing several drilled and tapped holes. Steel tubes with cone-shaped ends hold bolts tightly, allowing the tubes to be screwed into the node to create a building unit. Hundreds of these nodes and tubes go into structures such as the football stadium framework shown in Figure 87-8. There are many other systems used in modern building construction similar to the one described above.

KNOWLEDGE REVIEW

1. When was the elevator invented and by whom?
2. Describe the steps in completing a high-rise building.
3. Name three major structural components of a steel frame building.
4. Describe three ways of joining structural steel.
5. Describe how a cast-in-place skeleton is constructed.
6. What are *curtain walls?*
7. Describe a space frame system.
8. Build a model of a high-rise building.

Unit
88

Industrial and Commercial Construction

OBJECTIVES

After studying this unit, you should be able to:

- Name five kinds of industrial buildings.
- Explain why factories must be near transportation facilities.
- Name two types of warehouses.
- Name several kinds of maintenance shops.

KEY TERMS

prestressed beam   reinforced beam   unreinforced beam

Industrial construction includes mills, factories, warehouses, repair shops, maintenance facilities, and many other kinds of buildings needed to produce, store, and ship products. Commercial buildings are the offices, stores, shopping malls, and repair shops found in every city or town. Architects must design them with their special space requirements and functions clearly in mind (Figure 88-1). The typical industrial or commercial building has steel frame walls with an exterior cover of metal or some ceramic material.

Factories are designed to house the people, machines, and equipment used to manufacture products. They are usually single story buildings with very high ceilings so that conveyors, cranes, robots, ventilating ducts, and other special equipment can be installed over the large metal machines and manufacturing devices (Figure 88-2). Factories have concrete floors with special heavily reinforced concrete in areas that will hold extremely heavy equipment such as punch presses. Factories are generally built at the edge of cities or in rural areas where the price of land is relatively inexpensive. Important to any factory or mill is the availability of a means of transportation, including highways, railroads, airports, and ports for shipping overseas.

Fig. 88-1 This striking, dramatic shopping mall is an excellent example of modern steel construction. (Courtesy of Busch Industries, Inc.)

Fig. 88-2 A modern factory. (Courtesy of Cincinnati Milacron, Inc.)

Fig. 88-3 Large open spaces are needed for industrial warehouse operations. (Courtesy of GATX Logistics, Inc.)

Fig. 88-4 Industry provides safe workplaces for its employees. Note how this central work unit removes fumes, dust, and dirt. (Courtesy of The Lincoln Electric Company)

There are two kinds of warehouses; those that store the raw materials and those that contain finished products. Many warehouses have computer controlled equipment to move packages to correct locations for storage, and to retrieve products and deliver them to the shipping areas (Figure 88-3).

Maintenance shops are needed at airports, train terminals, and public works buildings to repair and maintain such vehicles as airplanes, trucks, buses, trains, and other transportation equipment. Many of these facilities have large enclosed welding booths, as well as air-exhaust equipment, to protect workers from dangerous fumes (Figure 88-4).

Mills and plants require special kinds of facilities depending on what is produced. Lumber mills have many kinds of metal machines for processing lumber and other wood products. Metal processing plants must be large to provide space for their special equipment.

Not all factories and stores are heavy steel structures. Smaller buildings, often constructed of a steel framework with an exterior cover of sheet metal, are used for light manufacturing, assembly, repair, and retail sales. Although many are well-designed and attractive, many are plain, utilitarian buildings inexpensive and easy to construct (Figure 88-5).

Concrete Frame Construction

Concrete beams, columns, wall panels, floors, and roof decks are commonly used in many buildings (Figure 88-6). These are, in fact, concrete castings. Concrete is weather- and fireproof, insect resistant, and durable. These features make the material ideal for exterior structures such as parking ramps.

The three common types of concrete beams are shown in Figure 88-7. The **unreinforced beam** has poor

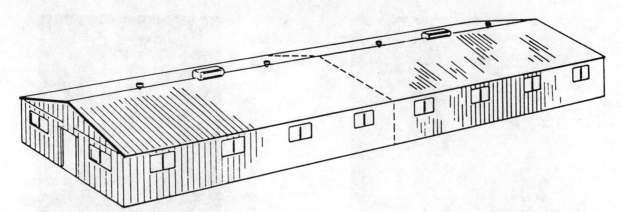

Fig. 88-5 Many small factories are durable, all-metal constructions.

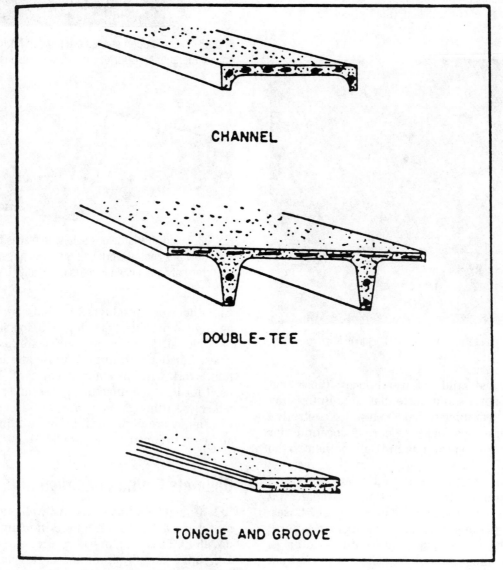

CHANNEL

DOUBLE-TEE

TONGUE AND GROOVE

Fig. 88-6 Three common types of concrete construction members.

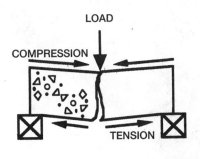

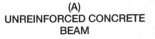

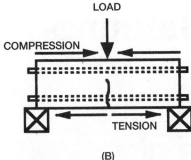

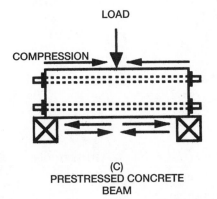

Fig. 88-7 Three types of concrete beams. Note how steel reinforcing rods affect their strength.

tension resistance and will eventually crack and fail under load. The **reinforced beam** has mild steel rods set in the form or mold prior to casting. These are better than the unreinforced beams, but are also subject to failure because the steel rods do not fully function until the beam cracks.

The **prestressed beam** as a load bearer is superior to the other types because it has a built-in compressive force at the beam bottom, and tension in the top. The prestressing (or pretensioning) process is illustrated in Figure 88-8. High tensile strength steel strands located in the casting form are stretched between abutments. Concrete is then poured into the form, and as it sets it bonds to the tensioned strands. When the concrete hardens to a proper strength, the strands are detached from the abutments. This prestresses the concrete by putting it under compression, creating a built-in resistance to loads. Prestressed beams, columns, and other structural members are manufactured in a factory and delivered to the construction site for erection.

KNOWLEDGE REVIEW

1. Describe the purposes of industrial buildings and list some examples.

2. Describe the purposes of commercial buildings and list some examples.

3. What are the functions of warehouses?

4. List some uses for all-metal utility buildings.

5. List some reasons for using concrete in construction.

6. Sketch and describe the three types of concrete beams.

7. Sketch and describe the method of making a prestressed concrete beam.

8. Visit some industrial and commercial buildings in your community. Write a report on the materials used in their construction.

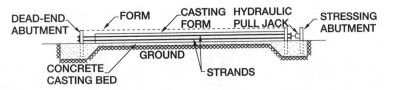

Fig. 88-8 A diagram of a prestressed concrete beam.

Unit 89

Public Works and Buildings

OBJECTIVES

After studying this unit, you should be able to:

- Name five major kinds of public buildings.
- List five major public works projects.
- Name two types of metals used in highway construction.
- Name two major parts of a bridge.
- List four kinds of bridges.
- List five kinds of dams.
- Describe the highest dam ever constructed in the United States.
- Name four uses for tunnels.
- Describe four types of buildings at airports.
- Name four kinds of utilities.

KEY TERMS

arch bridge	buttress dam	infrastructure	truss bridge
arch dam	earth-filled dam	rock-filled dam	utilities
beam bridge	gravity dam	suspension bridge	

Public works include roads, highways, bridges, dams, tunnels, and water systems, plus cable and computer networks. All of these are part of the **infrastructure** of our country that makes it possible for people to live, work, recreate, and move efficiently. Most of the projects are the responsibility of government, although some are privately funded. Public buildings are constructed like commercial and industrial buildings and include administration buildings, jails and prisons, libraries, museums, police and fire stations, schools, hospitals, laboratories, ball parks, stadiums, airports, train and bus stations, and repair and maintenance facilities (Figure 89-1).

Fig. 89-1 Airports are important public buildings which contribute to the progress of a community. (Courtesy of Kalamazoo/Battle Creek International Airport)

Fig. 89-2 Heavy equipment is needed to build roads and parking areas. (Courtesy of Kalamazoo/Battle Creek International Airport)

Fig. 89-3 A highway sign structure fabricated from aluminum extrusions. (Courtesy of Aluminum Company of America [ALCOA])

Roads and Highways

Roads and highways provide the ways for people to move overland from one part of the country to another. Products also are moved over roads and highways. The United States has forty thousand miles of interstate highways.

Roads are built of a variety of materials, including: natural soil, gravel, rock, tar, asphalt, portland cement (concrete), steel, and aluminum. The degree of difficulty in construction of roads or highways depends on the land and location. Usually some clearing is first required. Then large earth-moving equipment must clear the highway bed and do needed excavation (Figure 89-2). Temporary bridges are built where necessary. Roadbeds are constructed by bringing in gravel, rock, and soil to provide the base. Any necessary utilities, such as electrical wires, are placed alongside the roadbed. The finished roadway is covered with bituminous materials or concrete. Steel guardrails and aluminum highway signs complete the project (Figure 89-3). Inspectors make sure the road is built to specification.

Bridges

Bridges provide a way to cross waterways, valleys, roads, and other obstructions. There are many different types of bridges, including: beam, truss, arch, and suspension. A bridge may be fixed or movable (Figure 89-4).

Many different kinds of steel are used in bridges, including such alloys as nickel steel, silicon steel, and medium manganese steel. Aluminum is also used since it offers greater strength than structural steel, with one-third the weight. Aluminum is also more resistant to corrosion than steel. Aluminum can be used for all parts of a suspension bridge except the steel cables.

Precast concrete beams are also used for bridges. These beams contain reinforced bars of steel called *rebars*. Most bridges are composed of a combination of materials. For example, a bridge may have a superstructure of steel that rests on a substructure of concrete

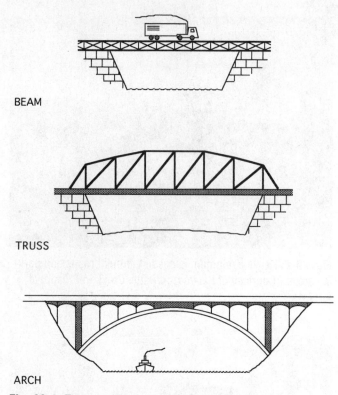

BEAM

TRUSS

ARCH

Fig. 89-4 Three common types of bridge structures.

Fig. 89-5 The Shasta Dam in California is a gravity dam over six hundred feet high. (Courtesy of American Concrete Institute)

abutments. The simplest bridge is the **beam bridge** in which horizontal members rest on supports. **Truss bridges** have a rigid framework of support members called trusses. **Arch bridges,** built in the shape of an arc, have been used for years to span deep canyons. Many of the early arch bridges were built of stone, but steel arcs are used today. The most famous of all large bridges are the suspension bridges. A **suspension bridge** has three major parts: towers, anchorages, and cables. Steel towers are found in most modern suspension bridges.

Dams

A dam is a large barrier placed across a river, stream, or channel. Dams are used to raise the water level behind the dam to generate electricity, control the flow of water for irrigation, and prevent flooding below the dam. The major parts of a dam are the dam structure itself, electrical power generating equipment, and spillways to control passage of floodwaters. The major types of dams include: earth-filled, rock-filled, gravity, arch, and buttress. The simplest dam is the **earth-filled** type. This dam depends on huge amounts of earth to hold back the water. In a **rock-filled dam,** over fifty percent of the dam's volume consists of dumped rock. **Gravity dams** are built so that the weight of the material provides the stability. Most gravity dams are built of concrete or steel-reinforced concrete (Figure 89-5).

The **arch dam** is one of the most important types. It is best utilized if it is built between canyon walls. The arc shape curves upstream to give the needed stability. In arch dams, the canyon walls absorb much of the pressure from the water behind it. Hoover (Boulder) Dam is one of the highest of these dams ever constructed. It was built in the Black Canyon of the Colorado River on the Arizona–Nevada border. It is 725 feet high and 1,244 feet long. The concrete base is 660 feet (two football fields) thick. It contains enough concrete (4,400,000 cubic yards) to build a two-lane highway from New York to San Francisco. The electrical power generated by the metal turbines of the dam is used in southern California (including Los Angeles), Arizona, and Nevada. The dam's reservoir, Lake Mead, irrigates more than one million acres of land and also serves for recreation.

Buttress dams are built with upstream sloping. There are very few buttress dams in the United States.

Tunnels

Tunnels are horizontal underground passageways for roads, highways, railroads, subways, electrical systems, and sewers. Tunnels are built under cities, through mountains, and under waterways. One famous tunnel is built under the English Channel, connecting France and England. The method of building tunnels depends on the material that must be removed. Soft rock tunnels are cut with metal saws and drills. Hard rock tunnels require blasting to loosen the materials. Tunnels are temporarily reinforced with wood beams. The completed tunnel has an inner lining of steel-reinforced concrete.

Fig. 89-6 City, as well as airport, fire stations are important community safety facilities. (Courtesy of Kalamazoo/Battle Creek International Airport)

Fig. 89-7 Control towers direct aviation ground and air traffic. (Courtesy of Kalamazoo/Battle Creek International Airport)

Airports

Airports combine heavy road construction with several kinds of buildings. Road construction includes runways, taxiways and aprons, parking lots, and access roads. All of the runways, taxiways, and aprons require extra thick steel-reinforced concrete. The buildings include passenger terminals, freight facilities, hangars, control towers, fire stations, and maintenance facilities (Figure 89-6). Each building must be designed and built for these special needs. Passenger and freight buildings are usually constructed of steel beams and trusses with an exterior of metal or ceramic materials. Control towers are built like high-rise buildings with windows all around the sides at the top level to give the traffic controllers a clear view of the entire airport (Figure 89-7).

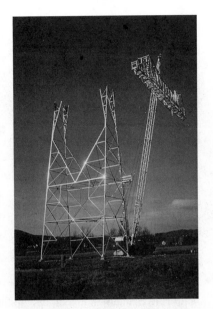

Fig. 89-8 A crane is raising a section of this aluminum electrical power transmission tower. (Courtesy of Aluminum Company of America [ALCOA])

Railroads

A railroad system requires tracks, trains, passenger stations, freight facilities, switching yards, and maintenance facilities. The tracks are made of steel fastened to wood crossbeams with heavy steel spikes. The engine, passenger, and freight cars are built of steel and aluminum.

Buses

Bus stations are usually located in the center of a city. A station may be a separate building or part of a larger structure. Buses are built primarily of steel and aluminum, with some copper.

Utilities

The water systems, sewer systems, telephone systems, and electrical systems that supply necessary services to the public are called **utilities.**

Water systems include pumping stations, filtering systems, and metal pipe that is laid underground. All buildings need water and are connected to the main lines. *Sewer systems* need similar facilities. *Telephone systems* include automated switching stations, lines, and facilities such as telephones and answering machines. Telephone lines are run either overhead or underground. Telephone towers are constructed of steel with the wires made of copper or fiberglass. *Electrical systems* include power stations for generating electricity either through various conventional means or through atomic power generation. Electricity is delivered to

Fig. 89-9 Three Rivers Stadium, home of the Pittsburgh Steelers, is an interesting metal structure. (Courtesy of American Iron & Steel Institute)

Fig. 89-10 Note the interesting structural details of the Eiffel Tower.

Fig. 89-11 This dramatic stainless steel sculpture is located in front of the fine arts building on a college campus. (Courtesy of the artist, Marcia Wood. "Prospect." Kalamazoo College, 1982. Photograph by David Curl)

in Paris, France. Its steel structure is striking and intricate, and its details invite close examination (Figure 89-10).

The welded stainless sculpture shown in Figure 89-11 stands near a college fine arts building. It was built for pure enjoyment, and to contribute to the art education of the students and the community. As you travel to our national capital in Washington, D.C., and to other cities in your state, you will find many other such structures honoring our presidents and other public figures and events.

KNOWLEDGE REVIEW

1. Name the construction projects that are part of the United States' infrastructure.

2. List five kinds of public buildings.

3. What are the major materials used in highways?

4. Name the two major parts of a bridge.

5. Name four kinds of bridges.

6. List five kinds of dams.

7. Describe the Hoover Dam.

8. What two countries in Europe are connected by a tunnel?

9. Name two main kinds of construction at an airport.

10. Describe some public structures or monuments you have seen.

11. Name two ways of generating electricity.

12. Design and build a bridge using pieces of wire cut from clothes hangers. Assemble the bridge with adhesive or soft solder.

customers through electrical lines that are either underground or overhead (Figure 89-8).

Public Structures

Various types of public works construction have been described and illustrated throughout this unit. However, many other public structures are to be found in all communities. For example, there are large and soaring football and baseball stadiums where fans enjoy and support their favorite teams (Figure 89-9). These feature reasonably comfortable seats, spacious restrooms and passageways, and well-placed eating facilities. They are durable structures, designed for safety and convenience.

The Eiffel Tower is an immense monument, rising over nine hundred feet to command a part of the skyline

SECTION 7

TRANSPORTATION SYSTEMS

A modern and efficient method of moving materials and machines to factories, and shipping their finished products, is a vital requirement of the manufacturing industries. This is achieved by a *transportation system* used to move people and cargo by trucks, trains, ships, and airplanes. There are two essentials of such a system: the **way,** or travel space; and the **mode,** or type of vehicle (Figure S7-1). For example, a truck and trailer travel on an interstate highway to move freight from city to city. The *way* is the highway and the *mode* is the truck and trailer. Similarly, airplanes move through established air routes, or *airways,* and materials are transported by ships over water routes, or *seaways* (Figure S7-2). The transportation industry refers to this system as *logistics,* meaning the planning, control, and management of material movement and storage.

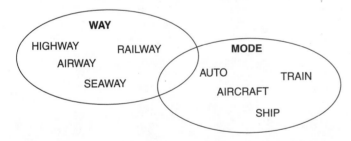

Fig. S7-1 The ways and modes of a transportation system.

Fig. S7-2 Materials and goods are moved by land, sea, and air transport. (Courtesy of GATX Logistics, Inc.)

Like any of the other systems described in this book, a transportation system is based upon the familiar input-process-output method illustrated in Figure S7-3. Here the material, human, and capital resources are the *inputs;* the planning, control, and management activities are the *process;* and the *output* is the actual movement of people and cargo by land, sea, and air.

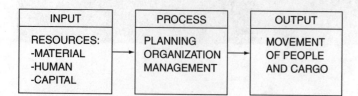

Fig. S7-3 The elements of a transportation system.

Unit 90

Land Transportation

OBJECTIVES

After studying this unit, you should be able to:

- Describe the essentials of a transportation system.
- Explain the transportation inputs, process, and outputs.
- Understand the basics of highway construction.
- Describe the types and purposes of the various highway vehicles.
- Describe the kinds and uses of off-road vehicles.
- Understand some alternate sources of automotive power.
- Understand the basics of railway construction.
- Describe the types and purposes of the various railway vehicles.

KEY TERMS

automobile	highway	passenger car	truck
diesel-electric engine	mode	railway	way
freight car			

Land transportation is exactly what the term implies; the moving of people and freight by cars, trucks, buses, and trains on land routes, such as highways and railways.

Highway Transportation

Probably no single invention has changed the United States as greatly as did the automobile (Figure 90-1). Think of all the ways in which cars, and other vehicles such as buses and trucks, affect your life. Perhaps you ride a bus to school. When your parents go to work or when you go on a vacation, the travel is probably by car. It is a quick, convenient form of transportation which has contributed to the growth of our cities and our nation.

The Highway

Road is the general term used to describe the prepared strip of land which provides a travel route for wheeled vehicles such as cars and trucks. *Streets* are the local roads traveled in cities. **Highways** are roads which connect cities and interstate systems. Modern highways are built from steel-reinforced concrete, portland cement, or bituminous concrete (blacktop) pavement, laid on a base of compacted stone and gravel (Figure 90-2). City streets are about thirty feet

Fig. 90-1 This MG roadster is an example of modern automotive transportation. (Courtesy of Gilmore Classic Car Club Museums)

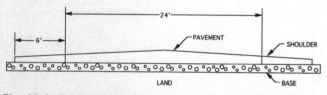

Fig. 90-2 Typical design for a paved road. The width of the driving surface and shoulders, and the pavement thickness, depend upon the kind of street or highway being constructed.

wide to accommodate two lanes of traffic and curbside parking. Two-lane highways are twenty-four feet wide with five- to eight-foot shoulders. Concrete and blacktop pavements are from six to twelve inches thick. Road thickness is determined by the weight and amount of highway traffic. This is calculated from the Equivalent Single Axle Load (ESAL) data collected from traffic counting devices, and from the freight truck weigh stations you see along the highways. Highways are generally paid for by gasoline taxes, and streets by local taxes.

Before the development of the automobile, travel was generally slow and difficult. The early roads in the United States were trails or unimproved roads used by horse-drawn carriages for both local and intercity travel. Many early settlements were on riverbanks because waterways were more convenient for transportation. The first long, hard-surfaced road was the Lancaster Turnpike in Pennsylvania, completed in 1795. It was surfaced with broken stones and gravel and extended sixty-two miles.

The first concrete road was built in Detroit in 1908. By 1924 there were thirty-one thousand miles of concrete roads in the United States. In 1925 a system of numbering highways was established. From 1924 until

Fig. 90-3 A map of the U.S. interstate highway system. East-west highways have even numbers, and north-south highways have odd numbers.

after World War II, truckers and automobile users realized that the highway system had to be improved. To help solve the problem of rapid highway transportation, the National System of Interstate and Defense Highways was established. Today there are over forty-two thousand miles of these superhighways. This system connects ninety percent of the United States cities that have populations over fifty thousand with multilane divided highways (Figure 90-3).

Highway Vehicles

Today there are more than 123 million automobiles and 35 million trucks and buses in the United States. They operate on nearly four million miles of streets and highways. The United States is a nation on wheels, and the automotive vehicle has become a necessity. Automobiles and trucks are the two most common of these transportation modes found on today's highways.

Automobiles

Automobiles are the personal vehicles that carry passengers on highways and streets. Modern cars are safe, convenient, and energy-efficient. There are many different kinds of automobiles, each designed to meet some personal transportation need. The most common are the two- and four-door sedans—the convertible coupe in Figure 90-4 is a special example. The minivan is a popular family car, providing both a comfortable ride and a convenient way of carrying things. Sports cars, such as the Plymouth Prowler shown in Figure 90-5, are exciting, powerful vehicles. These high-performance cars should be driven for thrill and joy, but safely. Buses are the vehicles that carry large numbers of people for local and long-distance travel (Figure 90-6).

Fig. 90-4 This striking Sebring convertible coupe exemplifies the concern for design, safety, and convenience in modern cars. (Courtesy of Chrysler Corporation)

Fig. 90-5 Sports cars are speedy, high-performance vehicles which should be driven responsibly and safely. (Courtesy of Chrysler Corporation)

Trucks

Trucks range in size from small half-ton pickups to large semitrailers that can carry thousands of pounds of freight (Figure 90-7). Large highway trucks are called tractors, and are constructed as metal shells to contain the engine and the driver compartments. These huge vehicles carry freight, animals, raw materials, food supplies, and liquids in specially designed bodies and trailers (Figure 90-8). Some tractor-trailer arrangements are designed for long-distance travel, often using two drivers for safety and efficiency. Some of the driver com-

Fig. 90-6 Buses such as this handsome model are used for both local and long-distance mass transportation. (Courtesy of Kalamazoo Metro Transit)

Fig. 90-7 Freight-hauling tractors and trailers are familiar sights along our nation's highways. (Courtesy of Aluminum Company of America [ALCOA])

partments have a sleeper cab so that one driver can rest while the other is at the wheel.

Off-Road and Specialized Vehicles

Other wheeled land transport includes off-road equipment such as bulldozers, graders, and gravel-haulers for highway construction; and home utility vehicles, such as lawn and garden tractors and snow-blowers. The many styles of bicycles and motorcycles described elsewhere in the book are another category. All-terrain recreational vehicles (ATVs) also are very popular with young people of driving age. These sporty, speedy three- and four-wheeled vehicles are designed for off-road travel over backwoods trails and sand dunes (Figure 90-9).

The wide range of specialized military equipment includes battle tanks, weapons and troop carriers,

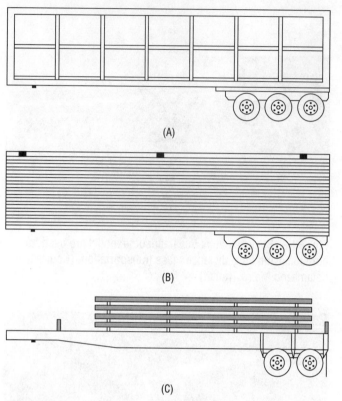

Fig. 90-8 Some typical freight-hauling trailers: (A) soft-side, (B) rigid-side, and (C) steel hauling flatbed.

Fig. 90-9 ATVs are rugged, sporty all-terrain vehicles. (Courtesy of Arctco, Inc.)

Fig. 90-10 The Humvee has replaced the Jeep as the military general-purpose vehicle. (Courtesy of U.S. Army)

ambulances, and the so-called High Mobility Multipurpose Wheeled Vehicle, or Humvee (Figure 90-10). This versatile four-wheel-drive cargo truck has a tough, durable, and lightweight aluminum and fiberglass body mounted on a steel channel frame. Its structure can be variously designed as a missile and armaments carrier, cargo hauler, or ambulance. One interesting feature is that the spare tire has been eliminated. Instead, the Humvee's tires have magnesium liners which permit flat-tire driving for distances up to thirty miles. Also, its modular construction simplifies field repairs.

Alternate Sources of Motor Power

The main problem with trucks and cars is that they are powered by internal combustion engines, which are a primary source of air pollution. Car engines

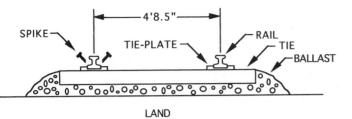

Fig. 90-11 Railway bed structure.

Fig. 90-12 Diesel-electric engines are a common type of railway locomotive used to pull passenger and freight cars. (Courtesy of Amtrak)

have become "cleaner" and more energy-efficient in recent years, and continue to improve. Catalytic converters, and the use of computers to control engine functions, have contributed to this improvement.

However, there has been an increasing interest in developing vehicles that use alternate sources of energy. One example is the *electric car,* which is powered by batteries that must be recharged periodically. The problems with electric vehicles include the inconvenience and cost of battery recharging; the weight of the batteries; and the space they require. The latter two factors reduce both the load capacity and the travel distance. Electric cars are therefore best suited to local city travel. Mail delivery vehicles would be a good application. *Solar cars* are another exciting possibility, and are described later in this unit.

Rail Transportation

This mode of transport involves moving goods and people in vehicles that run on steel rails. Therefore, the *way* is the railway or track, and the *mode* is the train.

The Railway

Railways are the steel track systems that provide the two hundred thousand miles of routes that freight and passenger trains travel. These ways include the numerous tunnels and bridges built to carry the tracks. The railway bed structure has several parts (Figure 90-11). The *ties* (or crossties) are either squared hardwood logs or steel-reinforced concrete castings eight feet, six inches long. They are laid on a fill of crushed rock called *ballast,* which is compressed, or tamped, to level and stabilize the land on which it is laid, and to provide drainage. The ties and the ballast, as well as the land on which they rest, form the *roadbed.* The ties are spaced about twenty-one inches apart, and support the rolled steel *rails* that the train wheels ride on. The rails sit on supporting forged steel *tie plates,* and are held by *spikes* driven through the plates into the wooden ties. Rails are held to concrete ties by special spring steel clips or special screws. The rails weigh from twenty to fifty pounds per running foot, and have a standard length of thirty-seven feet. Most railways are made of steel rail sections form-welded together for strength and durability, and to

prevent the familiar click-clack sounds. The welded stringers are a quarter mile long, and are preheated while laying to prevent buckling. The distance between the rails is four feet, eight and one-half inches, which is the standard gage in Canada, the United States, and Mexico, as well as other countries of the world.

Railway maintenance is a constant chore. Special inspection (detector) vehicles ride the rails to check for worn, cracked, or loose rails; broken ties; and washed-out ballast. Repair crews work continually to maintain a safe, fast, and reliable track system. In the railroad slang of times past, a wooden tie was called a *gandy,* and the track workers were called *gandy-dancers,* as they stepped from tie-to-tie in their work. Track maintenance used to be tiring, heavy work done by hand. It still is in some respects, but modern pneumatic hammers and tampers, rail cranes, and welders make it safer and easier.

Railway Vehicles

Locomotives and rail cars are the common rolling stock of the railroads. The steam engine, the first type used to pull cars, has been replaced by the modern **diesel-electric engine** (Figure 90-12). The diesel engine runs a generator which supplies power to electric motors located at the drive wheels. Electric locomotives generally pull the subway and elevated trains (els) for the rapid transport systems in cities. They derive their power from overhead lines or from a third rail laid beside the track. The gas turbine locomotive is also in wide use.

The material-hauling train vehicles (Figure 90-13) include **freight cars** (boxcars), which are sturdy wood and steel boxes on wheels used to haul packaged goods and merchandise. Some special insulated boxcars have mechanical *refrigeration* units to transport

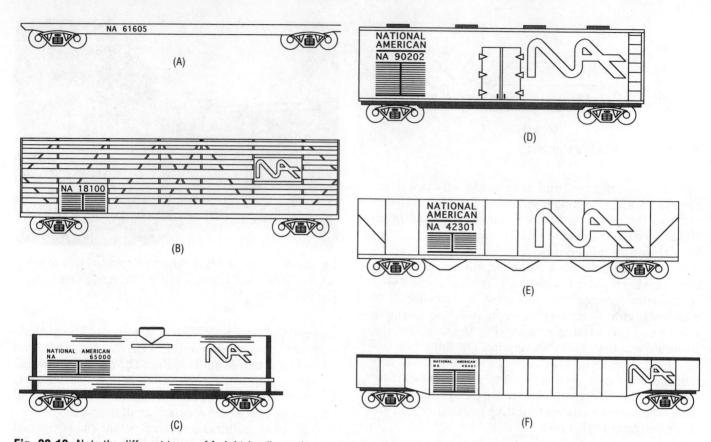

Fig. 90-13 Note the different types of freight-hauling train cars: (A) flat car, (B) stock car, (C) tank car, (D) refrigerator car, (E) hopper car, and (F) gondola.

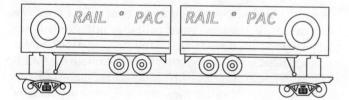

Fig. 90-14 A piggyback trailer-hauling car.

fresh produce, fruit, fish, and other perishables. *Hopper cars* are tough steel containers with high sides and sloped ends to hold coal, metal ores, and crushed stone. Hopper doors at the bottom of the car can be tripped for speedy and convenient unloading. Stainless steel *tank cars,* sometimes lined with plastic or porcelain, carry liquid chemicals, fuels, and crude oils. *Stock cars* have ventilated slat-sides and are used to transport cattle and other livestock. *Auto haulers* are designed with several floor levels to move new vehicles from automobile factories to distributors. *Gondolas* are open cars with short sides for hauling lumber, machinery, and large packed goods. *Flat cars* are open platforms which carry lumber, steel, machinery, and heavy equipment such as tractors and graders. Some models are fitted to carry freight trailers as a *piggyback* transport service (Figure 90-14).

The railway cartage vehicles described above are used for hauling specific freight and materials (Figure 90-15). The many models of railroad **passenger cars** are designed to carry people comfortably and safely. The best-known passenger carrier is the National Railroad Passenger Corporation, better known as Amtrak. This nationwide service operates about 250 intercity trains a day over twenty-four thousand miles of track, connecting some 525 communities (Figure 90-16). Amtrak has many long-distance routes running from coast to coast, as well as many regional lines serving passengers in the Northeast and West Coast travel corridors.

The passenger cars include coaches, lounge cars, diners, sleepers, and baggage cars. These attractive and convenient units provide passengers with safe and comfortable travel throughout the nation.

Experimental Locomotives

The need for faster and more efficient rail systems has led to some interesting vehicles. The linear induction motor (LIM) test car is a lightweight, high-speed vehicle designed for intercity rail travel. It operates at speeds up to 250 mph, is practically noiseless, and is nonpolluting. A structural diagram of this vehicle is shown in Figure 90-17.

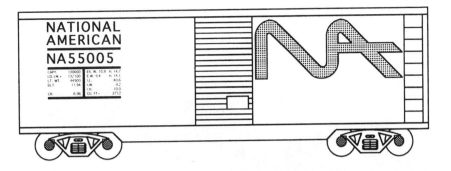

The markings on the side of a freight car contain important information. This insures that the right type of car is in the right place for the right load at the right time.

The markings on this boxcar indicate its size, weight, capacity, as well as other important facts. Here is an explanation of the complete markings.

CAPY. 130000 Car's nominal capacity is a 130,000 pound load.	EX. W. 10.8 H. 14.7 Width of car is 10'8" at a height of 14'7" above the rails.
LD. LMT. 132100 Load limit not to exceed 132,100 pounds.	E.W. 9.4 H. 14.1 Width of car at eaves is 9'4" at 14'1" above rails.
LT. WT. 44900 Car weight when empty.	I.L. 40.6 Inside length 40'6".
BLT. 11.94 Car built in November, 1994.	I.W. 9.2 Inside width 9'2".
CH. 6.96 Last shopping of car was in Chicago in June, 1996.	I.H. 10.0 Inside height 10'.
	CU. FT. 3712 Volume of car in cubic feet.

Fig. 90-15 How to read a freight car.

Fig. 90-16 A modern passenger train. (Courtesy of Amtrak)

Another experimental engine is the wheelless maglev (magnetic levitation), which operates on the principle that unlike magnetic poles attract, and like poles repel. Powerful electromagnets suspend (or levitate) and guide the train about ten millimeters above special flat rails. Similar magnets are used for propulsion. Also wheelless, the tracked air cushion research vehicle (TACRV) operates while suspended by air pressure over a guideway. A LIM supplies the power to move the TACRV at speeds up to 300 mph.

Research on these and other vehicles has led to quieter, more efficient, and faster rail travel. The new American Flyer is such an example (Figure 90-18). It will travel at 150 mph over reconstructed railbeds, and will enter service in 1999.

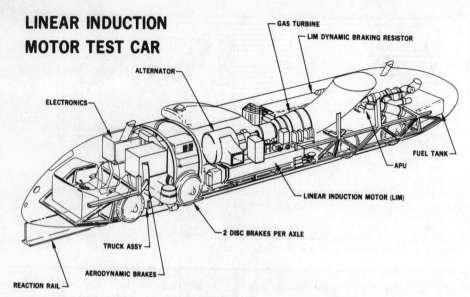

LINEAR INDUCTION MOTOR TEST CAR

GAS TURBINE

LIM DYNAMIC BRAKING RESISTOR

ALTERNATOR

ELECTRONICS

FUEL TANK

APU

LINEAR INDUCTION MOTOR (LIM)

2 DISC BRAKES PER AXLE

TRUCK ASSY

AERODYNAMIC BRAKES

REACTION RAIL

Fig. 90-17 Structural diagram of a LIM vehicle. (Courtesy of U.S. Department of Transportation)

Fig. 90-18 This striking new passenger train will enter service in 1999. (Courtesy of Amtrak)

IN FOCUS: *Solar Cars*

The Drexel University SunDragon Solar Racing Team won the 1994 Cross Continental category of the American Tour de Sol. The Tour de Sol is an annual race and exhibition of electric and solar vehicle technology, sponsored by the Northeast Sustainable Energy Association. The race began in New York City and ended in Philadelphia. SunDragon is a three-wheeled vehicle constructed almost entirely of composite materials (Figure 90-19). It weighs about 675 pounds, including the driver and batteries. The eight-square-meter solar panel contains eight hundred cells and provides the energy for the motor while racing, and for recharging the batteries at the end of the day. Eight 12-volt, deep-discharge, lead-acid batteries are used for energy storage. SunDragon is powered by a 7.5 kW (10 hp) brushless DC motor, and has a top speed of 110 km/h (75 mph).

Solar cars in the contest must be lightweight, energy-efficient vehicles, specially designed by students and faculty for this unique race. Another example, developed at Western Michigan University, illustrates the range of different styles possible in solar vehicle design (Figure 90-20). Much can be learned from these experimental cars about the potential for alternative power automobiles.

Fig. 90-19 The SunDragon prize-winning solar car. (Courtesy of National Instruments Corporation)

Fig. 90-20 The Sunseeker 95 is another interesting solar car design. (Courtesy of Western Michigan University)

KNOWLEDGE REVIEW

1. Describe the meanings of the transportation terms *ways* and *modes*.

2. What are the elements of a transportation system?

3. Prepare sketches of typical cross sections of a roadway and a railway.

4. Describe two alternative automotive power systems.

5. Prepare a bulletin board display of modern automobiles and their special features.

6. Prepare a research report on experimental railway locomotives.

Water Transportation

OBJECTIVES

After studying this unit, you should be able to:

- Recognize maps of important seaways of the world and American inland waterways.
- Distinguish among the various types of merchant vessels.
- Describe the unique functions of the container and the roll on/roll off ships.
- Describe some types of naval vessels.
- Explain the operational features of a hydrofoil and a hovercraft.
- Distinguish among the various types of recreational watercraft.

KEY TERMS

barge	displacement tonnage	icebreaker	sea lane
boat	gross tonnage	inland waterway	ship
canoe	hovercraft	jet boat	tanker
container ship	hydrofoil	patrol boat	vessel
deadweight tonnage			

Carrying people and materials by water is one of the oldest forms of transportation, dating back to the ancient Egyptians and Asians. The early European travelers in North America used boats and canoes to explore the waters of the Mississippi River valley. Water is still an important way of moving things between, as well as within, countries. Today, a wide range of water vehicles travel different kinds of waterways.

Waterways

Water vehicles, called **vessels,** travel on both oceans and inland waterways. The seas, rivers, and lakes are natural waterways, while humans have built many canals, such as the Panama, Suez, and New York State barge canals. The ocean waterways are called **sea lanes,** and they identify the many ship routes between nations (Figure 91-1). The oceans are vast and empty, but established sea lanes are necessary to direct traffic, provide weather information, and prevent collisions. Navigation on the seas is aided by radar, Loran, and other electronic guidance devices. The computer is also used for steering, direction, and speed controls.

The intracoastal waterway system skirts the American shores from Texas to Massachusetts, providing a safe and sheltered sea route. Hundreds of port cities are served by this system.

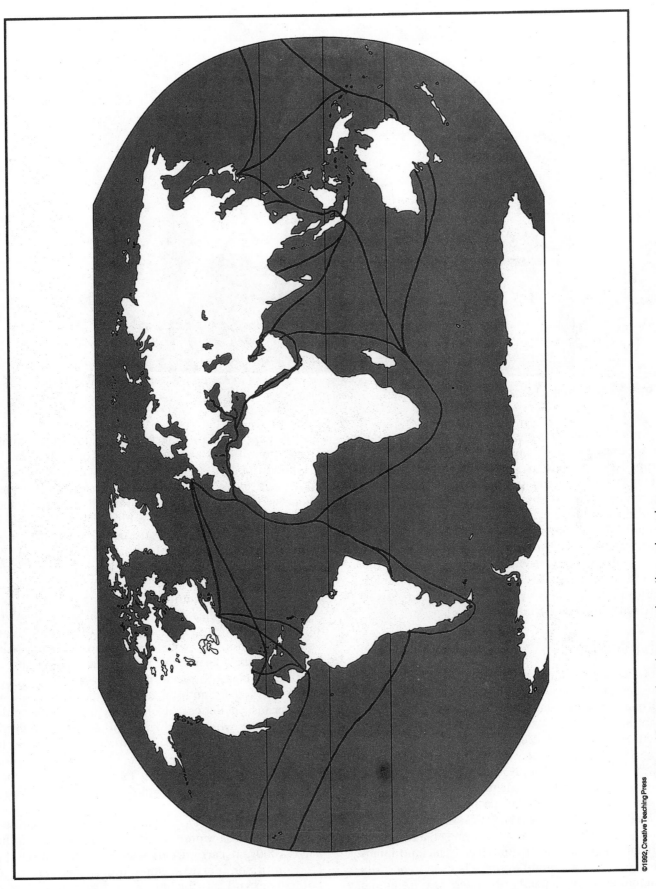

Fig. 91-1 Ocean-going ships travel sea lanes, such as those shown here.

513

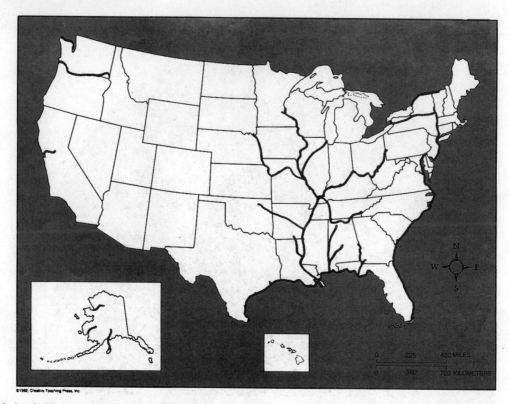

Fig. 91-2 Cargo is hauled between states on these inland and coastal waterways.

Inland waterways are the rivers, lakes, and canals used for travel within a country (Figure 91-2). The Mississippi River system is the most important in the United States—boats and barges can travel over eighteen hundred miles from Minneapolis to the Gulf of Mexico. The Illinois waterway, the Ohio River, and many other tributaries join the Mississippi to create a vast series of inland water routes. The Great Lakes/St. Lawrence Seaway system permits ocean-going ships to reach Canadian and American ports on the Great Lakes. Most of the American inland waterways are a combination of rivers, lakes, and constructed canals.

Water Vehicles

Generally speaking, a **ship** is a large, engine-powered vessel designed for navigation on the oceans. A **boat** is a small craft not designed for travel at sea. These distinctions are not hard and fast, for there are many types of these modern water vehicles. *Merchant ships* carry cargo and people. *Passenger ships* carry people to interesting parts of the world on ocean cruises. *Naval vessels* patrol and protect our nation's shores, and fight battles on foreign seas. *Recreational watercraft* provide people with the means of enjoying our numerous lakes and streams.

A ship's size is given by its tonnage, a term originally used to indicate how much a vessel could haul. Now the term **deadweight tonnage** is used to describe the actual carrying capacity of tankers and freighters,

and is expressed in long tons. One long ton equals 2,240 pounds, as contrasted to an ordinary ton of 2,000 pounds. This capacity includes crew, fuel, supplies, and cargo. **Gross tonnage,** a measure for passenger ships, is based upon the number of cubic feet of usable space. Each one hundred cubic feet equals one gross ton. Naval vessels use the quantity **displacement tonnage,** which equals the weight of the volume of water displaced by the ship as it floats. Boats and other small craft are measured by feet of length, such as a seventeen-foot racing canoe.

In ancient times, ships were powered by the wind or by humans rowing with oars. Over time, sailing vessels were improved, with stronger hulls, better sails, and by adding a keel to aid in stability and steering. A great advantage of the sailing ship was economy, because wind costs nothing. A main disadvantage was that because it depended on the wind, sailors could never be sure when the ship would arrive at its destination. Sailing ships were also difficult to maneuver in battle because of shifting winds. The use of the sailing ship reached its peak during the nineteenth century with the rugged, magnificent, and fast clipper ships (Figure 91-3).

The first successful steamboat in the United States was the Clermont, built by Robert Fulton in the early 1800s. It carried cargo and passengers on the Hudson River. Early ocean-going steamships used both steam and sail for power, as they often suffered

Fig. 91-3 Graceful clipper ships of the 1800s were the fastest of the sailing vessels, traveling at some twenty knots.

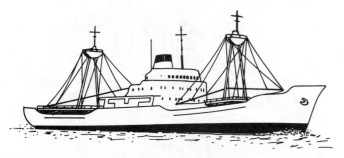

Fig. 91-4 A typical ocean-going merchant ship.

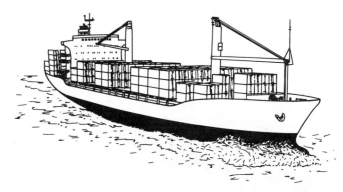

Fig. 91-5 A loaded container ship.

Fig. 91-6 Loading heavy equipment onto an RO/RO. (Courtesy of U.S. Army)

mechanical failures and had to use sails to finish their voyages. By the late 1860s, ships powered by steam were replacing the sailing ships. The first steamships were powered by reciprocating (piston) steam engines. Later, the steam turbine was developed, which was more efficient and used less fuel. Another advance was the use of the diesel engine, and today most ships are diesel powered.

The *Savannah* was the first nuclear-powered merchant ship. It went into service in 1962, and since that time many more such commercial and naval ships have been built. These boats are economical to operate and can travel great distances without refueling.

Merchant Ships

By far the greatest number of ships in use today carry cargo—everything from safety pins and grain to automobiles and large machinery. Since 1950, cargo shipping has changed greatly. Before then, cargo was carried inside the ship in large compartments called holds, which were loaded and unloaded by workers called stevedores (Figure 91-4). Since then, a new vessel called a **container ship** was developed. It was de-

signed to permit trailer-like containers to be stacked one on top of the other (Figure 91-5). They can therefore be transferred quickly and easily from the boat to a truck and driven to their final destination. This eliminates the handling of individual items on the dock and reduces both the cost and the time it takes to load and unload a ship. Special cranes on the docks load and unload the containers. A container ship may spend one day in port, as compared with six or more days for a ship handling noncontainerized cargo.

One interesting type of container ship is the *roll on/roll off* (RO/RO). In one application, cargo trailer vehicles are driven into the hold (or onto the deck) of the RO/RO, transported to their destination, and then driven off (Figure 91-6). Railcars also can be carried by this efficient water transport method.

The ships designed to carry liquid cargo are called **tankers,** and the large supertankers are about twelve hundred feet long and can haul two million barrels of crude oil. These huge ships are too large for ordinary port facilities. They must anchor outside the harbor to load and unload oil through pipes and hoses from on-shore storage tanks (Figure 91-7).

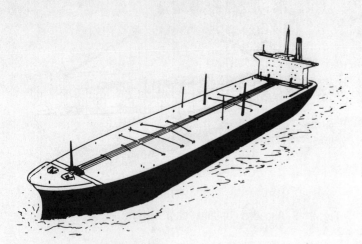

Fig. 91-7 Huge tankers carry crude oil and petroleum to and from ports around the world.

Fig. 91-8 Barge traffic on an inland river.

On inland waterways such as the Mississippi River system, petroleum products, coal, ore, and other bulk cargo are shipped by river **barges.** Powerful towboats push barges that are lashed securely together and then fastened to the towboat. (Tugboats pull barges, towboats push them.) (Figure 91-8.)

Passenger Ships

Before the invention of the airplane, the only way to cross the ocean was by ship. Passenger ships were then an important part of a shipping company fleet. Today most of the ocean-going passenger ships in service are cruise ships. There is also a need for fast, economical water transport between land areas separated by water. For example, ferry boats are used to transport people (and sometimes their cars) across rivers and lakes, and along coastal waterways. A unique craft used for short trips is the **hydrofoil** (Figure 91-9). This vessel actually rides above the water on fins called *foils,* propelled by jets of water from pumps driven by gas turbines. As the speed of the boat increases, the hull lifts out of the water. This effect reduces the water drag on the hull and permits a fast, smooth trip. The boat settles into the water at the dock to board passengers. One model of the hydrofoil has a cruising speed of fifty knots, even over twelve-foot waves.

A similar craft using a special propulsion system is the surface effect ship, or **hovercraft.** The hovercraft operates on a cushion of air which is created

Fig. 91-9 A hydrofoil vessel running on its three fins, or foils. (Courtesy of Aluminum Company of America [ALCOA])

Fig. 91-10 Drawing of a surface effect ship, showing how the stream of air moves to support the vessel.

under the craft by large fans. The air is pumped down from above and comes out the sides. This craft can operate on both land and water. Figure 91-10 shows the principle of the surface effect ship. The hovercraft can be used to transport either people or cargo for short distances. It can also be used as a short-distance land-sea military craft.

Naval Ships

Since the very beginning of water transportation, ships have been used for military purposes. Some famous naval battles were fought between wooden sailing ships, but modern naval vessels feature a combination of steam and nuclear power, and metal construction. The United States has many types of naval craft for many different uses. The most important duty of the Coast Guard is to protect life and property. Their **patrol boats** sail waterways or remain

Fig. 91-11 Patrol boat Point Camden. (Courtesy of U.S. Coast Guard)

Fig. 91-13 A rugged, durable, lightweight aluminum canoe is a popular recreational watercraft. (Courtesy of Aluminum Company of America [ALCOA])

Fig. 91-12 The aircraft carrier USS George Washington. (Courtesy of U.S. Navy; photograph by P/M Third Class Chris Vickers)

Fig. 91-14 This reliable jet boat is a speedy, maneuverable sport craft. (Courtesy of Arctco, Inc.)

on stand-by duty at stations to provide assistance to any boats that may be in danger (Figure 91-11). They also enforce sea laws by patrolling our coasts to intercept illegal vessels. **Icebreakers** clear pack ice from waterways, and track icebergs to provide safe passages for merchant and passenger ships.

The U.S. Navy has fleets of frigates, guided missile destroyers, heavy cruisers, battleships, submarines, and aircraft carriers to protect American interests throughout the world (Figure 91-12). Many of the vessels are nuclear powered. This energy source enables submarines, for example, to travel over sixty thousand miles without refueling, and remain underwater for long periods of time.

Recreational Craft

The oceans, lakes, and streams are traveled by thousands of people in thousands of pleasure boats. The **canoe** is one of the more popular, because it is easily portaged (carried) between lakes, is lightweight, and requires no other power source except human beings (Figure 91-13). **Jet boats** are fast and sporty, and use a battery to provide a powerful water jet for propulsion (Figure 91-14). Other recreational craft include: pontoon boats, sailboats, speedboats, and cruisers (Figure 91-15).

Fig. 91-15 Speedboats, sailboats, and cruisers are popular types of recreational craft.

KNOWLEDGE REVIEW

1. Draw some sea lanes and inland waterways on maps of the world and the United States.

2. How does a ship differ from a boat?

3. Name and describe the types of tonnage measurements used to describe a ship's size.

4. Describe container ships and the advantages they have over other cargo ships.

5. What is an RO/RO?

6. Describe the propulsion systems of the hydrofoil and the hovercraft. What are the advantages of these boats?

7. Prepare a research report on current vessels, including naval, merchant, and recreational.

8. Prepare a research report on modern sea navigation methods.

Unit
92

Aerospace Transportation

OBJECTIVES

After studying this unit, you should be able to:

- Understand the structure of the airway system.
- Describe the general types of aircraft in use today.
- Describe the purpose and operations of the Federal Aviation Administration.
- Know the difference between IFR and VFR.

KEY TERMS

air traffic control   airway   instrument flight rules visual flight rules

The carrying of people and cargo by aircraft is perhaps the most exciting and interesting form of transportation today. Huge airliners or fast military fighters climbing into the sky continue to be an awesome sight (Figure 92-1). It is the dream of many young people to one day pilot these magnificent machines. In this unit we shall explore the nature of the airways and the planes that travel them.

Fig. 92-1 The F1A 18 Marine jet fighter is an exciting military aircraft. (Courtesy of U.S. Navy; photograph by Chief Photographer's Mate Joseph Dorey)

The Airway

Like other forms of transportation, airplanes need paths or routes to follow. However, these **airways** generally have stricter control systems because of the nature of air travel. Since airplanes travel at great heights and speeds, their pilots seldom can see approaching air traffic. Instead, they must rely on their instruments—special radar and air communications—to avoid storms and other traffic. Airways are established routes on flight maps and charts. Useful airspace is considered to have a ceiling (top limit) of 75,000 feet. Its base is naturally not even, but follows the contours of the earth, changing above cities, mountains, and other obstructions, and reaching down to the ground at airports.

In the sky above the United States, there are more than 350,000 miles of federally designated airways

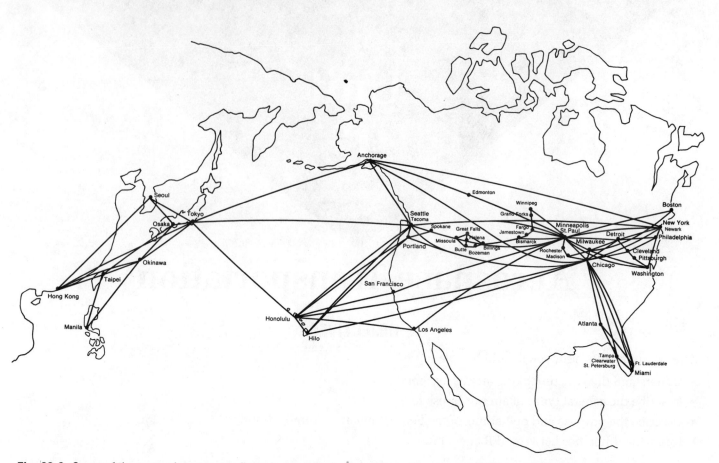

Fig. 92-2 Some of the many airway routes flown by aircraft are shown here. (Courtesy of Northwest Airlines)

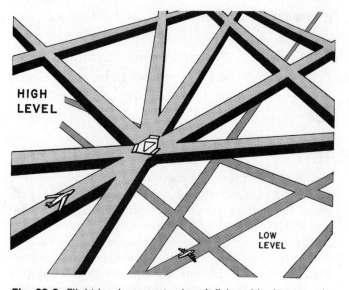

Fig. 92-3 Flight levels separate aircraft flying altitudes according to their direction of flight.

(Figure 92-2). These ways are divided into two different systems. The *low-level* system generally begins at 1,200 feet above the earth and goes up to, but does not include, 18,000 feet. The *high-level* system reaches from 18,000 feet to 45,000 feet (Figure 92-3). Airspace

above 45,000 feet is reserved for long distance, point-to-point flights.

Airplanes traveling on westerly headings must fly at even numbered altitudes. Those on easterly headings use odd numbered altitudes. Also, planes must fly at least one thousand feet above cities and five hundred feet above open country. Airways are the freeways of the sky, complete with an aerial version of signs, access roads, directional guides, and even parking places—those areas over airports known as *holding points*. Airplanes hold at these points when traffic is heavy, flying in an oval pattern so that they may be spaced in an orderly fashion before moving along the airways or into airports for landing.

Airway Vehicles

The first powered, sustained airplane flight took place at Kittyhawk, North Carolina, on December 17, 1903. There the Wright brothers flew an airplane some 120 feet, a flight which lasted twelve seconds. Since then, considerable progress has been made in the design and safety of aircraft. The Martin M-130 flying boat, called the China Clipper, was the world's first long range airliner. It entered service in 1935, carrying mail and passengers on Pan American Airways' transpacific

Fig. 92-4 This flying boat was used to carry people and cargo on early transpacific flights. (Courtesy of Pan American World Airways)

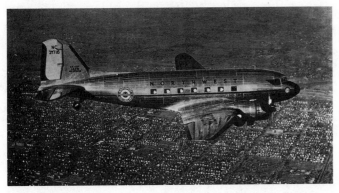

Fig. 92-5 The DC-3 was one of the important aircraft used in commercial aviation. Many are still in use. (Courtesy of Northwest Airlines)

Fig. 92-6 This huge Boeing 747 is one of the largest commercial jet airliners. (Courtesy of Northwest Airlines)

Fig. 92-7 A Boeing 747 cargo aircraft, with the nose door open for freight loading. Note that the cargo is contained on large loading pallets. (Courtesy of Northwest Airlines)

flights (Figure 92-4). The flying boat had a cruising speed of 130 mph and a range of 2,400 miles. (The *range* is the distance the airplane can fly without refueling.)

One of the most remarkable airplanes was the famous Douglas DC-3. It entered service in 1936, carried twenty-eight passengers, cruised at 185 mph, and had a range of about 330 miles (Figure 92-5). The rugged, dependable DC-3 is still flying cargo and people in many parts of the world. Following World War II, faster and larger airplanes with greater ranges were developed. All were driven by turbine- or piston-powered engines; jet engines were first used on military aircraft. The first commercial jet airliner was a British airplane, the Hawker-Siddley Comet, which flew in 1958. It traveled at a speed of 526 mph, carried eighty-one passengers, and had a range of 3,225 miles. This plane ushered in a new age in air travel. Since

that time, a number of modern jetliners have been designed. One of the more widely used is the Boeing 747 Jumbo Jet (Figure 92-6). The cruising speed of this airplane is 600 mph with a range of 10,000 miles. It can carry about five hundred passengers or 22,000 cubic feet of cargo.

While most airplanes carry both passengers and freight, there are models designed for cargo only. These have special doors and compartments to accommodate large items (Figure 92-7). Loading is done mechanically, and the cargo is placed in containers to speed up loading and unloading. Air freighting is very fast and dependable, and is growing in popularity.

There are many other types of aircraft designed for special uses. Light airplanes are used primarily for business and recreational travel (Figure 92-8). Both light and heavy helicopters serve as a means of passenger travel, and may also be used for fighting forest fires and for construction work.

Fig. 92-8 Light aircraft are used for business, commercial, and recreational travel. This flight instructor is working with a student pilot. (Courtesy of Western Michigan University)

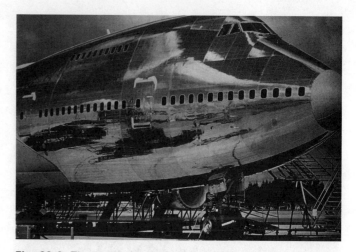

Fig. 92-9 The outer structure of the Boeing 747. (Courtesy of Aluminum Company of America [ALCOA])

Aircraft Structures

Airplanes are built of lightweight, tough alloys of aluminum and titanium, with composite materials also being used. One such composite is a graphite-epoxy compound as strong as titanium but much lighter. The structures are designed to be strong enough to withstand forces of flight such as wind and pressure. Figure 92-9 shows a typical aircraft structure covered with a thin, strong metal sheathing, or skin. The main parts of an aircraft appear in Figure 92-10, with the advanced composites applications clearly shown.

Much research is taking place on rocket engines and supersonic aircraft. Supersonic commercial aircraft, or SSTs, have been built as part of a joint French-British program. In addition, the aerospace research done by the National Aeronautics and Space Administration (NASA) has led to the familiar space vehicle missions which continue to excite people throughout the world (Figure 92-11). Much has been learned about flight, structure, power, and electronic guidance systems through NASA research.

Air Traffic Control

In the United States, the agency which controls the airways, aircraft, and air traffic is the Federal Aviation Administration (FAA). Its responsibilities begin at the drawing boards where aircraft are designed, and continue at the factories where they are made. It has authority over the people who dispatch the aircraft from airports, the crews who fly the planes, and the mechanics who maintain the aircraft. FAA responsibilities include the airspace, the navigation aids, the airway system, the eleven thousand airports in the United States, and the research needed to continually

improve the performance and safety of aircraft. One of its greatest responsibilities is to provide and maintain air traffic control.

All scheduled airline and military flights (and many general aviation flights) operate under **instrument flight rules** (IFR) regardless of weather conditions. This means that pilots must navigate by their onboard instruments—and according to instructions from FAA controllers—from takeoff to touchdown. Pilots file a flight plan before taking off, and are given a clearance that keeps them away from other planes flying IFR in the same area.

IFR operations are required when weather conditions fall below the minimum for cloud ceiling heights and visibility. In order to fly IFR, a civilian pilot must pass both a written and a flight test and receive an instrument rating from the FAA. When flying through "positive control" airspace (generally above 24,000 feet), pilots must fly under IFR regulations regardless of weather.

Pilots flying under **visual flight rules** (VFR) rely on their own sight and skill to avoid other aircraft, and must follow the rule: "See and be seen." Many local general aviation flights operate under VFR.

Air traffic in the United States operates under a "common system." This means that military and civilian aircraft are controlled by the same facilities. Traffic near an airport is controlled either by a military control tower or by one of the FAA airport control towers. Naval aircraft carriers also control air traffic near these and other vessels (Figure 92-12).

Traffic on the airways is controlled by Air Route Traffic Control Centers located throughout the country. Each center and tower handles traffic within its own area, using radar and communications equipment to keep aircraft moving safely. As the flight progresses, control is transferred from center to center and from center to tower.

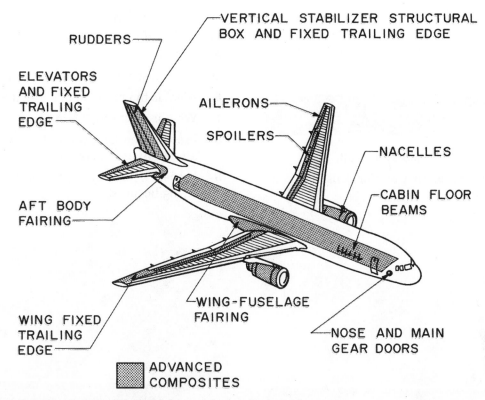

RUDDERS

VERTICAL STABILIZER STRUCTURAL BOX AND FIXED TRAILING EDGE

ELEVATORS AND FIXED TRAILING EDGE

AILERONS

SPOILERS

NACELLES

AFT BODY FAIRING

CABIN FLOOR BEAMS

WING FIXED TRAILING EDGE

WING-FUSELAGE FAIRING

NOSE AND MAIN GEAR DOORS

ADVANCED COMPOSITES

Fig. 92-10 The use of advanced composite materials on airframes is shown here.

Fig. 92-11 The beginning of a space shuttle mission. (Courtesy of NASA)

Fig. 92-12 This aircraft approach controller is at work aboard the aircraft carrier USS George Washington. (Courtesy of U.S. Navy; photograph by Airman Joe Hendricks)

In addition to control towers and Air Route Traffic Control Centers, there is a third air traffic facility called the Flight Service Station. It provides valuable information and other important services to civil aviators, air carriers, and military pilots. About 385 of these stations and combined station/towers are scattered around the nation, each covering an area of thousands of square miles. Flight Service Station specialists, who are experts on area terrain, and who provide preflight and inflight briefings, weather information, suggested routes, altitudes, and other information important to flight safety, staff these stations. If an airplane is overdue at its destination, the Station starts a search-and-rescue operation. If a pilot is lost or is having some trouble, it will give instructions and directions to the nearest emergency landing field.

Aircraft are an important part of our transportation system. They join with rail, land, and water vehicles to provide this nation with a fast, efficient, and safe way to move people and materials.

KNOWLEDGE REVIEW

1. Describe the airways used by aircraft.
2. Describe some of the early airplanes used in commercial travel.
3. List several responsibilities of the Federal Aviation Administration.
4. Explain the difference between IFR and VFR.
5. Explain some of the functions of air traffic control.
6. What is the function of a Flight Service Station?
7. What is meant by the *range* of an airplane?
8. Prepare a bulletin board display of some of the jet aircraft in use today.
9. Prepare a research report on the requirements for flying licenses.

PROJECT ACTIVITIES

Fig. P-1 Examples of candleholders made from I-beam scraps.

PROJECT 1: Structural Steel Candlesticks

Interesting candleholders can be made from I-beams, as shown in Figure P-1. See what other shapes can be found in the shop scrap bin. A hole of the proper size must be drilled to support the candle. Spray with satin lacquer, or use a gun bluing finish.

PROJECT 2: Steel Hammers

Some of the hammer shapes in Figure P-2 can be made on a lathe, and some will require grinding. They can be made smaller, for use as craft tools.

PROJECT 3: Fireplace Tools

The tool set in Figure P-3A combines gun-blued steel and teak wood for an interesting combination of materials. The brush can be purchased at a hardware store. Experiment with designs for forged poker ends. Figures P-3B–G show tool part details.

PROJECT 4: Fire Tool Handle

Shown in Figure P-4 is one of many other styles of handles which can be designed for the fire tools in Project 3. You may also wish to work out your own design.

PROJECT 5: Bird Feeder

Wire cloth and wood plugs are combined to make the functional, squirrel-proof feeder in Figure P-5A.

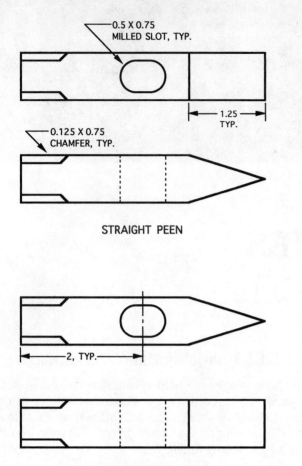

0.5 X 0.75
MILLED SLOT, TYP.

1.25
TYP.

0.125 X 0.75
CHAMFER, TYP.

STRAIGHT PEEN

2, TYP.

CROSS PEEN

Fig. P-2 Metalsmith hammers (0.75 × 0.75 × 4 tool steel).

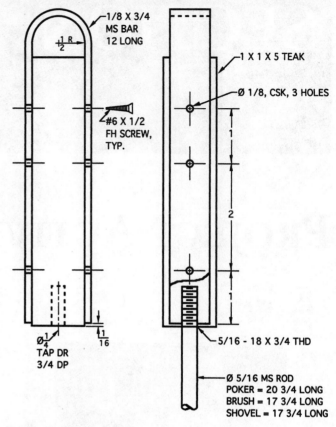

1/8 X 3/4
MS BAR
12 LONG

$\frac{1}{2}$ R

#6 X 1/2
FH SCREW,
TYP.

1 X 1 X 5 TEAK

Ø 1/8, CSK, 3 HOLES

1

2

1

Ø $\frac{1}{4}$
TAP DR
3/4 DP

$\frac{1}{16}$

5/16 - 18 X 3/4 THD

Ø 5/16 MS ROD
POKER = 20 3/4 LONG
BRUSH = 17 3/4 LONG
SHOVEL = 17 3/4 LONG

Fig. P-3B Handle details.

Fig. P-3A Fireplace tools.

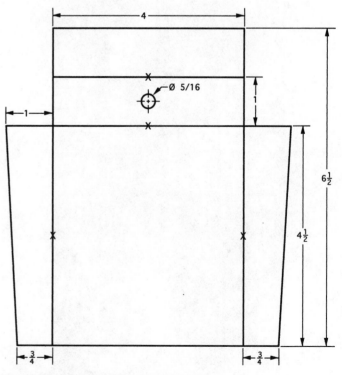

4

Ø 5/16

1

1

$6\frac{1}{2}$

$4\frac{1}{2}$

$\frac{3}{4}$

$\frac{3}{4}$

Fig. P-3C Shovel detail (20 ga. MS).

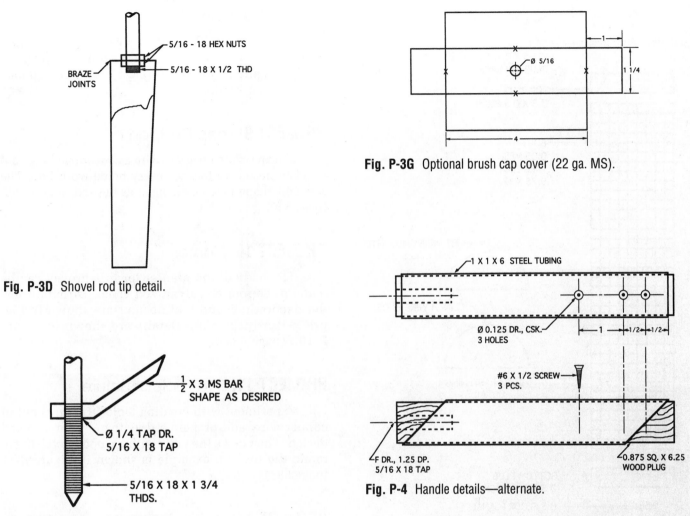

Fig. P-3D Shovel rod tip detail.

Fig. P-3G Optional brush cap cover (22 ga. MS).

Fig. P-3E Poker rod tip detail.

Fig. P-4 Handle details—alternate.

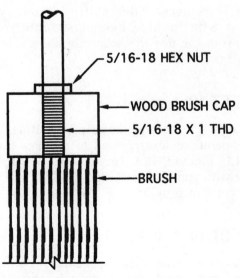

Fig. P-3F Brush rod tip detail.

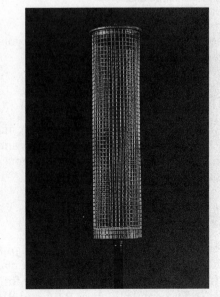

Fig. P-5A Bird feeder.

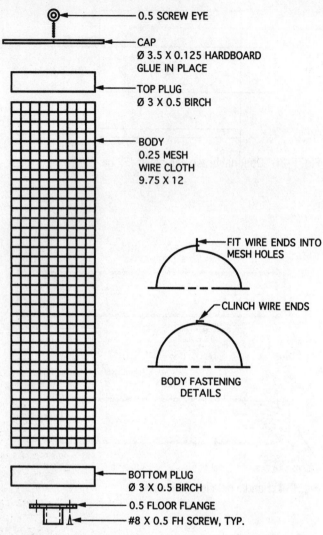

- → 0.5 SCREW EYE
- → CAP
  Ø 3.5 X 0.125 HARDBOARD
  GLUE IN PLACE
- → TOP PLUG
  Ø 3 X 0.5 BIRCH
- → BODY
  0.25 MESH
  WIRE CLOTH
  9.75 X 12
- → FIT WIRE ENDS INTO
  MESH HOLES
- → CLINCH WIRE ENDS

BODY FASTENING
DETAILS

- → BOTTOM PLUG
  Ø 3 X 0.5 BIRCH
- → 0.5 FLOOR FLANGE
- → #8 X 0.5 FH SCREW, TYP.

Fig. P-5B Bird feeder details.

The size and shape can be changed to suit your needs. Details are shown in Figure P-5B.

PROJECT 6: Putty Knife

A stiff yet flexible stainless steel blade, and a simple wood handle, make this an attractive and useful tool. Make it to metric dimensions to become familiar with this measuring system (Figure P-6).

PROJECT 7: Brass Pipe Vase

Short pieces of brass, copper, and steel pipe (scrap tailpipe ends from your local muffler shop make great workpieces) can be used to create a holder for fresh or dry flowers, or for pencils (Figures P-7A and B). Experiment with lathe-turned or sand-blasted designs. Use a conversion coat to give the piece a dark color, and then scratch an interesting design with a hacksaw blade.

PROJECT 8: Candleholder

Experiment with bending steel rods or bars into interesting shapes, and then make a holder cup for the candle (Figures P-8A and B)(page 530). The cup may be turned on a lathe.

PROJECT 9: Shop Dustpan

A heavy-duty dustpan, with a convenient long handle, is a necessary tool for every home workshop. The size and shape may be changed as desired (Figure P-9) (page 530).

PROJECT 10: Planter

The body of the planter for this project can be made of copper or galvanized sheet. Different border decorations and foot designs are shown in Figure P-10A (page 530). Details are shown in Figure P-10B (page 531).

PROJECT 11: Rod Table Structure

Experiment with creating table structures out of copper wire, and prepare a drawing of your selected design. Then make the table with a wood, metal, or ceramic tile top. An example is shown in Figure P-11 (page 531).

PROJECT 12: Candy Dish

Sketch some ideas for copper or brass candy dishes, make a pattern of your favorite, and trace it on the workpiece. After cutting it out, you can shape it with forming hammers. Draw your own foot design, or use the example shown in Figure P-12 (page 531).

PROJECT 13: Tool Set

Forging, machining, knurling, grinding, and polishing operations are required to produce the handy tool set in Figure P-13A (page 531). Color them with heat, or with gun bluing. Design a holder for the tools. Details are shown in Figure P-13B (page 532).

PROJECT 14: Belt Buckle

Shown in Figure P-14A (page 532) is one of many belt buckles you can design and make as an art metal activity. Use epoxy glue to adhere your favorite emblem to the buckle face, or create an etched pattern. Details are shown in Figure P-14B (page 532).

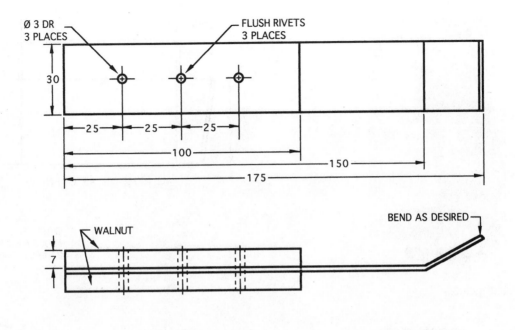

mm	INCHES
175	7
150	6
100	4
30	1.25
25	1
7	0.28

Fig. P-6 Putty knife details (0.50 stainless steel blade). Dimensions are in millimeters.

Fig. P-7A A vase made from a section of brass pipe with a deep-etched finish design.

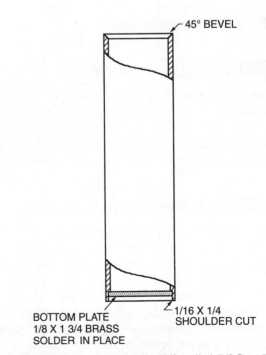

Fig. P-7B Brass pipe vase details (1/8 wall; 1 7/8 D × 7 H).

Fig. P-8A Steel candleholder.

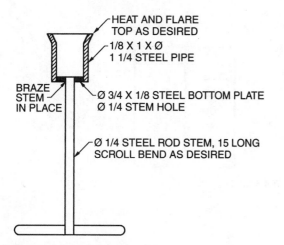

HEAT AND FLARE
TOP AS DESIRED

1/8 X 1 X Ø
1 1/4 STEEL PIPE

BRAZE
STEM
IN PLACE

Ø 3/4 X 1/8 STEEL BOTTOM PLATE
Ø 1/4 STEM HOLE

Ø 1/4 STEEL ROD STEM, 15 LONG
SCROLL BEND AS DESIRED

Fig. P-8B Candleholder details (CRS).

PROJECT 15: Figure Castings

Sketch some interesting animal shapes, similar to the one shown in Figure P-15 (page 532), and cut one out and trace it on a piece of 3/4-inch pine to create a casting pattern. Carve the surface to give it form. You can also consider antique pistols as pattern ideas.

PROJECT 16: Model Antique Cannon

An example of a model cannon to place on a study desk in your home is shown in Figure P-16 (page 533). If you wish to prepare a full-size drawing of it, the barrel is six inches long. You can work out the details, and change it as you wish. Library research can provide many other styles. Your class could prepare a casting match plate of the cannon and use it as a mass production project. The model can also be machined from brass bar stock.

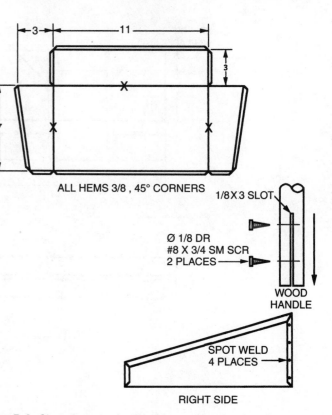

ALL HEMS 3/8 , 45° CORNERS

1/8 X 3 SLOT

Ø 1/8 DR
#8 X 3/4 SM SCR
2 PLACES

WOOD
HANDLE

SPOT WELD
4 PLACES

RIGHT SIDE

Fig. P-9 Shop dustpan details (26 ga. galvanized steel).

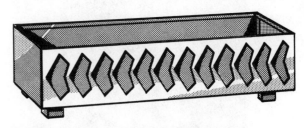

Fig. P-10A Two styles of sheet metal planter.

PROJECT 17: Boot Scraper

When working in the garden, on the farm, or playing in a field, your muddy boots can be cleaned easily with a handy scraper (Figures P-17A and B) (page 533). The two supports are forged.

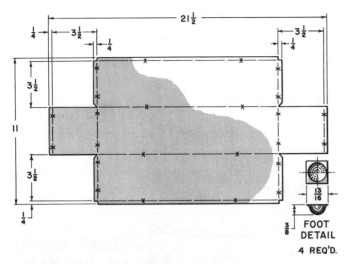

Fig. P-10B Planter layout details.

Fig. P-11 Rod table constructed of copper wire with a metal top.

PROJECT 18: Garden Hose Holder

The sturdy tool in Figure P-18A (page 533) can be used to hold a garden hose, a heavy-duty electrical extension cord, or a shop air hose. It is easy to make and convenient to use. Details are shown in Figure P-18B (page 533).

PROJECT 19: Garden Digger

One nice feature of the simple hoe shown in Figure P-19A (page 533) is that you can attach a short or a long handle to it, for use in a kneeling or standing position. It is simple to make, and the size and shape can be changed as desired. Experiment with similar garden tool designs. Details of the hoe are given in Figure P-19B (page 534).

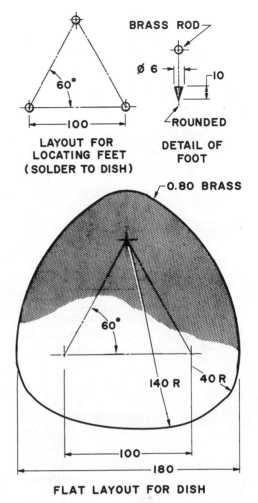

Fig. P-12 Candy dish details.

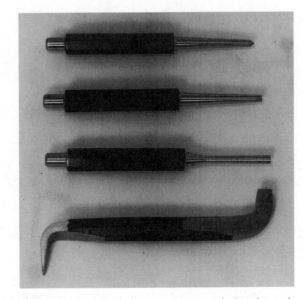

Fig. P-13A Tool set, including a center punch, hand punch, pin punch, and offset screwdriver.

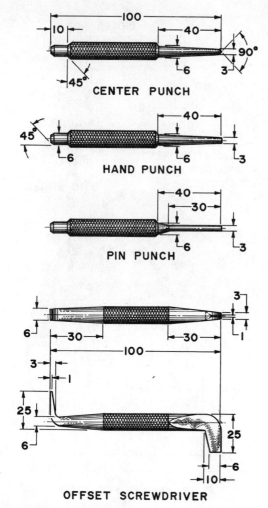

CENTER PUNCH

HAND PUNCH

PIN PUNCH

OFFSET SCREWDRIVER

Fig. P-13B Tool set details (ø 10 tool steel).

Fig. P-14A Belt buckle.

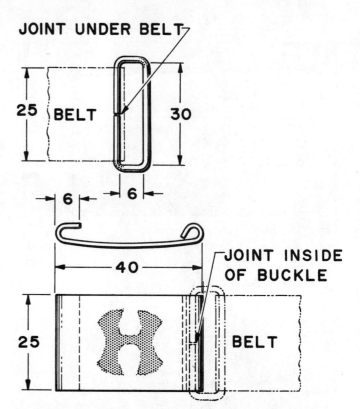

JOINT UNDER BELT

BELT

JOINT INSIDE OF BUCKLE

BELT

Fig. P-14B Belt buckle details. Components include: Two ø 2 brazing rod loops (65 long) and a 0.80 brass buckle (25 × 60). Dimensions are in millimeters.

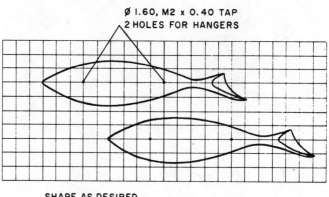

ø 1.60, M2 x 0.40 TAP
2 HOLES FOR HANGERS

SHAPE AS DESIRED.
BODY 20 mm AT THICKEST POINT
SQUARES 25 mm

Fig. P-15 Figure casting pattern.

PROJECT 20: Serving Fork and Knife

Shown in Figure P-20 (page 534) is one design for a serving set. Sketch some other ideas for different blade and fork shapes.

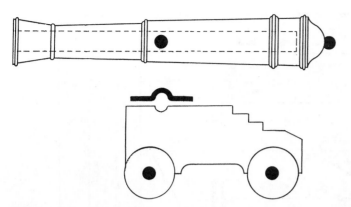

Fig. P-16 Model cannon illustration.

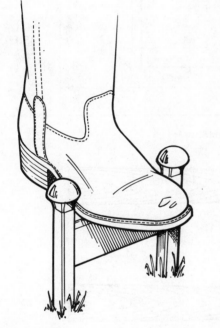

Fig. P-17A Boot scraper.

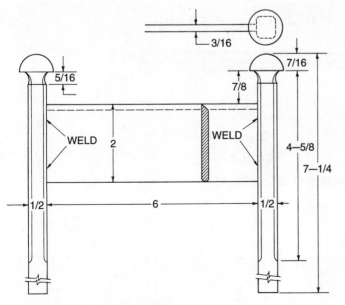

Fig. P-17B Boot scraper details.

Fig. P-18A Garden hose holder.

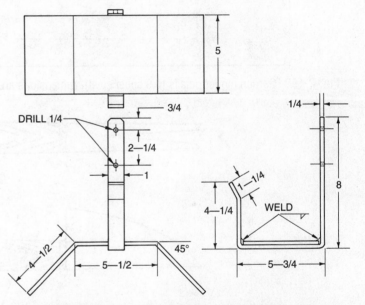

Fig. P-18B Garden hose holder details. The bracket is 1/4 × 1 × 18 H.R.S., and the apron is 5 × 14 1/2 × 11 ga. sheet steel.

Fig. P-19A Garden digger.

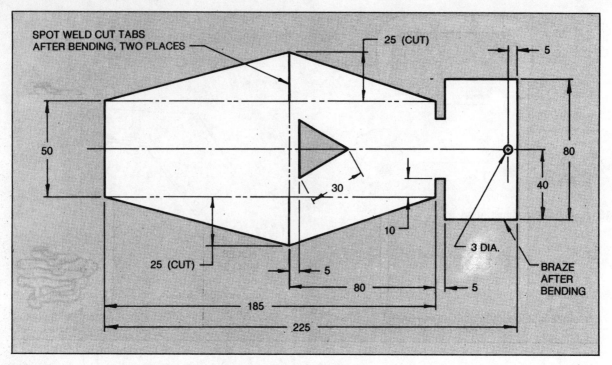

Fig. P-19B Garden digger details (0.9 sheet steel). Dimensions are in millimeters.

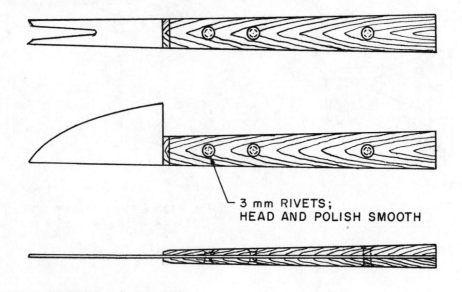

10 mm SQUARES

SCALE — 1:1

Fig. P-20 Serving fork and knife (1 mm stainless steel blades; walnut handles with oil finish).

Appendix

Customary-Metric Drill Sizes and Conversion Chart

Fractional Inch	Decimal Inch	Number or Letter	mm	Fractional Inch	Decimal Inch	Number or Letter	mm	Fractional Inch	Decimal Inch	Number or Letter	mm
1/64	0.0156				0.0610		1.55		0.1285	30	
	0.0157		0.4	1/16	0.0625				0.1299		3.3
	0.0160	78			0.0630		1.6		0.1339		3.4
	0.0165		0.42		0.0635	52			0.1360	29	
	0.0173		0.44		0.0650		1.65		0.1378		3.5
	0.0177		0.45		0.0669		1.7		0.1405	28	
	0.0180	77			0.0670	51		9/64	0.1406		
	0.0181		0.46		0.0689		1.75		0.1417		3.6
	0.0189		0.48		0.0700	50			0.1440	27	
	0.0197		0.5		0.0709		1.8		0.1457		3.7
	0.0200	76			0.0728		1.85		0.1470	26	
	0.0210	75			0.0730	49			0.1476		3.75
	0.0217		0.55		0.0748		1.9		0.1495	25	
	0.0225	74			0.0760	48			0.1496		3.8
	0.0236		0.6		0.0768		1.95		0.1520	24	
	0.0240	73		5/64	0.0781				0.1535		3.9
	0.0250	72			0.0785	47			0.1540	23	
	0.0256		0.65		0.0787		2.0	5/32	0.1562		
	0.0260	71			0.0807		2.05		0.1570	22	
	0.0276		0.7		0.0810	46			0.1575		4.0
	0.0280	70			0.0820	45			0.1590	21	
	0.0292	69			0.0827		2.1		0.1610	20	
	0.0295		0.75		0.0846		2.15		0.1614		4.1
	0.0310	68			0.0860	44			0.1654		4.2
1/32	0.0312				0.0866		2.2		0.1660	19	
	0.0315		0.8		0.0886		2.25		0.1673		4.25
	0.0320	67			0.0890	43			0.1693		4.3
	0.0330	66			0.0906		2.3		0.1695	18	
	0.0335		0.85		0.0925		2.35	11/64	0.1719		
	0.0350	65			0.0935	42			0.1730	17	
	0.0354		0.9	3/32	0.0938				0.1732		4.4
	0.0360	64			0.0945		2.4		0.1770	16	
	0.0370	63			0.0960	41			0.1772		4.5
	0.0374		0.95		0.0965		2.45		0.1800	15	
	0.0380	62			0.0980	40			0.1811		4.6
	0.0390	61			0.0981		2.5		0.1820	14	
	0.0394		1.0		0.0995	39			0.1850	13	

Fractional Inch	Decimal Inch	Number or Letter	mm	Fractional Inch	Decimal Inch	Number or Letter	mm	Fractional Inch	Decimal Inch	Number or Letter	mm
	0.0400	60			0.1015	38			0.1850		4.7
	0.0410	59			0.1024		2.6		0.1870		4.75
	0.0413		1.05		0.1040	37		3/16	0.1875		
	0.0420	58			0.1063		2.7		0.1890		4.8
	0.0430	57			0.1065	36			0.1890	12	
	0.0433		1.1		0.1083		2.75		0.1910	11	
	0.0453		1.15	7/64	0.1094				0.1929		4.9
	0.0465	56			0.1100	35			0.1935	10	
3/64	0.0469				0.1102		2.8		0.1960	9	
	0.0472		1.2		0.1110	34			0.1969		5.0
	0.0492		1.25		0.1130	33			0.1990	8	
	0.0512		1.3		0.1142		2.9		0.2008		5.1
	0.0520	55			0.1160	32			0.2010	7	
	0.0531		1.35		0.1181		3.0	13/64	0.2031		
	0.0550	54			0.1200	31			0.2040	6	
	0.0551		1.4		0.1220		3.1		0.2047		5.2
	0.0571		1.45	1/8	0.1250				0.2055	5	
	0.0591		1.5		0.1260		3.2		0.2067		5.25
	0.0595	53			0.1280		3.25		0.2087		5.3
	0.2090	4			0.3160	O		17/32	0.5312		
	0.2126		5.4		0.3189		8.1		0.5315		13.5
	0.2130	3			0.3228		8.2	35/64	0.5469		
	0.2165		5.5		0.3230	P			0.5512		14.0
7/32	0.2188				0.3248		8.25	9/16	0.5625		
	0.2205		5.6		0.3268		8.3		0.5709		14.5
	0.2210	2		21/64	0.3281			37/64	0.5781		
	0.2244		5.7		0.3307		8.4		0.5906		15.0
	0.2264		5.75		0.3320	Q		19/32	0.5938		
	0.2280	1			0.3346		8.5	39/64	0.6094		
	0.2283		5.8		0.3386		8.6		0.6102		15.5
	0.2323		5.9		0.3390	R		5/8	0.6250		
	0.2340	A			0.3425		8.7		0.6299		16.0
15/64	0.2344			11/32	0.3438			41/64	0.6406		
	0.2362		6.0		0.3445		8.75		0.6496		16.5
	0.2380	B			0.3465		8.8	21/32	0.6562		
	0.2402		6.1		0.3480	S			0.6693		17.0
	0.2420	C			0.3504		8.9	43/64	0.6719		
	0.2441		6.2		0.3543		9.0	11/16	0.6875		
	0.2460	D			0.3580	T			0.6890		17.5
	0.2461		6.25		0.3583		9.1	45/64	0.7031		
	0.2480		6.3	23/64	0.3594				0.7087		18.0
1/4	0.2500	E			0.3622		9.2	23/32	0.7188		
	0.2520		6.4		0.3642		9.25		0.7283		18.5
	0.2559		6.5		0.3661		9.3	47/64	0.7344		
	0.2570	F			0.3680	U			0.7480		19.0
	0.2598		6.6		0.3701		9.4	3/4	0.7500		
	0.2610	G			0.3740		9.5	49/64	0.7656		
	0.2638		6.7	3/8	0.3750				0.768		19.5
17/64	0.2656				0.3770	V		25/32	0.7812		
	0.2657		6.75		0.3780		9.6		0.7874		20.0
	0.2660	H			0.3819		9.7	51/64	0.7969		
	0.2677		6.8		0.3839		9.75		0.808		20.5
	0.2717		6.9		0.3858		9.8	13/16	0.8125		
	0.2720	I			0.3860	W			0.8268		21.0
	0.2756		7.0		0.3898		9.9	53/64	0.8281		
	0.2770	J		25/64	0.3906			27/32	0.8437		
	0.2795		7.1		0.3937		10.0		0.847		21.5
	0.2810	K			0.3970	X		55/64	0.8594		
9/32	0.2812				0.4040	Y			0.8661		22.0
	0.2835		7.2	13/32	0.4062			7/8	0.8750		
	0.2854		7.25		0.4130	Z			0.886		22.5
	0.2874		7.3		0.4134		10.5	57/64	0.8906		

Fractional Inch	Decimal Inch	Number or Letter	mm	Fractional Inch	Decimal Inch	Number or Letter	mm	Fractional Inch	Decimal Inch	Number or Letter	mm
	0.2900	L		27/64	0.4219				0.9055		23.0
	0.2913		7.4		0.4331		11.0	29/32	0.9062		
	0.2950	M		7/16	0.4375			59/64	0.9219		
	0.2953		7.5		0.4528		11.5		0.926		23.5
19/64	0.2969			29/64	0.4531			15/16	0.9375		
	0.2992		7.6	15/32	0.4688				0.9449		24.0
	0.3020	N			0.4724		12.0	61/64	0.965		24.5
									0.9531		
	0.3031		7.7	31/64	0.4844			31/32	0.9687		
	0.3051		7.75		0.4921		12.5				
	0.3071		7.8	1/2	0.5000				0.9843		25.0
	0.3110		7.9		0.5118		13.0	63/64	0.9844		
5/16	0.3125			33/64	0.5156			64/64	1.000		25.4
	0.3150		8.0								

Index

abrading, 5, 6, 146–49
abrasive cloth, hand
    polishing with, 154
abrasives, 35
  grinding with, 146–49
  on grinding wheels, 339–40
  kinds of, 147
  polishing with, 153–55
abrasive-stick dressers, 149
abrasive wheels, polishing
    with, 154
acid-resist enamel, 238, 239
acme screw thread, 159
acoustical properties, 32, 37
adhesion, 385
adhesive bonding, 206
adhesive films, 207
adhesives, 206–7
adjustable wrench, 170
advertising, 440–41
aerospace engineers, 28
aerospace transportation,
    519–24
air conditioning, installation
    of, 486
airless spraying, 405
air-lift hammers, 260
airplanes, 520–22
airports, construction of, 499
air traffic control, 522–24
airways, 519–20
airway vehicles, 520–22
alloys, 36, 39, 49
alloy steels, 44
all-purpose flat tool, 249
all-terrain vehicles (ATVs),
    505
aluminum, 50–51
aluminum anodizing,
    399–400

aluminum oxide, 147, 339
American National Thread
    System, 158
American Standard pipe
    threads, 159
analog computers, 454
anatomy, design for, 91
angle plate, 106
angular bends, 137–38,
    186–87
annealing, 232–33, 354
anode, 391, 397
anthropometry, design for, 90
antiquing, 395
anvils, 254
anvil tools, 254, 257–58
appendages, soldering, 212
application programs, 457
applied research
    scientists, 29
apprenticeship programs, 433,
    435
arc welding, 366, 374–78, 490
arch bridges, 498
arch dams, 498
architectural design, 488
articulated robots, 467
art metal, soldering, 241–43
art metalwork, 223–25
asphaltum varnish, 238, 239
assembling. *See* fastening
assembly drawings, 70
automated drafting, 460–61
automated storage and
    retrieval systems
    (AS/RS), 475
automatic operation, 413
automatic spraying, 405
automation, 413
automobiles, 504

back stick, 249
ball peen, 105
ball tool, 249
band machining, 296
bandsawing, 288, 296–302
band saws, 121
  basic operating techniques,
    300–302
  blades of, 298–99
  safety for, 299–300
  vertical and horizontal, 288,
    296
bar bender, 141
bar folder, 189–90
barges, 516
basements, construction of,
    483–84, 489
base metals, 49
base units, 62
basic oxygen furnace, 42
basic research scientists, 29
bastard file, 134–35
beading, 201–2
beading tool, 249
bead weld, 371–72
beam bridge, 498
beating down. *See* sinking
bed, of engine lathe, 304
beeswax, 238, 239–40
bench grinder, 148
bench metal, 113–15
bench shears, 120
bench vice, 105
bending, 7, 8, 136–39
  with brakes, 190, 192
  in hand-forging operations,
    256–57
  with machines, 140–45
  sheet metal, 185–88, 221
bending strength, 36

benefits, for workers, 435
beryllium, 51
bill of materials, 422, 445
binary system, 454
binding arbitration, 436
bits, 454
black antique finish, 395
black-iron rivets, 204
blacksmith's hammers, 254
blades, for saws, 117–18, 226,
    298–99
blanking, 218, 219
blast furnace, 39, 42
blind holes, drilling, 128
blowhorn stake, 185
blowpipe, 241
board-drop hammers, 260
boat, 514
bolt cutters, 120
bolts, 167, 490
bonds, 340
boring, 287, 322
box-and-pan brake, 190
box wrenches, 169–70
brakes, 190, 192
brass, 50
brazing, 10, 11, 373
brazing rod, 370
break lines, 68
brickwares, 34
bridges, construction of,
    497–98
Brinell hardness test, 358
brittleness, 35
broaching, 287
bronze, 50
bronze powder finish, 395
bronze welding. *See* brazing
brush finishing, 394–95
buffing, 11, 12, 147, 244–45